CHEMICAL ZOOLOGY

Volume IX

AMPHIBIA AND REPTILIA

Contributors to This Volume

ARNOLD F. BRODIE
R. A. COULSON
G. DANDRIFOSSE
WILLIAM H. DANTZLER
H. DENIS
HERBERT C. DESSAUER
M. GILLES-BAILLIEN
GERHARD G. HABERMEHL
WILFRIED HANKE
THOMAS HERNANDEZ
W. N. HOLMES
PAUL LICHT
GABRIEL M. MAKHLOUF
MARUS W. MUMBACH
WARREN S. REHM
FINDLAY E. RUSSELL
BRADLEY T. SCHEER
BOLLING SULLIVAN
A. R. TAMMAR
ALLAN R. THOMPSON

CHEMICAL ZOOLOGY

Edited by MARCEL FLORKIN

DEPARTMENT OF BIOCHEMISTRY
UNIVERSITY OF LIÈGE
LIÈGE, BELGIUM

and

BRADLEY T. SCHEER

DEPARTMENT OF BIOLOGY
UNIVERSITY OF OREGON
EUGENE, OREGON

Volume IX

AMPHIBIA AND REPTILIA

ACADEMIC PRESS New York and London 1974
A Subsidiary of Harcourt Brace Jovanovich, Publishers

ACADEMIC PRESS, INC.
111 Fifth Avenue, New York, New York 10003

United Kingdom Edition published by
ACADEMIC PRESS, INC. (LONDON) LTD.
24/28 Oval Road, London NW1

Library of Congress Cataloging in Publication Data

Florkin, Marcel.
Chemical zoology.

Includes bibliographies.
CONTENTS: v. 1. Protozoa, edited by G. W. Kidder.–
v. 2. Porifera, Coelenterata, and Platyhelminthes.–
v. 3. Echinodermata, Nematoda, and Acanthocephala.
[etc.]
1. Biological chemistry. I. Scheer, Bradley
Titus, Date joint author. II. Kidder, George
Wallace, Date ed. III. Title.
[DNLM: 1. Amphibia. 2. Reptiles. W1CH276 v. 9
1974/QL641 A526 1974]
QP514.F528 591.1'92 67-23158
ISBN 0–12–261039–3 (v. 9)

PRINTED IN THE UNITED STATES OF AMERICA

Contents

List of Contributors xi

Preface xiii

Contents of Other Volumes xv

Section I: AMPHIBIA

Chapter 1. **Biochemistry of Amphibian Development**

H. DENIS

I. Metabolism of Yolk during Oogenesis and Embryonic Development 4
II. Mitochondria in Oogenesis and Embryonic Development 8
III. Ribosomes in Oogenesis and Embryonic Development 12
References 19

Chapter 2. **Gastric Secretion in Amphibia**

GABRIEL M. MAKHLOUF AND WARREN S. REHM

I. Introduction 23
II. Chloride Transport 27
III. Separate Hydrogen Transport Mechanism 30
IV. Coupling of Chloride and Hydrogen Mechanisms 31
V. Electrical Aspects of Secretion 34
VI. Energetic Aspects of Secretion 37
VII. Ionic Requirements for Acid Secretion 42
VIII. Transport of Sodium, Potassium, and Water 43
References 46

Chapter 3. **Salt Balance and Osmoregulation in Salientian Amphibians**

Bradley T. Scheer, Marus W. Mumbach, and Allan R. Thompson

I. Ranges and Tolerance 51
II. Fluxes 52
III. Endocrine Controls 56
IV. Cellular Basis of Aldosterone Action 59
References 64

Chapter 4. **Bile Salts of Amphibia**

A. R. Tammar

I. Introduction 67
II. Caudata 68
III. Anura 69
IV. Significance of Amphibian Bile Salt Distribution 71
V. Amphibian Bile Salts to June 1971 74
References 76

Chapter 5. **Amphibian Hemoglobins**

Bolling Sullivan

I. Introduction 77
II. Hemoglobin Components 79
III. Hemoglobin Function 92
IV. Hemoglobin Structure 101
V. Hemoglobin Synthesis 111
VI. Concluding Comments 117
References 118

Chapter 6. **Endocrinology of Amphibia**

Wilfried Hanke

I. Structure and Function of Endocrine Glands 123
II. Regulation of Metabolic Processes 137

III. Regulation of Development and Metamorphosis 144
IV. Regulation of Gonadal Function and Reproduction 148
V. Regulation of Color Change 149
References 152

Chapter 7. **Venoms of Amphibia**

GERHARD G. HABERMEHL

I. Introduction 161
II. Urodela 164
III. Anura 167
IV. Conclusion 180
References 180

Section II: REPTILIA

Chapter 8. **Plasma Proteins of Reptilia**

HERBERT C. DESSAUER

I. Introduction 187
II. The Plasma Protein System 189
III. Specific Protein 193
References 213

Chapter 9. **Intermediary Metabolism of Reptiles**

R. A. COULSON AND THOMAS HERNANDEZ

I. Introduction 217
II. Metabolism of Carbohydrates 218
III. Metabolism of Amino Acids 224
IV. Intermediary Metabolism and Metabolic Rate 241
References 246

Chapter 10. **Digestion in Reptiles**

G. DANDRIFOSSE

I. Feeding Habits 249
II. Anatomical Considerations 251

III. Digestive Juice 252
IV. Mechanism of Digestion 260
References 270

Chapter 11. **Water and Mineral Metabolism in Reptilia**

William H. Dantzler and W. N. Holmes

I. Regulation of Urinary Excretion of Water and Ions 277
II. The Extrarenal Excretory System 312
References 334

Chapter 12. **Bile Salts in Reptilia**

A. R. Tammar

I. Introduction 337
II. Anapsida 337
III. Lepidosauria 339
IV. Crocodilia 342
V. Significance of Reptilian Bile Salt Distribution 343
VI. Reptilian Bile Salts to June 1971 346
References 350

Chapter 13. **Seasonal Variations in Reptiles**

M. Gilles-Baillien

I. Introduction 353
II. Seasonal Variations in Cold-Climate Reptiles 356
III. Seasonal Variations in Reptiles from Dry-Summer Regions 367
IV. Origin of Seasonal Variations 368
V. Concluding Remarks 373
References 374

Chapter 14. **Reptilian Hemoglobins**

Bolling Sullivan

I. Introduction 377
II. Hemoglobin Components 378

III. Hemoglobin Function 385
IV. Hemoglobin Structure and Synthesis 393
V. Concluding Comments 394
References 396

Chapter 15. **Endocrinology of Reptilia—The Pituitary System**

PAUL LICHT

I. Introduction 399
II. Growth Hormone (GH) and Prolactin (PL) 403
III. Gonadotropins and Reproduction 413
IV. Adrenocorticotropin (ACTH) and Corticoid Hormones 428
V. Thyrotropin (TSH) 437
VI. Concluding Remarks 442
References 444

Chapter 16. **Venoms of Reptiles**

FINDLAY E. RUSSELL AND ARNOLD F. BRODIE

I. Introduction 449
II. General Characteristics of Venom 450
III. Low Molecular Weight Venom Components 455
IV. Venom Enzymes 465
References 474

Author Index 479

Subject Index 503

List of Contributors

Numbers in parentheses indicate the pages on which the authors' contributions begin.

ARNOLD F. BRODIE (449), Department of Biochemistry, School of Medicine, University of Southern California, Los Angeles, California

R. A. COULSON (217), Department of Biochemistry, Louisiana State University School of Medicine, New Orleans, Louisiana

G. DANDRIFOSSE (249), Department of Biochemistry, University of Liège, Liège, Belgium

WILLIAM H. DANTZLER (277), Department of Physiology, College of Medicine, The University of Arizona, Tucson, Arizona

H. DENIS (3),* Department of Biochemistry, University of Liège, Liège, Belgium

HERBERT C. DESSAUER (187), Department of Biochemistry, Louisiana State University School of Medicine, New Orleans, Louisiana

M. GILLES-BAILLIEN (353), Department of General and Comparative Biochemistry, University of Liège, Liège, Belgium

GERHARD G. HABERMEHL (161), Institute of Organic Chemistry, Darmstadt Institute of Technology, Darmstadt, Germany

WILFRIED HANKE (123), Zoological Institute of the University (TH), Karlsruhe, Germany

THOMAS HERNANDEZ (217), Department of Pharmacology, Louisiana State University School of Medicine, New Orleans, Louisiana

W. N. HOLMES (277), Department of Biological Sciences, University of California, Santa Barbara, California

PAUL LICHT (399), Department of Zoology, University of California, Berkeley, California

GABRIEL M. MAKHLOUF (23), Department of Medicine, Medical College of Virginia, Richmond, Virginia

* Present address: Center for Molecular Genetics, Gif-sur-Yvette, France.

MARUS W. MUMBACH (51), Department of Biology, University of Oregon, Eugene, Oregon

WARREN S. REHM (23), The Medical Center, University of Alabama in Birmingham, University Station, Birmingham, Alabama

FINDLAY E. RUSSELL (449), Laboratory of Neurological Research, School of Medicine, University of Southern California, Los Angeles, California

BRADLEY T. SCHEER (51) Department of Biology, University of Oregon, Eugene, Oregon

BOLLING SULLIVAN (77, 377), Department of Biochemistry, Duke University Medical Center, Durham, North Carolina, and Duke University Marine Laboratory, Beaufort, North Carolina

A. R. TAMMAR (67, 337), Biochemistry and Chemistry Department, Guy's Hospital Medical School, London, England

ALLAN R. THOMPSON (51), Department of Biology, University of Oregon, Eugene, Oregon

Preface

Zoology is currently undergoing a period of transition in which chemical knowledge is progressively integrated with the more classic knowledge of morphology and systematics. Biochemical studies of species, as well as of higher taxa, open new disciplines to the zoologist and offer new viewpoints in considering problems of structure, function, development, evolution, and ecology. The biochemist has considerable opportunities for broadening his sphere of investigation because of the enormous selection of animal species available for study from which a great variety of compounds can be obtained and reactions observed. There are abundant prospects for fruitful collaboration between the biochemist and zoologist in studies in which the characteristics of the animal and the biochemical constituents and processes interact in significant ways.

Very often the initial obstacle in undertaking investigations in new fields is the complexity and scattered character of the literature. This treatise is aimed primarily at making it possible for zoologists and chemists, who have a limited knowledge of the literature in fields other than their own, to gain a valid impression of the present state of knowledge in chemistry and zoology and an introduction to the existing literature. Thus, we have invited research workers who have contributed significantly to problems involving combined chemical and zoological approaches to summarize the knowledge in their specific disciplines of interest and competence. The authors have been encouraged to be critical and synthetic and to include mention of gaps in knowledge as well as the established information.

The treatise is arranged by phyla, an arrangement which seemed most suitable for presenting chemical information of zoological significance and for bringing to the attention of chemists those aspects of biochemical diversity of greatest potential interest. Each section, dealing with a major phylum, is introduced by a discussion of the biology and systematics of the group. This is followed by chapters dealing with various aspects of the biochemistry of the group. In general, the authors of individual chapters have been given full freedom, within the limitations of space, to develop their assigned topic. We thought that in this way the reader would have the advantage of the author's personal experience in and attitude toward his field, and that this would more than compensate for any unevenness in coverage that might result.

We are grateful to Professor K. M. Wilbur for his help in the early planning of this treatise, to the authors for their cooperation and patience, and to the staff of Academic Press for their careful work.

MARCEL FLORKIN
BRADLEY T. SCHEER

Contents of Other Volumes

Volume I: PROTOZOA

Systematics of the Phylum Protozoa
John O. Corliss

Chemical Aspects of Ecology
E. Fauré-Fremiet

Carbohydrates and Respiration
John F. Ryley

Nitrogen: Distribution, Nutrition, and Metabolism
George W. Kidder

Lipid Composition, Nutrition, and Metabolism
Virginia C. Dewey

Growth Factors in Protozoa
Daniel M. Lilly

Transport Phenomena in Protozoa
Robert L. Conner

Digestion
Miklós Müller

The Chemistry of Protozoan Cilia and Flagella
Frank M. Child

Protozoan Development
Earl D. Hanson

Nucleic Acids of Protozoa
Manley Mandel

Carbohydrate Accumulation in the Protist—A Biochemical Model of Differentiation
Richard G. Pannbacker and Barbara E. Wright

Chemical Genetics of Protozoa
Sally Lyman Allen

Chemistry of Parasitism among Some Protozoa
B. M. Honigberg

AUTHOR INDEX—SUBJECT INDEX

Volume II

Section I: PORIFERA

The Sponges, or Porifera
Paul Brien

Skeletal Structures of Porifera
M. Florkin

Pigments of Porifera
T. M. Goodwin

Nutrition and Digestion
Raymond Rasmont

Composition and Intermediary Metabolism—Porifera
C. S. Hammen and Marcel Florkin

Chemical Aspects of Hibernation
Raymond Rasmont

Section II: COELENTERATA, CTENOPHORA

Introduction to Coelenterates
J. Bouillon

Pigments of Coelenterata
T. W. Goodwin

Chemical Perspectives on the Feeding Response, Digestion, and Nutrition of Selected Coelenterates
Howard M. Lenhoff

Intermediary Metabolism—Coelenterata
C. S. Hammen

The Chemistry of Luminescence in Coelenterates
Frank H. Johnson and Osamu Shimomura

Coelenterata: Chemical Aspects of Ecology: Pharmacology and Toxicology
C. E. Lane

Section III: PLATYHELMINTHES, MESOZOA

Introduction to Platyhelminthes
Bradley T. Scheer and E. Ruffin Jones

Nutrition and Digestion
J. B. Jennings

Intermediary Metabolism of Flatworms
Clark P. Read

Platyhelminthes: Respiratory Metabolism
Winona B. Vernberg

Growth Development and Culture Methods: Parasitic Platyhelminths
J. A. Clegg and J. D. Smyth

Chemical Aspects of the Ecology of Platyhelminths
Calvin W. Schwabe and Araxie Kilejian

Responses of Trematodes to Pharmacological Agents
Ernest Bueding

The Mesozoa
Bayard H. McConnaughey

AUTHOR INDEX—SUBJECT INDEX

Volume III

Section I: ECHINODERMATA

General Characteristics of the Echinoderms
Georges Ubaghs

Ionic Patterns
Shirley E. Freeman and W. P. Freeman

Feeding, Digestion, and Nutrition in Echinodermata
John Carruthers Ferguson

Carbohydrates and Carbohydrate Metabolism of Echinoderms
Philip Doezema

Lipid Metabolism
U. H. M. Fagerlund

Pigments in Echinodermata
T. W. Goodwin

Fertilization and Development
Tryggve Gustafson

Pharmacology of Echinoderms
Ragnar Fänge

Section II: NEMATODA AND ACANTHOCEPHALA

The Systematics and Biology of Some Parasitic Nematodes
M. B. Chitwood

The Biology of the Acanthocephala
Ivan Pratt

Skeletal Structures and Integument of Acanthocephala and Nematoda
Alan F. Bird and Jean Bird

Culture Methods and Nutrition of Nematodes and Acanthocephala
Morton Rothstein and W. L. Nicholas

Carbohydrate and Energy Metabolism of Nematodes and Acanthocephala
Howard J. Saz

Lipid Components and Metabolism of Acanthocephala and Nematoda
Donald Fairbairn

Nitrogenous Components and Their Metabolism: Acanthocephala and Nematoda
W. P. Rogers

Osmotic and Ionic Regulation in Nematodes
Elizabeth J. Arthur and Richard C. Sanborn

Chemical Aspects of Growth and Development
W. P. Rogers and R. I. Sommerville

The Pigments of Nematoda and Acanthocephala
Malcolm H. Smith

Pharmacology of Nematoda
J. Del Castillo

Chemistry of Nematodes in Relation to Serological Diagnosis
José Oliver-Gonzáles

Chemical Ecology of Acanthocephala and Nematoda
Alan F. Bird and H. R. Wallace

Gastrotricha, Kinorhyncha, Rotatoria, Kamptozoa, Nematomorpha, Nemertina, Priapuloidea
Ragnar Fänge

AUTHOR INDEX—SUBJECT INDEX

Volume IV: ANNELIDA, ECHIURA, AND SIPUNCULA

Systematics and Phylogeny: Annelida, Echiura, Sipuncula
R. B. Clark

Nutrition and Digestion
Charles Jeuniaux

Respiration and Energy Metabolism in Annelids
R. Phillips Dales

Respiratory Proteins and Oxygen Transport
Marcel Florkin

Carbohydrates and Carbohydrate Metabolism: Annelida, Sipunculida, Echiurida
Bradley T. Scheer

Nitrogen Metabolism
Marcel Florkin

Guanidine Compounds and Phosphagens
Nguyen van Thoai and Yvonne Robin

Annelida, Echiurida, and Sipunculida—Lipid Components and Metabolism
Manfred L. Karnovsky

Inorganic Components and Metabolism; Ionic and Osmotic Regulation: Annelida, Sipuncula, and Echiura
Larry C. Oglesby

Pigments of Annelida, Echiuroidea, Sipunculoidea, Priapulidea, and Phoronidea
G. Y. Kennedy

Growth and Development
A. E. Needham

Endocrines and Pharmacology of Annelida, Echiuroidea, Sipunculoidea
Maurice Durchon

Luminescence in Annelids
Milton J. Cormier

AUTHOR INDEX—SUBJECT INDEX

Volume V: ARTHROPODA Part A

Arthropods: Introduction
S. M. Manton

Arthropod Nutrition
R. H. Dadd

Digestion in Crustacea
P. B. van Weel

Digestion in Insects
R. H. Dadd

Carbohydrate Metabolism in Crustaceans
Lyle Hohnke and Bradley T. Scheer

Metabolism of Carbohydrates in Insects
Stanley Friedman

Nitrogenous Constituents and Nitrogen Metabolism in Arthropods
E. Schoffeniels and R. Gilles

Lipid Metabolism and Transport in Arthropods
Lawrence I. Gilbert and John D. O'Connor

Osmoregulation in Aquatic Arthropods
E. Schoffeniels and R. Gilles

Osmoregulation in Terrestrial Arthropods
Michael J. Berridge

Chemistry of Growth and Development in Crustaceans
Larry H. Yamaoka and Bradley T. Scheer

Chemical Aspects of Growth and Development in Insects
Colette L'Hélias

Author Index—Subject Index

Volume VI: ARTHROPODA Part B

The Integument of Arthropoda
R. H. Hackman

Hemolymph—Arthropoda
Charles Jeuniaux

Blood Respiratory Pigments—Arthropoda
James R. Redmond

Hemolymph Coagulation in Arthropods
C. Grégoire

Respiration and Energy Metabolism in Crustacea
Kenneth A. Munday and P. C. Poat

Oxidative Metabolism of Insecta
Richard G. Hansford and Bertram Sacktor

Excretion—Arthropoda
J. A. Riegel

Pigments—Arthropoda
T. W. Goodwin

Endocrines of Arthropods
František Sehnal

Chemical Ecology—Crustacea
F. John Vernberg and Winona B. Vernberg

Toxicology and Pharmacology—Arthropoda
M. Pavan and M. Valcurone Dazzini

Author Index—Subject Index

Volume VII: MOLLUSCA

The Molluscan Framework
Charles R. Stasek

Structure of the Molluscan Shell
C. Grégoire

Shell Formation in Mollusks
Karl M. Wilbur

Byssus Fiber—Mollusca
E. M. Mercer

Chemical Embryology of Mollusca
C. P. Raven

Pigments of Mollusca
T. W. Goodwin

Respiratory Proteins in Mollusks
F. Ghiretti and A. Ghiretti-Magaldi

Carbohydrates and Carbohydrate Metabolism in Mollusca
Esther M. Goudsmit

Lipid and Sterol Components and Metabolism in Mollusca
P. A. Voogt

Nitrogen Metabolism in Mollusks
Marcel Florkin and S. Bricteux-Grégoire

Endocrinology of Mollusca
Micheline Martoja

Ionoregulation and Osmoregulation in Mollusca
E. Schoffeniels and R. Gilles

Aspects of Molluscan Pharmacology
Robert Endean

Biochemical Ecology of Mollusca
R. Gilles

Author Index—Subject Index

Volume VIII: DEUTEROSTOMIANS, CYCLOSTOMES, AND FISHES

Section I: DEUTEROSTOMIANS

Introduction to the Morphology, Phylogenesis, and Systematics of Lower Deuterostomia
Jean E. A. Godeaux

Biochemistry of Primitive Deuterostomians
E. J. W. Barrington

Section II: VERTEBRATES (CRANIATA)

General Characteristics and Evolution of Craniata or Vertebrates
Paul Brien

A. Cyclostomes

Osmotic and Ionic Regulation in Cyclostomes
James D. Robertson

Endocrinology of the Cyclostomata
Sture Falkmer, Norman W. Thomas, and Lennart Boquist

B. Fishes

Biochemical Embryology of Fishes
A. A. Neyfakh and N. B. Abramova

The Muscle Proteins of Fishes
H. Tsuyuki

Plasma Proteins in Fishes
Robert E. Feeney and W. Duane Brown

Respiratory Function of Blood in Fishes
Gordon C. Grigg

Nitrogen Metabolism in Fishes
R. L. Watts and D. C. Watts

Respiratory Metabolism and Ecology of Fishes
Robert W. Morris

Electrolyte Metabolism of Teleosts—Including Calcified Tissues
Warren R. Fleming

Pigments of Fishes
George F. Crozier

Endocrinology of Fishes
I. Chester Jones, J. N. Ball, I. W. Henderson, T. Sandor, and B. I. Baker

Bile Salts in Fishes
A. R. Tammar

AUTHOR INDEX—SUBJECT INDEX

CHEMICAL ZOOLOGY

Volume IX

AMPHIBIA AND REPTILIA

SECTION I

AMPHIBIA

CHAPTER 1

Biochemistry of Amphibian Development

H. Denis

I. Metabolism of Yolk during Oogenesis and Embryonic Development 4
A. Structure of Yolk 4
B. Accumulation of Yolk during Oogenesis 6
C. Breakdown of Yolk during Embryonic Development 7
II. Mitochondria in Oogenesis and Embryonic Development 8
A. Structure and Activity of Mitochondria in Oocytes and Embryos .. 8
B. Mitochondrial DNA and RNA in Oocytes and Embryos 9
III. Ribosomes in Oogenesis and Embryonic Development 12
A. Structure and Synthesis of Ribosomes in Amphibia 12
B. Synthesis and Accumulation of Ribosomes during Oogenesis 14
C. Synthesis and Accumulation of Ribosomes during Embryonic Development 17
References 19

The amphibian egg is one of the largest cells that undergoes complete cleavage, i.e., gives rise to two identical daughter cells. The eggs of the Mexican axolotl, *Amblystoma mexicanum,* have a diameter of about 2 mm. Those of the South African clawed toad *Xenopus laevis* are much smaller (1–1.2 mm). The most common European frogs, *Rana esculenta* and *Rana temporaria,* lay eggs of intermediate size. The eggs of many other groups are larger than those of the amphibians, but they are much more loaded with yolk so that only a small part of the cell cleaves and actively participates in the elaboration of the embryo. Thanks to its large size and to its moderate load in yolk, the amphibian egg is the choice material for biochemical research on embryonic development.

As expected the egg of amphibians contains many times as much DNA, RNA, and protein as a somatic cell. How the oocyte accumulates these substances during its growth in the ovary and how the embryo later utilizes them is a matter of concern for the embryologist. A section through an uncleaved egg shows that its cytoplasm is loaded with organelles and inclusions that will play specific roles during embryonic development. The most numerous inclusions are the yolk platelets. These contain phosphorylated proteins which will supply the embryo with amino acids and inorganic phosphate. Other large and numerous inclusions are the lipochondria, or oil droplets. The yolk platelets and the lipochondria are distributed throughout the cytoplasm of the egg. Other bodies are confined mostly to the superficial layers of the cell. The pigment granules which lie immediately underneath the cell membrane protect the egg and especially its nucleus against solar radiation. The mitochondria that also lie at short distance from the surface make use of the oxygen which diffuses freely from outside to supply the egg and the embryo with energy. All the inclusions mentioned above can be observed with the light microscope. The electron microscope shows that the ground substance of the cytoplasm is filled with granules of glycogen and ribosomes.

In recent years, much information has been gathered on the process of yolk, mitochondria, and ribosome accumulation during oogenesis and on the fate of these organelles in embryonic development. The purpose of this chapter is to review the present state of knowledge in this field.

I. Metabolism of Yolk during Oogenesis and Embryonic Development

A. Structure of Yolk

The chemical composition of yolk in amphibians is quite simple. It almost entirely consists of two proteins called phosvitin and lipovitellin (Wallace, 1963b). Minor components of yolk are RNA (Panijel, 1950; Grant, 1953; Gross and Gilbert, 1956; Rounds and Flickinger, 1958), DNA (Ringle and Gross, 1962b; Wallace, 1963a; Baltus *et al.*, 1968), and carotenoids (Ringle and Gross, 1962b). RNA is probably a contaminant which binds to the yolk platelet during extraction (Wallace, 1963a). As indicated by its name, lipovitellin is a lipoprotein in which lipid makes up about 20% of the protein dry weight (Wallace, 1963b; Redshaw and Follett, 1971). Lipovitellin occurs as a dimer which splits into monomers at high pH (Wallace, 1965). Phosvitin is a phosphoprotein containing almost 10% of its dry weight as phosphorus (Wallace, 1963b; Redshaw and Follett, 1971). Lipovitellin has a much higher molecular weight (420,000) than phosvitin (32,000) (Wallace, 1963b). In yolk, there are two molecules of phosvitin for every dimer of lipovitellin (Wallace, 1963b). The amino acid composition of the yolk proteins of the South African clawed toad, *Xenopus laevis,* and of the American leopard frog, *Rana pipiens,* is given in Table I. Phosvitin has a very unusual chemical composition, since about 50% of its amino acids are represented by serine. As much as 72% of these serine residues are phosphorylated in *X. laevis* (Redshaw and Follett, 1971).

The yolk proteins are insoluble at neutral pH and moderate salt concentrations. They can, however, be easily brought into solution either

TABLE I
AMINO ACID COMPOSITION OF THE YOLK PROTEIN OF *Rana pipiens* AND *Xenopus laevis*

Amino acid	Content (residues/1000 residues)			
	Lipovitellin		Phosvitin	
	X. laevis[a]	*R. pipiens*[b]	*X. laevis*[a]	*R. pipiens*[b]
Glycine	53.4	51.5	28.4	59.5
Alanine	101.8	81.8	22.2	24.2
Valine	50.8	75.3	5.8	6.5
Leucine	85.1	87.6	14.0	10.1
Isoleucine	40.6	72.0	3.3	9.1
Serine	81.3	83.7	560.0	473.0
Threonine	56.4	46.9	7.2	21.2
½-Cystine	9.1	9.4	0	2.0
Methionine	20.1	26.1	0	1.0
Aspartic acid	87.7	87.0	49.8	63.6
Glutamic acid	138.2	107.2	101.4	99.3
Lysine	73.0	72.0	75.6	78.7
Arginine	56.4	48.2	59.6	62.5
Histidine	28.7	29.6	27.7	53.9
Tyrosine	29.6	27.7	9.4	14.6
Phenylalanine	42.3	34.2	7.2	2.0
Proline	50.3	48.5	29.8	18.1
Tryptophan	—	11.1	—	1.0

[a] From Redshaw and Follett (1971).
[b] From Wallace (1963b).

by bringing the pH below 4 or above 9 (Holtfreter, 1946; Gross and Gilbert, 1956) or by suspending them in concentrations of monovalent cations higher than 0.4 *M* (Ringle and Gross, 1962a; Wallace, 1963b; Wallace and Karasaki, 1963) or in concentrations of calcium ions higher than 3 m*M* (Essner, 1953; Gross, 1954; Flickinger, 1956; Gross and Gilbert, 1956; Ringle and Gross, 1962a). This allows the yolk proteins to be analyzed and purified by the standard biochemical techniques such as electrophoresis (Flickinger and Nace, 1952; Barth and Barth, 1954; Ringle and Gross, 1962b; Denis, 1964) and chromatography on TEAE-cellulose (Wallace, 1965; Redshaw and Follett, 1971).

Phosvitin and lipovitellin are present in yolk platelets in an almost dry state since the platelets contain only 30% water (Wallace, 1963a). This very low water content gives the yolk granule a much higher density than the remainder of the cell. Electron micrographic studies have shown that the yolk platelet has a crystal-like structure and is composed of electron-dense particles sometimes disposed in a regular array (Ward, 1962; Wartenberg, 1962; Karasaki, 1963; Lanzavecchia, 1965). These observations are generally interpreted as meaning that the molecules of phosvitin that are the most electron dense are arranged in a simple hexagonal lattice. The position of the lipovitellin molecules in this lattice cannot be determined with certainty. Several models were proposed in recent years to account for the mutual relationship of phosvitin and lipovitellin in yolk (Wallace and Dumont, 1968; Honjin *et al.*, 1965; Honjin and Nakamura, 1967; Leonard *et al.*, 1972). Studies with the electron microscope have further shown that the central crystal of the yolk platelet is surrounded with a superficial granular layer, both being contained within a simple limiting membrane (Ward, 1959; 1962; Wartenberg, 1962; Karasaki, 1963; Wallace and Karasaki, 1963; Lanzavecchia, 1965; Massover, 1971).

B. Accumulation of Yolk during Oogenesis

Yolk does not appear in the oocyte at the beginning of its growth. Oogenesis can be divided into two periods with respect to yolk accumulation (Panijel, 1951). During the first period, the cytoplasm of the oocyte remains free of yolk and lipochondria (Balinsky and Devis, 1963; Wischnitzer, 1966). During the second period, called vitellogenesis, the oocyte becomes loaded with yolk, lipochondria, pigment granules, and ribosomes (Balinsky and Devis, 1963; Wischnitzer, 1966). In *X. laevis*, yolk deposition begins when the oocyte has a diameter of about 400 μm (Balinsky and Devis, 1963). Yolk condensation first starts in the periphery of the oocytes (Wischnitzer, 1966). In several anuran species part of the yolk platelets appears within mitochondria (Ward, 1962; Wartenberg, 1962; Balinsky and Devis, 1963; Karasaki, 1963; Wischnitzer, 1966). This was never observed in urodeles (Wartenberg, 1962; Karasaki, 1963; Hope *et al.*, 1964).

The yolk proteins are not synthesized within the oocyte itself, but in the liver of the female (Flickinger, 1961, Rudack and Wallace, 1968). From the liver, the yolk proteins are secreted into the blood where they can easily be detected (Wallace and Jared, 1969; Redshaw and Follett, 1971). In the blood, phosvitin and lipovitellin are associated in one complex called vitellogenin, which consists of two molecules of phosvitin covalently bound to one lipovitellin dimer (Wallace, 1970). Vitellogenin contains 12% lipid and about 1.5% phosphorus. Its molecular weight is 460,000 (Follett and Redshaw, 1968; Wallace, 1970, Redshaw and Follett, 1971). Vitellogenin occurs in the blood not only of gravid females (Wallace and Jared, 1968a, 1969), but also of males that have been injected with estradiol-17β (Wallace and Jared, 1968b, 1969). From the blood, vitellogenin penetrates into the oocytes, presumably through the follicle cells (Hope *et al.*, 1963; Follett and Redshaw, 1968; Wallace and Jared, 1969). These cells are probably responsible for the selective uptake of vitellogenin from the blood (Wallace *et al.*, 1970). The molecular rearrangement by which soluble vitellogenin converts to insoluble phosvitin and lipovitellin is not yet known. Phosvitin was first thought to undergo further phosphorylation inside the oocyte. Wallace (1964) partially purified an enzyme (protein kinase, ATP protein phosphotransferase) from frog ovaries that apparently could transfer the terminal phosphate of ATP to phosvitin. More recently, however, Wallace *et al.* (1972) realized that phosvitin does not become phosphorylated in the oocyte, so that phosphorylation is not involved in the conversion of vitellogenin into phosvitin and lipovitellin.

C. Breakdown of Yolk during Embryonic Development

No visible breakdown of yolk occurs in embryos before neurulation (Bragg, 1939). But yolk certainly plays a role in embryonic development much before it begins to be digested. Yolk owes these morphogenetic properties to its physical rather than to its chemical characteristics. The vegetal part of the egg, which will develop into the endoderm and later the digestive tract, contains more yolk and less free cytoplasm than the animal part, which will give rise to the ectoderm (Pasteels, 1951). The consequence of this uneven distribution of yolk is that the vegetal hemisphere is denser than the animal one. Soon after fertilization, the egg becomes free from its envelopes and orients itself according to gravity. This rotation brings the animal (pigmented) part of the egg upward and the vegetal part downward. As a result, the egg acquires its bilateral symmetry. The dorsal blastoporal lip, which later organizes the embryo axis, will appear on the side of the egg that moves down during the rotation that follows fertilization (Ancel and Vintemberger, 1948). The determination of bilateral symmetry is therefore an epigenetic event. No biochemical explanation has so far been proposed for this phenomenon. Yolk might play a role in this process that is perhaps not entirely physical. An intimate mingling of the yolky cytoplasm with the superficial cytoplasm seems to be the prerequisite for the appearance of a blastoporal lip (Waddington, 1956). Such a mingling occurs in normal development as a consequence of the postfertilization rotation. It also occurs in experimental conditions involving a redistribution of the egg components through a forced rotation of the egg (Pasteels, 1951).

Breakdown of yolk first begins in the anterodorsal part of the embryo. It is concomitant with cell differentiation. The major part of yolk, which is located in endoderm cells, does not start to be digested before the embryo hatches. The process of yolk breakdown has been observed with the electron microscope (Karasaki, 1963). The yolk platelet first loses its superficial layer. The central crystal then cleaves into several parts and apparently gives rise to various types of membranes, particles and fibrils. Whether these formations represent degradation products of phosvitin and lipovitellin remains to be determined. Little is known about the biochemical mechanism of yolk breakdown. Large amounts of inorganic phosphate are split from yolk, probably by a phosphoprotein phosphatase (Barth and Barth, 1954; Flickinger, 1956), and this might contribute to the solubilization of the yolk proteins (Flickinger, 1956; Nass, 1956, 1962). It is generally believed that phosvitin and lipovitellin are degraded into amino acids. A proteolytic enzyme, cathepsin, was found by Deuchar (1958) to be associated with the yolk platelets of developing embryos, but this association might be an artifact.

II. Mitochondria in Oogenesis and Embryonic Development

A. Structure and Activity of Mitochondria in Oocytes and Embryos

The mature oocyte, which is a single cell, contains a very large number of mitochondria. These are mostly distributed near the cell surface (Balinsky and Devis, 1963; Wischnitzer, 1966). The mitochondria of eggs and oocytes have the same ultrastructure as those of somatic cells (Wischnitzer, 1966). Mitochondria multiply during the whole length of oogenesis. In previtellogenic oocytes of many species, a cluster of mitochondria called the Balbiani body or, erroneously, yolk nucleus can be observed even with the light microscope (Kemp, 1953; Wartenberg, 1962; Balinsky and Devis, 1963). When the oocyte enters vitellogenesis, the Balbiani body moves toward the oocyte surface, leaving behind small clusters of mitochondria (Balinsky and Devis, 1963).

Mitochondria from oocytes and eggs have the same properties as those from other tissues (Williams, 1965). A complete cytochrome system is present in egg mitochondria (Boell and Weber, 1955; Weber and Boell, 1955; Maggio and Ghiretti-Magaldi; 1958; Strittmatter *et al.*, 1960). In developing embryos, oxygen consumption increases severalfold from fertilization to feeding stage (Brachet, 1960). This increase is only moderate during cleavage but very fast after neurulation. Cytochrome oxidase activity remains constant in the embryo at least until neurulation, after which it it increases rapidly (Spiegelman and Steinbach, 1945; Boell and Weber, 1955; Petrucci, 1957, 1959; Boell *et al.*, 1971; Chase and Dawid, 1972). According to Chase and Dawid (1972) the cytochrome oxidase activity per mitochondrion is the same throughout embryonic development. The increase in enzyme activity in late embryos therefore reflects multiplication of the mitochondria.

There is no reason to believe that mitochondria might play a different role in oocytes and embryos than in somatic cells. The stock of mitochondria present in the egg becomes distributed among the developing organs and supplies them with energy through oxidative phosphorylation. The main substrate for energy production after gastrulation is glycogen (Brachet, 1960), whose amount per embryo begins to decline as soon as gastrulation starts (Brachet and Needham, 1935; Gregg, 1948). The energy needed for morphogenetic movements (gastrulation and neurulation) represents a very small proportion of the energy produced by respiration during early development (Brachet, 1944, 1960; Selman, 1958).

B. Mitochondrial DNA and RNA in Oocytes and Embryos

Oocyte and egg mitochondria contain small amounts of DNA, as do mitochondria of other tissues (Luck and Reich, 1964; Rabinowitz *et al.*, 1965). Since mitochondria are very numerous in the egg, they contain most of the egg DNA (Dawid, 1966). It has long been recognized that the DNA content of the amphibian egg is much higher than that of a somatic cell (Hoff-Jørgensen and Zeuthen, 1952; Sze, 1953; Gregg and Løvtrup, 1955; Kuriki and Okasaki, 1959; Baltus and Brachet, 1962). The latest measurements indicate that the egg of *X. laevis* contains from 300 to 500 times as much DNA as a somatic cell of the same species (Dawid, 1965). Various origins and functions were ascribed to egg DNA until the work of Dawid (1965, 1966). In *X. laevis* egg, mitochondrial DNA has a molecular weight of 10.6×10^6 and consists of double-stranded circular molecules with a contour length of about 5 μm (Wolstenholme and Dawid, 1967). Mitochondrial DNA from the urodeles *Amblystoma mexicanum* and *Necturus maculosus* is approximately 20% shorter than that of *X. laevis* and *R. pipiens* (Wolstenholme and Dawid, 1968). Although mitochondrial DNA has the same base composition (42% guanine + cytosine) as nuclear DNA, no homology could be detected between them (Dawid, 1965, 1966). During embryonic development, the amount of mitochondrial DNA per embryo remains constant until swimming stages, whereas bulk (nuclear) DNA increases very rapidly from the beginning (Fig. 1). This observation suggests that the number of mitochondria per embryo does not increase until late development.

Mitochondria of somatic tissues are endowed with partial autonomy and are able to synthesize DNA, RNA, and protein (Roodyn and Wilkie,

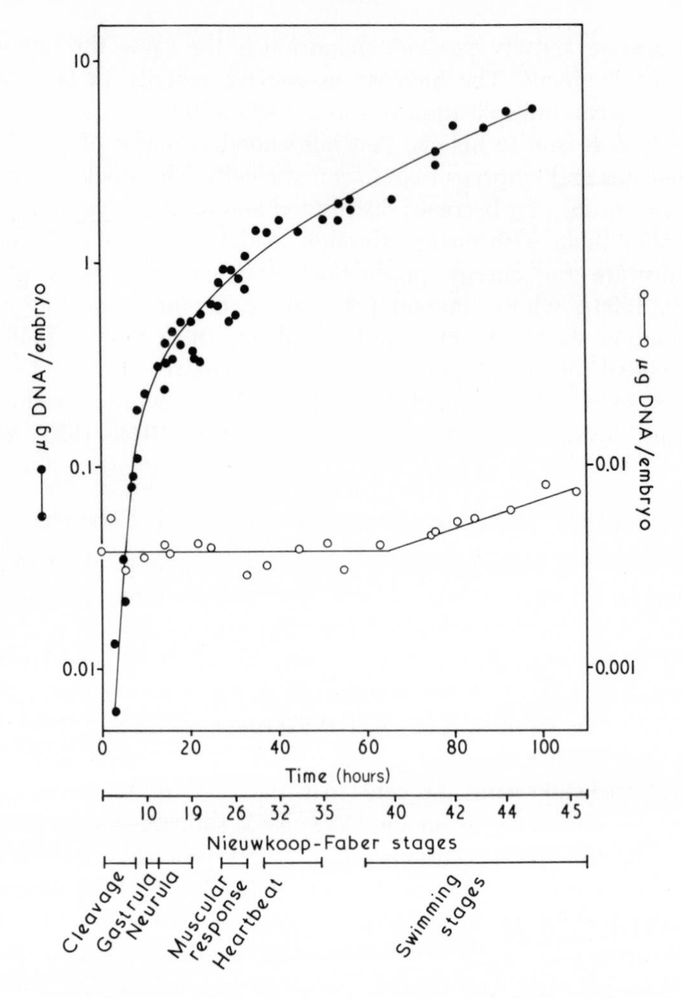

FIG. 1. Accumulation of bulk (●) and mitochondrial (○) DNA in developing embryos of *X. laevis*. Redrawn from Dawid (1965) and Chase and Dawid (1972) with permission.

1968; Borst and Kroon, 1969). The same is true for the mitochondria of the egg and oocytes. Synthesis of mitochondrial DNA in oocytes is certainly not coupled with that of chromosomal DNA, since mitochondria multiply during the whole length of oogenesis, whereas no synthesis of DNA occurs in the chromosomes (Brachet, 1960; Ficq, 1961). The same absence of coupling is observed in embryonic development (Fig. 1). A consequence of the autonomous multiplication of the mitochondria is that all mitochondria that are present in the adult derive from those of the egg. Definitive proof of maternal and cytoplasmic inheritance for mitochondrial DNA has recently been given by Dawid and Blackler (1971).

Mitochondrial DNA, which comprises 16,000 base pairs (Dawid *et al.*, 1971), contains enough information to code for about forty proteins of the same length as hemoglobin. However, no protein that is organized by mitochondrial DNA has so far been identified. The only well established function of mitchondrial DNA is to code for ribosomal and 4 S RNA (Dawid, 1971; Swanson and Dawid, 1970). In *X. laevis* ribosomal and 4 S RNA from egg mitochondria are complementary to 6.5% and to 3.5%, respectively of mitochondrial DNA (Dawid, 1971). Each molecule or circle of mitochondrial DNA, therefore, contains one sequence or gene for each species of ribosomal RNA and ten to twenty genes for 4 S RNA (Dawid, 1972). The mitochondrial ribosome is the smallest so far known. Its sedimentation coefficient is only 60 S, instead of 87 S for the cytoplasmic ribosome (Swanson and Dawid, 1970). It contains two RNA species with molecular weights of 5.3×10^5 and 3×10^5, respectively (Dawid and Chase, 1972) and no 5 S RNA (Dawid and Chase, 1971). Ribosomal RNA from mitochondria can be detected in eggs and embryos (Fig. 2). It makes up approximately 0.5% of the total RNA content of the egg (compare Figs. 2 and 4). Unlike the amount of mitochondrial DNA (Fig. 1), the amount of mitochondrial RNA per embryo increases from gastrula stage on, and both types of ribosomal RNA accumulate coordinately (Fig. 2). Synthesis of RNA in mitochon-

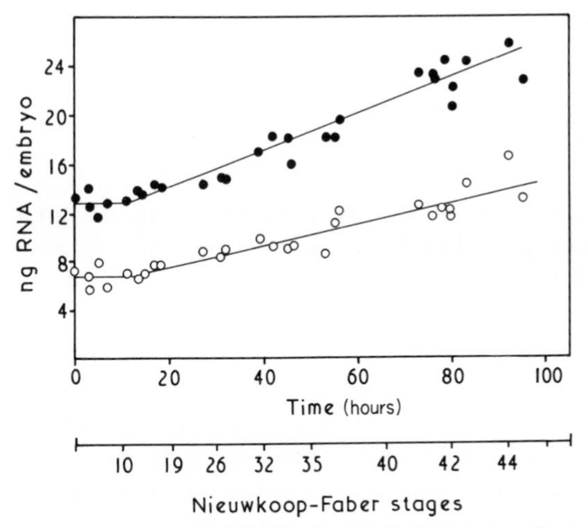

FIG. 2. Accumulation of mitochondrial ribosomal RNA in embryos of *X. laevis*. ● = large ribosomal RNA; ○ = small ribosomal RNA. Redrawn from Chase and Dawid (1972) with permission.

dria is therefore not under the same metabolic control as that of DNA (Chase and Dawid, 1972).

III. Ribosomes in Oogenesis and Embryonic Development

A. Structure and Synthesis of Ribosomes in Amphibia

The ribosome of the amphibian has the same structure as that of other multicellular organisms. Its sedimentation coefficient is about 80 S. It consists of two subunits sedimenting at 60 S and 40 S (Maden, 1971). The 60 S subunit contains two molecules of RNA with sedimentation coefficients of 28 S and 5 S and molecular weights of 1.5×10^6 and 4×10^4, respectively (Brown and Littna, 1964, 1966; Loening, 1968). A third RNA species sedimenting at 5.8–7.0 S has recently been discovered in the large ribosome subunits of somatic cells (Pène *et al.*, 1968; Sy and McCarty, 1970; King and Gould, 1970) and of amphibian oocytes (Denis and Mairy, 1972). It adheres to 28 S RNA by hydrogen bonds and probably derives from the same precursor as 28 S RNA. The smaller ribosomal subunit contains one molecule of RNA with a sedimentation coefficient of 18 S and a molecular weight of 7×10^5 (Brown and Littna, 1964; Loening, 1968). Proteins make up a little more than 50% of the ribosome mass in amphibians (Mairy and Denis, 1971; Ford, 1971a). There are about fifty different proteins in the large subunit of *X. laevis* and thirty in the small one (Ford, 1971a). Most of these proteins are present in one copy per ribosome, and no protein is apparently shared by the two subunits (Ford, 1971a). No definite function can be at present ascribed to the ribosomal proteins, and little is known about their arrangement within the ribosome.

In amphibians, as in other eukaryotes, all the major steps in ribosome formation occur in the nucleolus (Maden, 1971). The process begins with transcription of 40 S RNA, which is the common precursor to 28 S and 18 S RNA (Landesman and Gross, 1969; Loening *et al.*, 1969). Soon after transcription, the 40 S precursor is modified by methylating enzymes. It is then split into an 18 S and a 30 S molecule; 30 S RNA is further cleaved into 28 S RNA (Loening *et al.*, 1969). The 40 S precursor is about 10% longer than the finished molecules to which it gives rise. In the course of their maturation, 28 S and 18 S RNA associate with the ribosomal proteins and with 5 S RNA. Complete ribosomal subunits are released from the nucleolus into the nucleoplasm and then into the cytoplasm where protein synthesis takes place.

The nucleolar organizer which appears in metaphase cells as a secondary constriction of chromosome number 3 in *Amblystoma mexi-*

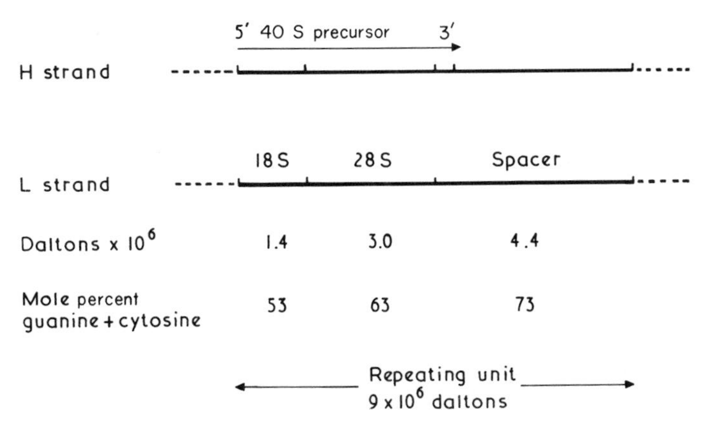

FIG. 3. Model of the repeating unit or gene coding for 28 S and 18 S RNA in *X. laevis*. Reproduced from Dawid *et al.* (1970) with permission.

canum (Callan, 1966) and of chromosome number 7 in *Bufo marinus* (Miller and Brown, 1969) is known to be the site of the structural genes for 28 S and 18 S RNA (Wallace and Birnstiel, 1966). In interphase cells, these genes are located within the nucleolus. The ribosomal genes were isolated in pure form from somatic cells of *X. laevis* by Brown and Weber (1968a) and Birnstiel *et al.* (1968). The ribosomal locus contains from 450 to 800 repeating units (Fig. 3) lying side by side (Wallace and Birnstiel, 1966; Brown and Weber, 1968b; Birnstiel *et al.* (1968). All repeating units are apparently identical to one another (Birnstiel *et al.*, 1969; Wensink and Brown, 1971). Each of them is a double-stranded stretch of DNA only half of which is transcribed into precursor 40 S RNA. The remainder of the repeating unit consists of "spacer" DNA which is not transcribed (Miller and Beatty, 1969a). As shown in Fig. 3, the spacer sequences have a different base composition from those that code for 28 S and 18 S RNA. Precursor 40 S RNA is transcribed from the DNA strand that has the higher density in cesium chloride gradients.

There are in *X. laevis* many more genes for 5 S RNA than for 28 S and 18 S RNA. The number of 5 S genes is 23,000 per haploid genome, according to Brown and Weber (1968a), and 9000 according to Birnstiel *et al.* (1972). The 5 S genes are certainly not linked to the 28 S and 18 S genes as they are in bacteria (Brown and Weber, 1968a). These genes have recently been isolated by Brown *et al.* (1972). They are apparently clustered and separated by nontranscribed spacers of low guanine + cytosine content. No information is available on the location nor on the degree of redundancy of the structural genes for ribosomal proteins.

B. Synthesis and Accumulation of Ribosomes during Oogenesis

The mature oocyte of *X. laevis* contains 4 μg of RNA (Brown and Littna, 1964). About 95% of this RNA is ribosomal in nature (Davidson *et al.*, 1964; Brown and Littna, 1966). The amount of ribosomal RNA present in one single egg is therefore at least 300,000 times as high as the amount found in a somatic cell. Ribosomes and ribosomal RNA do not accumulate at a constant rate during the whole length of oogenesis. Ribosome assembly is very slow in previtellogenic oocytes and proceeds much faster after the onset of vitellogenesis (Thomas, 1967, 1969).

To produce micrograms of ribosomal RNA, the oocyte must rely upon one single nucleus containing a tetraploid set of chromosomes. The chromosomal content of the oocyte throughout its growth is 12 pg of which only 0.05% or 0.006 pg is complementary to 28 S and 18 S RNA (Brown and Weber, 1968a). Even if the ribosomal genes were transcribed at maximum permitted rate from the beginning to the end of oogenesis, this would not be sufficient to produce 4 μg of RNA. How the oocyte nevertheless manages to synthesize so much ribosomal RNA was discovered independently by Gall (1968), Brown and Dawid (1968) and Evans and Birnstiel (1968). The oocyte amplifies its ribosomal genes at the very beginning of its growth period. The course of events can be reconstituted as follows (MacGregor, 1968; Gall, 1968). Between zygotene and pachytene stages, a very rapid synthesis of risobomal DNA takes place, and this DNA accumulates as an extrachromosomal cap containing 30 pg of DNA (MacGregor, 1968). This cap disperses at diplotene stage, and the amplified ribosomal genes are later found in the numerous nucleoli—500–700 in *A. mexicanum* (Callan, 1966); 1500 in *X. laevis* (Perkowska *et al.*, 1968)—that can be observed in growing oocytes. These nucleoli are not attached to the chromosomes (Callan, 1966). There are in each nucleolus several DNA-containing cores (Perkowska *et al.*, 1968), which are thought to be complete copies of an autosomal nucleolar organizer. From pachytene stage on, the oocyte is therefore at least 4000-ploid with respect to the nucleolar organizer and tetraploid with respect to the remainder of the genome. Miller and Beatty (1969a,b) isolated DNA-containing circular cores from oocyte nucleoli and published striking electron micrographs of the nucleolar "chromosomes" in the process of transcription. The nontranscribed spacer, the nascent 40 S precursor RNA, and the molecules of RNA polymerase could be recognized in these pictures. Until the end of oogenesis, each nucleolus synthesizes ribosomal RNA (Brown, 1966) and releases ribosomes into the cytoplasm as in somatic cells. When the oocyte reaches its maximum size, synthesis of ribosomal RNA stops, possibly through appearance of a specific repressor of the ribosomal genes (Crippa, 1970), the nucleoli gather in the center of the nucleolus (Callan, 1966) and disappear when the cell completes its first meiotic division. The amplified ribosomal genes are still present in the unfertilized egg, but they are never transcribed again (Brown and Dawid, 1968).

The 5 S RNA content of the egg is also much higher than that of a somatic cell (Brown and Littna, 1966). However, the oocyte does not amplify its 5 S genes (Brown and Dawid, 1968; Wegnez and Denis, 1972). It therefore contains far more 28 S and 18 S genes (2,000,000) than 5 S genes (100,000) but nevertheless accumulates 28 S, 18 S, and 5 S RNA in such a way that these molecules are present in equal numbers at the end of oogenesis (Brown and Littna, 1966; Mairy and Denis, 1971). Synthesis of 5 S RNA is not at all coordinate with that of 28 S and 18 S RNA (Mairy and Denis, 1971). The oocyte first accumulates large amounts of 5 S RNA, which is by far the most abundant RNA species in previtellogenic oocytes (Mairy and Denis, 1971; Ford, 1971b) (Fig. 4). At this stage of its growth, the oocyte contains at least one hundred fifty times as many molecules of 5 S RNA as molecules of 28 S RNA. A very small proportion of the 5 S RNA synthesized in previtellogenic oocytes is immediately included into the ribosomes, which are very few at that stage (Denis and Mairy, 1972; Wegnez and Denis, 1973). The 5 S RNA made in excess in small oocytes is stored partly in the cell sap and partly in ribonuclein particles sedimenting at 42 S (Ford, 1971b; Denis and Mairy, 1972). When the oocyte enters vitellogenesis, the 42 S particles disappear and the excess 5 S RNA synthesized earlier becomes included into the ribosomes (Mairy and Denis, 1972). Synthesis of 5 S RNA continues in vitellogenic oocytes, but this RNA represents a declining proportion of total RNA as the cell grows (Mairy and Denis, 1971) (Fig. 4). These observations strongly suggest that the oocyte switches on its 5 S genes during the whole length of its growth but does not fully utilize its amplified 28 S and 18 S genes until the onset of vitellogenesis. This causes the observed accumulation of 5 S RNA in small oocytes.

Almost nothing is known about the synthesis of ribosomal proteins during oogenesis. It is not even proved that these proteins are synthesized in the oocyte itself and not imported from the surrounding follicle cells. If the ribosomal proteins are indeed assembled within the oocyte, it will be interesting to see how the cell keeps the synthesis of these proteins in pace with that of ribosomal RNA.

It is important to know if there is any structural difference between the ribosomes of the egg and those of somatic cells. No comparison has yet been made as far as the ribosomal protein is concerned. The

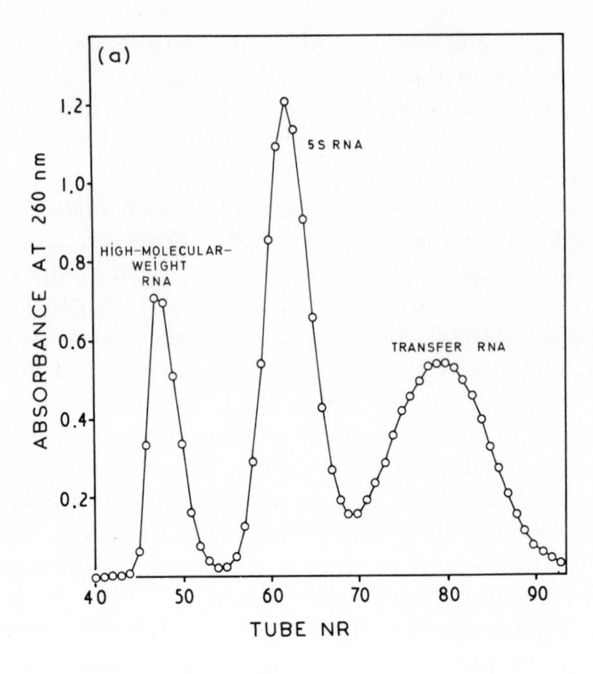

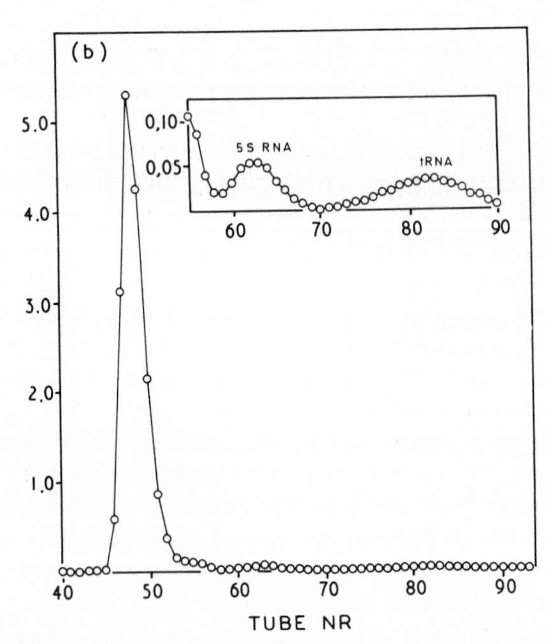

FIG. 4. Analysis on a column of Sephadex G-100 of the RNA extracted from previtellogenic oocytes (a) and from mature oocytes (b) of *X. laevis*.

largest two species of ribosomal RNA are likely to be identical in eggs and in somatic cells. Two lines of evidence support this interpretation. First, the amplified ribosomal genes are copied directly, perhaps through an RNA intermediate (Crippa and Tocchini-Valentini, 1971; Ficq and Brachet, 1971; Brown and Tocchini-Valentini, 1972; Mahdavi and Crippa, 1972), on the chromosomal genes at the beginning of the oocyte growth (Brown and Blackler, 1972). An alternative model proposed by Wallace *et al.* (1971), according to which the amplified genes would be copied on an episome transmitted from mother to daughter through the germ cell line, must be rejected (Brown and Blackler, 1972). Second, Dawid *et al.* (1970) carefully compared amplified ribosomal DNA from oocytes and ribosomal DNA from somatic cells and found them to be identical in all respects save in the degree of methylation of deoxycytidylic acid, which is higher in somatic ribosomal DNA. This difference most probably does not alter the template properties of ribosomal DNA. However, proof of the identity of 28 S and 18 S RNA in eggs and somatic cells will be gained only by partial or complete determination of the primary structure of these molecules.

A clear-cut difference between egg and somatic ribosomes was recently discovered by Denis *et al.* (1972) at the level of 5 S RNA. Somatic 5 S RNA of *X. laevis* differs from oocyte 5 S RNA by many physical and chemical properties (Denis *et al.*, 1972). These two molecules have a slightly different primary structure (Wegnez *et al.*, 1972; Ford and Southern, 1973). Oocyte 5 S RNA is also far more heterogeneous than somatic 5 S RNA. It follows that there are several types of 5 S genes in *X. laevis* cells. All 5 S genes are probably expressed in oocytes, whereas only part of them are expressed in somatic cells. Repression of the 5 S genes of oocyte type occurs at fertilization (Denis *et al.*, 1972). The observed differences between oocyte and somatic 5 S RNAs might confer different properties to the corresponding ribosomes. However, egg and oocyte ribosomes are able to correctly translate a somatic messenger that has been injected into the cell (Gurdon *et al.*, 1971; Lane *et al.*, 1971; Moar *et al.*, 1971). Until proof to the contrary appears, somatic and oocyte ribosomes must therefore be considered as functionally identical.

C. Synthesis and Accumulation of Ribosomes during Embryonic Development

Synthesis of RNA is very low during cleavage. It is strongly activated at the onset of gastrulation (Brown and Littna, 1964). At this stage, the nucleoli again become visible, and synthesis of 28 S and 18 S becomes detectable for the first time (Brown and Littna, 1964).

Failure to detect any synthesis of ribosomal RNA during cleavage is not, however, proof that the ribosomal genes are completely inactive before gastrulation (Emerson and Humphreys, 1970, 1971). The RNA content of the embryo does not increase immediately after initiation of ribosomal RNA synthesis (Fig. 5). This means that until the thirtieth hour of development, the amount of newly made RNA is negligible when compared to the amount that was present in the egg. When the embryo begins to feed (one hundredth hour of development), its RNA content is approximately doubled (Fig. 5). Since 95% of this RNA is ribosomal in nature (Brown and Littna, 1966), it follows that the ribosome stock of the embryo increases only twofold during development. It is therefore not astonishing that the embryo can live for a considerable period of time with the only ribosome stock of the egg. The anucleolate mutant of *X. laevis* discovered by Elsdale *et al.* (1958) is unable to synthesize any ribosomal RNA (Brown and Gurdon, 1964), because it has lost its ribosomal genes through a deletion (Wallace and Birnstiel, 1966), but nevertheless develops until the swimming stage. It degenerates and dies soon afterward (Wallace, 1960), presumably of protein deficiency.

Little information is available about the synthesis of the other ribosome components during embryonic development. Synthesis of ribosomal proteins is thought to start at gastrulation. Production of 5 S RNA commences at the same time as that of 28 S and 18 S RNA (Abe and Yamana, 1971) but is not coordinate with it, since the *Xenopus* embryo

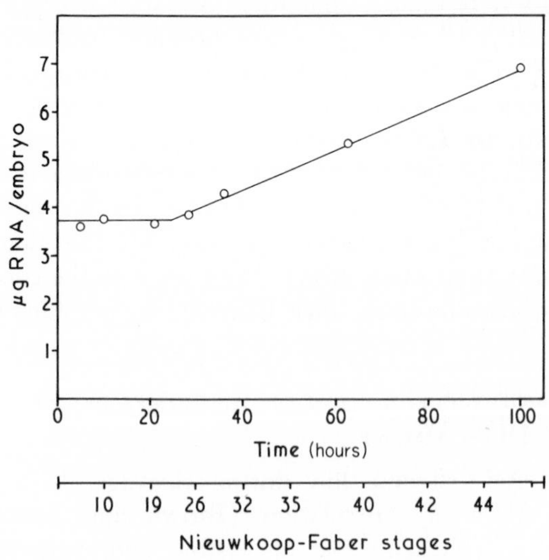

FIG. 5. Accumulation of RNA in developing embryos of *X. laevis*.

appears to synthesize more molecules of 5 S RNA than molecules of 28 S and 18 S RNA (Abe and Yamana, 1971). Only part of the 5 S molecules synthesized by the embryo would be incorporated into the ribosomes. The remainder is probably degraded. The overproduction of 5 S RNA in embryonic cells is not entirely unexpected, since there are in *Xenopus* fifty times as many 5 S genes as 28 S and 18 S ones (Brown and Weber, 1968a). It should be remembered, however, that all 5 S genes are not expressed in somatic tissues (Wegnez *et al.*, 1972; Ford and Southern, 1973).

REFERENCES

Abe, H., and Yamana, K. (1971). *Biochim. Biophys. Acta* **240**, 392.
Ancel, P., and Vintemberger, P. (1948). *Bull. Biol. Fr. Belg. Suppl.* **31.**
Balinsky, B. I., and Devis, R. J. (1963). *Acta Embryol. Morphol. Exp.* **6**, 55.
Baltus, E., and Brachet, J. (1962). *Biochim. Biophys. Acta* **61**, 157.
Baltus, E., Hanocq-Quertier, J., and Brachet, J. (1968). *Proc. Nat. Acad. Sci. U.S.* **61**, 469.
Barth, L. G., and Barth, L. J. (1954). "The Energetics of Development." Columbia Univ. Press, New York.
Birnstiel, M. L., Speirs. J., Purdom, I., Jones, K., and Loening, U. E. (1968). *Nature (London)* **219**, 454.
Birnstiel, M. L., Grunstein, M., Speirs, J., and Hennig, W. (1969). *Nature (London)* **223**, 1265.
Birnstiel, M. L., Sells, B. H., and Purdom, I. F. (1972). *J. Mol. Biol.* **63**, 21.
Boell, E. J., and Weber, R. (1955). *Exp. Cell Res.* **9**, 559.
Boell, E. J., D'Anna, T., Greenfield, P., and Petrucci, D. (1971). *J. Exp. Zool.* **178**, 151.
Borst, P., and Kroon, A. M. (1969). *Int. Rev. Cytol.* **26**, 107.
Brachet, J. (1944). "Embryologie chimique." Desoer, Liège.
Brachet, J. (1960). "The Biochemistry of Development." Pergamon, Oxford.
Brachet, J., and Needham, J. (1935). *Arch. Biol.* **46**, 821.
Bragg, A. N. (1939). *Biol. Bull.* **77**, 268.
Brown, D. D. (1966). *Nat. Cancer Inst., Monogr.* **23**, 297.
Brown, D. D., and Blackler, A. W. (1972). *J. Mol. Biol.* **63**, 75.
Brown, D. D., and Dawid, I. B. (1968). *Science* **160**, 272.
Brown, D. D., and Gurdon, J. B. (1964). *Proc. Nat. Acad. Sci. U.S.* **51**, 139.
Brown, D. D. and Littna, E. (1964). *J. Mol. Biol.* **8**, 669.
Brown, D. D., and Littna, E. (1966). *J. Mol. Biol.* **20**, 95.
Brown, D. D., and Weber, C. S. (1968a). *J. Mol. Biol.* **34**, 661.
Brown, D. D., and Weber, C. S. (1968b). *J. Mol. Biol.* **34**, 681.
Brown, D. D., Wensink, P. C., and Jordan, E. (1972). *J. Mol. Biol.* **63**, 57.
Brown, R. D., and Tocchini-Valentini, G. P. (1972). *Proc. Nat. Acad. Sci. U.S.* **69**, 1746.
Callan, H. G. (1966). *J. Cell Sci.* **1**, 85.
Chase, J. W., and Dawid, I. B. (1972). *Develop. Biol.* **27**, 504.
Crippa, M. (1970). *Nature (London)* **227**, 1138.
Crippa, M., and Tocchini-Valentini, G. P. (1971). *Proc. Nat. Acad. Sci. U.S.* **68**, 2769.

Davidson, E., Allfrey, V. G., and Mirsky, A. F. (1964). *Proc. Nat. Acad. Sci. U.S.* **52**, 501.
Dawid, I. B. (1965). *J. Mol. Biol.* **12**, 581.
Dawid, I. B. (1966). *Proc. Nat. Acad. Sci. U.S.* **56**, 269.
Dawid, I. B. (1971). *Carnegie Inst. Wash., Yearb.* **69**, 576.
Dawid, I. B. (1972). *J. Mol. Biol.* **63**, 201.
Dawid, I. B., and Blackler, A. W. (1971). *Carnegie Inst. Wash. Yearb.* **70**, 44.
Dawid, I. B., and Chase, J. W. (1971). *Carnegie Inst. Wash. Yearb.* **70**, 42.
Dawid, I. B., and Chase, J. W. (1972). *J. Mol. Biol.* **63**, 217.
Dawid, I. B., Brown, D. D., and Reeder, R. H. (1970). *J. Mol. Biol.* **51**, 341.
Dawid, I. B., Swanson, R. F., Chase, J. W., and Rebbert, M. (1971). *Carnegie Inst. Wash. Yearb.* **69**, 575.
Denis, H. (1964). *J. Embryol. Exp. Morphol.* **12**, 197.
Denis, H., and Mairy, M. (1972). *Eur. J. Biochem.* **25**, 524.
Denis, H., Wegnez, M., and Willem, R. (1972). *Biochimie.* **54**, 1189.
Deuchar, E. M. (1958). *J. Embryol. Exp. Morphol.* **6**, 223.
Elsdale, T. R., Fischberg, M., and Smith, S. (1958). *Exp. Cell Res.* **14**, 642.
Emerson, C. P., and Humphreys, T. (1970). *Develop. Biol.* **23**, 86.
Emerson, C. P., and Humphreys, T. (1971). *Science* **171**, 898.
Essner, E. S. (1953). *Protoplasma* **43**, 79.
Evans, D., and Birnstiel, M. L. (1968). *Biochim. Biophys. Acta* **166**, 274.
Ficq, A. (1961). *In* "Symposium on Germ Cells and Earliest Stages of Development," p. 121. Ist. Lombardo, Fondazione A. Baselli, Milano.
Ficq, A., and Brachet, J. (1971). *Proc. Nat. Acad. Sci. U.S.* **68**, 2774.
Flickinger, R. A. (1956). *J. Exp. Zool.* **131**, 307.
Flickinger, R. A. (1961). *In* "Symposium on Germ Cells and Earliest Stages of Development," p. 29. Ist. Lombardo, Fondazione A. Baselli, Milano.
Flickinger, R. A., and Nace, G. W. (1952). *Exp. Cell Res.* **3**, 393.
Follett, B. K., and Redshaw, M. R. (1968). *J. Endocrinol.* **40**, 439.
Ford, P. J. (1971a). *Biochem. J.* **125**, 1091.
Ford, P. J. (1971b). *Nature (London)* **233**, 561.
Ford, P. J., and Southern, E. M. (1973). *Nature New Biol.* **241**, 7.
Gall, J. (1968). *Proc. Nat. Acad. Sci. U.S.* **60**, 553.
Grant, P. (1953). *J. Exp. Zool.* **124**, 513.
Gregg, J. R. (1948). *J. Exp. Zool.* **109**, 119.
Gregg, J. R., and Løvtrup, S. (1955). *Biol. Bull.* **108**, 29.
Gross, P. R. (1954). *Protoplasma* **43**, 416.
Gross, P. R., and Gilbert, L. I. (1956). *Trans. N.Y. Acad. Sci.* [2] **19**, 108.
Gurdon, J. B., Lane, C. D., Woodland, H. R., and Marbaix, G. (1971). *Nature (London)* **233**, 177.
Hoff-Jørgensen, E., and Zeuthen, E. (1952). *Nature (London)* **169**, 245.
Holtfreter, J. (1946). *J. Exp. Zool.* **101**, 355.
Honjin, R., and Nakamura, T. (1967). *J. Ultrastruct. Res.* **20**, 400.
Honjin, R., Nakamura, T., and Shimasaki, S. (1965). *J. Ultrastruct. Res.* **12**, 404.
Hope, J., Humphries, A. A., and Bourne, G. H. (1963). *J. Ultrastruct. Res.* **9**, 302.
Hope, J., Humphries, A. A., and Bourne, G. H. (1964). *J. Ultrastruct. Res.* **10**, 547.
Karasaki, S. (1963). *J. Ultrastruct. Res.* **9**, 225.
Kemp, N. E. (1953). *J. Morphol.* **92**, 487.
King, H. W. S., and Gould, H. (1970). *J. Mol. Biol.* **51**, 687.

Kuriki, Y., and Okasaki, R. (1959). *Embryologia* **4**, 337.
Landesman, R., and Gross, P. R. (1969). *Develop. Biol.* **19**, 244.
Lane, C. D., Marbaix, G., and Gurdon, J. B. (1971). *J. Mol. Biol.* **61**, 73.
Lanzavecchia, G. (1965). *J. Ultrastruct. Res.* **12**, 147.
Leonard, R., Deamer, D. W., and Armstrong, P. (1972). *J. Ultrastruct. Res.* **40**, 1.
Loening, U. E. (1968). *J. Mol. Biol.* **38**, 355.
Loening, U. E., Jones, K. W., and Birnstiel, M. L. (1969). *J. Mol. Biol.* **45**, 353.
Luck, D. J. L., and Reich, E. (1964). *Proc. Nat. Acad. Sci. U.S.* **52**, 931.
MacGregor, H. C. (1968). *J. Cell Sci.* **3**, 437.
Maden, B. E. H. (1971). *Progr. Biophys. Mol. Biol.* **22**, 127.
Maggio, R., and Ghiretti-Magaldi, A. (1958). *Exp. Cell Res.* **15**, 95.
Mahdavi, V., and Crippa, M. (1972). *Proc. Nat. Acad. Sci. U.S.* **69**, 1749.
Mairy, M., and Denis, H. (1971). *Develop. Biol.* **24**, 143.
Mairy, M., and Denis, H. (1972). *Eur. J. Biochem.* **25**, 535.
Massover, W. H. (1971). *J. Ultrastruct Res.* **37**, 574.
Miller, L., and Brown, D. D. (1969). *Chromosoma* **28**, 430.
Miller, O. L., and Beatty, B. R. (1969a). *Science* **164**, 955.
Miller, O. L., and Beatty, B. R. (1969b). *J. Cell. Physiol.* **74**, *Suppl. 1*, 225.
Moar, V. A., Gurdon, J. B., Lane, C. D., and Marbaix, G. (1971). *J. Mol. Biol.* **61**, 93.
Nass, S. (1956). *Trans. N.Y. Acad. Sci.* [2] **19**, 118.
Nass, S. (1962). *Biol. Bull.* **122**, 232.
Panijel, J. (1950). *Biochim. Biophys. Acta* **5**, 343.
Panijel, J. (1951). "Etude cytochimique et biochimique de la gamétogenèse et de la fécondation chez la grenouille et l'Ascaris." Hermann, Paris.
Pasteels, J. (1951). *Bull. Soc. Zool. Fr.* **76**, 231.
Pène, J. J., Knight, E., and Darnell, J. E. (1968). *J. Mol. Biol.* **33**, 609.
Perkowska, E., MacGregor, H. C., and Birnstiel, M. L. (1968). *Nature (London)* **217**, 649.
Petrucci, D. (1957). *Acta Embryol. Morphol. Exp.* **1**, 105.
Petrucci, D. (1959). *Arch. Sci. Biol.* **43**, 25.
Rabinowitz, M., Sinclair, J., DeSalle, R., Haselkorn, R., and Swift, H. H. (1965). *Proc. Nat. Acad. Sci. U.S.* **53**, 1126.
Redshaw, M. R., and Follett, B. K. (1971). *Biochem. J.* **124**, 759.
Ringle, D. A., and Gross, P. R. (1962a). *Biol. Bull.* **122**, 263.
Ringle, D. A., and Gross, P. R. (1962b). *Biol. Bull.* **122**, 281.
Roodyn, D. B., and Wilkie, D. (1968). "The Biogenesis of Mitochondria." Methuen, London.
Rounds, D. E., and Flickinger, R. A. (1958). *J. Exp. Zool.* **137**, 479.
Rudack, D., and Wallace, R. A. (1968). *Biochim. Biophys. Acta* **155**, 299.
Selman, G. G. (1958). *J. Embryol. Exp. Morphol.* **6**, 448.
Spiegelman, S., and Steinbach, H. C. (1945). *Biol. Bull.* **88**, 254.
Strittmatter, C. F., Strittmatter, P., and Burdick, C. (1960). *Biol. Bull.* **119**, 341.
Swanson, R. F., and Dawid, I. B. (1970). *Proc. Nat. Acad. Sci. U.S.* **66**, 117.
Sy, J., and McCarty, K. S. (1970). *Biochim. Biophys. Acta* **199**, 86.
Sze, L. C. (1953). *Physiol. Zool.* **26**, 212.
Thomas, C. (1967). *Arch. Biol.* **78**, 347.
Thomas, C. (1969). *J. Embryol. Exp. Morphol.* **21**, 765.
Waddington, C. H. (1956). "Principles of Embryology." Allen & Unwin, London.
Wallace, H. (1960). *J. Embryol. Exp. Morphol.* **8**, 405.

Wallace, H., and Birnstiel, M. L. (1966). *Biochim. Biophys. Acta* **114,** 296.
Wallace, H., Morray, J., and Langridge, W. H. R. (1971). *Nature New Biol.* **230,** 201.
Wallace, R. A. (1963a). *Biochim. Biophys. Acta* **74,** 495.
Wallace, R. A. (1963b). *Biochim. Biophys. Acta* **74,** 505.
Wallace, R. A. (1964). *Biochim. Biophys. Acta* **86,** 286.
Wallace, R. A. (1965). *Anal. Biochem.* **11,** 297.
Wallace, R. A. (1970). *Biochim. Biophys. Acta* **215,** 176.
Wallace, R. A., and Dumont, J. N. (1968). *J. Cell Physiol. Suppl. 1,* **72,** 73.
Wallace, R. A., and Jared, D. W. (1968a). *Can. J. Biochem.* **46,** 953.
Wallace, R. A., and Jared, D W. (1968b). *Science* **160,** 91.
Wallace, R. A., and Jared, D. W. (1969). *Develop. Biol.* **19,** 498.
Wallace, R. A., and Karasaki, S. (1963). *J. Cell Biol.* **18,** 153.
Wallace, R. A., Jared, D. W., and Nelson, B. L. (1970). *J. Exp. Zool.* **175,** 259.
Wallace, R. A., Nickol, J. M., Ti Ho, and Jared, D. W. (1972). *Develop. Biol.* **29,** 255.
Ward, R. T. (1959). *Anat. Rec.* **134,** 651.
Ward, R. T. (1962). *J. Cell Biol.* **14,** 309.
Wartenberg, H. (1962). *Z. Zellforsch. Mikrosk. Anat.* **58,** 427.
Weber, R., and Boell, E. J. (1955). *Rev. Suisse, Zool.* **62,** 260.
Wegnez, M., and Denis, H. (1972). *Biochimie,* **54,** 1069.
Wegnez, M., and Denis, H. (1973). *Biochimie,* **55,** 1129.
Wegnez, M., Monier, R., and Denis, H. (1972). *FEBS Letters,* **25,** 13.
Wensink, P. C., and Brown, D. D. (1971). *J. Mol. Biol.* **60,** 235.
Williams, J. (1965). *In* "The Biochemistry of Animal Development" (R. Weber, ed.), Vol. 1, p. 13. Academic Press, New York.
Wischnitzer, S. (1966). *Advan. Morphog.* **5,** 131.
Wolstenholme, D. R., and Dawid, I. B. (1967). *Chromosoma* **20,** 445.
Wolstenholme, D. R., and Dawid, I. B. (1968). *J. Cell Biol.* **39,** 222.

CHAPTER 2

Gastric Secretion in Amphibia

Gabriel M. Makhlouf and Warren S. Rehm

I. Introduction 23
A. Scope of Review 23
B. *In Vivo* and *in Vitro* Secretion—A Comparison 24
C. Gastric Mucosa—Structure and Function 24
D. Cellular Origin of Secretion 27
II. Chloride Transport 27
A. Historical Prelude 27
B. Components of Chloride Transport 28
C. Chloride Carriers 28
III. Separate Hydrogen Transport Mechanism 30
IV. Coupling of Chloride and Hydrogen Mechanisms 31
V. Electrical Aspects of Secretion 34
A. Site of Gastric Potential Difference 34
B. Ionic Gradients and Gastric Potential Difference 35
VI. Energetic Aspects of Secretion 37
A. Energetic Requirements 37
B. Acid Secretion and Oxygen Consumption 38
C. Acid Secretion and Phosphate Metabolism 38
VII. Ionic Requirements for Acid Secretion 42
A. Sodium 42
B. Potassium 42
C. Calcium 42
VIII. Transport of Sodium, Potassium, and Water 43
A. Sodium Transport 43
B. Potassium Transport 43
C. Water Transport 44
References 46

I. Introduction

A. Scope of Review

Few studies have been concerned with the control of gastric secretion in Amphibia. Scattered observations—for example, the sensitivity of the mucosa to the mammalian antral hormone gastrin and the dependence of response to this hormone on the integrity of vagal pathways—suggest that regulatory mechanisms in amphibia resemble those in mammals (Kasbekar *et al.*, 1969; Morrissey and Wan, 1970; Grossman, 1967).

Most studies have been carried out *in vitro* and have been concerned largely with elucidating cellular mechanisms of secretion. This review will, therefore, focus on the analysis and interpretation of ionic transport and electrical activity in the isolated gastric mucosa of the frog and, to a lesser extent, the *Necturus.*

B. *In Vivo* and *in Vitro* Secretion—A Comparison

It is well to inquire at the outset how accurately *in vitro* secretion reflects secretion by the living animal. *In vitro* rates average 3–5 μeq cm^{-2} hour^{-1}; occasionally, higher rates are obtained, but these are still below *in vivo* rates. Morrissey and Wan (1970) report that the whole stomach of *Rana pipiens* with intact blood supply can secrete up to 200 μeq hour^{-1}, which is equivalent to 10–20 μeq cm^{-2} hour^{-1} of stretched fundic mucosa.

These comparisons provide a basis for interpreting certain differences observed *in vitro*. Carbonic anhydrase inhibitors such as Diamox inhibit mammalian *in vivo* secretion by about 90% but only inhibit amphibian *in vitro* secretion by 50% (Durbin and Heinz, 1958; Hogben, 1955b, 1966). This discrepancy suggests that the low rates observed *in vitro* may be near the uncatalyzed rate for the hydration of carbon dioxide. Recent studies support this view. When secretion is increased twofold (up to 9 μeq cm^{-2} hour^{-1}) by addition of fluorocarbon (a liquid with high oxygen capacity) to the bathing solutions, it becomes more susceptible to inhibition by Diamox (10^{-3} *M*) (Sachs *et al.,* 1970). However, the increase of secretion with fluorocarbon reported in this study has not been duplicated (R. L. Shoemaker, personal communication).

C. Gastric Mucosa—Structure and Function

The corpus of the stomach in Amphibia is histologically distinct from the distal portion, or antrum. Both portions are invaginated into branched tubular glands, but only the corpus glands are lined by granular, oxyntic, or acid-secreting cells. Some zymogen granules may be seen in the cytoplasm of these cells, but the peptic cells proper are mostly located in the lower esophagus. Oxyntic cells line the tubules, while mucous cells, also referred to as surface epithelial cells, line the pit region (Fig. 1). The striking difference between the amphibian and mammalian oxyntic cells is the absence of intracellular canaliculi in Amphibia (Ito, 1967). Instead, the apical (luminal) surface is thrown into folds and microvilli (Fig. 2). These are few in number, but they increase greatly during secretion only to disappear again when secretion is inhibited (Sedar, 1961a,b, 1962; Sedar and Forte, 1963). The increase during

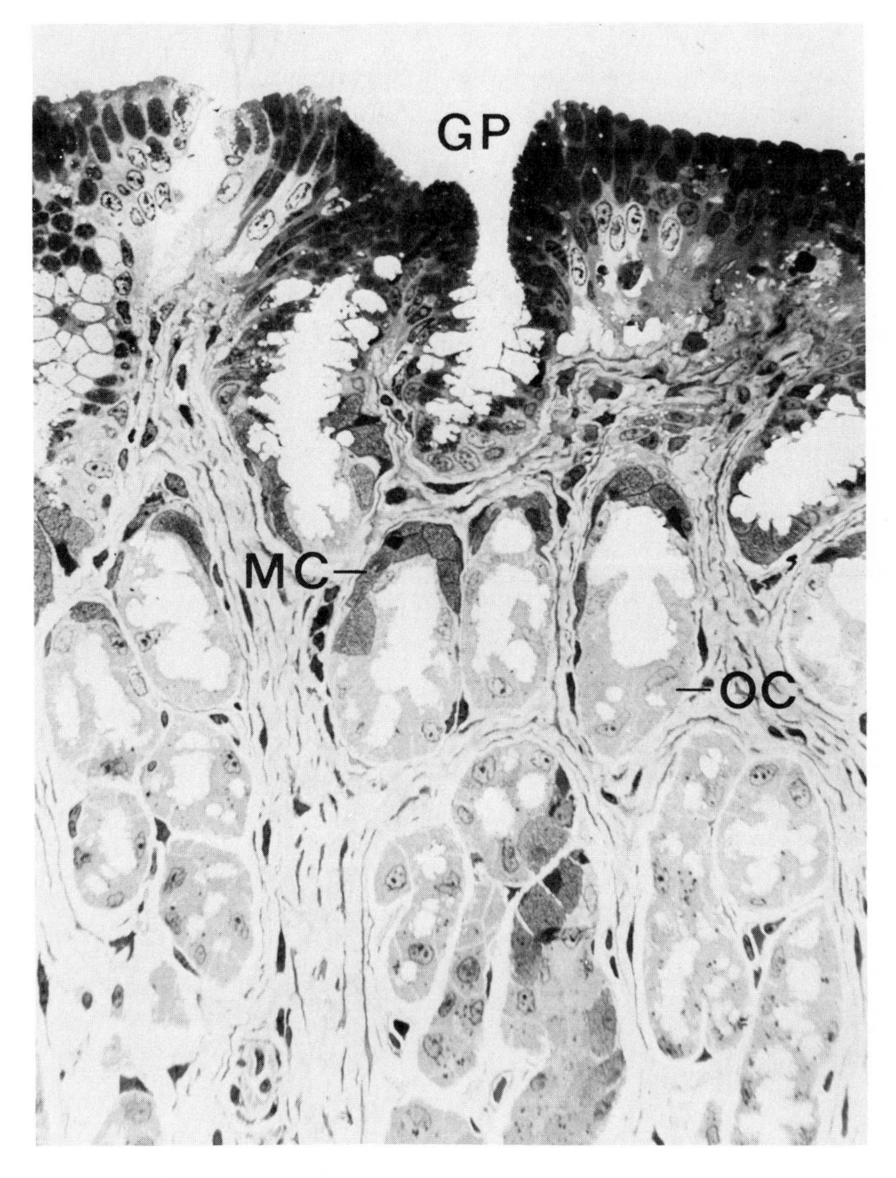

FIG. 1. Photomicrograph of the gastric mucosa of the fasted frog (*Rana catesbeiana*). GP = gastric pit, MC = mucous cells, OC = oxyntic cells. (Ito, 1971.)

secretion is thought to occur at the expense of the tubulovesicular system (smooth endoplasmic reticulum). This system of fine tubules is characteristic of oxyntic cells. It becomes abundant during stage XXIV of

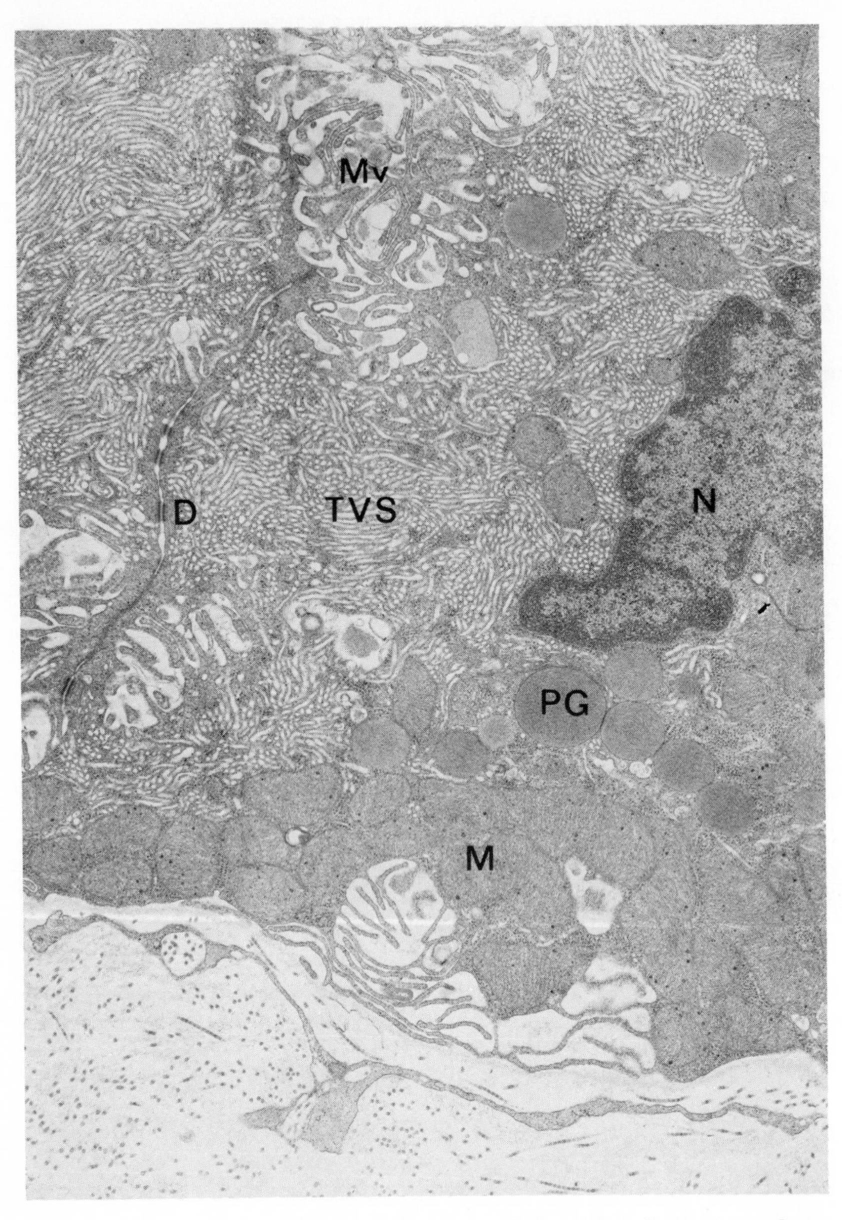

FIG. 2. Electronmicrograph of one oxyntic cell from fasted *Rana catesbeiana*. N = nucleus, PG = pepsin granules, M = mitochondria, D = desmosomes, TVS = tubulovesicular system (smooth endoplasmic reticulum), Mv = microvilli on apical (luminal) surface. Magnification × 11,000. (Ito, 1971.)

metamorphosis when acid secretion first appears (G. M. Forte *et al.*, 1969). During secretion, the tubules advance to the apex of the cell to be in dynamic continuity with the gastric lumen (Sedar, 1962; Forte, 1971).

Their behavior has given rise to the notion of a membrane replicating itself to fit the secretory requirements of the cell.

D. Cellular Origin of Secretion

The evidence which implicates the oxyntic cell of amphibia (or the parietal cell of mammals) as the site of acid secretion ranges as Davies (1959) aptly notes from "divine intervention to the use of Prussian blue." These cells contain large numbers of mitochondria (up to 20% of cell volume in the frog), a fact which is consistent with the high oxidative requirements for acid secretion. They also show proliferative changes of the luminal membrane which result in narrow channels capable of fulfilling the requirements for osmotic equilibration of secretion. Structural evidence such as this is favorable but not compelling. For example, the secretory cell of the salivary gland of the blowfly (*Calliphora erythrocephala* Meig) is a close morphological homolog of the gastric oxyntic cell in that it possesses a high complement of mitochondria and intracellular canaliculi with microvilli, but unlike the oxyntic cell, it does not develop detectable membrane changes when its secretion of isotonic potassium chloride increases sixtyfold (Ochsman and Berridge, 1970).

More reliable evidence can be derived from the combined study of structure and function. In the frog, where only two cell types exist, virtual denudation of the surface epithelium decreases but does not abolish acid secretion (Durbin, 1963–1964; Teorell and Wersall, 1945). A similar observation has been made in the mouse, but a more pertinent finding in mammals is that acid secretion and oxyntic cells appear simultaneously during gestation. The evidence derived from the use of acid–base indicators, though hallowed by continuous reference, is now considered of doubtful validity (Bradford and Davies, 1948; White *et al.*, 1956; Rehm, 1971).

II. Chloride Transport

A. Historical Prelude

Because the stomach concentrates acid by a factor of over one million, it seemed natural at first to assume that hydrogen ions were actively secreted, driving on their way a companion ion of opposite charge. Some early studies sought to demonstrate that anions moved passively across the gastric mucosa. Rehm (1950) pointed out that passive movement of chloride was incompatible with gastric electrophysiology, since chloride moved against both the concentration and electrical gradients. The argument was first enunciated for mammals but applied with equal validity to Amphibia in whom the stomach secretes isotonic hydrochloric acid and generates a potential difference of similar orientation (lumen negative). The advent of tracer techniques enabled Hogben (1951, 1952) to show by simultaneous flux and electrical measurements the capacity of the amphibian gastric mucosa to secrete chloride actively.

B. Components of Chloride Transport

Chloride transport can be separated into three components: active transport, exchange diffusion, and passive diffusion. Active transport is illustrated in the results of a typical experiment on the isolated frog mucosa (Table I) (Hogben, 1951). In the absence of an electrical or concentration gradient, a net flux of chloride from serosa to mucosa is observed which is equal to the algebraic sum of the hydrogen secretion rate J_H and the short circuit current I_{sc} (all expressed in μeq cm^{-2} hour^{-1}).

$$J_{sm}^{Cl} - J_{ms}^{Cl} = J_{net}^{Cl} = J_H + I_{sc} \quad (1)$$

The preponderance of net chloride flux over hydrogen rate indicates that the chloride mechanism is responsible for the orientation of the potential difference (lumen negative).

Exchange diffusion is present, since the unilateral flux of chloride against the direction of net transport, i.e., J_{ms}^{Cl} is greater than the total conductance of the mucosa. Evidently, most of this chloride traverses the mucosa in nonconducting form, bound to a membrane fragment or carrier. Some chloride is assumed to move by passive diffusion (Forte, 1969).

C. Chloride Carriers

The location of a chloride carrier on the mucosal (luminal) membrane is supported by two lines of evidence. Experiments in which the intracellular environment is allowed to exchange with isotopic chloride show

TABLE I

Chloride Transport by the Isolated Short Circuited Frog Gastric Mucosa[a]

All values in μEq cm^{-2} hr^{-1}

	Net Cl flux	Net charge transfer	
Serosa to mucosa	10.65	Current	3.05
Mucosa to serosa	6.38	H^+	1.20

[a] Hogben (1955b). Mean conductance: start, 3.21 mmho; end, 2.28 mmho.

that equilibration is more rapid across the luminal surface (Cotlove and Hogben, 1956). If the same surface is exposed to solutions of low chloride content, the chloride fluxes in both directions are reduced (Heinz and Durbin, 1958). This is taken to mean that the chloride fluxes are coupled by sharing the same carrier: the concentration of chloride in the mucosal solution dictates the supply of carrier (and companion chloride ion) to the inner surface of the luminal membrane, thus regulating serosal-to-mucosal flux.

The presence of a second chloride carrier located on the serosal membrane is indicated by the following evidence. For every hydrogen secreted, an equivalent base is formed within the cell (Davies, 1951; Teorell, 1951). Since the hydrogen and hydroxyl conductances of the serosal membrane are close to zero, neither the exit of a hydroxide ion nor the entry of a hydrogen ion across the serosal membrane could effect neutralization of cellular contents (Sanders and Rehm, 1970). This is more likely brought about by the operation of the carbon dioxide system and the production of bicarbonate ions at a rate approximately equal to the hydrogen secretory rate. However, the ionic conductance of the serosal membrane to bicarbonate is also close to zero, so that bicarbonate must exit from the cell by a neutral mechanism (Rehm and Sanders, 1972). Chloride enters in equivalent amounts, and the requirements of electroneutrality are such as to restrict its entry mainly to a neutral mechanism, the simplest being a carrier-mediated chloride: bicarbonate exchange. However, entrance of a chloride with a cation (sodium or potassium) via a carrier and exit of bicarbonate with the same cation via the same or a different carrier could account for the observations.

Other findings support the view of carrier-mediated chloride movement across the serosal membrane. Experiments in which the resistance of the serosal membrane is increased by addition of 1 m*M* barium to the serosal fluid support the view of a neutral chloride mechanism at the serosal membrane (Pacifico *et al.*, 1969). Despite the rise in resistance, hydrogen rate is not altered; calculations show that due to the high resistance of the membrane, chloride and bicarbonate could not cross as free ions (i.e., via conductance channels).

It might be argued that an active chloride:bicarbonate mechanism at the serosal membrane would do away with the need for a carrier-mediated chloride mechanism on the luminal membrane despite the independent evidence adduced above for the latter. On this basis, however, it would be difficult to explain the orientation of the potential difference or the fact that thiocyanate inhibits hydrogen but not active chloride transport. It should be pointed out that unlike bicarbonate, chloride can also move across the serosal membrane as a free ion; the total ionic conductance of the membrane being equal to the sum of potassium and chloride conductances (see below) (Harris and Edelman, 1964; Spangler and Rehm, 1968).

The chloride carriers impart a degree of specificity to chloride transport. The chloride mechanisms, however, can transport other monovalent anions (Davies and Forte, 1963; Durbin *et al.*, 1969; Hogben and Green, 1958). Some, like bromide and nitrate, sustain acid secretion while others, like thiocyanate, do not. The order of the efficacy of the halogen series in sustaining acid secretion is bromide, chloride, and iodide (Durbin, 1963–1964).

The exact steric and charge requirements for the operation of the carriers remain largely unknown. Covalent linkage is not likely but electrostatic binding may prevail. It is of interest that large anions, and also nonelectrolytes, in the mucosal solution reduce considerably the flux of chloride from serosa to mucosa (Durbin *et al.*, 1964).

III. Separate Hydrogen Transport Mechanism

Heinz and Durbin (1959) obtained the first clear evidence for a separate hydrogen transport mechanism after replacing chloride in the bathing solution with sulfate, a nontransportable anion. Partial replacement with sulfate reduces the utilization of chloride and brings about a proportionate decrease in acid rate and short circuit current. Complete replacement markedly decreases the potential difference in the resting mucosa but reverses its sign in the secreting mucosa (Rehm *et al.*, 1963). The change in orientation of the potential difference during secretion suggests that an independent electrogenic hydrogen pump can operate in the absence of net anion transport. If the potential difference during secretion arises from the electromotive force of an electrogenic hydrogen mechanism, then inhibition of secretion should result in a concurrent decrease in potential difference. This, in fact, is the case, and a precise linear relationship can be demonstrated between hydrogen rate and potential difference (Fig. 3), thus establishing the validity of the electrogenic nature of the hydrogen pump under these conditions (Rehm and Lefevre, 1965).

Isethionate is also an effective substitute for chloride and has the advantage over sulfate in being monovalent (Forte *et al.*, 1963; Kaneko-Mohammed and Hogben, 1964). Gluconate and glucuronate, however, fail to sustain hydrogen secretion, which displays, in this instance, an obligatory dependence on the presence of a minimal concentration of chloride (about 5 m*M*) (Durbin, 1963–1964).

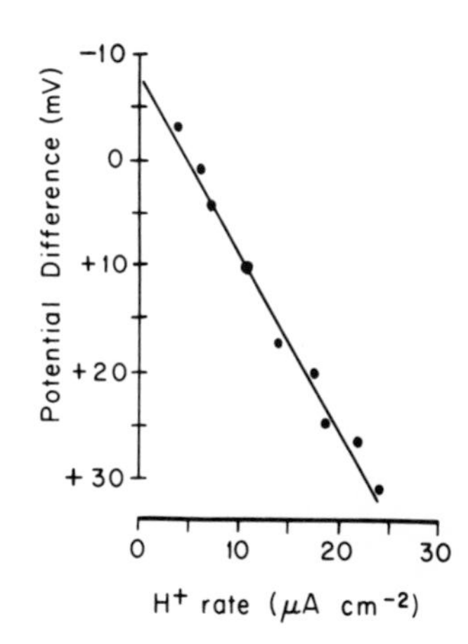

FIG. 3. Inverse linear relation between the hydrogen rate and the potential difference in sulfate solutions (Rehm and Lefevre, 1965).

An independent Cl mechanism is easily demonstrable in the presence of thiocyanate; acid secretion is inhibited and the equality embodied in Eq. (1) reduces to

$$J_{\text{net}}^{\text{Cl}} = I_{\text{sc}} \tag{2}$$

Thus far, the evidence indicates that under specific conditions two separate mechanisms (pumps), each capable of generating an electromotive force by the net transfer of electrical charge can be distinguished: a monovalent anion pump capable of operating in the absence of hydrogen secretion and a hydrogen pump capable of operating in the absence of net anion transport. Normally, however, the two mechanisms are coupled electrically and biochemically. The nature of this linkage is discussed below.

IV. Coupling of Chloride and Hydrogen Mechanisms

Two types of mechanisms may be considered (Rehm, 1967). One is a unitary, neutral mechanism in which both hydrogen and chloride are secreted at the same site by a single mechanism (Fig. 4a). The other is a separate site or electrogenic mechanism in which chloride and hydrogen (and thus charge) are transported at separate sites across the membrane (Fig. 4b). In this latter instance, the separate hydrogen and chloride mechanisms are coupled electrically.

In the case of the unitary mechanism, the potential difference would arise indirectly as a result of ion gradients between the cell and bathing fluids produced by the transport mechanism. In the electrogenic scheme, the hydrogen and chloride mechanisms would constitute electric batteries, each with its own electromotive force (emf) and resistance. These

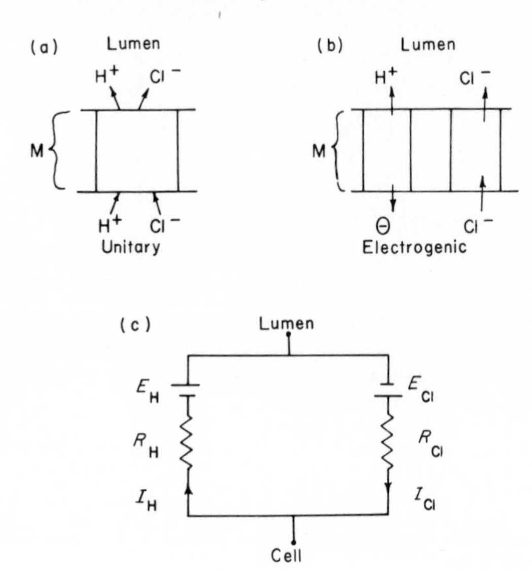

FIG. 4. The two principal schemes proposed for hydrochloric acid secretion. A unitary scheme (a), where a set of biochemical reactions results in the simultaneous secretion of hydrogen and chloride; there is no net transport of charge across the membrane and the resistance is expected to be high. An electrogenic or separate site scheme (b) with hydrogen and chloride secreted at separate sites; there is a net transport of charge across the membrane at each site with electrical coupling between sites. The equivalent circuit for the electrogenic scheme is shown in c.

emfs would be present in the absence of hydrogen and chloride gradients across the membrane, but their magnitudes would be a function of the chloride and hydrogen gradients across the respective mechanisms (Rehm, 1965). An equivalent circuit for the electrogenic scheme is shown in Fig. 4c. The potency of the mechanisms is given by the ratio of emf to resistance, i.e. E_A/R_H and E_{Cl}/R_{Cl} for hydrogen and chloride, respectively.

It is possible that the hydrochloric acid mechanism is neither completely electrogenic nor completely neutral and that the degree of electrogenicity may not be constant (Rehm, 1964). The ability to dissociate the chloride and hydrogen mechanisms alluded to earlier (addition of thiocyanate, substitution of sulfate for chloride) would seem to favor a separate site theory. However, the results of short circuiting the mucosa raise some challenging problems. For a group of mucosae in the steady-state, no significant change in the short circuit current can be detected despite increasing hydrogen rates (Durbin and Heinz, 1958). On the basis of this and other observations, Durbin and Heinz concluded that net chloride transport was the product of two separate chloride mechanisms; a unitary mechanism for hydrochloric acid and a parallel independent "electromotive" chloride mechanism. According to this scheme, no change in the short circuit current is expected in going from the resting to the secreting state and vice versa. We agree with this reasoning.

However, later studies on individual mucosae showed significant transient increases in short circuit current with the onset of secretion and transient increases with thiocyanate inhibition of secretion (Rehm, 1962). In agreement with the observations of Durbin and Heinz (1958), the steady state short circuit current during secretion did not vary from that in the resting state. Furthermore, there were frequent occasions in which the hydrogen rate could be made to increase by increasing the dose of histamine without change in short circuit current (Rehm, 1962). Within the framework of the separate site theory, this makes it necessary to postulate biochemical coupling of the separate electrogenic mechanisms in addition to the obvious electrical coupling. It is conceivable that if biochemical coupling is sufficiently tight, equal increments in hydrogen and chloride rates would leave the short circuit current fixed. A scheme of tightly coupled electrogenic mechanisms, however, would be difficult to differentiate experimentally from the scheme of Heinz and Durbin (1958). One approach might involve the study of relationship between electrical resistance and hydrogen rate. In contrast to a neutral scheme, a separate site electrogenic scheme demands a low resistance. In the scheme of Durbin and Heinz (1958) the resistance is not expected to vary with the hydrogen rate. Experimentally, however, the resistance decreases as the hydrogen rate increases, a finding which favors the electrogenic scheme (Rehm, 1965; Noyes and Rehm, 1970).

It is worth noting also that implicit in the scheme of Durbin and Heinz is an assumption of a two-limb equivalent circuit, one limb for the chloride electrogenic mechanism and a parallel limb for the hydrogen electrogenic mechanism. A more realistic construct should include circuits for both the luminal and serosal membranes (Rehm and Sanders, 1972); the short circuit current I_{sc} in this instance is given by

$$I_{sc} = \frac{E_0}{R_{Cl} + R'_{Cl}} - \frac{R_{Cl} I_H}{R_{Cl} + R'_{Cl}} \tag{3}$$

where E_0 and R_{Cl} are, respectively, the emf and resistance of the luminal chloride mechanism, R'_{Cl} the resistance of the chloride channels in the serosal membrane, and I_H the hydrogen rate expressed in milliamperes per square centimeter. The best estimates indicate that R'_{Cl} is substantially greater than R_{Cl} (Spangler and Rehm, 1968; Kidder and Rehm, 1970; Noyes and Rehm, 1970; unpublished data of J. A. Wright and W. S. Rehm). It follows after substitution of reasonable values in Eq. (3) that the short circuit current I_{sc} would change only slightly during marked changes in hydrogen rate and, if R_{Cl} were negligible compared to R'_{Cl}, would be virtually independent of the hydrogen rate. The simple circuit implicit in the scheme of Durbin and Heinz, however, assumes that R'_{Cl} is zero so that Eq. (3) becomes

$$I_{sc} = \frac{E_0}{R_{Cl}} - I_H \tag{4}$$

According to Eq. (4), the short circuit current should decrease by an amount equal to the increase in hydrogen rate. The more realistic equivalent circuit and estimates of resistance thus obviate objections to the separate site electrogenic theory based on the scheme of Durbin and Heinz.

V. Electrical Aspects of Secretion

A. Site of Gastric Potential Difference

The isolated amphibian gastric mucosa maintains a potential difference such that the lumen is negative with respect to the serosa (average —30 mV in the frog and —20 mV in *Necturus*). The oxyntic cell, which is mainly responsible for ion and water transport, is also mainly responsible for electrical activity. Forceful blotting of the mucosa removes the surface epithelial cells but spares the oxyntic cells in the deeper regions, resulting in a lower potential difference without significant alteration in short circuit current (Durbin, 1963, 1967). Unlike acid, the short circuit current is not affected by leakage of ions through a damaged surface.

Villegas (1962) studied the potential drops across the serosal and mucosal membranes separately with microelectrodes. The oxyntic cell, marked electropheretically with dye for later identification, shows a drop of 11 mV (cell positive to lumen) across the luminal membrane and of 18 mV (serosa positive to cell) across the serosal membrane.

Recent studies of *Necturus* surface cells (unmarked but punctured under direct vision) show a potential difference of 29 mV (cell negative to lumen) across the luminal membrane and of 46 mV (cell negative to serosa) across the serosal membrane (Shoemaker *et al.*, 1969, 1970). Smaller potential differences are observed in surface antral cells. The changes of potential observed on passing current across the mucosa or on addition of mecholyl to the bathing solutions indicate that surface cells are the site of at least one electromotive force.

Both on developmental and functional grounds, it seems likely that all gastric cells possess an active chloride mechanism responsible in part for the transmucosal potential difference. This mechanism is present in the tadpole stomach prior to the development of acid secretion and is partly responsible for the orientation of the potential difference (J. G. Forte *et al.*, 1969). An identical mechanism is retained in the mature antrum (Shoemaker, 1967). It is also retained in the mature fundus, where it is coupled to a distinctive hydrogen mechanism in oxyntic cells. The event is heralded morphologically by the appearance during stage XXIV of metamorphosis of numerous mitochondria and tubulovesicles (smooth endoplasmic reticulum) and, biochemically, by the development of high oxidative metabolism and a sensitivity to anoxia (Durbin, 1968; Villegas, 1964; Forte, 1969).

B. Ionic Gradients and Gastric Potential Difference

A full description of electrical activity in the gastric mucosa must take into account the contribution of ionic gradients to the potential across the separate cellular interfaces.

1. *Serosal Permeability*

The chloride mechanism on the luminal membrane maintains a low intracellular chloride, while a sodium pump on the serosal membrane maintains a high intracellular potassium and a low sodium. This distribution of ions permits the development of diffusion potentials across a permeable serosal membrane.

It was noted earlier that studies on the effect of changing the ionic composition of the serosal solution show that the conductivity of the mucosa is equal to the sum of the chloride and potassium conductances (Harris and Edelman, 1964; Spangler and Rehm, 1968). Thus, when either chloride or potassium are varied at constant product ($K \times Cl$ = constant), the change in potential difference for a tenfold change in concentration is about 56 mV, this being the magnitude predicted on the basis that the total conductance of the serosal membrane is equal to the sum of the chloride and potassium conductances. J. G. Forte *et al.* (1969) arrived at a similar estimate (57 mV) for the tadpole stomach prior to the development of acid secretion.

Other findings support this view; changing the concentration of hydrogen, hydroxide, bicarbonate, or sodium on this side does not induce a significant change in potential (Rehm, 1967). Assuming chemical equilibrium to prevail for potassium chloride across the serosal membrane, the potential drop of 18 mV (serosal fluid positive to cell) across this membrane shown by Villegas (1962) would be satisfied by potassium and chloride concentrations in the continuous phase of the cell of about 8 and 40 m*M*, respectively. The value of 40 m*M* for chloride is consistent with most calculations of intracellular concentration for this ion, but the value for potassium is much too low to satisfy average intracellular potassium concentration. However, there is some evidence that most of the potassium is compartmentalized and that consequently the residual potassium concentration in the continuous phase of that cell is low (Rehm *et al.*, 1966). It seems, therefore, probable that both potassium and chloride are distributed across the serosal membrane at or near electrochemical equilibrium.

2. *Mucosal (Luminal) Permeability*

The results of changes in ionic composition of the mucosal fluid are more difficult to interpret. Replacing chloride with an impermeant anion (sulfate or isethionate) results in a variable increase in potential difference (lumen more negative) (Rehm, 1965; Rehm *et al.*, 1963). Replacing sodium with potassium (i.e., increasing potassium concentration) yields a response like that shown on the left side in Fig. 5 (Rehm, 1968). The initial rise in potential difference (lumen more negative) indicates a greater conductance of the mucosal membrane to potassium than to sodium. However, the potential difference reaches a peak in about 2 minutes and then slowly declines to about the original level. This type of response suggests a feedback mechanism tending to maintain a con-

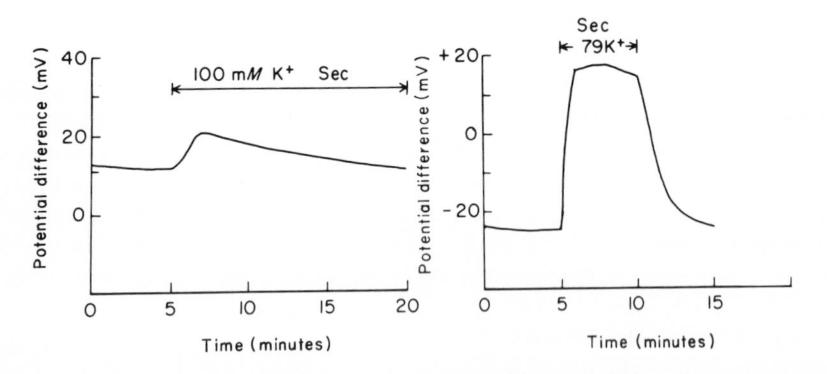

FIG. 5. Effect on the transmucosal potential difference of increasing potassium concentration in the luminal solution. *Left:* A change of potassium from 4 m*M* to 100 m*M* in chloride solutions. *Right:* A change of potassium from 4 m*M* to 79 m*M* in sulfate solutions. Positive potential difference indicates serosa is positive to lumen in external circuit.

stant potential difference in the face of changing ionic composition. It is of interest that a marked elevation of potassium on the mucosal side (or serosal side) results in a decrease in net chloride transport (Hogben, 1968). Since the hydrogen rate is not appreciably changed by elevation of potassium on the mucosal side, it appears that the decrease in net chloride transport is balanced by an approximately equal increase in potassium transport from mucosa to serosa.

In chloride-free solutions (sulfate replacing chloride on both sides) changes from sodium to potassium in the mucosal fluid result in a large and sustained increase in potential difference (Fig. 5, right side) (Davis *et al.*, 1963, 1965). This more profound effect on potential difference in chloride-free solutions is also seen with other cations. Thus, a change in hydrogen concentration on the mucosal side from pH 7 to pH 2 has almost no effect on the potential difference, but a change to pH 1 increases the potential difference by 13 mV in chloride solutions and by 24 mV in chloride-free (sulfate) solutions (Harris and Edelman, 1964). The increase of potential difference may be due to passive conductance of the luminal membrane to hydrogen or due to a change in the emf of the electrogenic hydrogen mechanism (Rehm, 1972).

Microelectrode studies of *Necturus* surface epithelial cells confirm the heightened response with respect to sodium (Makhlouf *et al.*, 1969). Addition of sodium to a luminal solution containing choline sulfate increases mostly the potential across the luminal membrane. Addition of sodium to the serosal side is as usual without effect. Similar responses can also be elicited from antral surface cells.

It seems reasonable to conclude that in chloride solutions the total conductance of the mucosal membrane is given by the sum of the conductances of the chloride and hydrogen electrogenic mechanisms (g_{Cl} and g_H in Fig. 4c) and that the conductance of all other ions is relatively low. This view of the mucosal membrane accords well with the fact that the amphibian stomach absorbs but little Na and does not easily dissipate secreted hydrogen, the leak or passive H^+ conductance being very low.

VI. Energetic Aspects of Secretion

A. Energetic Requirements

A comparison of the minimum energy required to concentrate hydrogen ions in gastric secretion (8600 cal/mole for 0.12 *N* hydrochloric acid in the frog) and the maximum energy derived from glucose oxidation (114,000 cal/mole oxygen utilized) yields only a gross estimate of energy conversion by the gastric mucosa, since a calculation of energy based on concentration and electrical gradients ignores the internal resistance of the system (Crane and Davies, 1949).

B. Acid Secretion and Oxygen Consumption

The quantitative relation which must exist between the amount of acid secreted and metabolic agents consumed (for example oxygen, ATP) has been widely used to evaluate models linking metabolism and secretion. Models of the redox type (Conway and Brady, 1948; Crane *et al.*, 1947; Rehm, 1950; Robertson, 1960) have stressed the derivation of hydrogen ions from hydrogen-bearing substrates. One version invokes irreversible donation of hydrogen atoms to the pump and electrons to oxygen, setting an upper limit of four to the number of hydrogen ions transported for every mole oxygen consumed. Other versions, free from the limitation imposed by irreversibility, can account for higher stoichiometric ratios (Rehm, 1950; Alonso and Harris, 1965; Davies and Ogston, 1950; Hogben, 1960). A full account of this aspect is given in a forthcoming review by one of us (Rehm, 1972).

Accurate estimates of stoichiometric efficiency are hampered by the cellular heterogeneity of the mucosa and the complexity of its transport processes. It is generally agreed that for single mucosae, as for groups of mucosae, the ratio of hydrogen ions secreted to moles of oxygen consumed is around 2 (Bannister, 1965; Davenport, 1952, 1957; Forte and Davies, 1964; Villegas and Durbin, 1960). Davies and co-workers (Crane and Davies, 1951; Davies, 1957; Forte and Davies, 1963, 1964) have singled out the frequency with which higher ratios are met, particularly when the increment or decrement in hydrogen rate associated with stimulation by histamine or inhibition by thiocyanate is compared with the corresponding change in oxygen consumption (i.e. $\Delta H/\Delta O$). If attention is focused on chloride transport given as the sum of hydrogen rate and short circuit current, the mean ratio of chloride ions transported to moles of oxygen consumed is 3.8 with 40% of the cases above 4. It is, of course, possible that such high ratios are an artifact of the *in vitro* state where hydrogen rates are low and where, therefore, the short circuit current represents an excessive fraction of the total transport rate. Recent studies on canine mucosal flaps *in vivo* show that at high secretory rates, the hydrogen:oxygen ratio is close to 2 and only marginally higher for chloride (Moody, 1968).

C. Acid Secretion and Phosphate Metabolism

1. ATP and Acid Secretion

There is convincing evidence that ATP can furnish the energy to drive transport in a number of systems. For example, injection of ATP into the squid axon restores active sodium transport following inhibition by dinitrophenol (Caldwell *et al.*, 1960). Intracellular ATP is known to drive the sodium pump of the red blood cell (Glynn, 1957). So far, similar experiments have not been feasible for the gastric muscosa. Attempts to restore secretion in the presence of dinitrophenol or during anoxia by adding massive amounts of ATP to the serosal solution have not met with success (Rehm and Lefevre, 1965; Sanders and Rehm, 1971). The reverse, in fact, occurs with ATP markedly inhibiting acid secretion. Evidence in favor of ATP involvement is derived from the close positive correlation between mucosal ATP levels and hydrogen rates (Forte, 1966; Forte *et al.*, 1965) (Fig. 6). The correlation persists during anoxia.

Both Durbin (1968) and Forte *et al.* (1965) have noted that while the onset of anoxia is accompanied by a rapid fall in acid rate, there persists a finite hydrogen rate, which fades but slowly during the first hour. Durbin further noted that during this period, the decay of ATP

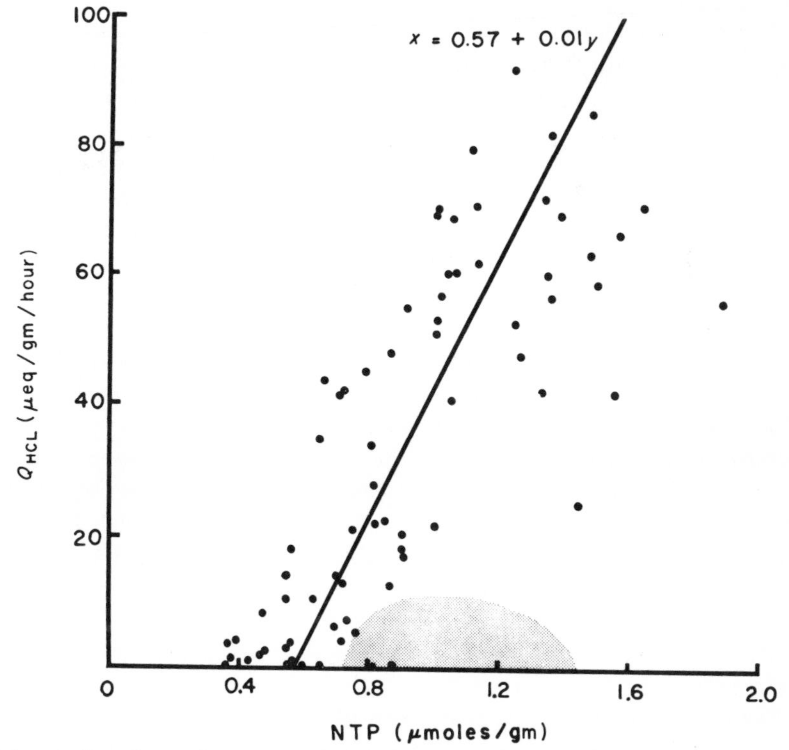

FIG. 6. Correlation between acid rate (Q_{HCl}) and nucleotide triphosphate levels (NTP) in frog gastric mucosa. Shaded area covers data obtained following inhibition by thiocyanate. (Forte *et al.*, 1965.)

was relatively slow, indicating that anaerobic mechanisms could generate sufficient ATP to sustain for a time a low rate of secretion.

It is possible that ATP may not furnish energy directly to the transport mechanism, but instead reverse the flow in the electron transport system (a well established finding in mitochondria) with a hydrogen mechanism of the redox type (Rehm, 1972).

2. *ATP and Thiocyanate*

It is noteworthy that a potent inhibitor such as thiocyanate should have no effect on the levels of phosphorylated metabolites in both mucosa and gastric mitochondrial preparations (Forte *et al.,* 1965). The finding recalls an early observation by Davies and Terner (1949) that thiocyanate has little effect on oxygen consumption. Forte *et al.* (1965) showed in effect that thiocyanate decreases the utilization of ATP but has no effect on its production, a fact which assumes added interest in the light of the discovery by Kasbekar and Durbin (1965) of a thiocyanate-inhibited gastric adenosine triphosphatase.

3. *Gastric ATPase and ATP-driven Models of Secretion*

The presence of ATPase provides a means of coupling ATP-bound energy to ion transport. Two properties single out the gastric ATPase from the ubiquitous cation-stimulated ATPases; it is stimulated by bicarbonate and other bases and inhibited by thiocyanate. Both properties have been noted in some gut ATPases and may indeed be shared by ATPases of other secretory tissues (Turbeck *et al.,* 1968).

On the basis of their discovery, Durbin and Kasbekar (1965) advanced a hypothesis for the utilization of ATP in the secretion of both hydrogen and chloride and the generation of electrical activity (Fig. 7). In its details, their hypothesis recalls the forced exchange hypothesis of Hogben (1955a,b). It postulates the formation of an oriented complex composed of enzyme, chloride, and carbonic acid. ATP reacts with the complex resulting in marked lowering of the pK of carbonic acid–enzyme complex. Hydrogen dissociates from the complex and moves into the lumen. The hydrolysis of ATP reorients the complex in a manner which results in the movement of chloride into the lumen and bicarbonate into the cell. Carbonic acid is presumed to derive from the hydration of carbon dioxide under the influence of carbonic anhydrase. Histochemical evidence confirms that carbonic anhydrase is found in the apical membrane of the oxyntic cell (Hansson, 1968).

Kasbekar and Durbin (1965) considered that certain steps in the sequence could be bypassed. In the absence of chloride (for example, sulfate substitution), the sequence would be the same except that chlo-

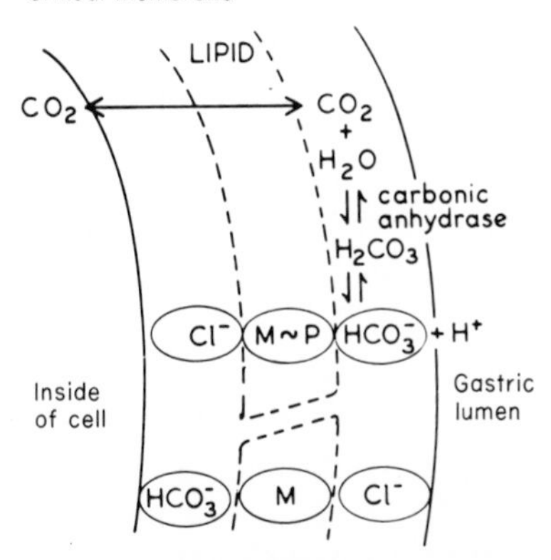

FIG. 7. Model for the utilization of ATP in the production of acid and the generation of short circuit current. Steps in anion translocation are shown from top to bottom. M ~ P refers to phosphorylated carrier. (Durbin and Kasbekar, 1965.)

ride would not enter the cycle, hence hydrogen would enter the lumen and bicarbonate the cell. There would now be a net transport of charge across the membrane which would constitute an electrogenic hydrogen pump. For secretion to proceed, there would have to be a return circuit—sodium and potassium moving from lumen into the cell and/or anions (sulfate) moving from cell to lumen.

If the chloride site is not specific, then other anions, including thiocyanate, can compete with chloride for attachment to it. This explains how thiocyanate can be actively transported and support electrical activity. Its inhibitory action results, presumably, from its attachment to the site normally occupied by bicarbonate. The latter, in fact, is known to protect ATPase from inhibition by thiocyanate.

The requirement for two sites predicted by Schlesinger *et al.* (1955) on other grounds was deduced in this hypothesis from the dual action of thiocyanate. The hypothesis allows for the possible dissociation and resultant electrogenicity of hydrogen and chloride mechanisms but assumes that they are normally coupled electrically and biochemically through dual sites located on the catalytic carrier enzyme.

4. *Adenosine 3′,5′-Monophosphate*

Recent studies (Alonso and Harris, 1965; Harris *et al.*, 1969) indicate that the adenylcyclase–cyclic AMP system may play a role in the sequence of metabolic intracellular events leading to acid secretion. Harris and co-workers have shown that exogenous cyclic AMP can initiate secretion from the resting mucosa. Secretion is also stimulated by agents such as methylxanthines, which increase the mucosal content of cyclic AMP. The details of this system, particularly those which might link it to the gastric ATP–ATPase system remain to be clarified.

VII. Ionic Requirements for Acid Secretion

As indicated earlier, the requirements for chloride or transportable monovalent anion depends on the nature of the substituting anion. The requirement for cations is more specific, particularly for potassium and calcium.

A. Sodium

Davenport (1963) has shown that mucosal sacs bathed in solutions containing choline chloride or mannitol with tris buffer and incubated under oxygen pressure maintained secretion at a somewhat reduced rate. He concluded on this basis that the major function of sodium was osmotic. Later studies showed that, for reasons unknown, the requirement for sodium is more evident in bicarbonate buffer, though even then total elimination of sodium does not abolish secretion. In the frog, two out of three mucosae maintain acid secretion but do not develop a potential difference (Sachs *et al.*, 1966). In the *Necturus*, the mucosae also secrete, but the potential difference inverts (Shoemaker *et al.*, 1967). This is characteristic of the operation of an independent hydrogen transport mechanism. The absence of sodium recalls in some respects the action of Diamox in uncoupling the hydrogen and chloride mechanisms. Sodium thus appears to be necessary for the operation of the chloride mechanism and for its efficient coupling to the hydrogen mechanism.

B. Potassium

Complete elimination of potassium from the bathing solutions leads to drastic effects; secretion ceases, resistance rises, and potential difference gradually decreases to zero (Davis *et al.*, 1963; Rehm *et al.*, 1966). Rubidium but not cesium reverses these effects. Since the potassium content of the mucosa falls only slightly, these effects are thought to be due to potassium loss from a labile compartment where its presence is necessary for secretion. This compartment is readily accessible and secretion can be restored by addition of potassium to either the mucosal or serosal solutions (Davenport and Abbrecht, 1965; Rehm *et al.*, 1966).

C. Calcium

Washing the mucosa with calcium-free solutions increases the resistance and decreases hydrogen rate (first phase); this is followed by a marked reduction of resistance and a cessation of secretion (second phase) (Jacobson *et al.*, 1965; Schwartz *et al.*, 1967). Forte and Nauss (1963) attributed the second phase to chelation of calcium from junctional complexes and the abnormal entry of serosal bicarbonate into the lumen.

Addition of calcium to the serosal side reverses both phases, while addition of calcium to the mucosal side reverses the second phase and only partially the first. The diverse effects on the two sides indicate a greater permeability of the serosal membrane to calcium. Barium, magnesium, and strontium are not effective in restoring function.

VIII. Transport of Sodium, Potassium, and Water

A. Sodium Transport

The greater part of sodium transport in the mature stomach is passive. The directional asymmetery in open circuit is largely accounted for by the potential difference (Crane and Davies, 1949). In short circuit, the ratio of unidirectional sodium fluxes is close to unity, consistent with about 90% of sodium moving passively (Hogben, 1951, 1955a). The residual flux is attributed to active transport from mucosa to serosa. This represents only a minor fraction of the acid rate (about 5%).

Sodium transport is more evident during development (J. G. Forte *et al.*, 1969). A ouabain-sensitive sodium pump located on the serosal membrane accounts for a part of the potential difference in the tadpole stomach. A similar pump accounts for a part of the potential difference in the mature antrum (Shoemaker, 1967).

Sodium transport in mammalian mucosae is reported to exceed acid secretion, but this evidence must be set against the poor function of these preparations which secrete at a minute fraction of *in vivo* rates (2–4 μeq cm^{-2} hour^{-1} versus 200 μeq cm^{-2} hour^{-1} in denervated canine pouches and even more in intact pouches) (Kitahara *et al.*, 1969). Poor function, which could result from improper oxygenation, is reported to enhance sodium transport (K. J. Obrink, personal communication). If improper oxygenation were to impair the integrity of the luminal membrane, which normally is impermeable to sodium and other cations, the serosal sodium pump, which normally operates from cell interior to serosa, could elicit a significant transepithelial sodium transport.

B. Potassium Transport

The transport of potassium, like that of sodium, is largely passive (Hogben, 1955a, 1966; Villegas, 1963b). The flux ratio under short circuit conditions is close to unity (0.9). The residual net flux from serosa to mucosa parallels the acid rate and is sensitive to the same inhibitors (amytal, dinitrophenol, thiocyanate) (Sachs and Pacifico, 1968). The correlation between the secretion of hydrogen and potassium suggests a link between them and points to the oxyntic cell as their common site of origin. It must be recalled, however, that the secretion of acid is associated with the greater part of osmotic water movement. The possibility exists, therefore, that the transport of potassium may not result from its presumed link to the transport of hydrogen but from the transit of osmotic water through a labile intracellular potassium compartment.

C. Water Transport

1. Water of Secretion

The normality of acid which accumulates in frog mucosal sacs incubated in isotonic saline approaches occasionally the cation concentration of the bathing solution (Davies, 1948, 1951; Davies and Longmuir, 1948). If account is taken of all secreted ions, including sodium and potassium, secretion is found to be isosmotic to the bathing solution (Fig. 8) (Makhlouf, 1971). Changes in osmolarity of the serosal solution induced by addition or withdrawal of sodium chloride are accompanied by parallel changes in the osmolarity of secretion. Changes induced by addition of glucose or substitution of glucose for sodium chloride are countered by secretion of an osmotic equivalent of ions.

These observations clearly establish that water has the same chemical potential in the bulk solutions on either side of the membrane. Water must therefore move in response to osmotic gradients created within or close to the surface of the membrane by the net transport of solutes (Curran, 1960; Diamond, 1971; Patlak *et al.*, 1963). The reflection coefficient of the luminal membrane is probably close to unity. The mem-

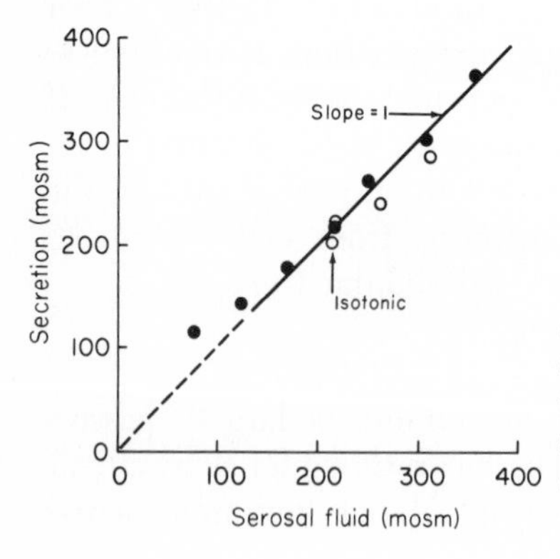

Fig. 8. Relation between the osmolarity of serosal solution and the osmolarity of gastric secretion *in vitro*. Solid circles = upon addition or withdrawal of sodium chloride from serosal solution; open circles = upon addition of glucose or substitution of glucose for sodium chloride. (Makhlouf, 1971.)

brane itself is thrown into folds which multiply greatly during secretion. The arrangement results in the formation of forward channels which are open to the tubular lumen and where osmotic equilibration can be effected.

2. Osmotic Properties of the Mucosa

The gastric mucosa may be described in terms of its effective pore area. In the frog, 93% of this area is occupied by pores with radii of 2.5 Å and 7% with radii of 60 Å (Durbin *et al.*, 1956; Villegas, 1963a). The intact canine mucosa is occupied by a similar set of pores (Altamirano and Martinoya, 1966). In both species, the distribution is highly skewed in favor of the smaller pores but is probably continuous between the two extremes. The larger pores may represent transient defects resulting from cell turnover or from experimental damage.

The mucosa responds to external osmotic gradients by net water flow in the expected direction (Durbin *et al.*, 1956; Villegas, 1963b). Equivalent gradients, however, elicit greater flow in the direction of the lumen (Makhlouf, 1972). The opposite is true of the frog small intestine (Loeschke *et al.*, 1970). Osmotic rectification in these structures, it would seem, predominates in the direction of active transport.

Water flow is near linear for gradients up to 200 mosm (Durbin *et al.*, 1956). Beyond that level, water flow diminished progressively (G. M. Makhlouf, unpublished observations). The deviation from linearity is also seen in canine mucosa and results probably from the behavior of tubules and canaliculi during water flow (Altamirano and Martinoya, 1966). The entry or exit of water from these restricted spaces alters their composition before the change becomes manifest in the bulk solutions. The osmotic gradient to which the oxyntic cells are exposed is thus altered. These cells are presumed to be responsible for mucosal osmotic permeability, a view which is favored by the finding that histamine increases osmotic permeability at the same time as it increases the apical surface of the oxyntic cells and stimulates secretion (Villegas, 1963b).

Another consequence of tubular geometry is that isotonic acid solutions placed in the lumen of the stomach are rendered transiently hypotonic (Berkowitz and Janowitz, 1966; Teorell, 1940, 1947; Terner, 1949). According to Rehm *et al.* (1970), the sequence of events is as follows: As sodium exchanges for hydrogen, the greater mobility of hydrogen renders the blind end of the tubules transiently hypertonic and obligates the entry of water. As a result of water entry, the bulk solutions become hypotonic. Recent studies by one of us (Makhlouf and Yau, 1973) confirm that isotonic acid solutions accumulate water, and the extent of accumulation appears to be sufficient to account fully for the resultant hypotonicity.

ACKNOWLEDGMENTS

The authors wish to thank Dr. S. Ito for Figs. 1 and 2, hitherto unpublished; Dr. J. G. Forte for Fig. 6; Drs. R. P. Durbin and D. K. Kasbekar for Fig. 7; and Dr. C. A. M. Hobgen for Table I.

REFERENCES

Alonso, D., and Harris, J. B. (1965). *Amer. J. Physiol.* **208**, 18.
Altamirano, M., and Martinoya, C. (1966). *J. Physiol.* (*London*) **184**, 771.
Bannister, W. H. (1965). *J. Physiol.* (*London*) **177**, 429.
Berkowitz, J. M., and Janowitz, H. D. (1966). *Amer. J. Physiol.* **210**, 216.
Bradford, N. M., and Davies, R. E. (1948). *Biochem. J.* **42**, xii.
Caldwell, P. C., Hodgkin, A. L., Keynes, R. D., and Shaw, T. I. (1960). *J. Physiol.* (*London*) **152**, 561.
Conway, E. J., and Brady, T. G. (1948). *Nature* (*London*) **162**, 456.
Cotlove, E., and Hogben, C. A. M. (1956). *Fed. Proc., Fed. Amer. Soc. Exp. Biol.* **15**, 41.
Crane, E. E., and Davies, R. E. (1949). *Biochem. J.* **45**, xxiii.
Crane, E. E., and Davies, R. E., (1951). *Biochem. J.* **49**, 169.
Crane, E. E., Davies, R. E., and Longmuir, N. M. (1947). *Nature* (*London*) **159**, 468.
Curran, P. F. (1960). *J. Gen. Physiol.* **43**, 1137.
Davenport, H. W. (1952). *Fed. Proc., Fed. Amer. Soc. Exp. Biol.* **11**, 715.
Davenport, H. W. (1957). *In* "Metabolic Aspects of Transport Across Cell Membranes" (Q. R. Murphy, ed.), pp. 295–302. Univ. of Wisconsin Press, Madison.
Davenport, H. W. (1963). *Amer. J. Physiol.* **204**, 213.
Davenport, H. W., and Abbrecht, P. H. (1965). *Amer. J. Physiol.* **208**, 528.
Davies, R. E. (1948). *Biochem. J.* **42**, 609.
Davies, R. E. (1951). *Biol. Rev.* **26**, 87.
Davies, R. E. (1957). *In* "Metabolic Aspects of Transport Across Cell Membranes" (Q. R. Murphy, ed.), pp. 277–293. Univ. of Wisconsin Press, Madison.
Davies, R. E. (1959). *Amer. J. Dig. Dis.* [N.S.] **4**, 173.
Davies, R. E., and Forte, J. G. (1963). *In* "Pathophysiology of Peptic Ulcer" (S. C. Skoryna, ed.), pp. 3–22. McGill Univ. Press, Montreal.
Davies, R. E., and Longmuir, N. M. (1948). *Biochem. J.* **42**, 621.
Davies, R. E., and Ogston, A. G. (1950). *Biochem. J.* **46**, 324.
Davies, R. E., and Terner, C. (1949). *Biochem. J.* **44**, 377.
Davis, T. L., Rutledge, J. R., and Rehm, W. S. (1963). *Amer. J. Physiol.* **205**, 873.
Davis, T. L., Rutledge, J. R., Keesee, D. C., Bajandas, F. J., and Rehm, W. S. (1965). *Amer. J. Physiol.* **209**, 146.
Diamond, J. M. (1971). *Fed. Proc., Fed. Amer. Soc. Exp. Biol.* **30**, 6.
Durbin, R. P. (1963). *Fed. Proc., Fed. Amer.* **22**, 663.
Durbin, R. P. (1963–1964). *J. Gen. Physiol.* **47**, 735.
Durbin, R. P. (1967). *In* "Handbook of Physiology" (Amer. Physiol. Soc., J. Field, ed.), Sect. 6, Vol. II, pp. 879–888. Williams & Wilkins, Baltimore, Maryland.

Durbin, R. P. (1968). *J. Gen. Physiol.* **51**, 233.
Durbin, R. P., and Heinz, E. (1958). *J. Gen. Physiol.* **41**, 1035.
Durbin, R. P., and Kasbekar, D. K. (1965). *Fed. Proc., Fed. Amer. Soc. Exp. Physiol.* **24**, 1377.
Durbin, R. P., Frank, H., and Solomon, A. K. (1956). *J. Gen. Physiol.* **39**, 535.
Durbin, R. P., Kitahara, S., Stahlmann, K., and Heinz, E. (1964). *Amer. J. Physiol.* **207**, 1177.
Durbin, R. P., Way, L. W., and Kasbekar, D. K. (1969). *Biophys. J.* **9**, A268.
Forte, G. M., Limlomwongse, L., and Forte, J. G. (1969). *J. Cell Sci.* **4**, 709.
Forte, J. G. (1966). *Fed. Proc., Fed. Amer. Soc. Exp. Biol.* **24**, 1382.
Forte, J. G. (1969). *Amer. J. Physiol.* **216**, 167.
Forte, J. G. (1971). *In* "Membranes and Ion Transport" (E. E. Bittar, ed.), Vol. III, pp. 111–165. Wiley (Interscience), New York.
Forte, J. G., and Davies, R. E. (1963). *Amer. J. Physiol.* **204**, 812.
Forte, J. G., and Davies, R. E. (1964). *Amer. J. Physiol.* **206**, 218.
Forte, J. G., and Nauss, A. H. (1963). *Amer. J. Physiol.* **205**, 631.
Forte, J. G., Adams, P. H., and Davies, R. E. (1963). *Nature (London)* **197**, 874.
Forte, J. G., Adams, P. H., and Davies, R. E. (1965). *Biochim. Biophys. Acta* **104**, 25.
Forte, J. G., Limlomwongse, L., and Kasbekar, D. K. (1969). *J. Gen. Physiol.* **54**, 76.
Glynn, I. M. (1957). *Progr. Biophys.* **8**, 241.
Grossman, M. I. (1967). *In* "Handbook of Physiology" (Amer. Physiol. Soc. J. Field, ed.), Sect. 6, Vol. II, pp. 835–863. Williams & Wilkins, Baltimore, Maryland.
Hansson, H. P. J. (1968). *Acta Physiol. Scand.* **73**, 427.
Harris, J. B., and Edelman, I. S. (1964). *Amer. J. Physiol.* **206**, 769.
Harris, J. B., Nigon, K., and Alonso, D. (1969). *Gastroenterology* **57**, 377.
Heinz, E., and Durbin, R. P. (1958). *J. Gen. Physiol.* **41**, 101.
Heinz, E., and Durbin, R. P. (1959). *Biochim. Biophys. Acta* **31**, 246.
Hogben, C. A. M. (1951). *Proc. Nat. Acad. Sci. U.S.* **37**, 393.
Hogben, C. A. M. (1952). *Proc. Nat. Acad. Sci. U.S.* **38**, 13.
Hogben, C. A. M. (1955a). *Amer. J. Physiol.* **180**, 641.
Hogben, C. A. M. (1955b). *In* "Electrolytes in Biological Systems" (A. M. Shanes, ed.), pp. 176–204. Amer. Physiol. Soc., Washington, D.C.
Hogben, C. A. M. (1960). *Amer. J. Med.* **29**, 726.
Hogben, C. A. M. (1966). *Fed. Proc., Fed. Amer. Soc. Exp. Biol.* **24**, 1353.
Hogben, C. A. M. (1968). *J. Gen. Physiol.* **51**, 240s.
Hogben, C. A. M., and Green, N. D. (1958). *Fed. Proc., Fed. Amer. Soc. Exp. Biol.* **17**, 72.
Ito, S. (1967). *In* "Handbook of Physiology" (Amer. Physiol. Soc., J. Field, ed.), Sect. 6, Vol. II, pp. 705–741. Williams & Wilkins, Baltimore, Maryland.
Ito, S. (1971). Unpublished electronmicrographs.
Jacobson, A., Schwartz, M., and Rehm, W. S. (1965). *Amer. J. Physiol.* **209**, 134.
Kaneko-Mohammed, S., and Hogben, C. A. M. (1964). *Amer. J. Physiol.* **207**, 1173.
Kasbekar, D. K., and Durbin, R. P. (1965). *Biochim. Biophys. Acta* **105**, 472.
Kasbekar, D. K., Ridley, H. A., and Forte, J. G. (1969). *Amer. J. Physiol.* **216**, 961.

Kitahara, S., Fox, K. R., and Hogben, C. A. M. (1969). *Amer. J. Dig. Dis.* [N. S.] **14**, 221.
Loeschke, K., Bentzel, C. J., and Csaky, T. Z. (1970). *Amer. J. Physiol.* **218**, 1723.
Makhlouf, G. M. (1971). Gastroenterology **60**, 784.
Makhlouf, G. M. (1972). *Fed. Proc. Fed. Amer. Soc. Exp. Biol.* **31**, 827.
Makhlouf, G. M., and W. M. Yau (1973). *Fed. Proc. Fed. Amer. Soc. Exp. Biol.* **32**, 394.
Makhlouf, G. M., Shoemaker, R. L., and Sachs, C. (1969). *Proc. Biophys. Congr., 3rd, 1969* p. 79.
Moody, F. G. (1968). *Amer. J. Physiol.* **215**, 127.
Morrissey, S. M., and Wan, B. (1970). *Comp. Biochem. Physiol.* **34**, 507.
Noyes, D. H., and Rehm, W. S. (1970). *Amer. J. Physiol.* **219**, 184.
Ochsman, J. L. and Berridge, M. J. (1970). *Tissue Cell* **2**, 281.
Pacifico, A. D., Schwartz, M., Mackrell, T. N., Spangler, S. G., Sanders, S. S., and Rehm, W. S. (1969). *Amer. J. Physiol.* **216**, 536.
Patlak, C. S., Goldstein, D. A., and Hoffman, J. F. (1963). *J. Theor. Biol.* **5**, 426.
Rehm, W. S. (1950). *Gastroenterology* **14**, 401.
Rehm, W. S. (1962). *Amer. J. Physiol.* **203**, 63.
Rehm, W. S. (1964). *In* "Transcellular Membrane Potentials and Ion Fluxes" (F. M. Snell, ed.), p. 64. Gordon & Breach, New York.
Rehm, W. S. (1965). *Fed. Proc., Fed. Amer. Soc. Exp. Biol.* **24**, 1387.
Rehm, W. S. (1967). *Fed. Proc., Fed. Amer. Soc. Exp. Biol.* **26**, 1303.
Rehm, W. S. (1968). *J. Gen. Physiol.* **51**, 250s.
Rehm, W. S. (1972). *In* "Greenberg's Metabolic Pathways" (L. E. Hokin, ed.), Vol. 6, p. 187, Academic Press, New York.
Rehm, W. S. (1972). *Arch. Intern. Med.* **129**, 270.
Rehm, W. S., and Lefevre, M. E. (1965). *Amer. J. Physiol.* **208**, 922.
Rehm, W. S., and Sanders, S. S. (1972). *In* "Gastric Secretion" (G. Sachs and E. Heinz, eds.), Academic Press, New York.
Rehm, W. S., Davis, T. L., Chandler, C., Gohmann, E., Jr., and Bashirelahi, A. (1963). *Amer. J. Physiol.* **204**, 233.
Rehm, W. S., Sanders, S. S., Rutledge, J. R., Davis, T. L., Kurfees, J. F., Keesee, D. C., and Bajandas, F. J. (1966). *Amer. J. Physiol.* **210**, 689.
Rehm, W. S., Butler, C. F., Spangler, S. G., and Sanders, S. S. (1970). *J. Theor. Biol.* **27**, 433.
Robertson, R. N. (1960). *Biol. Rev. Cambridge Phil. Soc.* **35**, 231.
Sachs, G., and Pacifico, A. (1968). *Fed. Proc., Fed. Amer. Soc. Exp. Biol.* **27**, 748.
Sachs, G., Shoemaker, R., and Hirschowitz, B. I. (1966). *Proc. Soc. Exp. Biol. Med.* **123**, 47.
Sachs, G., Clark, L. C., and Makhlouf, G. M. (1970). *Proc. Soc. Exp. Biol. Med.* **134**, 694.
Sanders, S. S., and Rehm, W. S. (1970). *Physiologist* **13**, 300.
Sanders, S. S., and Rehm, W. S. (1971). *Biophys. Soc. Abstr.* **11**, 79a.
Schlesinger, H. S., Dennis, W. H., and Rehm, W. S. (1955). *Amer. J. Physiol.* **183**, 75.
Schwartz, M., Kashiwa, H. K., Jacobson, A., and Rehm, W. S. (1967). *Amer. J. Physiol.* **212**, 241.

Sedar, A. W. (1961a). *J. Biophys. Biochem. Cytol.* **9**, 1.
Sedar, A. W. (1961b). *J. Biophys. Biochem. Cytol.* **10**, 47.
Sedar, A. W. (1962). *Ann. N. Y. Acad. Sci.* **99**, 9.
Sedar, A. W., and Forte, J. G. (1963). *Anat. Rec.* **145**, 283.
Shoemaker, R. L. (1967). Ph.D. Thesis, University of Alabama, Tuscaloosa.
Shoemaker, R. L., Hirschowitz, B. I., and Sachs, G. (1967). *Amer. J. Physiol.* **212**, 1013.
Shoemaker, R. L., Makhlouf, G. M., and Sachs, G. (1969). *Biophys. J.* **9**, A263.
Shoemaker, R. L., Makhlouf, G. M., and Sachs, G. (1970). *Amer. J. Physiol.* **219**, 1056.
Spangler, S. G., and Rehm, W. S. (1968). *Biophys. J.* **8**, 1211.
Teorell, T. (1940). *J. Gen. Physiol.* **23**, 263.
Teorell, T. (1947). *Gastroenterology* **9**, 425.
Teorell, T. (1951). *J. Physiol. (London)* **114**, 267.
Teorell, T., and Wersall, R. (1945). *Acta Physiol. Scand.* **10**, 243.
Terner, C. (1949). *Biochem. J.* **45**, 150.
Turbeck, B. O., Nedegaard, S., and Kruse, H. (1968). *Biochim. Biophys. Acta* **163**, 354.
Villegas, L. (1962). *Biochim. Biophys. Acta* **64**, 359.
Villegas, L. (1963a). *Biochim. Biophys. Acta* **75**, 131.
Villegas, L. (1963b). *Biochim. Biophys. Acta* **75**, 377.
Villegas, L. (1964). *Biochim. Biophys. Acta* **88**, 227.
Villegas, L., and Durbin, R. P. (1960). *Biochim. Biophys. Acta* **44**, 612.
White, T. D., Swigart, R. H., and Rehm, W. S. (1956). *Amer. J. Physiol.* **184**, 453.

CHAPTER 3

Salt Balance and Osmoregulation in Salientian Amphibians*

Bradley T. Scheer, Marus W. Mumbach, and Allan R. Thompson

I. Ranges and Tolerances ... 51
II. Fluxes ... 52
III. Endocrine Controls ... 56
IV. Cellular Basis of Aldosterone Action ... 59
References ... 64

The control of salt and water balance in frogs and toads is perhaps better understood than that in any other group of animals outside the Mammalia. Adolph and Adolph (1925) reviewed literature on water balance extending back as far as 1795. Anurans have also provided the classic tissues—skin, bladder, stomach (Makhlouf and Rehm, this volume)—for the study of active ion transport by and endocrine control of osmotic permeability of multicellular membranes. Bentley (1971) has recently reviewed the subject of the present chapter, and we shall consider here recent and especially significant publications and our own recent unpublished work in the light of his account.

I. Ranges and Tolerance

There is considerable variation among species of Salientia in tolerance of dehydration and of exposure to saline solutions. *Rana cancrivora,* but not other species of Ranidae, tolerate prolonged exposure to seawater, and more terrestrial amphibians, as *Scaphiopus couchi,* enter into a state of estivation when deprived of sources of liquid water in air of low relative humidity. Our observations (unpublished) with *Rana esculenta* and *R. pipiens* show that the highest concentration of sodium chloride tolerated during prolonged immersion in pure sodium chloride solutions is 150 m*M*/liter. Mean values of plasma sodium concentration in our studies were near 110 m*M*/liter in frogs kept continually immersed in media at sodium concentrations between 0 and 50 m*M*/liter; above

* Original unpublished work cited was supported by a grant AM03539 from the National Institutes of Health, United States Public Health Service.

this last value, plasma concentrations increased parallel with the concentration of the medium up to the highest level tolerated, 135–150 m*M*/liter. Our tentative conclusion is that the tolerance limit is determined by the ability of the regulatory mechanisms of the frog to maintain levels of plasma sodium concentration below 150 m*M*/liter.

In general, nearly all of the osmotic activity of anuran plasma is due to sodium chloride, with minor contributions from other salts and from colloids, except in hypertonic media or under other dehydrating conditions. In these latter circumstances, the osmotic concentration of the plasma is increased by substantial amounts of urea. The urea concentration in plasma of *R. cancrivora* is 40 m*M*/liter in fresh water, and increases to 400 m*M*/liter in seawater (Gordon *et al.*, 1961). Gordon (1962) and Tercafs and Schoffeniels (1962) report increases in blood urea in *Bufo viridis* in saline media, and we (Scheer and Markel, 1962) found that plasma urea increases from 0.3 m*M*/liter in fresh water to 1 m*M*/liter in animals kept in 150 m*M* sodium chloride. The increased urea levels are due to urine retention, with increased urea synthesis (Balinsky *et al.*, 1961) and decreased urea elimination (Scheer and Markel, 1962).

II. Fluxes

Aside from the osmotic role of urea production and retention, osmotic regulation in anurans is largely a matter of the regulation of plasma salt concentration, and this regulation must be accomplished through control of sodium pumps in the skin, nephron tubule, and bladder wall; through control of the permeability of membranes to water; and through behavioral mechanisms. Adolph (1927b) concluded that the urine could not be an important regulatory channel for salts, since the urine is always hypotonic to the blood, and our results (unpublished) with *R. esculenta* and *R. pipiens* confirm his and those of others (Ridley, 1964a) (*R. catesbeiana*) in showing that the concentration of sodium in the bladder urine is always less than that in plasma collected at the same time. Except in media with less than 20 m*M* sodium per liter, we find that urinary sodium concentration is always less than that of the medium.

Most estimates of urine elimination (Adolph, 1927a,b) have utilized ligation of the cloaca or cannulation for urine collection, and these procedures may cause disturbance in the normal physiological processes (Potts and Evans, 1967). We measured urine elimination rates by following the elimination of an isotopic indicator (^{24}Na, Glofil-^{125}I*, or inulin-^{14}C)

* Glofil (© Abbott Laboratories) is sodium iothalamate labeled with a radioisotope of iodine.

and converted these rates into rates of sodium elimination using values obtained by analysis of bladder urine immediately after a measurement. Mean values of urine elimination rate for *R. pipiens* are near 10 μl/gm/hour for animals in dilute media (less than 20 m*M* sodium per liter). The elimination rate falls to zero in media 100 m*M* sodium per liter and higher. Urinary efflux of sodium is appreciable (Table I). Greenwald (1971) has attempted to measure urinary sodium loss under similar conditions, but his estimates are based on assumption of skin loss as 10% of total loss, and other seemingly unjustified assumptions. Our estimates for one series of observations on *R. pipiens* are given in Table I; clearly, urinary loss, though appreciable, is never more than 20% of total loss computed from isotope efflux. Urine elimination, as Greenwald (1971) notes, is generally periodic, with periods of an hour or more between urinations. The rates of elimination of water and salts in the urine depend on a number of variables—glomerular filtration rate, rate of sodium uptake from the nephron tubule, permeability of the tubule to water, and rate of sodium uptake from the bladder and permeability of the bladder wall. The results presented in Table II show that acclimation to salt solution (135 m*M*/liter) does not alter glomerular filtration rate appreciably, but has a marked influence on urine elimination rate, evidently due to reabsorption of water from the urinary tract in concentrated media, and the addition of water, probably by osmosis, to the urine in dilute media.

The importance of urine retention by frogs under dehydrating conditions was noted by Darwin (1896) and is reviewed by Bentley (1966). Our results show that a single emptying of the bladder (by catheterization) results in a lower plasma [Na], by comparison with frogs in which

TABLE I
SODIUM EFFLUX FROM FROGS (*R. pipiens*) ACCLIMATED TO SODIUM CHLORIDE SOLUTIONS[a]

Medium sodium (m*M*/liter)	Plasma sodium (m*M*/liter)	Total sodium efflux (μ*M*/gm/hour)	Urinary sodium efflux (μ*M*/gm/hour)
0.5	125	0.38	0.0025
1.3	140	0.49	0.0092
10.8	133	0.57	0.117
99.8	144	0.78	0.060
135.1	154	0.96	0

[a] Values are means of 5 animals.

TABLE II
URINARY TRACT FUNCTION IN FROGS (*R. pipiens*) ACCLIMATED TO SODIUM CHLORIDE SOLUTIONS[a]

Function	Tap water	135 m*M* NaCl	Probability of a larger difference[b]
Clearance (μl plasma/gm body weight/hour)	22.2	23.8	<2.5%
Elimination (μl/gm body weight/hour)	57.2	7.6	<2.5%
Urine/plasma ratio	0.6	3.7	<0.1%

[a] Inulin = ^{14}C used as indicator.
[b] Calculated by analysis of variance.

the bladder remains full for several hours. The volume of dilution of sodium (sodium space or extracellular fluid volume) is also greater in "full-bladder" animals, presumably because the bladder contents form a part of the sodium space of the animal. Passive exchange of water and sodium with plasma, and active uptake of sodium across the bladder wall are important parts of this mechanism (Adolph, 1927c; Uranga and Quintana, 1958; Ruibal, 1962a,b).

Bentley (1969) observed influx of water across anuran skins to range from 8 to 28 μl/cm²/hour, which is of the same order of magnitude as our value of 10 μl/gm/hour for urinary efflux (all weights are body weights). Bentley (1971) states, "Amphibians do not drink, except in certain non-physiological situations." He also notes the results of Alvarado and Moody (1970) to the effect that bullfrog (*R. catesbeiana*) tadpoles depleted of salts (in distilled water) drink the medium at a rate of about 33% of the body weight each day. Krogh (1939) quoted Overton (1904) to the effect that frogs kept in solutions more concentrated than 140 m*M* drink the medium. We have measured drinking rates in *R. pipiens* using polyvinylpyrollidine labeled with ^{131}I as indicator and find that these frogs drink the medium at all concentrations, from a drinking rate of 0.4 μl/gm/hour in dilute media (0.5 m*M* sodium per liter) to one of 9.3 μl/gm/hour in concentrated media (135 m*M*/liter). In media of high salt concentrations, the result is an increase in intake of sodium, but our studies of sodium efflux from the gut suggest that all the salt taken into the gut from saline media is eliminated from the gut, and the intake of water compensates for osmotic and urinary water loss across the skin. Intestinal elimination of sodium was estimated by administering labeled polyvinylpyrollidine by stomach tube and measuring the elimination of the label; analysis of intestinal contents at the end of an observation period permitted calculation of sodium efflux rates from the intestine of animals acclimated to 135 m*M* sodium per liter at 1.3 μM/gm/hour, essentially equivalent to the rate of intake by drinking.

The total sodium efflux rates given for *R. pipiens* in Table I are typical of a large number of observations we have made in the last decade (unpublished). Values for *R. esculenta* under the same experimental conditions are not significantly different. Our calculation of total efflux from measurements of isotope efflux utilized an efflux coefficient relating the efflux rate to size of the internal pool. The efflux coefficient did not vary significantly with concentration of the medium, and the increased efflux rates in more concentrated media result from a considerable increase in the internal sodium pool of frogs kept in such media. This increase results from an increase in plasma [Na], but to a larger extent from the increase in the sodium dilution volume (sodium space), from a value of 40% of the body weight in media less concentrated than 20 m*M*/liter to a value of 66.6% body weight in 135 m*M*/liter. This is an important, and previously unnoticed, aspect of acclimation to concentrated media; frogs in concentrated media are noticeably edematous during acclimation in comparison with frogs kept in dilute media. The possibility that the bladder contents constitute a part of the sodium space and that urine is retained in concentrated media may also account for some of the difference in sodium space.

In frogs (*R. pipiens*) acclimated to various concentrations of sodium chloride, the active transport mechanism of the skin is important in influx of sodium. The accepted role of this mechanism is that of uptake of salt to replace that lost in the urine or by diffusion efflux. Our recent studies with *R. pipiens* confirm the observations previously reported with isolated skin (Maetz, 1959; Bishop *et al.*, 1961; Myers *et al.*, 1961; Gordon, 1962) that prolonged exposure of frogs to saline solutions results in a decrease of the active sodium transport rate across the isolated skin. Our studies (unpublished) show that the development of this phenomenon is progressive and requires 5–10 days of continuous exposure. Isotopic flux measurements on whole animals permitted calculation of an influx coefficient, the proportionality coefficient relating influx rate to size of external sodium pool. The value of the influx coefficient decreases with increasing concentration of medium (Table III). Interpretation of the influx coefficient as an index of active transport rate is justified by the fact that in our studies with *R. pipiens*, influx of sodium must occur against an electrochemical diffusion gradient in all media. The plasma of the intact animal is electrically positive (about 30 mV) with respect to the external medium in all the media we studied, and the

TABLE III
VALUES OF INFLUX COEFFICIENTS FOR SODIUM IN WHOLE FROGS (*R. pipiens*) ACCLIMATED TO SODIUM CHLORIDE SOLUTIONS

Medium concentration (m*M* Na/liter)	Influx coefficient (gm^{-1} $hour^{-1}$)
0.5	8.17
1.2	3.33
12	2.60
104	1.53
137	0.13

plasma also contains a higher concentration of sodium than does the medium, except in the most concentrated media.

Under natural conditions, frogs have behavioral means for regulation of water balance. Dehydrated anurans are known (Bentley, 1971) to be able to absorb water across the skin from moist sand, as well as from water. In a preliminary study, M. W. Mumbach (unpublished) observed that frogs (*R. pipiens*) given a choice of a platform out of water and immersion in water, spent roughly half of a 24-hour period out of the water. When the bladder was emptied before beginning a period of observation, the frogs remained in the water for nearly all of the first 24-hour period in the experimental chamber. This suggests to us, following Steen (1929) and Ruibal (1962a,b), that evacuation of the bladder depletes the total body water supply.

III. Endocrine Controls*

The first evidence for a hormonal effect on amphibian water balance was provided by Brunn (1921), when he showed that extracts of mammalian neurohypophysis injected into frogs caused a dramatic increase in weight. Thirty years of studies of the "Brunn effect" by many investigators ultimately led to the demonstration by Heller (1941, 1950) that there is a specific "amphibian water balance hormone." The neurohypophysial hormones of amphibians have since been identified as arginine vasotocin (8 Arg oxytocin) and mesotocin (8 Ile oxytocin) (Bentley, 1971), a pattern characteristic of the Choanichthyes and of reptiles, at opposite ends of the accepted phylogenetic sequence leading to and from Amphibia. The amphibian neurohypophysial principles have the combined effects of increasing permeability to water of skin, bladder,

* This section is by Marus W. Mumbach.

and probably renal tubular membranes, and of stimulating increased rates of active uptake of sodium across these membranes. Vasotocin has a glomerular antidiuretic action, decreasing the glomerular filtration rate (Maetz, 1963). Bentley (1969) reports that immersion of frogs and toads in saline solution results in release of vasotocin into the circulation, and we (Ridley, 1964b; F. R. Nelson and R. W. Wise, unpublished) have noted a distinct change in staining properties of a region in the preoptic nucleus when frogs (*R. pipiens*) are acclimated to salt solutions. The permeability effects and antidiuretic effects of vasotocin release may well account for the increase in extra-cellular fluid volume (sodium space) and plasma sodium which we have noted. Ridley (1964a) showed that [Na] increases in cerebrospinal fluid of bullfrogs exposed to saline solutions.

The corticosteroid aldosterone has been demonstrated in several species of amphibians (Bentley, 1971), but has otherwise been identified only in reptiles, birds, and mammals among the vertebrates. The known mechanism of action of this hormone, demonstrated in many instances on isolated skins and bladders of frogs and toads, is to stimulate an increase in the rate of active transport of sodium (Bentley, 1971). Our earlier results (Myers *et al.*, 1961; Bishop *et al.*, 1961) showed that treatment of whole animals with vasotocin could reverse the effects of removal of the adenohypophysis or of immersion in salt solutions on the transport properties of the isolated skin. The implication that the anterior lobe of the pituitary exerts some control on the secretion of aldosterone by the interrenal glands of frogs has thus far not been confirmed. Bentley (1971) discusses this question at length, but misinterprets some of the results from our laboratory. Ridley (1964a) found no change in plasma [Na] of bullfrogs after removal of the adenohypophysis, and his and our earlier (Myers *et al.*, 1961; Bishop *et al.*, 1961) studies concerned only the properties of the isolated skin. In our more recent studies (unpublished) of sodium fluxes with the whole animal, we find no differences in sodium flux through any channel between intact individuals of *R. esculenta* and individuals after adenohypophysectomy. Ridley's (1964a) study on *R. catesbeiana* and ours (unpublished) on *R. esculenta* both demonstrated lower [Na] in urine of hypophysectomized animals than in unoperated animals kept in low salt concentrations, but our data are not sufficient to permit inferences concerning the cause of this difference.

On the basis of available information, we can formulate a hypothetical interpretation of the actions of aldosterone and of vasotocin in relation to the environmental situations frequently encountered by frogs and toads (Fig. 1). In fresh water, the urinary tract is of prime importance

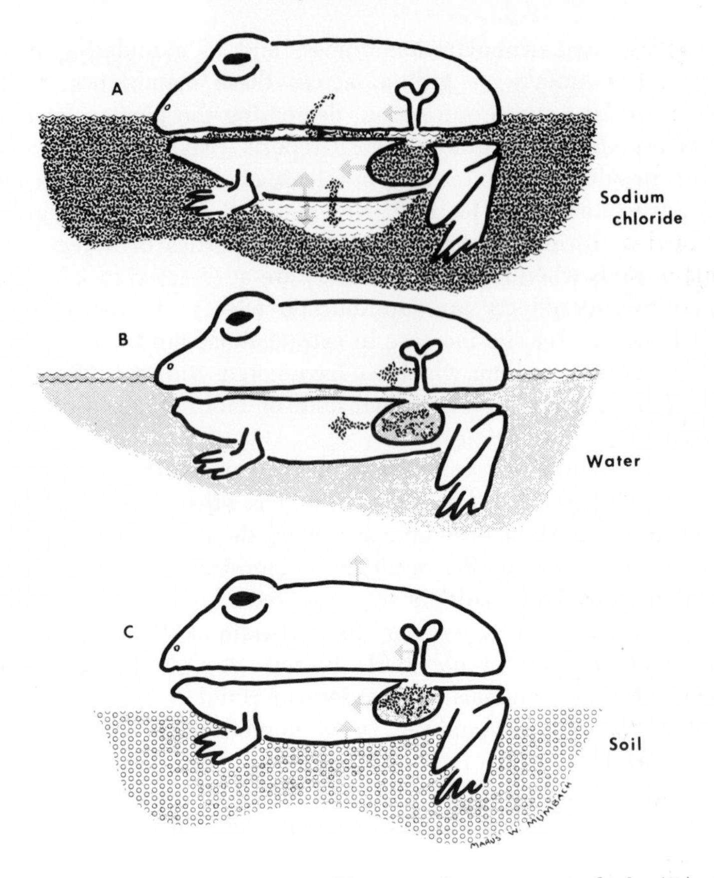

FIG. 1. (A) Aquatic dehydration: aldosterone low, vasotocin high. (B) Aquatic hydration: aldosterone high, vasotocin low. (C) Terrestrial dehydration: vasotocin high.

in the elimination of water taken in by drinking and by osmosis. In this function, a relatively high glomerular filtration rate, supplemented by reabsorption of sodium from the renal tubule and bladder, and low osmotic influxes from the medium and from the bladder contents are probably assured by low levels of vasotocin and relatively high levels of aldosterone. The water lost during a period of terrestrial life can be restored either by a return to fresh water, or (McClanahan and Baldwin, 1969) by absorption of water stored in the urinary bladder and uptake from moist earth or sand. Such uptake is presumably facilitated by high levels of vasotocin, which increases the osmotic permeability of the skin and bladder membranes, and by increasing active sodium transport, which facilitates osmosis as well.

In the situation, not outside the experience of many species of frogs, where water is only available from pools of brackish water of salinity near the plasma level, our results and those of others investigating saline media offer a basis for further hypotheses. As for the terrestrial situation, a high level of vasotocin facilitates reabsorption of water from the urinary tract. A high level of aldosterone would be disadvantageous in supplementing the effect of vasotocin on sodium uptake from the medium. The gradual decrease in active sodium transport rate across the skin during a prolonged exposure to saline media supports the idea of a continuing decrease in rate of aldosterone secretion, leading to a reduced plasma level. Water is conserved by suppression of urination, and any osmotic loss to the medium is restored by drinking. For maximum biological interest, further study of regulatory mechanisms in Salientia may well include, along with treatment of the animal as a convenient system for the study of physiological problems, some consideration of the integration of this system internally, and externally as a part of an ecosystem.

Recent measurements (M. W. Mumbach, unpublished, 1972) of aldosterone levels in plasma of *R. pipiens* using radioimmune assay support some of the predictions made above. The aldosterone level in plasma of frogs kept in distilled water is 10–14 ng/ml, and only traces of the hormone are present in frogs kept in 150 m*M* sodium chloride.

IV. Cellular Basis of Aldosterone Action*

For many years, epithelial membranes from amphibians have been used as model systems for studying the mechanisms of transepithelial transport of water and solutes. Frog skins and toad bladders provide large areas of transporting tissue which remain viable *in vitro* for many hours. The experimenter has the immense advantage of being able to manipulate the composition of the solutions bathing both surfaces of the tissue, whereas with most cell preparations only the external medium is accessible. In addition to performing the active transport of salts, principally sodium, the tissues also respond to the hormones aldosterone and antidiuretic hormone (ADH). It is perhaps fair to state that most of what is known of the action of these hormones has first been observed in these tissues. Leaf (1965) holds that the urinary bladder of the toad is a functional analog of the mammalian kidney tubule. Due to space limitations, this review will be limited to recent progress in elucidating the mechanism of action of aldosterone.

* This section is by Allan R. Thompson.

Crabbé (1961) was the first to demonstrate an effect of aldosterone on active sodium transport *in vitro.* Toad urinary bladders treated with aldosterone responded by approximately doubling the short circuit current after a lag period of several hours following administration of hormone. This latent period was unaffected by the concentration of aldosterone. Tissue glycogen content decreased, but tissue water content was unaffected. Bladders from animals that had been preconditioned to aldosterone by injection of hormone the day previous to experimentation failed to respond *in vitro.*

Subsequent experimentation on the mode of action of the hormone has been mainly concerned with two major questions: Does aldosterone exert its effect through the *de novo* synthesis of protein? And does the observed stimulation of sodium transport result from a lowering of the mucosal permeability barrier or does it result from a stimulation of the active transport "pump" mechanism? The early evidence relating to these questions has been reviewed by Sharp and Leaf (1966). The main points will be summarized here, and their review should be consulted for references.

The inhibitors of protein synthesis, actinomycin D and puromycin, block the response of the toad bladder to aldosterone. The time lag between the administration of hormone and the first appearance of the stimulation of sodium transport indicates that the hormone acts as an intermediate in some biochemical pathway which has as its net effect the stimulation of sodium flux. Radioautography of toad bladders showed that virtually all the label assimilated after treatment by aldosterone-^{3}H was localized in the nuclei of the mucosal cells (Edelman *et al.*, 1963; Porter *et al.*, 1964). Exposure to aldosterone for a brief period, followed by removal of the hormone still produced the increased sodium transport after the normal lag period. Tissues treated with uridine-^{3}H showed an increase in total tissue ribonucleic acid. Radioautography of similarly treated bladders showed that virtually all of the label appeared in the nuclei of mucosal cells. Despite these results, it was not possible to demonstrate any increase in the amount of labeled amino acids incorporated into protein in aldosterone–treated bladders.

Recently, Rousseau and Crabbé (1971) reexamined the question of incorporation of precursors into RNA and proteins. They found that nuclear RNA synthesis and uridine uptake were stimulated by aldosterone. They failed to detect any increase in cytoplasmic RNA or any changes in the rate of turnover of messenger RNA. In addition, polysomal protein synthesis patterns, ribosomal processing, and protein turnover rates were unaffected by aldosterone.

Even though Edelman (1972) has demonstrated the existence of specific aldosterone-binding proteins from rat kidney, the evidence supporting increased protein synthesis as the sole route of action of aldosterone remains circumstantial. This may be due to the use of methods of detection that are too crude to detect the changes in levels of protein that must be synthesized in order to stimulate sodium transport. This may not be surprising, since the total amount of material which may be necessary to affect membrane processes is most likely quite small. It has been estimated that less than 1% of the total surface area of the toad bladder is involved in transport processes (Gatzy and Clarkson, 1965; Cuthbert and Painter, 1969).

In an effort to assess the relative contribution of different metabolic pathways, Kirchberger *et al.* (1971) have examined the evolution of $^{14}CO_2$ from glucose labeled in the one and six positions. After aldosterone treatment, they found that more label appearing in the evolved carbon dioxide was derived from glucose labeled in the six position than from the one position. It was concluded that the hexose monophosphate shunt was inhibited by aldosterone. These workers allowed 2 hours to pass between the time of labeled glucose administration and recovery of the labeled carbon dioxide. Goodman *et al.* (1971) found that if the time course of appearance of $^{14}CO_2$ was followed after addition of aldosterone there was a very high rate of recovery from glucose labled in the one position. This increase began within 20 minutes after the administration of aldosterone and reached a peak after 40 minutes. After this time, the rate of evolution from glucose-1-^{14}C decreased relative to the rate from glucose-6-^{14}C. Thus, it appears that the hexose monophosphate shunt is stimulated for a short period, followed by an inhibitory period. In order to examine this further, the pattern of lipogenesis was followed. It was found that the incorporation of uniformly labled glucose into lipid was altered. These findings have led this group to propose that the mode of action of aldosterone may be mediated through increased phospholipid synthesis. They suggest that the results indicating a mode of action through protein synthesis reflect the induction of an enzyme with a rather short half-life. The enzyme would be expected to be involved in the synthesis of phospholipid, which would be responsible for the changes in membrane permeability. It has been thought for some time that the protein synthesized in response to aldosterone treatment has a half-life of about 40 minutes (Sharp and Leaf, 1966).

It does not seem likely that all the accumulated evidence in favor of the induction of protein synthesis is in error. The failure of investigators to demonstrate conclusively the synthesis of a protein involved in the transport process leads us to believe that the hypothesis put forward by Goodman and his co-workers (1971) should be carefully explored.

The appearance of the fluid mosaic model of cell membrane structure (Singer and Nicolson, 1972) which asserts that phospholipids rather than proteins are the major structural component of membranes, makes this hypothesis all the more intriguing.

There are two possible physiological means by which the hormone may stimulate active sodium transport: (1) by increasing the permeability of the mucosal border to sodium thus increasing the size of the active transport pool, or (2) it may stimulate the pump mechanism directly by increasing the energy supply available for active transport. Sharp and Leaf (1964) found that aldosterone-treated bladders had a higher concentration of radioactive sodium than did paired control bladders. This result is expected if the effect of aldosterone is to lower the apical permeability barrier to sodium without increasing the activity of the pump. It was also found that the increased oxygen consumption associated with aldosterone treatment was dependent on the presence of sodium in the medium bathing the mucosal surface of the bladder. Fimognari *et al.* (1967) found that the aldosterone effect was dependent on the presence of glucose, β-hydroxybutyrate, oxaloacetate, or acetoacetate. Other Krebs cycle intermediates were without effect. They concluded that aldosterone functions by stimulating one or more steps in the Krebs cycle between condensing enzyme and α-ketoglutarate dehydrogenase.

Aldosterone has been shown to lower the electrical resistance across the toad bladder (Civan and Hoffman, 1971). Bladders treated with aldosterone and then incubated until the metabolic substrates were depleted responded to the addition of pyruvate by increasing short circuit current without changing total transepithelial resistance. Such results are consistent with the idea that aldosterone has its major site of action at the mucosal border of the bladder.

Experiments that rely on the measurement of total tissue radioactivity are open to some criticism. The major point that can be raised is that since the bladder contains a physiologically inert supporting layer under the mucosal cell layer, this layer may contain most of the radioactivity thought to be in the transport pool (Sharp and Leaf, 1966). This criticism has been at least partially answered by the development of techniques that allow the isolation of sheets of epithelial cells (Gatzy and Berndt, 1968). Handler and others (1972a) found that aldosterone and ouabain caused an increase in epithelial cell sodium, while amiloride, a diuretic though to act only on the mucosal surface of toad bladder (Biber, 1971), caused a decrease in cell sodium. These results have been confirmed by others (Leaf and MacKnight, 1972; Handler *et al.*, 1972b). These findings are consistent with the previously reported findings on whole tissue and lend support to the hypothesis that aldosterone exerts its effect at the outer permeability barrier.

Lipton and Edelman (1971) were unable to demonstrate any changes in epithelial cell sodium, potassium, or chloride after aldosterone treatment. They concluded that aldosterone has a dual site of action. Their results require that aldosterone stimulate both the rate of entry across the mucosal border and the sodium pump to the same extent. Snart (1972) has found a two-stage dose-response effect with aldosterone. In addition, a two-stage response to temperature change was also observed. These data were taken to indicate that the major effect of aldosterone is to increase the mucosal permeability, with an additional, but secondary stimulation of the pump, which is independent of the increase in the size of the sodium pool.

This hypothesis is supported by a comparison of the effects of aldosterone and insulin on sodium transport in the toad bladder (Crabbé, 1972). It was found in bladders treated with hormones and then incubated for several hours to deplete endogenous substrate that only aldosterone-treated bladders responded to further additions of glucose. It seems to us that the accumulated evidence indicates that the major effect of aldosterone is to lower the mucosal permeability barrier with a secondary direct stimulation of the pump or an increase in the availability of energy-yielding substrates. The recently devised techniques for determining the rate of uptake of sodium at the outer barrier of the frog skin (Rotunno *et al.*, 1970; Biber and Curran, 1970) may provide useful quantitative tools for determining the relative contributions of increased sodium permeability and direct stimulation of the pump mechanism.

The major share of the work on aldosterone has been done on toad bladder, with relatively little on either the frog or toad skin. Porter (1971) found that toad skin responded in a manner similar to bladder. The short-circuit current showed an increase after the usual lag period. Hviid-Larsen (1970) noted that aldosterone treatment initiated sloughing of the stratum corneum in toad skin. After the completion of the sloughing process, the short circuit current increased by about 300%. If the effects of aldosterone on toad skin are followed for 10 or more hours, a rather distinct pattern of response is noted (Hviid-Larsen, 1972). The time course of the response can be resolved into three phases: (1) an initial stimulatory period which is characterized by an increase in active sodium transport, (2) a decreased transepithelial resistance, and (3) a decreased potential difference. Following the initial stimulatory period is a refractory period during which the hormonal stimulation of sodium transport disappears. This is accompanied by an increase in potential difference and resistance. The refractory period is accompanied by a sloughing of the stratum corneum. This process takes about 5 hours to complete. After the completion of the sloughing process, the final stimulatory period begins. This period corresponds to the initial stimulatory period in physiological parameters.

While the question of the mode of action of aldosterone is still unanswered, recent progress indicates that a reasonable understanding of cellular events underlying the response to this hormone is within reach in the not too distant future. We would like to suggest that the phospholipid synthesis hypothesis discussed earlier should receive careful attention. It may be suggested that a fruitful approach to the question of the ultimate membrane events surrounding the action of aldosterone may well lie in the use of specific molecular probes, such as the use of glutaraldehyde as recently applied fruitfully by Eggena (1972) to examine the mode of action of vasopressin.

REFERENCES

Adolph, E. F. (1927a). *J. Exp. Zool.* **47**, 1–30.
Adolph, E. F. (1927b). *J. Exp. Zool.* **49**, 321–351.
Adolph, E. F. (1927c). *Amer. J. Physiol.* **81**, 315–324.
Adolph, E. F., and Adolph, P. E. (1925). *J. Exp. Zool.* **43**, 105–149.
Alvarado, R. H., and Moody, A. (1970). *Amer. J. Physiol.* **218**, 1510–1516.
Balinsky, J. B., Cragg, M., and Baldwin, E. (1961). *Comp. Biochem. Physiol.* **3**, 236–244.
Bentley, P. J. (1966). *Biol. Rev. Cambridge Phil. Soc.* **41**, 225–316.
Bentley, P. J. (1969). *J. Endocrinol.* **43**, 359–369.
Bentley, P. J. (1971). "Endocrines and Osmoregulation." Springer-Verlag, Berlin and New York.
Biber, T. U. L. (1971). *J. Gen. Physiol.* **58**, 131–144.
Biber, T. U. L., and Curran, P. F. (1970). *J. Gen. Physiol.* **56**, 83–99.
Bishop, W. R., Mumbach, M. W., and Scheer, B. T. (1961). *Amer. J. Physiol.* **200**, 451–453.
Brunn, F. (1921). *Z. Gesamte Exp. Med.* **25**, 170.
Civan, M. M., and Hoffman, R. E. (1971). *Amer. J. Physiol.* **220**, 324–328.
Crabbé, J. (1961). *J. Clin. Invest.* **40**, 2103–2110.
Crabbé, J. (1972). *J. Steroid Biochem.* **3**, 229–235.
Cuthbert, A. W., and Painter, E. (1969). *Brit. J. Pharmacol.* **35**, 29–50.
Darwin, C. R. (1896). "Naturalist's Voyage around the World." Appleton, New York.
Edelman, I. S. (1972). *J. Steroid Biochem.* **3**, 167–172.
Edelman, I. S., Bogoroch, R., and Porter, G. A. (1963). *Proc. Nat. Acad. Sci. U.S.* **50**, 1169–1176.
Eggena, P. (1972). *J. Gen. Physiol.* **59**, 519–533.
Eliassen, E., and Jørgensen, C. B. (1951). *Acta Physiol. Scand.* **23**, 143–151.
Fimognari, G. M., Porter, G. A., and Edelman, I. S. (1967). *Biochim. Biophys. Acta* **135**, 89–99.
Gatzy, J. T., and Berndt, W. O. (1968). *J. Gen. Physiol.* **51**, 770–784.
Gatzy, J. T., and Clarkson, T. W. (1965). *J. Gen. Physiol.* **48**, 647–671.

Goodman, D. B. P., Allen, J. E., and Rasmussen, H. (1971). *Biochemistry* **10**, 3825–3831.
Gordon, M. S. (1962). *J. Exp. Biol.* **39**, 261–270.
Gordon, M. S., Schmidt-Nielsen, K., and Kelly, H. M. (1961). *J. Exp. Biol.* **38**, 659–678.
Greenwald, L. (1971). *Physiol. Zool.* **44**, 149–161.
Handler, J. S., Preston, A. S., and Orloff, J. (1972a). *Amer. J. Physiol.* **22**, 1071–1074.
Handler, J. S., Preston, A. S., and Orloff, J. (1972b). *J. Steroid Biochem.* **3**, 137–141.
Heller, H. (1941). *J. Physiol. (London)* **100**, 125–141.
Heller, H. (1950). *Experientia* **6**, 368–376.
Hviid-Larsen, E. (1970). *Acta Physiol. Scand.* **79**, 453–461.
Hviid-Larsen, E. (1972). *J. Steroid Biochem.* **3**, 11–20.
Jørgensen, C. B. (1947). *Nature (London)* **160**, 872.
Kirchberger, M. A., Chen, L. C., and Sharp, G. W. G. (1971). *Biochim. Biophys. Acta* **241**, 861–875.
Krogh, A. (1939). "Osmotic Regulation in Aquatic Animals." Cambridge Univ. Press, London and New York.
Leaf, A. (1965). *Ergeb. Physiol., Biol. Chem. Exp. Pharmakol.* **56**, 216–263.
Leaf, A., and MacKnight, A. D. C. (1972). *J. Steroid Biochem.* **3**, 237–245.
Lipton, P., and Edelman, I. S. (1971). *Amer. J. Physiol.* **221**, 733–741.
McClanahan, L., and Baldwin, R. (1969). *Comp. Biochem. Physiol.* **28**, 381–389.
Maetz, J. (1959). *In* "La méthode des indicateurs nucléaires dans l'étude des transports actifs d'ions" (I. Coursaget, ed.), pp. 185–196. Pergamon, Oxford.
Maetz, J. (1963). *Symp. Zool. Soc. London* **9**, 107–140.
Myers, R. M., Bishop, W. R., and Scheer, B. T. (1961). *Amer. J. Physiol.* **200**, 444–450.
Overton, E. (1904). *Verh. Phys.-Med. Ges. Wurzburg* **36**, 277–295.
Porter, G. A. (1971). *Gen. Comp. Endocrinol.* **16**, 443–451.
Porter, G. A., Bogoroch, R., and Edelman, I. S. (1964). *Proc. Nat. Acad. Sci. U.S.* **52**, 1326–1333.
Potts, W. T. W., and Evans, D. H. (1967). *Biol. Bull.* **133**, 411–425.
Ridley, A. R. (1964a). *Gen. Comp. Endocrinol.* **4**, 481–485.
Ridley, A. R. (1964b). *Gen. Comp. Endocrinol.* **4**, 486–491.
Rotunna, C. A., Villalonga, F. A., and Cereijido, M. (1970). *J. Gen. Physiol.* **55**, 716–735.
Rousseau, G., and Crabbé, J. (1971). *Eur. J. Biochem.* **25**, 550–559.
Ruibal, R. (1962a). *Physiol. Zool.* **35**, 133–147.
Ruibal, R. (1962b). *Physiol. Zool.* **35**, 218–223.
Scheer, B. T., and Markel, R. A. (1962). *Comp. Biochem. Physiol.* **7**, 289–297.
Sharp, G. W. G., and Leaf, A. (1964). *Nature (London)* **202**, 1185–1188.
Sharp, G. W. G., and Leaf, A. (1966). *Physiol. Rev.* **46**, 593–633.
Singer, S. J., and Nicolson, G. L. (1972). *Science* **175**, 720–731.
Snart, R. S. (1972). *J. Steroid Biochem.* **3**, 129–136.
Steen, W. B. (1929). *J. Exp. Zool.* **43**, 215–220.
Tercafs, R. R., and Schoffeniels, E. (1962). *Life Sci.* **1**, 19–24.
Uranga, J., and Quintana, G. (1958). *C. R. Soc. Biol.* **152**, 1826–1827.

CHAPTER 4

Bile Salts of Amphibia

A. R. Tammar

I. Introduction 67
II. Caudata 68
III. Anura 69
A. Bufonidae 69
B. Ranidae 70
C. Other Anurans 70
IV. Significance of Amphibian Bile Salt Distribution 71
V. Amphibian Bile Salts to June 1971 74
References 76

I. Introduction

Bile salts are secretory products of vertebrate livers and are involved in the digestion and absorption of fat. Cholesterol is assumed to be the universal precursor of bile salts (although there is a certain amount of evidence from one amphibian species to the contrary); hence, bile salts are steroidal in nature. The final products of cholesterol catabolism are either bile alcohols or bile acids. Bile salts themselves are derivatives of these alcohols or acids. Bile alcohols are esterified with one sulfate group (two are encountered occasionally, but not in Amphibia) and bile acids form peptide links (conjugate) with either taurine or glycine. This chapter will concern itself with alcohol sulfates and taurine conjugated acids, as glycine conjugation is not encountered below Mammalia.

Primary bile salts are synthesized by the liver directly from cholesterol, and secondary bile salts are those that have been modified by intestinal microorganisms and returned to the liver by the enterohepatic circulation. They are either resecreted by the liver along with the primary bile salts or further modified by liver hydroxylation systems prior to resecretion.

The subject of species differences in bile salts has been extensively reviewed by Haslewood most recently in 1967 (Haslewood, 1967a,b). In this context, C_{24}-5β acids are regarded as advanced and all other bile salts as more primitive. Table I at the end of the chapter has a complete list of references to work on individual species.

II. Caudata

The biles of three species of this order have been investigated. The giant Japanese salamander *Megalobatrachus japonicus* contains the sulfate of 5α-cholestane-3α,7α,12α,26,27-pentol, 5α-cyprinol (Fig. 1) as well as allocholic and allochenodeoxycholic acids. The C_{24} acids are said to be unconjugated. In addition, small amounts of 3α,7α,12α-trihydroxy-5α-cholestan-26-oic and 5α-cholestan-26-oic acids have been found. The newt *Diemyctylus pyrrhogaster* also contains 5α-cyprinol sulfate and also the sulfate of 5α-cholestane-3α,7α,12α,25,26-pentol 5α-bufol (Fig. 2). Recent investigations (Hoshita *et al.*, 1967c) have revealed no bile acid conjugates in *D. pyrrhogaster* bile, although Okasaki (1944) claimed to have isolated small quantities of both cholic and deoxycholic acids. *Salamandra salamandra* bile contains the 24-sulfate of 25,26-dihomo-5α-cholane-3α,7α,12α,24ξ,26-pentol, 5α-ranol (Fig. 3) and tauroallocholate. Thus, the very few caudatan bile salts examined so far, although exclusively derivatives of 5α-cholestane, show wide divergence of side chain oxidation pattern even within one family.

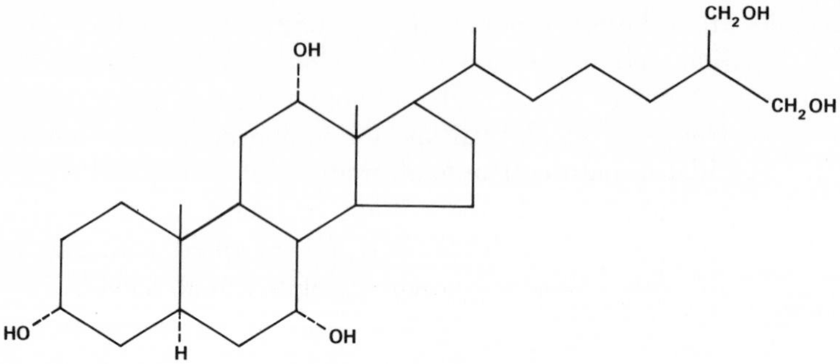

FIG. 1. 5α-Cyprinol.

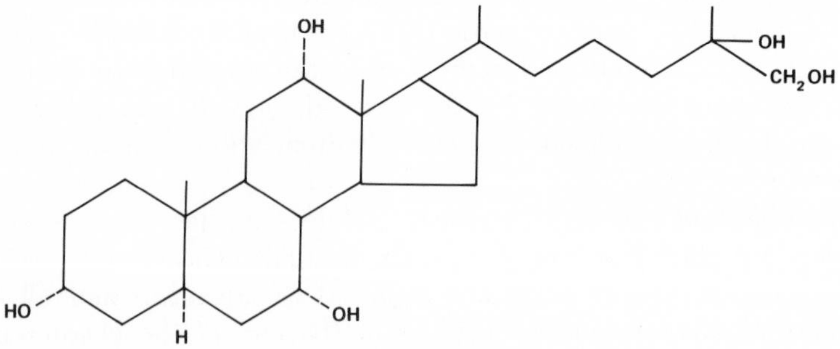

FIG. 2. 5α-Bufol.

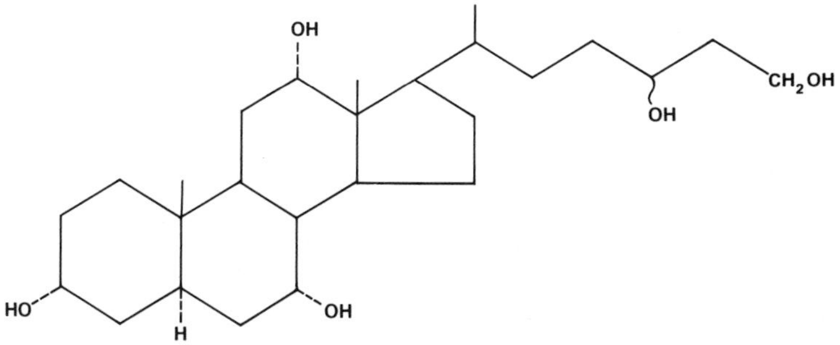

FIG. 3. 5α-Ranol.

III. Anura

A. BUFONIDAE

In contrast to caudatans, and indeed ranids, the bufonids examined show a marked similarity in their biles in that all four species apparently have 5β-bufol sulfate as the principal bile salt. The four species are *Bufo bufo, B. marinus, B. regularis,* and *B. vulgaris japonicus* (= *B. bufo formosus*). Only the last of these has been thoroughly investigated, and in addition to 5β-bufol sulfate its bile contains 3α,7α,12α-trihydroxy-5β-cholest-23-en-26-oic and 3α,7α,12α-trihydroxy-5β-cholest-22-ene-24-carboxylic acids (Fig. 4). This latter acid is curious because of its C_{28} formula, and it is difficult to see how it could have been derived from cholesterol. Indeed, intraperitoneal injection of *B. v. japonicus* with cholesterol-4-^{14}C led to the formation of labeled 3α,7α,12α-trihydroxy-5β-cholestan-26-oic acid, 5β-cholestane-3α,7α,12α,26-tetrol, and 5α-cholestane-3α,7α,12α,26-tetrol [the only 5α bile alcohol reported from a toad (Hoshita *et al.*, 1965)] but not to radioactive 3α,7α,12α-trihydroxy-5β-cholest-23-en-26-oic and 3α,7α,12α-trihydroxy-5β-cholest-22-ene-24-carboxylic acids. Phytosterols have been found in *B. v. japonicus* skin (Hüttel and Behringer, 1937), and subsequent analysis (Morimoto, 1966) has shown that they are C_{27} steroids (93%), β-sitosterol (4%), and campesterol (3%).

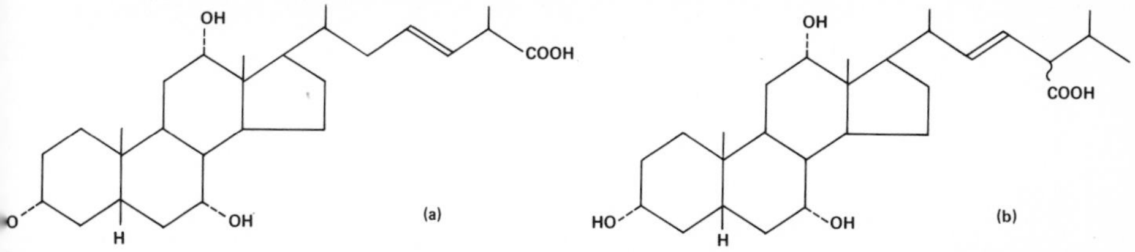

FIG. 4. (a) 3α,7α,12α-Trihydroxy-5β-cholest-23-en-26-oic acid; (b) 3α,7α,12α-trihydroxy-5β-cholest-22-ene-24-carboxylic acid.

None of these appears at first sight to be a likely precursor of $3\alpha,7\alpha,12\alpha$-trihydroxy-5β-cholest-22-ene-24-carboxylic acid, and it could be that the true precursor is ergosterol, a substance much favored for this role by Japanese workers in the 1930s.

Bufonid biles then are characterized by the presence of 5β-bufol and other 5β-steroid acids, some of which are unlikely to have been derived from cholesterol. The nature of conjugation of toad bile salts is not fully known, since biles from apparently healthy toads contain detectable quantities of unconjugated material, apparently a proportion of the bile acids.

B. Ranidae

The biles of six species of this family have been analyzed, and all of them are quite different. The ranids thus do not display the homogeneity normally associated with bile salts from within one of the lower taxonomic groups. Four of these species appear to have only, or chiefly, one bile salt, all primitive (see Table I). *Rana catesbeiana* and *R. nigromaculata,* however, have bile salt mixtures which, although mainly primitive, also contain taurocholate. No member of this family appears to have 5β-bufol, which may reasonably be regarded as characteristic of the Bufonidae. The bewildering diversity of bile salt types exhibited by the Ranidae has been attributed by Haslewood (1967b) to the fact that this group is actively evolving. This may be so, but many more species must be investigated before it can be established with certainty.

C. Other Anurans

The only other anuran species that has been investigated is *Discoglossus pictus,* which appears to have the taurine conjugate of $3\alpha,7\alpha,12\alpha$-trihydroxy-5β-cholestan-26-oic acid (Fig. 5) as its principal or sole bile salt.

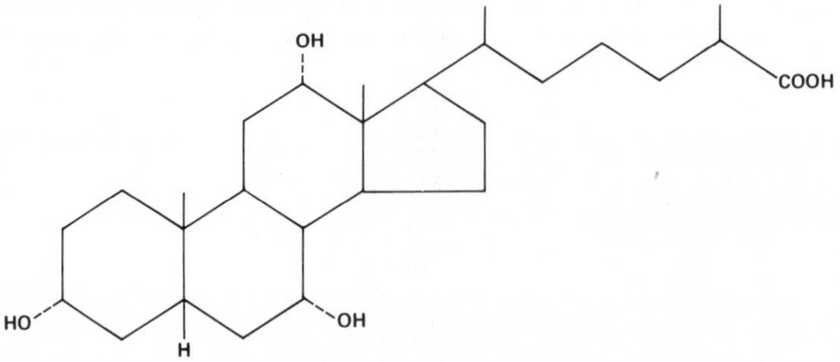

OH
COOH
HO
OH
H

Fig. 5. $3\alpha,7\alpha,12\alpha$-trihydroxy-5β-cholestan-26-oic acid.

IV. Significance of Amphibian Bile Salt Distribution

Chapter 15 of Volume VIII of this work correlates the various piscine bile salts with intermediates on the main route of mammalian cholic acid biosynthesis from cholesterol. Although it is possible to do the same for the bile salts already mentioned in this chapter, the end result is less satisfactory. This is because the mutations necessary to produce the different amphibian bile salts do not appear to have coincided with the taxonomic groupings of morphologists. Furthermore, there is evidence that certain toad bile salts are derived from sterols other than cholesterol and nothing is known of the routes whereby they are produced. No cecilian bile salts have yet become available for examination.

Consideration of amphibian bile salts shows that they fall into two steroid categories—5α and 5β. The changes that take place in the steroid nucleus to transform cholesterol into 5α-cholestane-3α,7α,12α-triol or 5β-cholestane-3α,7α,12α-triol are shown in Fig. 6. Caudatans, with exclusively 5α bile salts, appear not to possess the 5β-reductase which converts 7α,12α-dihydroxy-4-cholesten-3-one (Fig. 6d) into 7α,12α-dihydroxy-5β-cholestan-3-one (Fig. 6f). Toads, on the other hand, appear not to possess the 5α-reductase which transforms 7α,12α-dihydroxy-4-cholesten-3-one (Fig. 6d) into 7α,12α-dihydroxy-5α-cholestan-3-one (Fig. 6e). Frogs appear to possess both enzymes.

Modifications to the steroid C_8 side chain which result in the formation of C_{24} acids are shown in Fig. 7, where a through e represent the steps in mammalian cholic acid biosynthesis. Hydroxylation at C-26 of 5α(or 5β)-cholestane-3α,7α,12α-triol (Fig. 7a) affords 27-deoxy-5α(or 5β)-cyprinol (Fig. 7b) a key intermediate, since hydroxylation at C-27 affords 5α(or 5β)-cyprinol (Fig. 7g) and hydroxylation at C-25 affords 5α(or 5β)-bufol (Fig. 7h). Some frogs have a 27-hydroxylation system, all toads seem to have a 25-hydroxylation system, and caudatans appear to have both.

Oxidation of 27-deoxy-5α(or 5β)-cyprinol gives 3α,7α,12α-trihydroxy-5α(or 5β)-cholestan-26-oic acid (Fig. 7c), which has been found in the 5β-form in some frogs and in the 5α-form in *Megalobatrachus japonicus*. Hydroxylation of the C_{27} acid at C-24 gives 3α,7α,12α,24-tetrahydroxy-5α(or 5β)-cholestan-26-oic acid (Fig. 7d), which in cholic (and presumably allocholic) acid biosynthesis is further oxidized to the 24-ketone (Fig. 7e), which itself undergoes thiolytic cleavage to form cholic acid (Fig. 7f) and ultimately taurocholate. This route, which is well established for mammalian cholic acid biosynthesis (Danielsson and Tchen, 1968), may also be inferred for allocholic acid biosynthesis in *M. japonicus*. Amimoto (1966b) injected *M. japonicus* with tritiated

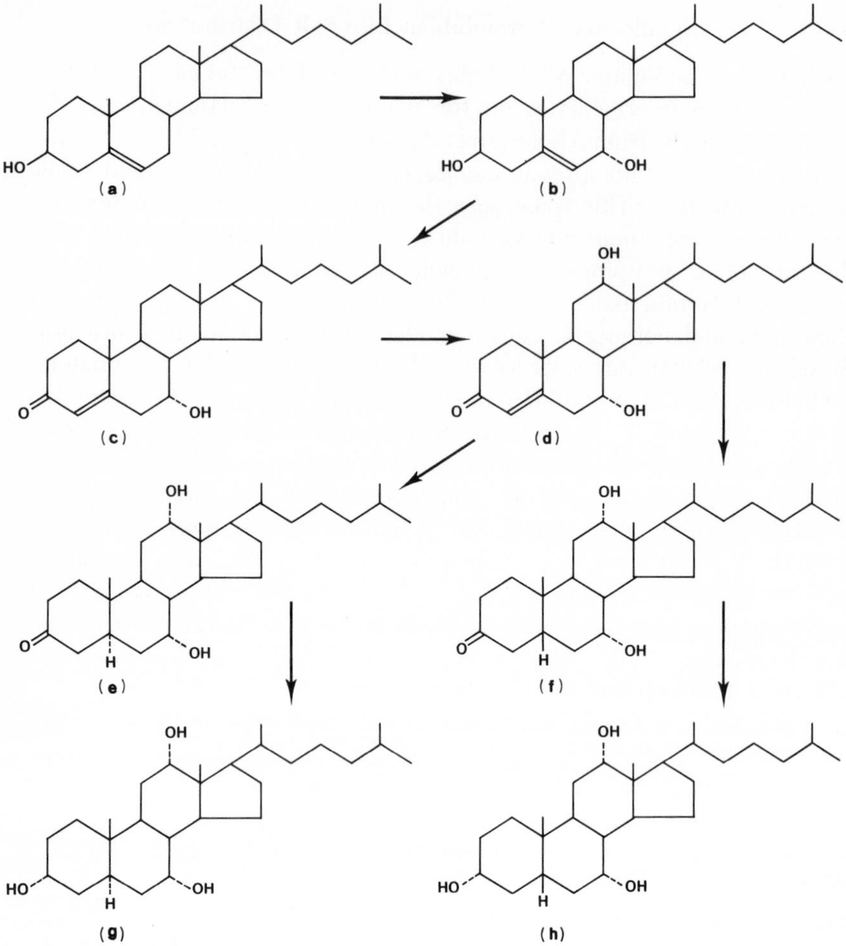

FIG. 6. Modifications to the steroid nucleus in cholic acid biosynthesis.

5α-cholestane-3α,7α,12α,26-tetrol. Subsequent analysis of the bile afforded labeled 5α-cholestane-3α,7α,12α,26,27-pentol, 3α,7α,12α-trihydroxy-5α-cholestan-26-oic acid, and allocholic acid.

Injection of tritiated 5β-cholestane-3α,7α,12α-triol (Fig. 7a) into *R. catesbeiana* (Betsuki, 1966) resulted in the formation of 3α,7α,12α-trihydroxy-5β-cholestan-26-oic acid (Fig. 7c), 26-deoxy-5β-ranol (Fig. 7i), 5β-ranol (Fig. 7j), and cholic acid (Fig. 7f). Decarboxylation of 3α,7α,12α,24-tetrahydroxy-5β-cholestan-26-oic acid (Fig. 7d) would afford 26-deoxy-5β-ranol, 26-hydroxylation of which would give 5β-ranol. If this mechanism of ranol formation, which Betsuki (1966) favors, be

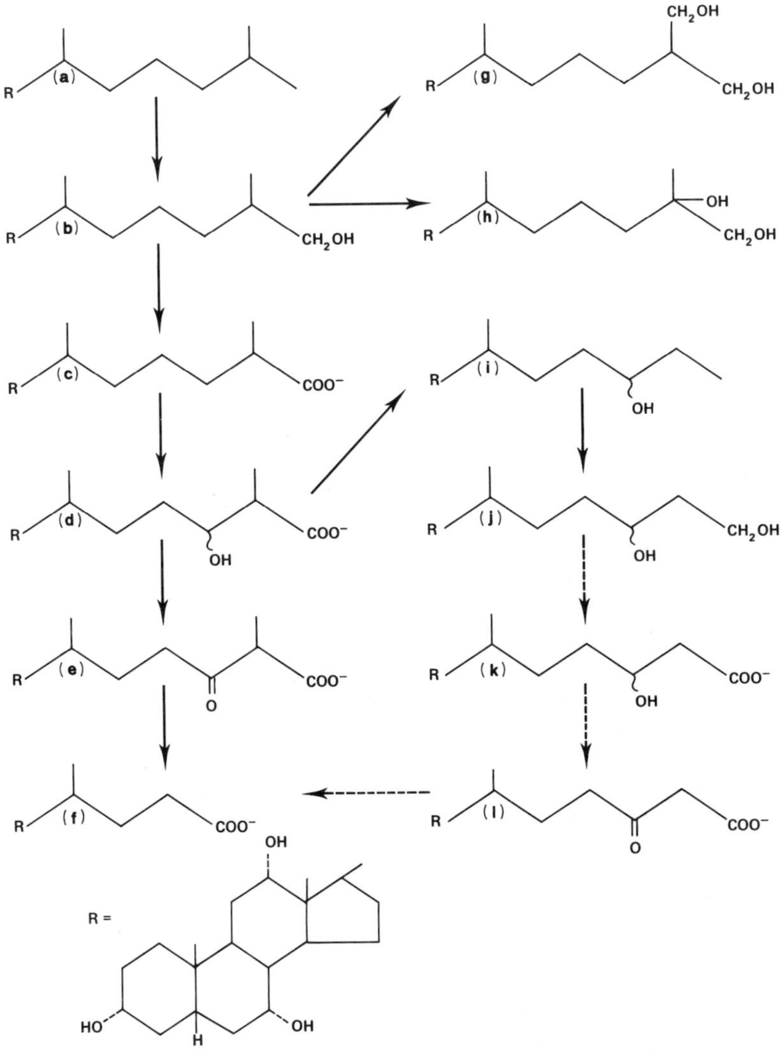

FIG. 7. Side-chain modifications in cholic acid biosynthesis.

correct, then the decarboxylating enzyme system is restricted in its occurrence to certain ranids and *S. salamandra*. Danielsson and Kazuno (1964) have shown that the bile fistula rat is capable of converting 5β-ranol to cholic acid in a yield of at least 40%. A route going through intermediates k and 1 (Fig. 7). was proposed for the conversion, and it is possible that this is the route whereby *R. catesbeiana* produces cholic acid.

Thus, the bile acids or alcohols characteristic of a particular amphibian species can be regarded either as intermediates in the mammalian cholic acid biosynthetic pathway which cannot be further metabolized in that species or, alternatively, as intermediates so modified as to be removed from the pathway. The former type is illustrated by $3\alpha,7\alpha,12\alpha$-trihydroxy-5β-cholestan-26-oic acid (Fig. 7c) found in *Rana ridibunda* and *Discoglossus pictus* and the latter type by the cyprinols and bufols (Fig. 7g and h).

The small number of amphibian species studied to date has brought to light a variety of species differences in bile salts that, apart from the Bufonidae, appear to exhibit no pattern consistent with already existing morphologically based taxonomic groupings. More exhaustive researches into amphibian bile salts, as yet unpublished, are currently being undertaken in this Department. An investigation of about 30 hitherto unreported anuran biles shows that biles of species of the genus *Bufo* are consistent with those listed in Table I and all have a proportion of unconjugated bile acids. The species of frogs examined exhibit the wide diversity of bile salts already encountered and there is, as yet, no clarification of the present confused scene.

V. Amphibian Bile Salts to June 1971

Table I, which follows, lists all amphibian bile salts examined up to June 1971. No mention is made of the nature of the conjugation of any of the bile alcohols or acids in column 2, but the reader may assume that in the bile all acids (with exceptions in bufonid bile) are taurine conjugated and that all the alcohols occur as sulfate esters.

TABLE I
AMPHIBIAN BILE SALTS EXAMINED UP TO JUNE 1971

Species	Acid or alcohol found	Reference
Caudata		
Cryptobranchidae		
Megalobatrachus japonicus	5α-cyprinol; 3α,7α,12α-trihydroxy-5α-cholestan-26-oic, allocholic, and allochenodeoxycholic acids	Amimoto (1966a,b); Hoshita *et al.* (1967a)
Salamandridae		
Diemyctylus pyrrhogaster	5α-cyprinol; 5α-bufol; ? cholic acid	Okasaki (1944); Hoshita *et al.* (1967c)
Salamandra salamandra	5α-ranol; allocholic acid	Haslewood (1967b), also unpublished
Anura		
Bufonidae		
Bufo vulgaris japonicus	5β-bufol; 3α,7α,12α-trihydroxy-5β-cholest-23-en-26-oic and 3α,7α,12α-trihydroxy-5β-cholest-22-ene-24-carboxylic acids	Hayakawa (1953); Okuda *et al.* (1962); Hoshita *et al.* (1967b)
Bufo marinus *Bufo bufo* *Bufo regularis*	5β-bufol	Haslewood (1967b)
Ranidae		
Rana catesbeiana	5α-ranol; 5β-ranol; 25(*R*)- and 25(*S*)-3α,7α,12α-trihydroxy-5β-cholestan-26-oic and cholic acids	Mabuti (1941); Kazuno *et al.* (1963, 1965a)
Rana nigromaculata	5β-cyprinol; 25(*R*) and 25(*S*)-3α,7α,12α-trihydroxy-5β-cholestan-26-oic and cholic acids	Komatsubara (1954); Kazuno *et al.* (1965b)
Rana ridibunda	3α,7α,12α-trihydroxy-5β-cholestan-26-oic acid; probably 5β-cyprinol	Haslewood (1967b), also unpublished
Rana temporaria	5α-ranol	Haslewood (1952, 1964)
Rana esculenta	5β-cyprinol and 5β-ranol (minor amounts)	G. A. D. Haslewood (unpublished)
Rana pipiens	5β-ranol	G. A. D. Haslewood (unpublished)
Discoglossidae		
Discoglossus pictus	3α,7α,12α-trihydroxy-5β cholestan-26-oic acid	G. A. D. Haslewood (unpublished)

REFERENCES

Amimoto, K. (1966a). *J. Biochem.* (*Tokyo*) **59**, 340.

Amimoto, K. (1966b). *Hiroshima J. Med. Sci.* **15**, 225; *Chem. Abstr.* **67**, 115002c.

Betsuki, S. (1966). *J. Biochem.* (*Tokyo*) **60**, 411.

Danielsson, H., and Kazuno, T. (1964). *Acta Chem. Scand.* **18**, 1157.

Danielsson, H., and Tchen, T. T. (1968). *In* "Metabolic Pathways" (D. M. Greenberg, ed.), 3rd ed., Vol. 2, pp. 117–168. Academic Press, New York.

Haslewood, G. A. D. (1952). *Biochem. J.* **51**, 139.

Haslewood, G. A. D. (1964). *Biochem. J.* **90**, 309.

Haslewood, G. A. D. (1967a). *J. Lipid Res.* **8**, 535.

Haslewood, G. A. D. (1967b). "Bile Salts." Methuen, London.

Hayakawa, S. (1953). *Proc. Jap. Acad.* **29**, 279 and 285; *Chem. Abst.* **48**, 13766i.

Hoshita, T., Sasaki, T., Tanaka, S., Betsuki, S., and Kazuno, T. (1965). *J. Biochem.* (*Tokyo*) **57**, 751.

Hoshita, T., Amimoto, K., Nakagawa, T., and Kazuno, T. (1967a). *J. Biochem.* (*Tokyo*) **61**, 750.

Hoshita, T., Okuda, K., and Kazuno, T. (1967b). *J. Biochem.* (*Tokyo*) **61**, 756.

Hoshita, T., Hirofuji, S., Nakagawa, T., and Kazuno, T. (1967c). *J. Biochem.* (*Tokyo*) **62**, 62.

Hüttel, R., and Behringer, H. (1937). *Hoppe-Seyler's Z. Physiol. Chem.* **245**, 175.

Kazuno, T., Masui, T., Nakagawa, T., and Okuda, K. (1963). *J. Biochem.* (*Tokyo*) **53**, 331.

Kazuno, T., Masui, T., and Okuda, K. (1965a). *J. Biochem.* (*Tokyo*) **57**, 75.

Kazuno, T., Betsuki, S., Tanaka, Y., and Hoshita, T. (1965b). *J. Biochem.* (*Tokyo*) **58**, 243.

Komatsubara, T. (1954). *Proc. Jap. Acad.* **30**, 614 and 618; *Chem. Abstr.* **50**, 386f.

Mabuti, H. (1941). *J. Biochem.* (*Tokyo*) **33**, 131.

Morimoto, K. (1966). *Hiroshima J. Med. Sci.* **15**, 145; *Chem. Abstr.* **67**, 18859e.

Okasaki, Y. (1944). *J. Biochem.* (*Tokyo*) **36**, 65 and 77.

Okuda, K., Hoshita, T., and Kazuno, T. (1962). *J. Biochem.* (*Tokyo*) **51**, 48.

CHAPTER 5

Amphibian Hemoglobins

Bolling Sullivan

I. Introduction 77
II. Hemoglobin Components 79
A. Anura 82
B. Urodela 89
C. Apoda 92
III. Hemoglobin Function 92
A. Anura 93
B. Urodela 98
IV. Hemoglobin Structure 101
A. Anura 101
B. Urodela 103
C. Sequence Studies and Their Structure–Function Correlations 104
V. Hemoglobin Synthesis 111
A. Seasonal Variation and Phenylhydrazine Treatment 111
B. Sites of Hemoglobin Synthesis 113
C. The Larval to Adult Transition 115
VI. Concluding Comments 117
References 118

I. Introduction

Studies of human hemoglobins have provided us with the amino acid sequences of the various polypeptide chains, the three-dimensional structures of the oxygenated and deoxygenated conformational derivatives, and a genetic scheme for the evolution of protein molecules. Sufficient data are available to generate specific models that explain the cooperative interaction of subunits, the regulation of function by allosteric effectors, and the genetic control of synthesis in normal and diseased states. Hemoglobin variants in man have helped to define the functional properties of human hemoglobin as well as to provide a model system for the study of genetic disorders. Perhaps a thorough knowledge of one hemoglobin molecule is sufficient in that all of its significant properties can be deduced. Perhaps, but that is not likely. Hemoglobin is one of the most variable molecules known. It is clearly adapted to serve the needs of many species, and it is this variation in structure and function that many people find so fascinating. Some of these variations have properties quite unlike human hemoglobin, yet they illustrate facets of hemoglobin function that are obscure in human hemoglobin. Certainly a thorough knowledge of other hemoglobins will give us a perspective we badly need. A recent review of primate hemoglobins (Sullivan, 1971) has shown that some aspects of the structure, function, and genetic control of human hemoglobins are limited to man and his closest relatives. Are human hemoglobins specifically, and primate hemoglobins in general, typical or very specialized forms? To answer questions such as these we need perspective. Human and chimpanzee hemoglobins are identical, yet these species are always placed in different genera and usually in different families. On the other hand two species of frogs in the genus *Rana* have hemoglobins which appear to differ by 25%. That's perspective, or lack of it.

Comparative studies of hemoglobins outnumber those of any other protein molecule. Interest in amphibian hemoglobins was sparked by McCutcheon's finding that the oxygen-binding properties of tadpole hemoglobin differ significantly from those of adult frog hemoglobin (McCutcheon, 1936). He correctly attributed these differences to "the type and properties of the hemoglobin." Larval and adult bullfrog (*Rana catesbeiana*) hemoglobins and the mechanisms involved in their ontogenetic switch have been the most studied among amphibian hemoglobins. That is natural, since this change is comparable in part to the fetal–adult hemoglobin transition in man and offers a model system for experimental studies of differentiation. Other studies have shown, however, that amphibian hemoglobins may be quite helpful in elucidating properties common to all hemoglobins. Certainly, amphibian hemoglobins are interesting in their own right, because they do illustrate the plasticity of molecular architectures in response to adaptive pressures. Amphibians clearly illustrate a stage in the transition from aquatic to terrestrial life. Certain properties of their hemoglobins may also illustrate this transition, but it is likely that the more recent adaptations of amphibians to various environments will obscure any broad generalizations. Amphibian hemoglobins surely contain much useful phylogenetic information encoded in their amino acid sequences which would be of great interest to those interested in the taxonomy and phylogeny of amphibians. Whether or not technology and the priorities of modern civilization will allow a significant amount of this information to be extracted remains to be seen.

This review will attempt to summarize critically what is known about the structural and functional properties of amphibian hemoglobins and how their synthesis is controlled. By necessity, the data often will be compared with what is known about human hemoglobins (most recently summarized by Huisman and Schroeder, 1970; Antonini and Brunori, 1971). Such a comparison naturally will reveal the inadequacies of our present data on amphibian hemoglobins, but it is hoped that it will also stimulate and direct future studies of these interesting protein molecules.

II. Hemoglobin Components

Hemoglobins occur in enucleated erythrocytes in most mammalian species. They occur in nucleated erythrocytes in most other vertebrates. Hemoglobin normally occurs in the cytoplasm, but has been found in the nucleus of frog (Davies, 1961) and newt (Tooze and Davies, 1963) erythrocytes. Hemoglobin is erratically distributed among invertebrate groups and occurs either intracellularly or extracellularly. X-ray crystallographic studies have shown that the three-dimensional structures of the globin chains from *Chironomus* (insect) (Huber *et al.*, 1971), *Glycera* (annelid) (Padlan and Love, 1968), and lamprey (cyclostome) (Hendrickson and Love, 1971) are very similar to those of human α and β chains. Amino acid sequence studies of these molecules confirm the apparent homology at the primary level (these and other globin sequences are summarized in Dayhoff, 1972). Comparison of the amino acid sequences and atomic structures of various myoglobins supports the theory that myoglobins and hemoglobins are descendants from a common gene (Ingram, 1961). The apparent unity in origin and three-dimensional structure among various globin chains is accompanied by enormous structural and functional diversity. This diversity is apparent even among the amphibia, a morphologically conservative group.

All amphibians have nucleated erythrocytes. These elliptical cells range in size from 36.3×62.5 μ in *Amphiuma* to lower limits near 12.2×19.5 μ as in *Hyla crucifer* (Table I). It should be pointed out that a distinct gap in size exists between erythrocytes of urodeles and those of anurans and also between amphibians and all but the lowest vertebrates. The adaptive significance of these large erythrocytes in urodeles is unknown. The hemoglobin concentration in amphibian erythrocytes is variable and usually less than in mammalian cells. Enucleated erythrocytes also occur in numerous urodeles and anurans, particularly those adapted to a terrestrial environment, but the proportion enucleated is quite variable and usually less than 10%. Enucleated erythrocytes outnumber normal erythrocytes in the urodele *Batrachoseps attenuatus* (Szarski and Kosmos, 1965). Enucleated erythrocytes consume less oxygen than nucleated ones and are therefore more efficient oxygen carriers, but they must be replaced more often. If enucleation occurs after a period of normal circulation, it may serve to prolong the useful life

TABLE I
ERYTHROCYTE DIMENSIONS AND HEMOGLOBIN CONCENTRATION IN UREDELES AND ANURANS

Species	Erythrocyte dimensions (μ)	Hemoglobin concentration (gm/100 ml blood)	References
Urodeles			
Amphiuma means	36.3 × 62.5	9.4	Wintrobe, 1934
Cryptobranchus alleganiensis	21.0 × 40.5	13.3	Wintrobe, 1934
Necturus maculosus	28.2 × 52.8	4.6	Wintrobe, 1934
Dicamptodon ensatus (adult)	29.4 × 51.4	4.4	Wood, 1971
Ambystoma maculatum	22.5 × 38.7	—	Szarski and Czopek, 1966
Taricha granulosa	20.5 × 37.5	—	Szarski and Czopek, 1966
Plethodon cinereus	17.0 × 32.5	—	Szarski and Czopek, 1966
Desmognathus quadramaculatus	21.7 × 34.2	—	Szarski and Czopek, 1966
Eurycea bislineata	16.5 × 34.7	—	Szarski and Czopek, 1966
Anurans			
Bufo americanus	12.5 × 19.7	—	Szarski and Czopek, 1966
Hyla crucifer	12.2 × 19.5	—	Szarski and Czopek, 1966
Rana catesbeiana	15.3 × 28.4	7.8	Wintrobe, 1934

of the erythrocyte (Szarski and Czopek, 1965). The life span of nucleated erythrocytes in *Bufo marinus* is 700–1400 days compared to the 120 day span of enucleated human erythrocytes (Altland and Brace, 1962). Other amphibians have equally long-lived erythrocytes (Carmena, 1971a).

The major hemoglobin component in hemolyzates from humans (*Homo sapiens*) is a tetramer composed of equal amounts of two types of polypeptide chains. Each chain has a molecular weight (m.w.) of about 16,000 and the tetramer m.w. is 64,500. The two types of polypeptide chain (designated α and β) have heme at the active site (which binds one molecule of molecular oxygen) and differ in length by five residues (141 residues in the α chain and 146 residues in the β chain). The α and β chains from *H. sapiens* are coded at separate, nonlinked loci. Similarities in the amino acid sequences of the α and β chains indicate a genetic relationship between the two loci (Ingram, 1961). One chain locus (presumed to be the β) arose from the α chain locus by complete gene duplication followed by translocation and differentiation (for discussions of gene duplication, see Dixon, 1966; Watts and

Watts, 1968). Human hemolyzates also contain a minor hemoglobin component whose m.w. is 64,500. This tetramer contains two α and two δ chains. The δ chains are very similar to β chains in amino acid sequence, differing at only ten positions. The δ chain locus arose by total gene duplication of the β chain locus (Ingram, 1961) and the δ chain locus is linked tightly to the β chain locus.

Red cells from fetal human blood contain a different major hemoglobin, fetal hemoglobin. This hemoglobin tetramer has a m.w. of 64,500 and contains two α and two γ chains. The γ chain, like the β and δ chains, contains 146 amino acids. Although the γ chain shares sequence homology with both the α and β chains, it is believed to have derived from the β chain locus by gene duplication (Ingram, 1961). Several other tetramers containing two types of polypeptide chains occur in very young human embryos. Apparently, these are the major components prior to three months' gestation (Huehns *et al.*, 1964). These tetramers often contain an aditional type of chain designated ϵ. The genetic relationship of the ϵ chain to the other chains is not known, but it is interesting that it seems to pair with both α-like and γ-like chains in human embryos. In addition to these aforementioned components, various minor components occur with regularity in adult and fetal hemolyzates. These usually amount to less than 10% of the hemoglobin present and for most their genetic and synthetic origins are obscure.

Although *H. sapiens* has a single major adult hemoglobin component, most species of vertebrates have multiple components. Presumably these result from duplications of the α- and β-chain loci and recombination (usually) into all possible heterotetramers ($\alpha_2{}^A\beta_2{}^A$, $\alpha_2{}^B\beta_2{}^A$, $\alpha_2{}^A\beta_2{}^B$, $\alpha_2{}^B\beta_2{}^B$, etc.). These multiple components are found in all normal individuals. The number of hemoglobins in hemolyzates from various amphibians has been determined by a variety of techniques. The most widely used technique is electrophoresis. This technique, which has undergone a series of refinements, will separate molecules according to their charge and to some extent, depending on the supporting medium used, according to their size and shape or, roughly, molecular weight. Originally, paper electrophoresis was used. It was replaced by starch-gel electrophoresis which has superior resolving power. More recently, discontinuous polyacrylamide gel electrophoresis and isoelectric focusing have been used. These new techniques have even greater resolving power and should be used in the future along with starch-gel electrophoresis. The correct interpretation of electrophoretic experiments often requires that the oxidation state of the iron in hemoglobin and the degree of polymerization of hemoglobin (monomer, dimer, tetramer, octamer, . . . , polymer) be known. Many workers have been unaware of this, with the result that their data are often confusing and contradictory. Although new chains are still being discovered in man (multiple α chains, Abrahamson *et al.*, 1970; Brimhall *et al.*, 1970; multiple γ chains, Schroeder *et al.*, 1968, 1972; Huisman *et al.*, 1969) the electrophoretic and chromatographic properties are well defined. This is not true for other species, as will be clear from our discussion of amphibian hemoglobins.

A. Anura

The sixteen living families of anurans contain 218 genera and approximately 2600 species (Porter, 1972). Although morphologically similar, they differ widely in their abilities to tolerate heat, desiccation, and salt. Anuran species abound in the tropics, but some species reach Tierra del Fuego, South Africa, and Tasmania in the south and the Arctic Circle in the north. Hemolyzates from very few species have been studied in any detail.

1. *Adult Components*

It is apparent from the studies to be described below that many species of frogs and toads have geographically variable hemolyzates. Furthermore, many if not most species have more than one hemoglobin component in their hemolyzates. These two facts have made it almost impossible to resolve the many studies of frog hemoglobins into a comprehensible picture. Clearly, they have made it impossible to divide the studies into normally occurring and polymorphic components, so I have chosen to lump the two categories. In the future it is hoped that authors will state the origin of their samples and determine whether or not the animals used were polymorphic for hemoglobin. Wild-trapped specimens are to be preferred over commercial lots of unknown origin. The term polymorphism is often incorrectly applied to electrophoretic data; it correctly describes genetic variation within a species and should not be applied to a species simply because its hemolyzates contain multiple components found in all individuals. Hemoglobins from many species of amphibians polymerize by disulfide bond formation (Riggs *et al.*, 1964; Trader and Frieden, 1966). Polymerized hemoglobin has altered electrophoretic properties and can mimic polymorphism. Treatment of hemolyzates with mercaptoethanol or dithiothreitol at pH values above 7.0 should reduce the disulfide bridges and decrease many artifacts seen in electrophoretic experiments.

Discoglossidae: Hemolyzates from *Bombina bombina* and *B. variegata* contain two hemoglobin components, as determined by paper electrophoresis (Marchlewska-Koj, 1963).

Pipidae: In a study of protein synthesis in nucleated erythrocytes, Maclean *et al.* (1969) chromatographed hemolyzates from *Xenopus laevis* on columns of carboxymethylcellulose. Three fractions were observed and eluted in the following order: F-1 (43%), F-2 (37%), and F-3 (6%). The resolution was not particularly good, but there do seem to be two major and one minor components. These authors also noted that unless mercaptoethanol was added to the hemoglobin solutions, much hemoglobin remained at the top of the column. This was interpreted as indicating that *Xenopus* hemoglobin polymerizes. Later work (Maclean and Jurd, 1971a) indicates there are two components (A_1 and A_2). A_1 accounts for about 95% of the hemolyzate. Perhaps one of the components observed earlier represented a polymerized fraction. Hemoglobins from *X. l. laevis, X. l. victorianus,* and their hybrids are indistinguishable (Maclean and Jurd, 1972).

Pelobatidae: Marchlewska-Koj (1963) using paper electrophoresis found two hemoglobins in hemolyzates from *Pelobates fuscus.*

Leptodactylidae: Bertini and Rathe (1962) examined the electrophoretic patterns of hemolyzates from twelve leptodactylid species by paper electrophoresis. Only *Telmatobius hauthali, T. oxycephalus, Leptodactylus bufonius, Pleurodema bibroni, P. nebulosa,* and *Calyptocephalella gayi* had single-banded hemoglobin patterns. Two-banded hemoglobin patterns were found in hemolyzates from *Eleutherodactylus discoidalis, Leptodactylus chaquensis, L. luticeps, L. podicipinus,* and *Pleurodema tucumana.* Three-banded hemoglobin patterns were observed for *Leptodactylus ocellatus* and *Pleurodema bufonina.* Disk gel electrophoresis of *L. fallax* hemolyzates resolved two components (Gatten and Brooks, 1969).

Hylidae: Dessauer *et al.* (1957) reported two hemoglobin bands in hemolyzates from *Hyla cinerea* examined by paper electrophoresis. Similar results were reported by Marchlewska-Koj (1963) for *H. arborea.* Additional paper electrophoretic studies by Bertini and Rathe (1962) also revealed two hemoglobin bands in *H. trachytorax* and *Phyllomedusa sauvagei.* Dessauer and Nevo (1969) examined hemolyzates from 850 *Acris crepitans* and *A. gryllus* collected in twenty of the United States. Each species showed two hemoglobins in starch-gel electropherograms. The patterns were species specific. Although the hemoglobins of all *A. crepitans* populations were electrophoretically similar, the electrophoretic patterns of their globin chains (after removal of the heme groups) differed and fell into three population groups: (1) plains group, (2) delta group, and (3) Appalachian group.

Bufonidae: Fox *et al.* (1961) examined hemolyzates from *Bufo valliceps, B. fowleri,* and their hybrids by starch-gel electrophoresis. Each species shows a single distinct band, the hybrid toads showed both parental bands. Single bands were observed in hemolyzates from *B. bufo, B. viridis,* and *B. calamites* by Marchlewska-Koj (1963). Bertini and Rathe (1962) subjected hemolyzates from six species and subspecies of *Bufo* to paper electrophoresis. Each hemolyzate (*B. arenarum, B. marinus, B. paracnemis, B. spinulosus, B. granulosus major, B. g. fernandezae*) contained a single component. Kurata and Arakawa (1963) found that hemolyzates from *B. vulgaris* contain a single hemoglobin band on paper electrophoresis. Guttman (1967) studied hemolyzates from *B. regularis* and *B. rangeri.* Starch-gel electrophoresis revealed the amazing total of eleven different components arranged into eight phenotypes among thirteen specimens of *B. regularis* collected at El Mahalla el Kubra, Egypt. Individuals had from two to four individual components. With one exception, a specimen having an additional fast-moving component, all *B. rangeri* hemolyzates contained a single hemoglobin band. Examination of six species of the *B. americanus* group showed extensive hemoglobin polymorphism (Guttman, 1969). The numbers of each species examined are in parentheses. *Bufo americanus* (1), *B. hemiophrys* (1), *B. houstonensis* (2), and *B. microscaphus* (4) hemolyzates were all monomorphic, single banded and species specific. *Bufo terrestris* (10) and *B. woodhousei* (12) hemolyzates were polymorphic, showing four and five phenotypes, respectively. In both samples there is a noticeable lack of heterozygotes. Brown and Guttman (1970) studied the hemoglobins of *B. arenarum* (4), *B. spinosus* (5), and their natural (1) and artificial (1) hybrids. Both species were polymorphic for hemoglobin, but again no heterozygotes were found. Furthermore, the artificial and natural hybrid individuals had different phenotypes; neither showed more than a single band. Guttman (1972) reports that hybrid individuals between the following *Bufo* species did show both parental hemoglobin bands: *B. houstonensis* × *B. valliceps, B. woodhousei* × *B. houstonensis, B. woodhousei* × *B. valliceps,* and *B. garmani* × *B. regularis.* Hybrid individuals showing only one parental band were: *B. perplexus* × *B. bocourti, B. speciosa* × *B. arenarum,* and *B. arenarum* × *B. valliceps.* Several crosses produced hybrid individuals with a hemoglobin band intermediate between the parental bands: *B. perplexus* × *B. leutkeni, B. coccifer* × *B. valliceps,* and *B. marinus* × *B.*

spinulosus. In the first two crosses, hemolyzates from hybrid individuals contained one hybrid-specific band and one parental band. In addition to examining hybrid specimens, Guttman (1972) also has examined hemolyzates from 37 species of *Bufo*. Of these 19 were polymorphic at the hemoglobin loci: *B. garmani, B. punctatus, B. kellogi, B. speciosa, B. cognatus, B. marmoreus, B. perplexus, B. valliceps, B. canaliferus, B. coniferus,* and *B. granulosus*; other polymorphic species have been described above. Introgressive hybridization can account for a few cases of polymorphism, but not the majority. Considering that less than five individuals were examined in many species, the high frequency of polymorphism in the genus *Bufo* must be regarded as a low estimate. It is interesting to note that although many species of *Bufo* hybridize in the wild (and many more can be artificially hybridized) their hemoglobins are most often species specific. From the breeding data one might hypothesize that although the species are behaviorally isolated and differentiated, they are not genetically differentiated. However, the hemoglobin data lead to the opposite conclusion, at least as concerns the genetic differentiation.

Ranidae: Paper electrophoretic studies of hemolyzates from *Rana tigrina* showed two hemoglobin bands (Bhown, 1965). Paper electrophoresis of hemolyzates from *R. temporaria, R. terrestris,* and *R. esculenta* showed single components (Marchlewska-Koj, 1963). Electrophoretic studies of hemolyzates from *R. grylio* demonstrated four fractions (Trader and Frieden, 1966). A population study of *R. pipiens* by Gillespie and Crenshaw (1966) revealed two hemoglobin bands. On starch-gel electrophoresis at pH 8.6, the faster band was the less concentrated. Frogs from the piedmont and coastal plain of Maryland were compared to specimens from Wisconsin. Hemoglobin polymorphism was evident in all populations. Manwell (1966) resolved hemolyzates from *R. pipiens* into approximately three bands, a slow major band and two faster minor bands. Starch-gel electrophoresis of *R. esculenta* hemoglobin revealed the presence of three major and one minor hemoglobin. Carboxymethylcellulose chromatography separated the three major components (Tentori *et al.*, 1965). The minor component appears to be polymorphic in English populations (Manwell and Baker, 1970). Only two major components were resolved by Chauvet and Acher (1971). Christomanos (1967) reported two hemoglobins in *R. ridibunda*.

Hemolyzates from *R. catesbeiana*, the bullfrog, have been examined by a number of workers. Hemolyzates from this species readily polymerize, and this has caused some confusion. Starch-gel electrophoresis of bullfrog hemolyzates revealed the presence of two, three, or four components (Manwell, 1966; Baglioni and Sparks, 1963). Hamada *et al.* (1964b) applied *R. catesbeiana* hemolyzates to a carboxymethylcellulose column and resolved the hemoglobins into a major and two minor components. Stratton and Wise (1967) resolved four components in bullfrog hemolyzates. Using polyacrylamide-gel electrophoresis Moss and Ingram (1968a) separated bullfrog hemolyzates into five fractions. The major fraction was preceded by three minor fractions and followed by one minor fraction. The major band represented the polymerized fraction (Riggs, *et al.*, 1964). Similar results were obtained by Just and Atkinson (1972). Aggarwal and Riggs (1969) have fractionated bullfrog hemolysates into four fractions by carboxymethylcellulose chromatography. The first fraction, component A, represents less than 1% of the total hemolyzates. The amount of component B varies from less than 5% to 35% of the hemolyzate in different individuals. Component C, the third to elute, is the major fraction, accounting for from 65% to almost 100% of the total hemolyzate in various bullfrogs. Component D was not quantified but does not appear to represent more than 1–3% of the hemolyzate. These results agree rather well with those of Moss and Ingram, except that the minor component just ahead of the major component was not resolved by Aggarwal and Riggs (1969). Polymerization was restricted to component C. Although the data are far from satisfactory, it appears that those *Rana* species examined to date have one and sometimes two major hemoglobins. The most common pattern is a slowly electrophoresing major band with two more rapidly migrating minor bands. There is both intra- and interspecific variation in the amount of the major components present. Compare, for example, the results obtained by Moss and Ingram (1968a) with those of Maniatis and Ingram (1971b).

2. *Larval Components*

Maclean and Jurd (1972) report that *Xenopus laevis* tadpoles have two distinct hemoglobins and that they share no common chain with adult frogs. Tadpole hemoglobin F_1 is the major component and the ratio F_1/F_2 increases with the age of the tadpole. Kurata and Arakawa (1963) studied hemolyzates from larval and adult forms of *Bufo vulgaris*. They were able to differentiate the two hemolyzates by paper electrophoresis. Larval hemoglobin migrated more slowly and both larval and adult hemolyzates contained a single type of hemoglobin. Dessauer *et al.* (1957) reported that on paper electrophoresis tadpole (larval) hemoglobin migrates more slowly than hemoglobin from adult *Rana pipiens*. Manwell (1966) confirmed these findings for *R. pipiens* and also reported variation in hemoglobin patterns for individual tadpoles. Elzinger

(1964) has reported that individual tadpoles of *R. pipiens* nearing metamorphosis have hemoglobin components that do not match electrophoretically the components of premetamorphic tadpoles or adult frogs. Elzinger (1964) and Manwell (1966) also found variation in different populations of *R. pipiens* tadpoles and concluded that tadpole hemolyzates were composed of five to six hemoglobin components. Variation in hemoglobin patterns of tadpoles was not paralleled by variation in adult hemoglobin patterns, indicating few if any common polypeptide chains.

Hamada *et al.* (1964b) were able to resolve tadpole (*R. catesbeiana*) hemolyzates into three components by carboxymethyl-cellulose chromatography. Hemolyzates contained one major and two minor components. Baglioni and Sparks (1963) resolved tadpole hemolyzates from *R. catesbeiana* into three components by starch-gel electrophoresis and concluded that tadpoles and adults share no common hemoglobin fractions. Trader and Frieden (1966) separated hemolyzates from *R. grylio* tadpoles into three or four fractions. Electrophoresis in starch-gel showed that no fractions were identical to adult fractions. Moss and Ingram (1968a) resolved hemolyzates from *R. catesbeiana* tadpoles into four bands by disk gel electrophoresis. The single major and three minor bands migrated more rapidly toward the anode than did the major component of frog hemoglobin. These workers noted variation in the number of minor bands in different tadpole populations. No systematic differences were observed between 2 and 16 gm tadpoles. Tadpoles obtained from a dealer in Connecticut were consistently different in their hemoglobin patterns from tadpoles obtained from a dealer in North Carolina.

Preparative disk gel electrophoresis separated tadpole hemolysates into four fractions whose relative amounts were 6, 20, 70, and 4%. Some of the minor bands stained differently with benzidine. Stratton and Wise (1967) resolved five bands by disk electrophoresis of *R. catesbeiana* tadpole hemolyzates. Aggarwal and Riggs (1969) chromatographed tadpole (*R. catesbeiana*) hemolyzates on columns of carboxymethylcellulose into at least six different components. Three of these components were themselves heterogeneous. When some of the components were remixed, additional bands were generated. The first fraction eluted from their column was a dimer, and these workers suggest that chromatography perturbs the dissociation equilibrium and allows individual chains of one or more tetrameric components to be separated. These chains exist in free solution as dimers. Moss and Ingram (1968a) noticed dissociation of tadpole hemoglobin on columns of Sephadex G-100. Aggarwal and Riggs (1969) chromatographed tadpole hemolyzates from several sources and the chromatograms showed few similarities. Wise (1970) electrophoresed tadpole hemolyzates on polyacrylamide gels and found two bands for *R. grylio* and four to five bands for *R. catesbeiana*. The latter species showed considerable variation from individual to individual.

Using immunological techniques, Wise (1970) was unable to distinguish between the various tadpole fractions. Aggarwal and Riggs (1969) were able to hybridize one hemoglobin component of the frog with one of the tadpole components. This produced some new electrophoretic components, which might explain why Herner and Frieden (1961), Manwell (1966), and Elzinger (1964) have all found components in metamorphosing tadpoles which were not equatable to components in tadpoles or adult frogs. It does not explain the unique components found in metamorphosed frogs that as tadpoles had been sensitized to adult frog hemoglobin (Maniatis *et al.*, 1969). On the other hand, Just and Atkinson (1972) found no unique hemoglobins in metamorphosing tadpoles. Their results do show different hemoglobin paterns in hemolyzates from tadpoles of different stages, all prior to the appearance of adult hemoglobin.

Although early reports indicated that ranid tadpoles and frogs shared hemoglobin components or at least one polypeptide chain (Hamada and Shukuya, 1966), later reports do not support this hypothesis. Electrophoretic analyses of isolated frog and tadpole globin chains in dissociating solvents such as 8 *M* urea have indicated few if any common chains (Baglioni and Sparks, 1963; Stratton and Wise, 1967; Moss and Ingram, 1968a). Antibodies prepared to frog (*R. catesbeiana*) hemoglobin or hemolyzates do not cross react with tadpole hemoglobins (Stratton and Wise, 1967; Wise, 1970; Maniatis and Ingram, 1971b). Antibodies to tadpole hemoglobin do not cross react with adult hemoglobin. The same is true of larval and adult hemoglobins from *B. vulgaris* (Kurata and Okada, 1968). The amino acid compositions of isolated globin chains from frogs and tadpoles are dissimilar (Aggarwal and Riggs, 1969).

In summary, we can state only that frog and tadpole hemoglobins share no common antigenic determinants. Electrophoretic and amino acid composition studies also support the hypothesis that frog and tadpole hemoglobins are dissimilar. Unfortunately, we cannot yet define the correct number of hemoglobin components nor their proportions in any tadpole species. Many species appear to be highly polymorphic, but only a limited number of species have been examined.

3. *Embryonic Components*

Kurata and Arakawa (1963), Kurata *et al.* (1965), and Kurata and Okada (1968) reported the presence of embryonic hemoglobins in toad embryos (*Bufo vulgaris japonicus*) during early development before erythropoiesis by primitive erythroblasts. The hemoglobin, which was prepared from developing eggs, had the spectral properties of hemoglobin but differed on electrophoresis from both tadpole and adult hemoglobins. Embryonic hemoglobin, like adult toad hemoglobin, was not resistant to alkali denaturation. Larval hemoglobin was alkaline resistant. Kurata and Arakawa (1963) also report that embryonic hemoglobin is lacking in *Rhacophorus sehlogelis* var *arborea*. Later studies (Kurata and Okada, 1968) have shown that antisera to adult toad hemoglobin will not cross react with larval or embryonic hemoglobin. Antisera to larval hemoglobin does not cross react with adult or embryonic hemoglobin. Using fluorescein isothiocyanate-labeled antisera to embryonic hemoglobin, these workers were able to associate the hemoglobin with a dense layer of small particles that surrounds the yolk platelets. No structural studies of these interesting embryonic hemoglobins have been made nor have their functional properties been defined. However, unpublished data of Yoneyama and Sugita quoted by Kurata and Okada (1968) indicates that embryonic hemoglobins are monomeric.

B. Urodela

Salamanders are the least specialized group of amphibians. The 316 species are arranged into 8 families and occur in wet areas of the temperate and tropical regions.

1. Adult Components

Salamander hemoglobins have seldom been studied. Dessauer *et al.* (1957) examined hemolyzates from seven species (including six families) by paper electrophoresis and found a single band in all families except the Salamandridae, which showed two bands. Similar studies by Marchlewska-Koj (1963) revealed one to three bands in *Triturus* species and two bands in *Salamandra salamandra*. Starch-gel electrophoresis of hemolyzates from thirty-eight *Siren* (*S. intermedia nettingi*, *S. i. texana*, and *S. lacertina*) revealed from two to four hemoglobin components in each species or subspecies (Guttman, 1965). Some variation from sample to sample was encountered, but this was not attributable to polymerization of hemoglobin. Wood (1971) found one major and two minor components in hemolyzates of *Dicamptodon ensatus*. Newcomer (1967) found a single hemoglobin band by starch-gel electrophoresis in hemolyzates from *Ambystoma annulatum* (in one locality sampled more than one band was observed, see Section II, B, 3). Although Amiconi *et al.* (1970) did not report on the electrophoretic properties of *Ambystoma tigrinum tigrinum* and Axolotl *Ambystoma mexicanum* hemolyzates, they did state that the hemoglobins used in their oxygen-binding experiments represented the "fast" fraction and accounted for 80% of the total hemoglobin. Thus, more than one band must occur in this species, and this has been confirmed by acrylamide gel electrophoresis (Gahlenbeck and Bartels, 1970; Maclean and Jurd, 1971b). Shontz (1968), using starch-gel electrophoresis, noted single hemoglobin bands in hemolyzates from *Desmognathus fuscus* and *D. monticola,* but more complex banding patterns in hemolyzates from *D. ochrophaeus* and *D. quadramaculatus.* In the latter two species, there appeared to be one major band and one or more minor bands. Brown and DeWitt (1970) report that *Triturus viridescens* has a single major band and one very minor band in its hemolyzate (studied by disk gel electrophoresis). Sorcini *et al.* (1970) using starch-gel electrophoresis found a single hemoglobin band in some *Triturus cristatus,* but others showed additional bands (see Section II, B, 3).

Taketa and Nickerson (1973) have studied the electrophoretic properties of hemolyzates from *Cryptobranchus alleganiensis, Necturus maculosus,* and *Hynobius tsuensis.* A single component of identical mobility was found in hemolyzates from 9 *C. a. alleganiensis* and 15 *C. a. bishopi.* Three *Necturus* and 6 *Hynobius* had two hemoglobins, and the patterns were species specific.

Reference to Table II indicates that very little of the diversity among salamanders has been sampled in spite of the fact that North America is one of the two centers of radiation for this group of amphibians. Furthermore, there have been almost no attempts to fractionate components in hemolyzates containing several components nor to rigorously

TABLE II
NUMBERS AND DISTRIBUTION OF SALAMANDERS [ORDER URODELA (CAUDATA)][a]

Family	Genera	Species	Distribution
Hynobiidae	5	30	Asia
Cryptobranchidae	2	1	Eastern Asia
		2	Eastern North America
Sirendiae	2	3	Southeastern North America
Proteidae	2	6	Europe and eastern North America
Salamandridae	15	42	Eurasia and North America
Amphiumidae	1	3	Southeastern North America
Ambystomidae	4	32	North America to Mexico
Plethodontidae	23	183	Europe (one species); North, Central, and South America

[a] From Brame (1967).

demonstrate that only a single component is present in any species. Future studies should be aimed not only at examining additional species, but also at better defining the hemolyzates from any one species.

2. *Larval Components*

Although it has been known for 20 years that larval and adult frogs have different hemoglobins, parallel studies in salamanders are quite recent. Amiconi *et al.* (1970) have shown that the oxygen-binding properties of hemoglobins from neotenic and metamorphosed axolotl (*Ambystoma mexicanum*) are slightly different, the neotenic form has a hemoglobin with a lower oxygen affinity. Similar results were reported by Gahlenbeck and Bartels (1970), who in addition indicated that there was a change in the electrophoretic properties. However, other workers have denied electrophoretic differences between the neotenic and adult forms of the axolotl (Marchlewska-Koj, 1964; Edwards and Justus, 1969). Maclean and Jurd (1971b) have shown that there is a change in electrophoretic pattern between tadpoles and neotenic adults of *A. mexicanum,* but there is no change between the neotenic adult and the metamorphosed adult. Similar larval to adult changes have been reported for *Dicamptodon ensatus,* a western salamander (Wood, 1971), and *Ambystoma maculatum* (Edwards and Justus, 1969). Larval hemolyzates usually contain a major and one or more minor components. The latter may disappear after metamorphosis. Brown and DeWitt (1970) examined *Triturus viridescens* larvae, elfs, and adults. In this species, two transformations occur, the first from the aquatic larva to the terrestrial elf, and the second from the terrestrial elf to the aquatic adult. However, changes in respiratory proteins occur only during the larval to elf transformation. Disk electrophoretic studies show that the elf and the adult both have the same hemoglobin pattern, one major band and one minor band. However, the larva has a pattern showing two major bands, one of which is unique to the larvae and the other corresponds to the minor adult band.

Wade and Rose (1972) were unable to discern any differences in the electrophoretic properties of hemolyzates from larval and transformed *Ambystoma tigrinum.*

3. *Polymorphism*

Hemoglobin polymorphism has been reported in *Ambystoma annulatum* (Newcomer, 1967) and *Triturus cristatus* (Sorcini *et al.,* 1970). In addition, Newcomer (1967) reports (unpublished data) that hemoglobin polymorphism also was found in *Ambystoma opacus* and *Dicamptodon ensatus* among the thirteen ambystomid species he examined.

Polymorphic hemoglobin alleles in *A. annulatum* populations (twenty-three individuals examined) were localized in Washington County, Arkansas. They were not observed among the small samples from Adair County, Oklahoma, or Stone County, Missouri. The allelic distribution gave a good Hardy–Weinberg fit. Wood (1971) also found hemoglobin polymorphism in his study of larval and adult *D. ensatus*. Sorcini *et al.* (1970) observed two alleles in *T. cristatus* populations (116 examined from unidentified sources). By characterization of the terminal amino acids on each chain, they were able to determine that the variant was at the α chain locus. These authors did not compare their allelic frequencies with those expected from random mating. I have made such calculations from their data and find a paucity of heterozygotes in their sample ($X^2 = 3.74$; probability $\simeq 0.05$).

Studies of the distribution of normal and polymorphic hemoglobins in salamander hemolyzates have been few in number and rather limited in scope. Those species examined do not seem to have complex hemolyzates as do many other species of amphibians and fishes. Considering the enormous increase in DNA content reported for salamanders, one would have guessed just the opposite.

C. Apoda

Two families containing approximately 150 species of caecilians occur in the Old and New World tropics. Although these animals are often available from animal dealers, to my knowldge none of their hemolyzates have been examined.

III. Hemoglobin Function

The functional properties of hemoglobins have recently been reviewed by Antonini and Brunori (1971). Although mainly concerned with human hemoglobin, this book covers the subject in great detail and is heartily recommended. When discussing hemoglobin function, one commonly refers to the P_{50}, or oxygen pressure at which half the binding sites are occupied; the Bohr effect, which today is interpreted as the pH dependence of P_{50}; and n or heme–heme interaction, which is a measure of the cooperativity of oxygen-binding by a multiunit molecule. For human and most vertebrate hemoglobins, P_{50} is in the range 10–30 mm Hg, Bohr effects are -0.40 to -0.70 and n is 2–3. The exact values depend upon temperature, solvent, and concentrations of allosteric effectors.

This review is concerned primarily with macromolecular interactions rather than a comparative physiological approach. Physiological adaptations of respiration are reviewed in Foxon (1964). More recent references are Guimond and Hutchison (1972), Emilio and Shelton (1972), Howell *et al.* (1970), Johansen (1972), Johansen *et al.* (1970), Jones (1972), Lenfant and Johansen (1967, 1972), Reeves (1972), and Romer (1972). Although little effort is made here to relate the *in vitro* functional properties of hemolyzates to respiratory adaptations, it is a recognized goal. On the other hand, a comparative study of oxygen binding by bloods in the physiological region is limited in scope and necessarily obscures many fundamental differences in structure and function. With the recognition that organic phosphates are strong modifiers of oxygen affinity, we are a step closer to translating *in vitro* properties into *in vivo* functional adaptations.

A. Anura

Hall (1966) studied the functional properties of blood from *Bufo marinus.* The Bohr effect (—0.28) was about half the value found for human blood, but the oxygen affinity and temperature dependence were similar to values obtained with mammalian bloods. M. L. Coates (unpublished) has confirmed the low Bohr effect of *B. marinus* hemolyzates and studied the effects of organic phosphates on this hemoglobin. Stripping the hemoglobin of phosphates only slightly increases its oxygen affinity. Examination of the phosphate content of erythrocytes indicates considerable quantities of adenosine triphosphate (ATP) and 2,3-diphosphoglyceric acid (DPG) (ATP/Hb $\simeq$ 0.2; 2.3-DPG/Hb $\simeq$ 0.7). Leggio and Morpurgo (1968) have studied the effects of temperature and pH on hemoglobin solutions prepared from *Bufo bufo* hemolyzates. Their results are rather curious for they find that at high (45°C) and low (4°C) temperatures the Bohr effect vanishes. From these observations, they hypothesize that active metabolism fails at low temperatures due to a lack of oxygen and the animal is forced to hibernate. Although these authors were aware of methemoglobin formation, this must contribute to experiments done at higher temperatures. I have found it extremely difficult, if not impossible, to do accurate, reproducible oxygen-binding experiments at 45°C. Perhaps if one includes an enzymatic or nonenzymatic methemoglobin reduction system, good experiments would be possible. It is more difficult to hypothesize why the Bohr effect disappeared at low temperatures. It should be pointed out, however, that this leads to a nonlinear Arrhenius plot. One could put forth a mechanism to explain these results, but it would be elaborate and would depend upon multiple components and conformational changes. Finally, it should be pointed out that hibernation has not evolved in response to lowered metabolic levels, but to a lack of food in the environment. Animals have adapted to lower temperatures (as many marine species have done) when there was an ample supply of food.

It was the work of McCutcheon (1936) and later of Riggs (1951) that clearly showed that tadpole and frog hemoglobins were not similar in function. Their oxygen-binding properties were shown to be quite different; tadpole hemoglobin lacks a Bohr effect and has a high oxygen affinity, while frog hemoglobin has a "normal" Bohr effect and lowered affinity. Although this work was done on *Rana catesbeiana*, it has been shown that erythrocytes from *Rana clamitans* tadpoles also lack a Bohr effect and have a high oxygen affinity (Manwell, 1966). Hamada *et al.* (1964b) have repeated the earlier work on *R. catesbeiana* adults and tadpoles with similar results.

Brunori *et al.* (1968) studied the oxygen-binding properties of *R. esculenta* hemolyzates and purified hemoglobin components (Fig. 1). In general, the two purified components had properties similar to those of the whole hemolyzate, but both components had higher oxygen affini-

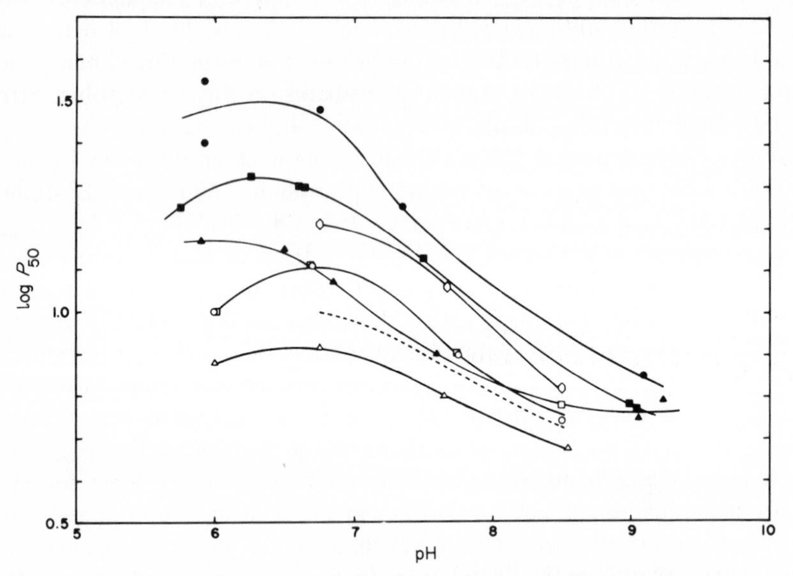

FIG. 1. Oxygen affinity of whole frog hemolyzates and their isolated fractions as a function of pH. Closed symbols refer to hemoglobins from *R. esculenta* (Brunori *et al.*, 1968): Solid circle = total hemolyzate, solid square = component I, solid triangle = component II. Open symbols refer to hemoglobins from *R. catesbeiana* (Aggarwal and Riggs, 1969): Circle = total hemolyzate, hexagon = component B, triangle = component C, square = 1:2 mixture of B and C (as was original hemolysate). The broken line is the calculated curve for 1:2 mixture of B and C assuming no interaction. Temperature = 20–23°C, dilute phosphate or borate buffers.

ties than the hemolyzate. Similar results have been obtained by others (Sullivan and Nute, 1968; B. Sullivan, unpublished work on turtle hemoglobins), probably because small effector molecules such as DPG, ATP, or inositol hexaphosphate (IHP) are lost during purification. Component II in their preparation had a greater oxygen affinity than Component I (Fig. 1). The value of n (1.4–2.9) was pH dependent, being lower at lower pH values. Kinetic studies (Brunori *et al.*, 1968) showed that the velocity constants for the dissociation of oxygen do not parallel the Bohr effect as they do in mammalian hemoglobins. The velocity constants of enzymatically modified hemoglobins (carboxypeptidase digested) were much more similar to unmodified hemoglobins than is true for mammalian hemoglobins.

There is some evidence that intracellular metabolites may play a seasonal role in regulating the oxygen affinity, similar to the way that DPG modulates the affinity of human hemoglobin from individuals acclimated to different altitudes. Gahlenbeck and Bartels (1968) report the oxygen affinity of blood from cold-adapted frogs differs from the affinity of blood from warm adapted individuals. These workers (Gahlenbeck and Bartels, 1970) also noted a difference in the oxygen affinity of blood from neotenic and metamorphosed *Ambystoma mexicanum,* although most workers have denied any electrophoretic differences between the two forms. As mentioned earlier, purified hemoglobins from frogs always have an increased oxygen affinity, implying that some allosteric effector has been removed in the purification.

The effects of organic phosphates on bullfrog and tadpole (*R. catesbeiana*) hemoglobins have recently been reported (Araki *et al.*, 1971). Passage of hemolysates through gel filtration columns (i.e., stripping away of allosteric effectors) greatly increases the oxygen affinity of both hemolyzates. ATP, IHP, and DPG were found in erythrocytes of frogs and tadpoles (Table III) and when added to stripped hemolyzates resulted in a decrease in the oxygen affinity. However, IHP,

TABLE III
ORGANIC PHOSPHATES IN *R. catesbeiana* ERYTHROCYTES[a]

	Concentration (mM)			
Stage	Hemoglobin	ATP	IHP	2,3-DPG
Adult	3.5	3.01	0.29	1.5
Tadpole	4.84	6.02	0.48	3.13

[a] From Araki *et al.* (1971).

which has the greatest effect on most mammalian hemoglobins, had no effect on adult frog hemolyzates. Since it occupies the largest volume of the three effectors, we can only guess that it may be sterically prevented from reaching the binding site (see additional comments in Section IV, C). Interestingly, stripped frog hemolyzates never regain the low oxygen affinity of unstripped samples, even in saturating amounts of ATP or DPG. The interaction of organic phosphates with frog hemoglobin does not affect heme–heme interaction, at least at pH 7.0.

As mentioned earlier, *R. catesbeiana* hemoglobin polymerizes to octameric and larger molecules. Apparently this has no effect on the oxygen-binding properties (Riggs, 1966), although Trader and Frieden (1966) came to the opposite conclusion. Aggarwal and Riggs (1969) have suggested that Trader and Frieden separated several of the components as well as polymerized from unpolymerized hemoglobin. It has been shown that components differ in their oxygen affinities (Fig. 1). Aggarwal and Riggs (1969) found that the oxygen affinity of adult *R. catesbeiana* hemolyzates is increased at hemoglobin concentrations below 0.5% by dialysis or by stripping small molecules from hemoglobin by passing the hemolyzate through Sephadex G-25. Again, the addition of DPG only partially reversed the increased oxygen affinity caused by stripping. These authors also report that dialysis is just as effective as stripping in increasing the oxygen affinity at pH 8, but much less effective at pH 7. Reaction of sulfhydryl groups with iodoacetamide blocks polymerization (Riggs *et al.*, 1964) but does not affect the oxygen-binding properties (Aggarwal and Riggs, 1969). When iodacetic acid reacts with human hemoglobin, the oxygen-binding properties are changed.

Isolation of the individual components and measurement of their oxygen-binding properties shows that component C has a higher affinity for oxygen and a smaller Bohr effect than component B (Fig. 1). When the components are remixed, these workers claim there is interaction between the components. The oxygen affinity is not that calculated from the properties and proportions of the mixed components, but is shifted to the right (i.e., has a lower affinity). The remixed components and the original dialyzed hemolyzate have similar properties. These experiments are open to some criticism. These authors state "however, a small difference exists between the data for hemolysates dialysed successively against barbital and phosphate buffers." According to my calculations, the difference is not small at all, but amounts to 0.25 log units, which is greater than the entire Bohr effect for component C.

Part of the trouble may be attributable to the fact these workers have performed their experiments at a protein concentration dangerously close to the point at which oxygen affinity is strongly dependent upon protein concentration (see their Fig. 5). Another point is that comparable log P_{50} values for material dialyzed versus phosphate or barbital buffers (i.e., the material used for the oxygen equilibrium measurements of components B, C, mixture, and control) are lower than the log P_{50} values of stripped hemolyzates (stripped, pH 6.84, log P_{50} = 1.22; mixture of phosphate and barbital dialyzed hemolyzate pH $\simeq$ 6.84, log $P_{50} \leq 1.1$). These experiments illustrate how very difficult it is to take apart a hemolyzate and put it back together again. The oxygen affinities of purified components seldom sum to the oxygen affinity of the hemolyzate, probably because of methemoglobin formation and loss of small effector molecules. Whereas it is certainly possible that components dissociate and form hybrid tetramers—tetramers that cannot be observed during electrophoresis because their equilibria are perturbed—this explanation remains to be proved for amphibian hemoglobins.

The oxygen-binding properties of *R. catesbeiana* tadpole hemoglobin have been studied by Riggs (1951, 1964), Hamada *et al.* (1964b), and Aggarwal and Riggs (1969). Although it is generally stated that tadpole hemoglobin lacks a Bohr effect, the more precise data of Aggarwal and Riggs (1969) indicate there is a small Bohr effect (—0.10 compared to —0.28 for the adult frog), and A. Riggs (private communication) has indicated that the presence or absence of a Bohr effect depends upon the buffer used in the oxygen-binding experiments. Dialysis does not affect the oxygen affinity of tadpole hemoglobin. Oxygen affinities of erythrocytes or hemolyzates from different tadpole species are quite similar (*R. clamitans* $\simeq$ 4.8 mm Hg, 24°C; *R. grylio* $\simeq$ 5.0 mm Hg, 30°C; and *R. catesbeiana* $\simeq$ 4.0 mm Hg, 20°C). Oxygen equilibria of individual tadpole hemoglobin components have not been attempted. Because tadpole hemoglobin dissociates readily, its oxygen affinity is strongly concentration dependent (Riggs, 1964).

Tadpole (*R. catesbeiana*) erythrocytes contain significant amounts of IHP, ATP, and 2,3-DPG (Table III). Stripped hemolyzates respond to these organic phosphates and in the order given. The addition of phosphates to stripped hemolyzates can restore the original oxygen affinity (Araki *et al.*, 1971).

Araki *et al.* (1973) have reported that tadpole (*Rana catesbeiana*) hemolyzates have a large Bohr effect in the presence of fortyfold molar excess of ATP or IHP. No Bohr effect is observed for fresh hemolyzates, stripped hemolyzates, or hemolyzates to which 2,3-DPG has been added. Unfortunately, the data reported by these authors are internally inconsistent. For example, the log P_{50} value at pH 7.0 and 1 mM ATP is about 1.0 at 24°C and 0.57 at 20°C. The difference in temperatures used for the two sets of experiments cannot account for the large differences in oxygen affinity. At pH 8.5, the difference between comparable experiments is as large or larger. The finding of n values over 4 is perplexing. The finding that ATP at pH values above 7.7 increases the oxygen affinity compared to stripped hemolyzates is also puzzling. Apparently, although this is not too clear, numerous substances (such as chloride ions, inorganic phosphates, the intracellular complement of anions, and 2,3-DPG) can decrease the oxygen affinity but independently of pH. ATP and IHP produce pH-dependent decreases or increases in oxygen affinity. These data are quite inconsistent with our notions of how anions interact with human and other mammalian hemoglobins and should be repeated.

B. Urodela

A variety of workers have examined the oxygen-binding properties of salamander blood or hemoglobins. Lenfant and Johansen (1967) compared the oxygen-binding properties of whole blood from *Necturus maculosus, Amphiuma tridactylum,* and *Rana catesbeiana* in an attempt to show the changes accompanying adaptation to terrestrial environments. At physiological pressures of carbon dioxide and a temperature of 20–22°C, the P_{50}, or half saturation pressure, was 14.5 mm Hg for *Necturus* blood and 27 mm Hg for *Amphiuma* blood. The Bohr effects were −0.13 and −0.21, respectively. Wood (1971) studied the oxygen-binding properties of hemolyzates and purified hemoglobins from larvae of adults of *Dicamptodon ensatus.* Hemoglobins from the two forms were identical. They showed moderate heme–heme interactions ($\simeq$ 2.7), rather low oxygen affinity and a definite response to ATP. ATP increased during metamorphosis and resulted in a decrease in the oxygen affinity of adult blood. The Bohr effect was small (−0.13 in adults). Coates and Metcalfe (1971) reported a slight difference in oxygen-binding properties between blood of *Taricha granulosa* and *T. rivularis.* They interpret these differences in terms of the relatively recent adaptation by *T. rivularis* to a less aquatic environment. Groups of *Taricha* were also acclimatized to different aquatic and terrestrial environments, but these acclimations were without effect on the oxygen-binding properties of whole blood. The half saturation values under physiological conditions were 31.2 and 34.5 mm Hg at 20°C for *T. granulosa and T. rivularis,* respectively. Heme–heme interaction values were 2.2–2.3 and the Bohr effects were −0.18 (*T. granulosa*) and −0.13 (*T. revularis*).

The oxygen-binding properties of whole blood from salamanders shows one unusual feature, the Bohr effect is consistently −0.2 or less whereas in mammals it is nearer —0.6 (Fig. 2). The Bohr effect varies from species to species, but in most vertebrates it is in the range —0.3 to —0.7. Morpurgo *et al.* (1970) reported a reverse Bohr effect in hemolyzates from *Triton* (*Triturus*) *cristatus.* Only individuals with the "slow" hemoglobin phenotype were used. The reversal of the Bohr effect was temperature dependent, having a negative slope (normal) at 4°C and a positive slope (reversed) at 30°C. Furthermore, the switch in Bohr effects occurred between 3 and 10°C in animals acclimated to 5°C and between 20 and 30°C in animals acclimated to 30°C. These workers also noted that if the hemoglobin was stripped prior to performing the oxygen-binding experiments (stripping involved passage through a column of Sephadex to remove low molecular weight molecules), then the Bohr effect remained reversed at all temperatures. Although it is certainly possible that these experiments accurately reflect the properties of *Triturus* hemoglobin, I suspect that methemoglobin formation, that demon of oxygen-binding experiments, has complicated their results. Methemoglobin formation occurs at elevated temperatures and low pH values. It tends to occur more rapidly in oxygen-binding experiments involving nonmammalian hemoglobins; in fact, the more primitive the vertebrate (phylogenetically speaking) the more susceptible is its hemoglobin to oxidation. Methemoglobin usually increases the oxygen affinity and lowers the heme–heme interaction value. At the higher temperatures used in the experiments of Morpurgo *et al.* (1970), more oxidation would occur, and this appears to reverse the normal Bohr effect (since the Bohr effect is small anyway, it would not take much change in oxygen affinity to reverse it). Passage of hemoglobin down a column would increase the methemoglobin content, thus reversing the Bohr effect at all temperatures. The apparent change in thermostability of hemoglobin from animals acclimated to 30°C remains a puzzle, however, and deserves further investigation.

The recent study by Amiconi *et al.* (1970) is extremely exciting. These authors studied the binding of oxygen by purified hemoglobin from *Ambystoma t. tigrinum* and the Axolotl *Ambystoma mexicanum* as a function of pH and temperature. They found that the shape of the oxygen-binding curve was independent of protein concentration, pH, and temperature. The Hill plot (from which the heme–heme interaction value is obtained) had a slope of 2.0 near 50% saturation. The apparent heat of oxygenation (ΔH) was slightly pH dependent, but in general only half the value reported for most other hemoglobins. The pH dependence of the oxygen affinity was unique among animal hemoglobins (Fig. 2). The oxygen affinity was maximum at pH 7.2–7.6 and decreased slightly

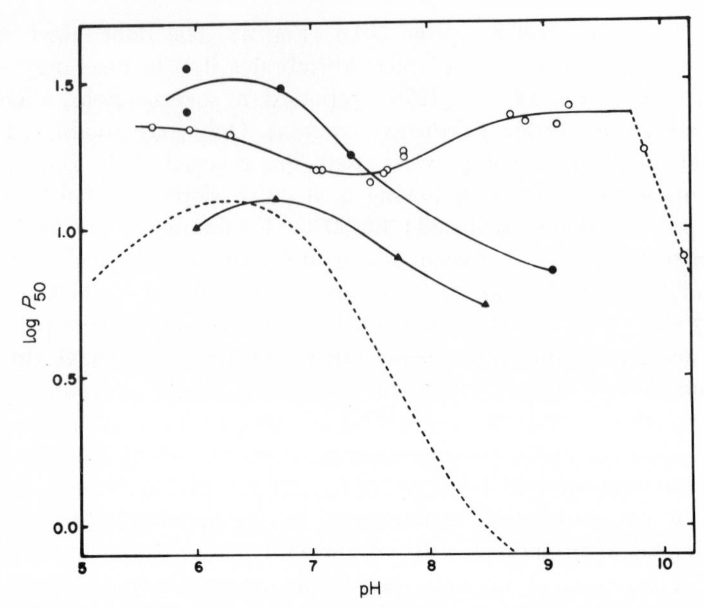

FIG. 2. Oxygen affinity of frog, salamander, and human hemolyzates as a function of pH. Solid circle = *Rana esculenta* (Brunori *et al.*, 1968), solid triangle = *Rana catesbeiana* (Aggarwal and Riggs, 1969), open circle = *Ambystoma trigrinum* (Amiconi *et al.*, 1970), broken line = Homo sapiens (Sullivan, 1967). Temperature = 20–23°C, dilute phosphate or borate buffers.

at both higher and lower pH values. Hemoglobins from both species behaved identically, the neotenic form of *A. mexicanum* had a hemoglobin with a slightly higher oxygen affinity. Apparently, there is an additional Bohr group with a pK near 9. These authors suggest that it may be an α or ϵ-amino group. It is not clear whether there is an acid Bohr group in this hemoglobin, but it appears there may not be one. The kinetics of the dissociation of oxygen from *Ambystoma* hemoglobin were also anomalous. The first-order rate constant as a function of pH does not parallel the Bohr effect as it does in all mammalian hemoglobins which have been studied. This may necessitate a reexamination of the kinetic interpretation of the Bohr effect.

The oxygen-binding properties of *Cryptobranchus* and *Necturus* hemolyzates were studied by Taketa and Nickerson (1973). The former has a low oxygen affinity (20 mm Hg at 20°C) and no Bohr effect, whereas the latter has a high affinity (5 mm Hg at pH 7.4 and 20°C) which is pH dependent. Heme–heme interaction values were 1.7 for *Necturus* and 2.7 for *Cryptobranchus* hemolyzates in 0.14 *M* phosphate buffers.

IV. Hemoglobin Structure

Hemoglobin structural work is periodically updated in the "Atlas of Protein Sequence" (Dayhoff, 1972). Reference should also be made to the crystallographic studies of Perutz and his colleagues, which for the most part are published in *Nature*.

A. Anura

1. *Association, Dissociation, and Polymerization*

Although most vertebrate hemoglobins are tetramers of 64,000 m.w. and contain two types of polypeptide chains in equal proportions, there is a growing list of exceptions or modifications of this pattern. Svedberg and Hedenius (1934) observed hemoglobin molecules in amphibian hemolyzates with sedimentation coefficients near 7.3 S in addition to the normally sedimenting species (4.7 S). These polymerized hemoglobins were found in hemolyzates from *Salamandra maculosa, Bufo viridis,* and *B. valliceps,* but not in hemolyzates from *Rana temporia.* In hemolyzates from *B. valliceps,* an additional component was observed to sediment at 12.5 S. These workers noted that the amount of polymerized hemoglobin increased with time. Trader *et al.* (1963) observed polymerized components in *Rana catesbeiana, R. grylio,* and *Xenopus laevis* hemolyzates and noted that these polymerized components appeared after metamorphosis of tadpoles. Riggs *et al.* (1964) showed that polymerization occurs *in vitro* after hemolysis in *R. catesbeiana* and several turtle species and that sulfhydryl reagents prevent polymerization. Polymer can be dissociated by mercaptoethanol, a disulfide-reducing agent. Tadpole hemoglobin does not polymerize. Polymerized hemoglobin usually accounts for 60–80% of the adult bullfrog hemolyzate. Polymerization does not affect the oxygen-binding properties of *R. catesbeiana* hemoglobin (Riggs, 1966). In those species whose hemoglobins polymerize, polymerization can occur *in vivo* if the methemoglobin reducing system in erythrocytes fails (Sullivan and Riggs, 1964). Methemoglobin formation is rather common among tadpoles (Manwell and Baker, 1970, p. 186). Trader and Frieden (1966) found that the rate of polymer formation parallels the disappearance of reactive sulfhydryl groups. Polymerization was observed to occur in hemolyzates from *R. catesbeiana, R. grylio,* and *X. laevis* but not in hemolysates from *Rana pipiens* or *Leptodactylus pentadactylus.* Loss of sulfhydryl groups in polymerizing hemolyzates amounted to 2 per 65,000 m.w.

Riggs (1964) has reported that the electrophoretically fast tadpole (*R. catesbeiana*) hemoglobin is in dissociation equilibrium with its α

and β chains. More recently, Aggarwal and Riggs (1969) determined the sedimentation rates for the isolated components from *R. catesbeiana* tadpole hemolyzates. One of the six components examined had a $s_{20,w}$ of 2.4, indicating that the tetramer was dissociating into smaller subunits. When this fraction was mixed with any other tadpole fraction, the 2.4 S material disappeared. Why this should happen is not entirely clear (see also Riggs and Bownds, 1963). Sedimentation studies of hemolyzates from *Rana esculenta* revealed no polymer (Tentori *et al.*, 1965). However, later studies (Elli *et al.*, 1970) have shown that *R. esculenta* hemoglobin reversibly polymerizes in the deoxygenated form. The polymer (apparently an octamer) accounts for about 75% of the hemolyzate and depolymerizes to tetramers when reoxygenated. The polymer is not dissociated by dilution. Furthermore, amphibian hemoglobins do not dissociate in acetate buffers as do mammalian hemoglobins. Kurata and Okada (1968) have reported that hemoglobins in toad eggs are monomeric.

2. *Amino Acid Composition Studies*

A number of workers have reported the amino acid composition of frog hemolyzates or purified hemoglobins (Christomanos *et al.*, 1965; Christomanos, 1967, 1972; Tentori *et al.*, 1965; Hamada and Shukuya, 1966; Trader and Frieden, 1966; Aggarwal and Riggs, 1969). Because of the confusion that surrounds the isolation and characterization of hemoglobins in different hemolyzates from adults and tadpoles, it seems rather useless to try to summarize the amino acid compositional data. Most fractions contain all the amino acids, but some or most tadpole fractions lack cysteine. Naturally, there is disagreement on this point also. Most fractions contain significant amounts of isoleucine, an amino acid missing in some mammalian species of hemoglobin. In general, adult *R. catesbeiana* hemoglobins seem to have more cysteine and methionine and less tryptophan and isoleucine than do tadpole hemoglobins. Perhaps the most remarkable thing about the amino acid compositional data is that all fractions, both from adults and tadpoles, are rather similar.

3. *End Group Studies*

Hamada, Shukuya, and Kaziro (1964a) and Hamada and Shukuya (1966) reported that adult *Rana catesbeiana* has glycine on the N-terminus of both the α and β chains. Following glycine is leucine on the α chain and serine on the β chain. Tadpole hemoglobin has the following N-termini: α chain, Gly-Leu; β chain, Val-Ala. Carboxypeptidase digestion revealed histidine as C-terminal on the α chains of adult and tadpole hemoglobin and alanine and glutamic acid as the C-terminal residues of tadpole and frog β chains, respectively. Christomanos (1967) reported threonine and traces of glycine at the N-terminus of *Rana ridibunda* hemoglobin. Christomanos *et al.* (1965) reported N-terminal glycine followed by serine or methionine in *Rana esculenta* hemoglobin. Further studies on this species by Tentori *et al.* (1967) have revealed both components I and II have two types of chain (one of which is shared, Tentori *et al.*, 1967) whose carboxytermini are Tyr-Arg and Tyr-His. These workers found four residues of N-terminal glycine per tetramer (for component I and for component II). N-terminal analyses by Moss and Ingram (1968a) yielded valine for *R. catesbeiana* tadpole globin and glycine for frog globin. However, yields were only about 50%, which indicated a possible blocked end group.

Elzinger (1964) found only 0.93 moles of valine per tetramer of tadpole hemoglobin and 2.1 moles of glycine per tetramer of frog globin. DeWitt and Ingram (1967) have found that half the amino-terminal end groups of tadpole and frog hemoglobin are acetylated. It is unclear why other workers have found four end groups per tetramer (Hamada *et al.*, 1964a; Hamada and Shukuya, 1966; Tentori *et al.*, 1967). Sequence studies by Chauvet and Acher (1968) of *R. esculenta* hemoglobin have revealed N-terminal glycine followed by serine in the β chain and N-terminal acetylalanine in the α chain. Compared to human β chains, *Rana* β chains lack the first six residues of the amino terminus. However, toad (*Bufo bufo*) hemoglobin (one purified component) has Val-His-Leu at the amino terminus of its β chain, which is identical to human hemoglobin. The α chain begins Ala-Leu and is acetylated. The C-terminal residues of both chains are identical to human (Caffin *et al.*, 1969).

B. Urodela

It has always been assumed that the molecular weights and subunit organizations of salamander hemoglobins are similar to human hemoglobin, namely a tetramer of molecular weight 64,000, which is composed of equal amounts of two different polypeptide chains (α and β). However, studies by Riggs *et al.* (1964) and Sullivan and Riggs (1967) have shown polymerized hemoglobins among Amphibia and Reptilia, and recent sedimentation studies by Elli *et al.* (1970) have underscored the danger of making broad assumptions. Axolotl *Ambystoma mexicanum* hemoglobin sediments as a single symmetrical peak with $s_{20,w}$ of $\sim$4.5 in the pH range 5–9. Human hemoglobin behaves similarly. However, when *A. mexicanum* hemoglobin is deoxygenated, it sediments as two components at pH values below 7. The major component has an $s_{20,w}$ of $\sim$6.8 and the minor component sediments with an $s_{20,w}$ of $\sim$4.5

The major and minor components refer to sedimenting species, not electrophoretic species. Above pH 7, only one peak was observed, $s_{20,w}$ ~4.5 or slightly higher. The polymerization was completely reversible simply by addition of the ligand, oxygen. This polymerization is concentration dependent and disappears below 2 mg/ml. Interestingly, the shape of the oxygen-binding curve remains the same over a protein concentration of 0.2–5.0 mg/ml, indicating that polymerization is not affecting subunit interaction (Amiconi *et al.*, 1970). The main electrophoretic component (about 80% of the hemolyzate) behaved similarly to the whole hemolyzate in its polymerization properties. Svedberg and Hedenius (1934) observed polymerized hemoglobin in hemolysates from *Salamandra maculosa.* Normal molecular weights were observed for larval and adult hemolyzates from *Dicamptodon ensatus* (Wood, 1971). Hemolyzates from *Triturus cristatus* did not polymerize under the same experimental conditions in either its oxygenated or deoxygenated state (Elli *et al.*, 1970). These studies underscore the need for more comparative studies and the surprises they are likely to uncover.

No amino acid sequence studies have been reported for salamander hemoglobins but some are in progress. Sorcini *et al.* (1970) reported that the carboxy-terminal amino acids from *T. cristatus* hemoglobin are Tyr-Arg for one chain and Tyr-His for the other chain. These sequences match the terminal sequences of human α and β hemoglobin chains. The partial amino acid sequences of the α and one of two β chains found in *Taricha granulosa* are shown in Figs. 4 and 5 (M. Coates, R. T. Jones, P. Stenzel, and B. Brimhall, personal communication, 1973). These are discussed in the next section.

Cryptobranchus hemolyzates, which show a single component on electrophoresis, contain six reactive sulfhydryl groups per tetramer, whereas *Necturus* hemolyzates (two components) contain six to eight reactive sulfhydryl groups per tetramer (Taketa and Nickerson, 1973). Tryptic peptide maps of *Cryptobranchus* and *Necturus* hemolyzates are different but show numerous homologies (Taketa and Nickerson, 1973).

C. Sequence Studies and Their Structure–Function Correlations

Amino acid sequence studies are regularly summarized by Dayhoff (1972).

1. Anura

The β chain from the major hemoglobin fraction in *Rana esculenta* hemolyzates has been completely sequenced by Chauvet and Acher (1968, 1970a,b,c, 1972). When compared to human β chain (Fig. 3), frog β chain contains sixty-one substitutions and is six residues shorter

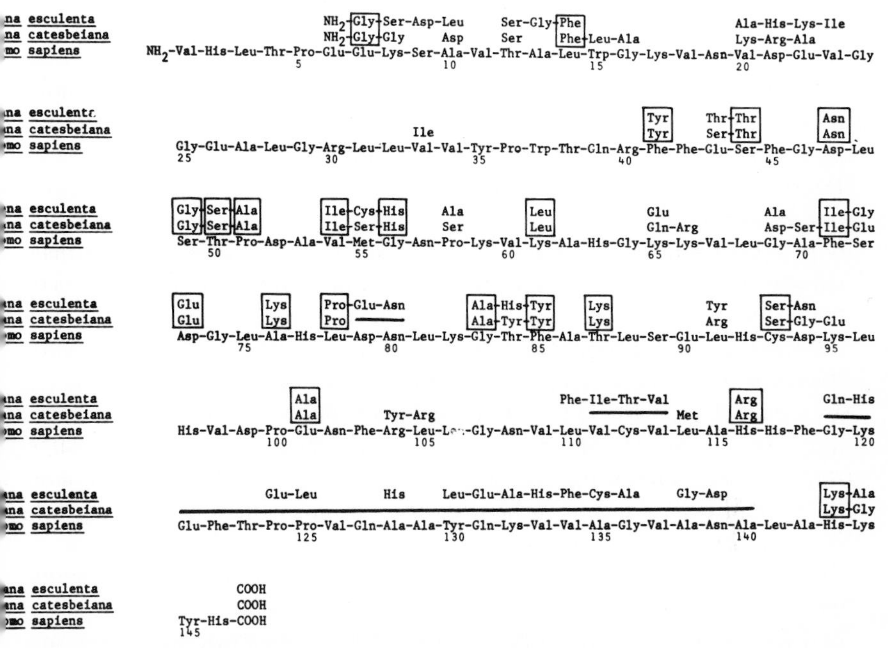

FIG. 3. Comparison of the sequences of β chains from ranid frogs and man. The sequence of *R. esculenta* β chain is complete (Chauvet and Acher, 1972). The sequence of *R. catesbeiana* β chain is incomplete as indicated by the dark horizontal bars (T. O. Baldwin and A. F. Riggs, personal communication, 1973). These ranid frog β chains are six residues shorter than human β chains. Only those positions in frog β chains that differ from human β chains are shown.

in length. The missing residues are the first six residues from the N-terminal end. Comparison of these substitutions with the X-ray model (Perutz *et al.*, 1968) and the substitutions observed in other species (Perutz *et al.*, 1965, 1968) reveals the following patterns of substitution. Among the nine positions invariant in other globin chains, only one substitution occurs (β-132, Lys → Ala, a two step mutation). The role of this lysine residue in hemoglobin function is unknown but it appears to be positioned near the N-terminal residue and may help to position it for DPG, IHP, or ATP binding. It is interesting to note that of the thirty-three positions classified as internal by Perutz *et al.* (1965), only eight are substituted in the frog chain, and only one substitution involves a change between unlike amino acids (β-133, Val → His, which may restore the positive charge in this area). The surface residues are most susceptible to variation and in the frog β chain twenty-five of the forty-three positions so classified by Perutz *et al.* (1965) are substituted. Those residues which interact with the heme group remain rather invariant (only four of twenty are substituted). There are eighteen residues on the β chains which are involved in the contact plane α_1,β_1 (Perutz *et al.,* 1968). As these authors have pointed out, substitutions are not uncommon along this plane of contact, but are relatively rare along the plane of contact α_1,β_2 which involves nine residues on the β chain. In the frog chain there are eight β-chain substitutions along the α_1,β_1 plane and only one along the α_1,β_2 plane.

It is very difficult to determine the exact significance of many of these substitutions, but, there are two very important changes in frog β chains. It has been mentioned previously that half of the amino-terminal groups are acetylated (DeWitt and Ingram, 1967). These occur on the α chain in *Rana esculenta* (Chauvet and Acher, 1968). Since the α-amino group of the N-terminal residue of the α chain contributes about a quarter of the Bohr protons (Perutz *et al.,* 1969), the Bohr effect should be reduced in frog hemoglobin. It has a value of —0.28 compared to —0.6 to —0.7 for human hemoglobin. Approximately half of the Bohr protons come from the C-terminal histidine (146) of the β chain. In deoxyhemoglobin, this residue forms a salt bridge with aspartic acid 94 of the β chain. In the two ranid frogs, this aspartic acid is replaced by asparagine or glycine. Although this may not diminish the Bohr effect, it certainly could by "causing" the histidine to deprotonate in the deoxy conformation. There is some evidence that the proton is not released until tyrosine 145 is expelled from between the F and H helices (see Perutz, 1970). Also, aspartic acid 94 is thought to donate some of the acid Bohr protons; thus, this acid (or negative) Bohr effect should be diminished. The acid Bohr value for frog (+0.36) is considerably lower than that of human hemoglobin (+0.65). The other residue implicated in the acid Bohr effect, namely the α-carboxyl group of α 141 arginine, can no longer bond to the α-amino group of the N-terminus of the complementary α chain because the N-terminus is acetylated. It is unclear whether or not this residue is available to solvent in deoxyhemoglobin. If histidine 146 is no longer contributing significantly to the Bohr effect, this might explain why its removal with carboxypeptidase had only a small effect on the functional properties (Brunori *et al.,* 1968).

T. O. Baldwin and A. F. Riggs (personal communication, 1973) have recently completed much of the sequence of the β chain of hemoglobin component B from *Rana catesbeiana* (Fig. 3). They are uncertain as to the exact nature of residues 79, 80, 111–113, and 119–140. Because of these uncertainties, the following comparisons are restricted to 113 positions. The β chain of *R. catesbeiana* like that of *R. esculenta* is shortened on the N-terminal end by six residues. The two species share twenty-two identities not found in human β chain (boxed in Fig. 3).

The most surprising feature of the comparison is that the two frog β chains are so very different. These differences are summarized in Table IV. Thus, at 28 of 113 positions, the chains are different, and it requires 35 mutations to make this transition. Differences of this magnitude when extrapolated to chains of 146 residues are greater than those found at the ordinal level in mammals. One explanation might be that the β chains from the two species are not strictly homologous; that is, they represent products of separate β_A and β_B loci. Duplications of the β chain locus are not uncommon (Sullivan, 1971), yet still the differences are extraordinarily large. Another plausible explanation is that members of the genus *Rana* diverged at approximately the same time as did the orders of mammals, thus allowing equal years of time for hemoglobin differentiation. However, the earliest ranid fossil dates to the Miocene (Porter, 1972) which is considerably more recent than the period of mammalian radiation. It will be most interesting to see the β chain sequences from additional *Rana* species and from other hemoglobin components from these species.

Arnone (1972) has provided a detailed map of the binding site of DPG. Beta chain interactions with DPG involve the α-amino group of N-terminal valine (β_1), the protonated ring of histidine (β_2), the ϵ amino groups of lysine (β_{82}), and the protonated ring of histidine (β_{143}). In the frog β chains, histidine 143 is replaced by lysine, which probably interacts with the negative phosphate groups. The most notable change is the N-terminal deletion including His-2. Shortening of the β chain is known to diminish the effect of DPG on sheep (Benesch *et al.*, 1968; Bunn, 1971) and abnormal human hemoglobins (J. Bonaventura, C. Bonaventura, M. Brunori, G. Amiconi, and E. Antonini, personal communication, 1973). Not surprisingly, then, the effect of DPG on frog hemoglobin is quite small (Araki *et al.*, 1971).

TABLE IV
MUTATION AND SEQUENCE DIFFERENCES AMONG FROG AND HUMAN β CHAINS[a]

	Rana esculenta	*Rana catesbeiana*	*Homo sapiens*	
Rana esculenta		21/7/0	30/11/1	Class of mutation
Rana catesbeiana	35(28)		34/13/0	1 step/ 2 step/ 3 step
Homo sapiens	55(42)	60(47)		
	Total number of mutations or (sequence differences)			

[a] Based on a comparison of 113 residue positions.

It is apparent that many of the unusual features of oxygen-binding by frog (*Rana*) hemoglobins can be directly correlated with sequence differences. As such, frog hemoglobins will undoubtedly prove to be useful probes of hemoglobin function. Tadpole hemoglobin, with its unusual Bohr effect, should be particularly illuminating.

As outlined earlier, the distribution of polymerizing oxyhemoglobins is erratic among the Amphibia. Similar results have been described for turtle hemoglobins (Sullivan and Riggs, 1967). If each occurrence of polymerization involves the same (one or more) cysteine group on one chain, the erratic distribution implies either that the sulfhydryl group varies in its availability and/or it is being substituted by other residues. If we assume the reactive sulfhydryl group is on the β chain, can it be located? Chauvet and Acher (1972) have located the two residues of cysteine at positions 55 and 135. Both are in helices, and although the cysteine at position 55 appears to be the more available residue, neither appears to protrude outward.

Hemoglobin from *R. esculenta* does not appear to polymerize via disulfide bond formation. However, hemoglobin from *R. catesbeiana* does, but Baldwin and Riggs have not located all the cysteine residues. There is no cysteine at β_{55}, which eliminates that position from consideration. However, there are two cysteine residues in the unsequenced segment 119–140. One is probably located at residue 132 as it is the *R. esculenta* chain. Residue positions 119–126 seem to be exposed, and should the second cysteine occur in this region, it might be the residue responsible for polymerization. On the other hand, Baldwin (1971) has presented some evidence that component B (*R. catesbeiana*), which does not polymerize, contains a β chain similar to component C, which does polymerize. The α chains of these two components appear dissimilar and contain cysteine residues. This implies that α chain cysteine residues are responsible for polymerization. Of course, polymerization might involve an intermolecular α–β disulfide bridge. Utilizing existing data, it is not possible to locate the cysteine residue(s) responsible for hemoglobin polymerization in amphibians and reptiles. In the mouse (Bonaventura, 1969; Bonaventura and Riggs, 1967), the cysteinyl residue is β_{13}. In two abnormal human hemoglobins that polymerize, the cysteinyl residues allowing polymerization are in different segments of the sequence (β_9 in Hemoglobin Pôrto Alegre and β_{83} in Ta-li) (Bonaventura and Riggs, 1967; Blackwell *et al.*, 1971). Apparently, polymerization can be effected by cysteine substitutions at numerous positions throughout the sequence.

2. *Urodela*

The partial sequence of the α chain of hemoglobin from the newt *Taricha granulosa* is shown in Fig. 4 (M. Coates, R. T. Jones, P. Stenzel,

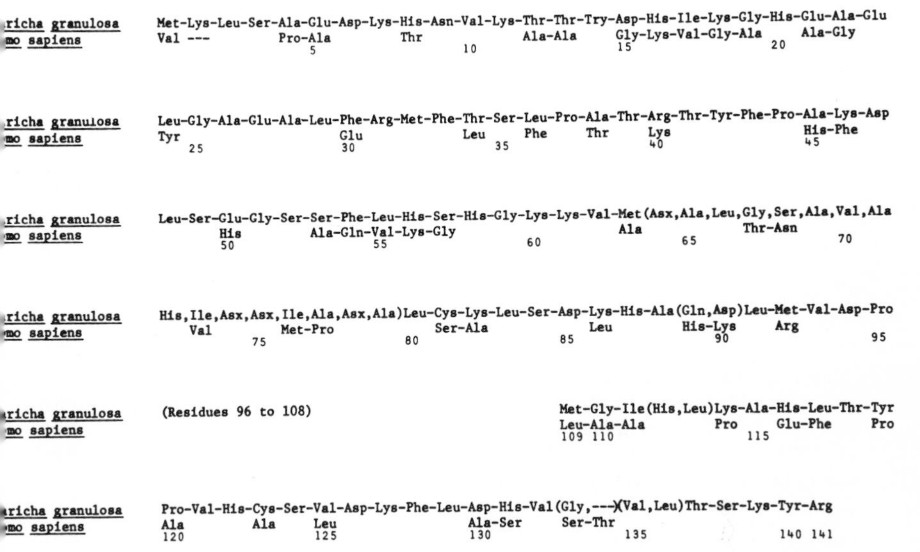

FIG. 4. The partial sequence of the α chain from the newt, *Taricha granulosa* (M. Coates, R. T. Jones, P. Stenzel, and B. Brimhall, personal communication, 1973). Those residues within parentheses have not been sequenced, but have been positioned to maximize homology with the human α chain. Differences between human and *Taricha* α chains are indicated on the lower line, identities are not. Residues 96–108 are to be sequenced. There are 142 instead of 141 residues because human α chain does not have anything corresponding to *Taricha* residue 2. However, residue 134 or 135 appears to be missing in *Taricha*.

and B. Brimhall, personal communication, 1973). This amphibian chain appears to contain a single insertion (residue 2) and to have a single deletion (residue 134 or 135), which together preserve the length at the normal 141 residues. Should subsequent work locate the presumed deleted residue, then one might suggest that valine at position 1 in the human chain and lysine at position 2 in the newt chain are indeed homologous, and that the methionine residue might represent an uncleaved portion of the initiator complex (*N*-formylmethionine). Because this sequence is fragmentary and because there are no other amphibian sequences with which to compare it, it seems premature to search for other peculiarities in the sequence.

The partial sequence of the β chain of hemoglobin from *T. granulosa* is shown in Fig. 5 (M. Coates, R. T. Jones, P. Stenzel, and B. Brimhall, personal communication, 1973). There are at least two β chains and one α chain in this species. Although only slightly more than one-third of the sequence is known, still it shows interesting similarities to the β chains of ranid frogs. Identities at positions 41, 50, 54, and 56 seem to unite (phylogenetically) the amphibian β chains. However, serine

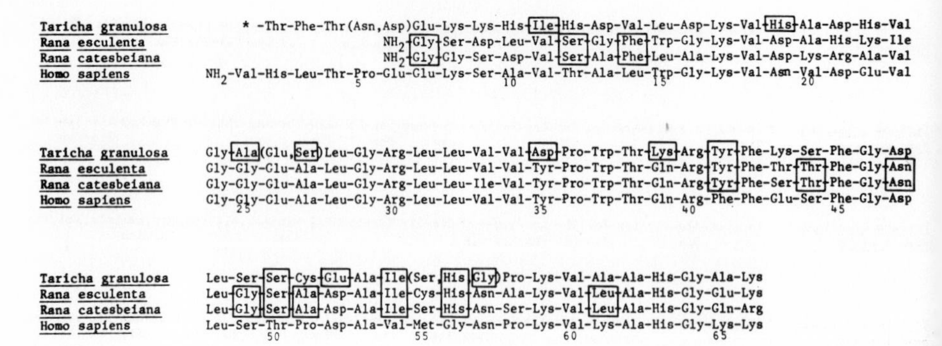

FIG. 5. The partial sequence of one of the β chains from the newt *Taricha granulosa* (M. Coates, R. T. Jones, P. Stenzel, and B. Brimhall, personal communication, 1973). The N-terminal amino acid is blocked, probably by an acetyl group, and does not respond to end-group analysis. Residues within parentheses have not been ordered, except by compositional analysis and homology with other vertebrate β chains. The sequence is compared to the β chains of ranid frogs and man. Boxed residues may be particularly indicative of phylogenetic relationships and are discussed in the text.

at position 50 commonly ocurs among other mammalian β chains and is not a distinguishing feature of amphibian β chains. The substitution of histidine for glycine at position 56 is particularly important, because it is unique among vertebrate β chains and because it is a two-step mutation. Isoleucine at position 54 and tyrosine at position 41 seem to be limited to amphibian and mouse β chains. They probably represent parallel mutations but do serve as convenient markers.

Substitutions at positions 7, 12, 14, 19, 44, 47, 49, 51, and 61 appear to group anuran β chains. With the exception of alanine at position 51, which is present early in the evolution of mammalian β chains and is subsequently replaced by proline; and leucine at position 61, which occurs also in avian hemoglobins, these substitutions appear to be valid markers. The phenylalanine at position 14 in amphibian β chains is also common to ungulate β chains, undoubtedly a parallel mutation. As mentioned earlier, the shortening of the ranid β chain by six residues does not appear to have antedated the division of *Bufo* and *Rana*. However, it will probably be an excellent generic or familial marker.

Substitutions at positions 11, 25, 27, 35, 39, 52, and 57 appear to group urodeles. Other residues may also be common to this assemblage, but cannot be recognized at this time. The two step mutation at position 57 (Asn $\rightarrow$ Gly) looks particularly promising as a marker.

The functional properties of hemolyzates from *T. granulosa* were described by Coates and Metcalfe (1971). The unusual property of this hemolyzate seems to be a low Bohr effect (which appears common to all urodele hemolyzates). Wood (1971) reported that hemolyzates from adult *Dicamptodon* were rather insensitive to ATP. The blocked α amino group on the N-terminus of the β chain together with the substitution of threonine for histidine at position 2 in the β chain would both serve to decrease the Bohr effect and lower the sensitivity to organic phosphates. However, it is unlikely that these substitutions account for all the functional changes. Other residues responsible for unique functional properties observed of hemolyzates from several species of urodeles remain to be determined.

V. Hemoglobin Synthesis

The most recent review of hemoglobin synthesis is by Maclean and Jurd (1972). These authors tend to wander from their subject, and some of their statements are clearly incorrect, however, they have united a diverse group of references into a useful form. Ingram (1972) has noted several widespread features of erythropoiesis in amphibians and higher vertebrates and Frieden and Just (1970) discuss the hormonal control of hemoglobin transition in amphibians.

Rosse *et al.* (1964) showed that sublethal exposure of *Rana pipiens* to hypoxia did not stimulate erythropoiesis. However, bleeding adults of about one-third of their blood volume did stimulate erythropoiesis, and serum from bled frogs stimulated erythropoiesis in untreated frogs.

Harrington and Becker (1973) have reported that mature erythrocytes (*Rana pipiens?*) are stimulated by weak electrical signals. The morphological changes resemble dedifferentiation. New proteins are synthesized, but no attempts were made to characterize the properties of these products. They do seem to differ from the complement of proteins (hemoglobins) found in mature cells.

A. Seasonal Variation and Phenylhydrazine Treatment

Flores and Frieden (1968) were the first to quantitate effects of phenylhydrazine treatment (which lyses red cells) in amphibians. They noted a greater than 99% drop in the hemoglobin content and erythrocyte numbers in phenylhydrazine-treated tadpoles and adults of *Rana catesbeiana.* Similar effects on tadpoles have been noted by DeWitt *et al.* (1972). Tadpoles survived and began to synthesize hemoglobins. Adult frogs did not survive, but adult *Rana pipiens* treated in a similar manner did survive. Apparently, the severity of the response is less in *R. pipiens* (Meints, 1969). This latter worker also noted enlarged spleens in responding animals, but erythrocyte maturation and hemoglobin synthesis appeared to take place in the peripheral circulation.

DeWitt *et al.* (1972) and Flores and Frieden (1972) have studied the details of the recovery process and the susceptibility of different stages to phenylhydrazine treatment. Flores and Frieden (1972) noted the following order of susceptibility: froglets more susceptible than prometamorphic tadpoles more susceptible than adult frogs more susceptible than metamorphic climax tadpoles more susceptible than triiodothyronine-treated tadpoles. The rate of regeneration of erythrocytes was: metamorphic climax tadpoles greater than triiodothyronine-treated tadpoles greater than adult frogs greater than froglets greater than prometamorphic tadpoles. The order of erythrocyte regeneration is consistent with the hypothesis that phenylhydrazine treatment also damages the liver. Thus, tadpoles, whose erythrocytes mature in the liver, would be most susceptible. One interesting observation was that in those animals treated with triiodothyronine and phenylhydrazine, adult hemoglobin was found in stage XVIII tadpoles, or before the development of bones. This suggests that adult erythrocytes may also be synthesized elsewhere than in the bone (e.g., the spleen). No unusual hemoglobins were found in any animals that had been treated with phenylhydrazine.

DeWitt *et al.* (1972) found a few proerythroblasts and basophilic erythroblasts appearing in the peripheral circulation 4 days after treatment with phenylhydrazine. These same erythrocyte precursors are found in the liver prior to their appearance in the peripheral circulation and the spleen. This evidence points to the liver as the site of maturation of these cells. These cells are clearly undergoing mitosis as opposed to mature tadpole erythrocytes, which show little or no mitotic activity. Sixteen days after phenylhydrazine treatment, these immature erythroid cells are still undergoing mitosis and are actively synthesizing hemoglobins, as evidenced by their high content of polysomes. In all phenylhydrazine-treated tadpoles, the protein synthesized during the recovery phase was tadpole hemoglobin (DeWitt *et al.*, 1972; Maniatis and Ingram, 1972). Thyroxine treatment (which induces metamorphosis) of tadpoles recovering from phenylhydrazine-induced anemia resulted in the appearance of adult frog hemoglobin at the time expected for non-anemic tadpoles.

Adults of *Xenopus laevis* treated with phenylhydrazine begin to synthesize significant amounts of HbF_1, a fetal component. In addition to the synthesis of the normal adult component (HbA) up to 58% of newly synthesized hemoglobin is HbA_2, which normally occurs as a minor component accounting for about 5% of the hemolyzate. Fluorescein-labeled antibody studies show that the cells synthesizing HbF_1 are also synthesizing HbA_2. Thus, anemia (induced by phenylhydrazine or bleeding) induces the synthesis of new components in otherwise normal adult erythrocytes (Maclean and Jurd, 1971a).

B. Sites of Hemoglobin Synthesis

It has been amply demonstrated by a number of workers that there is a transition in hemoglobin components that accompanies metamorphosis. This transition might be accomplished in a number of ways. First, mature nucleated erythrocytes retain some synthetic capacity and the transition might be accomplished by turning off tadpole hemoglobin production and initiating adult hemoglobin synthesis in tadpole erythrocytes. Alternatively, tadpole erythrocytes might be replaced by erythrocytes synthesizing adult frog hemoglobin. These cells could originate in the same tissue as tadpole erythrocytes or they might originate in another tissue or organ. It is necessary to differentiate between these alternatives before one can intelligently probe control mechanisms operative in natural and hormone-induced metamorphosis.

If the hemoglobin transition is accomplished by inhibiting tadpole hemoglobin synthesis and initiating frog hemoglobin synthesis, then one might expect to find both hemoglobins in the same erythrocytes during the transition period. Wood (1971) found intermediate electrophoretic patterns for hemolyzates from neotenic salamanders that had developed gonads. Hollyfield (1966b), DeWitt (1968), Moss and Ingram (1968b), and Benbassat (1970) were able to distinguish *Rana pipiens* or *R. catesbeiana* tadpole and frog erythrocytes by a number of morphological and physiological characteristics. Rosenberg (1970) electrophoresed hemoglobin from single cells and concluded that both forms do not occur in the same cell. Maniatis and Ingram (1971c) using *R. catesbeiana* employed immunofluorescent techniques to determine that erythrocytes contain tadpole or adult frog hemoglobin, but not both. However, Jurd and Maclean (1969, 1970) using similar techniques concluded that in *Xenopus laevis*, up to 25% of the erythrocytes in metamorphosing animals contained both fetal and adult hemoglobins. In *Xenopus* unlike *Rana*, the switch seems to occur within a single cell line (see also Maclean and Jurd, 1971a). The possible occurrence of embryonic and larval hemoglobins in the same cell has not been investigated but seems unlikely. Thus, in contrast to the fetal to adult switch in man which takes place in single erythrocytes (Betke and Kleihauer, 1958; Kleihauer *et al.*, 1967), for some species larval and adult amphibian hemoglobins occur in separate erythrocytes.

Rosenberg (1972) has recently fused tadpole and frog erythrocytes

(*R. catesbeiana*) and found a preferential synthesis of frog hemoglobin in the binucleate cells. This was interpreted as indicating that frog erythrocytes contained a substance that represses tadpole hemoglobin synthesis. However, several alternative interpretations of her experiments exist. The most crucial point considers the use of leucine-^{14}C incorporation as an indicator of hemoglobin synthesis. Aggarwal and Riggs (1969) have shown that some tadpole hemoglobin components contain only two-thirds as many leucines as frog components. This might account for the apparent preferential synthesis of frog hemoglobin. It seems to me that the utilization of cysteine-^{14}C as well as leucine-^{14}C would be useful in synthesis experiments, because the former amino acid is found in adult but not tadpole hemoglobin (*R. catesbeiana, R. grylio*). At any rate, it seems clear now that the tadpole to frog hemoglobin transition takes place in separate cells in *Rana* and that the control mechanism of interest probably occurs at the tissue level, although of course, tadpole hemoglobin synthesis is being repressed (or not induced) in frog cells.

Since tadpole and adult frog hemoglobins do not coexist intracellularly, it seems likely that cells containing them will be synthesized by different tissues or organs. Jordan and Spiedel (1923a,b) reported that the spleen is the site of erythropoiesis in adult *R. catesbeiana* and that bone marrow is also active in the spring. In tadpoles, they found the intertubular regions of the kidney to be active. During metamorphosis (hormone induced), they noted a switch in erythropoiesis from the kidney to the spleen. This erythropoietic transition has been thoroughly investigated by Maniatis and Ingram (1971a,b). They found that in premetamorphic tadpoles and thyroxin-treated tadpoles the liver is the site of erythropoiesis. In the adult frog, erythropoiesis takes place in the bone marrow. However, Carmena (1971b) concluded that hemoglobin synthesis in adult *R. montezumae* takes place primarily in the liver but also in the spleen and bone marrow. Although one can pinpoint those cells actively making hemoglobin, it is more difficult to pinpoint their embryological origin. Nevertheless, by combining data from the experiments of Maniatis and Ingram (1971a) with data from transplantation experiments carried out by Hollyfield (1966a) (and earlier data reviewed in Wilt, 1967), we can construct the following tentative picture of hemoglobin synthesis in *R. catesbeiana* (and *R. pipiens?*), realizing, of course, that there may be species differences. Embryonic hemoglobin probably occurs in cells originating in the blood islands. Tadpole hemoglobin is synthesized in cells originating in the pronephric and mesonephric tissue but maturing in the liver. Adult frog erythrocytes mature in bone marrow and may originate there also.

It has recently been reported by Broyles and Frieden (1973) that one major tadpole (*Rana catesbeiana*) component seems to be synthesized in the kidney while the other two major components are synthesized in the liver. These results indicate that erythropoiesis in amphibians differs significantly from erythropoiesis in higher vertebrates and should provide a stimulus for future studies.

C. The Larval to Adult Transition

With these thoughts in mind, let us now turn our attention to the synthetic events that accompany the larval to adult hemoglobin transition. It should be mentioned at the outset that there is disagreement among workers, primarily as to the timing of the larval to adult hemoglobin switch. This may result from the use of several species (*R. grylio, R. catesbeiana, R. pipiens*), as well as genetic variation in hemoglobin structure, which apparently is most prevalent in tadpoles. Also, it is known that there is significant seasonal variation in the ability of tadpoles and frogs to synthesize hemoglobin (Maclean *et al.*, 1969; Friedmann *et al.*, 1969; Meints and Carver, 1972; Harris, 1972). Temperature is also an important factor (Cline and Waldmann, 1962). Complicating the picture even more is the fact that many workers have found it convenient to induce metamorphosis with thyroxine or triiodothyronine. The extent to which induced metamorphosis mimics natural metamorphosis remains to be determined. Certainly there are some significant differences in hemoglobin synthesis. Also, it should be noted that when Maniatis *et al.* (1969) inhibited the synthesis of adult frog hemoglobin by sensitizing tadpoles to adult hemoglobins, new hemoglobin components appeared that were neither adult nor larval. The genetic origins and chemical compositions of these interesting hemoglobins have not been studied.

Red cells from premetamorphic tadpoles (stages V to XI) rapidly incorporate labeled precursors into tadpole nucleic acid and hemoglobin (Theil, 1967; McMahon and DeWitt, 1968; Benbassat, 1970). As tadpoles continue to mature (prometamorphic, stages XII to XIX), hemoglobin synthesis decreases (Moss and Ingram, 1965; Benbassat, 1970). During metamorphic climax (stage XX to XXV), hemoglobin and nucleic acid synthesis increase and adult hemoglobin appears (Hamada *et al.*, 1966; Theil, 1967, 1970; Moss and Ingram, 1968b; Just and Atkinson, 1972). Although Benbassat (1970) found some adult hemoglobin synthesis at the end of metamorphic climax, erythrocytes of froglets still contained predominantly tadpole hemoglobins. In *R. pipiens*, he was unable to demonstrate adult hemoglobins in froglets. These results are in conflict with those of all other workers who found adult hemoglobin occurring at stage XXII and the transition complete within about 2 weeks (or at stage XXIV). Benbassat (1970) and Theil (1967) have reported that premetamorphic tadpoles synthesize a considerable amount (about 10%?) of nonhemoglobin protein in erythrocytes. Later tadpole stages do not synthesize significant amounts of nonhemoglobin protein. Jurd and Maclean (1970) have shown that tadpoles of *Xenopus laevis* in metamorphic climax show only traces of adult hemoglobin and young adults retain a large proportion of tadpole type hemoglobin. In *Ambystoma mexicanum* the transition from larval to adult hemoglobin types is gradual and begins soon after all four legs appear in the larval form.

Grasso and Woodard (1967) determined that no loss of DNA occurs during erythrocyte development in *Triturus viridescens*. They also observed DNA synthesis and mitosis in hemoglobinized cells. Just and Atkinson (1972) were able to demonstrate a temporal separation between DNA and protein synthesis. Using *R. catesbeiana,* they found DNA synthesis begins at stage XX, is maximum at stage XXII, and then declines. Protein synthesis starts at stage XXII and is maximum at stage XXIV, after which it declines.

Although hormone-induced metamorphosis appears to mimic natural metamorphosis, Just and Atkinson (1972) have noted important differences. Approximately 10 days after the injection of triiodothyronine, protein synthesis increases and the hemoglobin content of tadpole erythrocytes increases. The hemoglobin level is significantly higher than in controls by day 16 and remains high through day 32. Although protein synthesis increases, most hemoglobin synthesized is of the larval type. Interestingly, Moss and Ingram (1965, 1968b) and Theil (1967) noted a significant drop in protein synthesis *in vitro* 8 days after hormone treatment. Thus, thyroid hormone administration can partially mimic normal metamorphosis. DNA synthesis occurs shortly after treatment, and it is followed by the appearance of immature erythrocytes which contain measurable amounts of adult hemoglobin. However, only a low level of adult hemoglobin is produced, and labeled precursors are actively incorporated into tadpole hemoglobin. In normal metamorphosis in *Rana,* almost all hemoglobin synthesis results in adult hemoglobin production. Although adult hemoglobin production is stimulated by thyroid hormone injections, tadpole hemoglobin production is also increased, and tadpole erythrocytes are not removed from the circulation. Thus, thyroxine is involved in the control of the larval to adult hemoglobin transition, but other factors must also be operative. DeWitt (1971) has reported a difference in methionyl and arginyl transfer RNAs in larval and adult bullfrogs. Other possible control factors have not been studied.

VI. Concluding Comments

It is clear that amphibian hemoglobins offer a fertile field for further investigation. The sequence studies, although fragmentary at present, should serve to stimulate additional structural and functional studies. Obviously, frog and salamander hemoglobins have several peculiar oxygen-binding properties that might offer real insight into the structure–function relationships of hemoglobins in general. Chemical modification studies of these unusual hemoglobins should prove quite interesting.

Most if not all amphibian species have multiple hemoglobins. These seem to differ in their functional properties and may exist to provide the animal with a more constant oxygen supply over a wide range of body temperatures. Accompanying the complex hemoglobin pattern in amphibian hemolyzates is a fairly high incidence of genetic polymorphism. The problems associated with genetic polymorphisms must be resolved before additional studies on ranid species are meaningful. Also, the genetic control of hemoglobin structures in *Bufo* should be thoroughly investigated in order to understand the unusual patterns of hemoglobin inheritance observed in this genus.

The role of organic and inorganic phosphates in regulating oxygen affinity has been studied very little in amphibians. Clearly DPG, ATP, and IHP either singly or in some combination, may control oxygen affinity in amphibian bloods. What levels of phosphate exist in hibernating, cold-adapted, and warm-adapted frogs? Are all hemoglobin components phosphate sensitive? Are there geographical and/or altitudinal variations in phosphate levels? What affinity-determining factors other than phosphates are operative in adult *Rana catesbeiana*? How have the unusual structural features of the β chain modified phosphate binding?

Hemoglobin synthesis during the tadpole to adult transition will continue to demand attention. First, however, we must resolve or confirm what appear to be basic differences in hemoglobin synthesis between amphibian species (i.e., between *Rana* and *Xenopus*). Second, it seems desirable to work with pure lines as are now available from the University of Michigan. Using genetically homogeneous populations of *R. pipiens* has several advantages, not the least of which is that the hemoglobin does not seem to polymerize. When the events of hemoglobin transition during natural and hormone-induced metamorphosis are agreed upon, then we can more clearly attack the actual control mechanisms involved in this ontogenetic switch.

As mentioned throughout this review, only a small number of species have been studied in any detail, and these are species from an even smaller fraction of the available phylogenetic assemblages. As our horizons widen, so do the experimental possibilities. So also does our understanding of the applicability of our present generalizations. Past research, though often conflicting, has illuminated a number of interesting avenues for future research. These will certainly lead to a better understanding not only of amphibians and their many adaptations, but also of man and his own specializations.

ACKNOWLEDGMENTS

We wish to thank the following persons who allowed us to incorporate their unpublished data into this chapter: T. O. Baldwin, J. and C. Bonaventura, B. Brimhall, M. Coates, R. T. Jones, and A. F. Riggs.

REFERENCES

Abramson, R. K., Rucknagel, D. L., Shreffler, D. C., and Saave, J. J. (1970). *Science* **169,** 194.

Aggarwal, S. J., and Riggs, A. (1969). *J. Biol. Chem.* **244,** 2372.

Altland, P. D., and Brace, D. C. (1962). *Amer. J. Physiol.* **203,** 1188.

Amiconi, G., Antonini, E., Brunori, M., Sorcini, M., and Tentori, L. (1970). *Int. J. Biochem.* **1,** 582.

Antonini, E., and Brunori, M. (1971). "Hemoglobin and Myoglobin in their Reaction with Ligands." Amer. Elsevier, New York.

Araki, T., Kajita, A., and Shukuya, R. (1971). *Biochem. Biophys. Res. Commun.* **43,** 1179.

Araki, T., Kajita, A., and Shukuya, R. (1973). *Nature (London) New Biol.* **242,** 254.

Arnone, A. (1972). *Nature (London)* **237,** 146.

Baglioni, C., and Sparks, C. E. (1963). *Develop. Biol.* **8,** 272.

Baldwin, T. O. (1971). Ph.D. Thesis, University of Texas, Austin.

Benbassat, J. (1970). *Develop. Biol.* **21,** 557.

Benesch, R., Benesch, R. E., and Yu, C. I. (1968). *Proc. Nat. Acad. Sci. U.S.* **59,** 526.

Bertini, F., and Rathe, G. (1962). *Copeia, No. 1,* p. 181.

Betke, K., and Kleihauer, E. (1958). *Blut* **4,** 241.

Bhown, A. S. (1965). *Indian J. Exp. Biol.* **1,** 272.

Blackwell, R. Q., Liu, C. S., and Wang, C. L. (1971). *Biochim. Biophys. Acta* **243,** 467.

Bonaventura, J. (1969). *Biochem. Genet.* **3,** 239.

Bonaventura, J., and Riggs, A. (1967). *Science* **158,** 800.

Brame, A. H. (1967). *J. S. W. Herpetol. Soc.* **2,** 1.

Brimhall, R., Hollan, S., Jones, R. T., Koler, R. D., and Szelenyi, J. G. (1970). *Clin. Res.* **18,** 184.

Brown, E. R., and DeWitt, W. (1970). *Comp. Biochem. Physiol.* **35,** 495.

Brown, L. E., and Guttman, S. I. (1970). *Amer. Midl. Natur.* **83,** 160.

Broyles, R. H., and Frieden, E. (1973). *Nature (London) New Biol.* **241,** 207.

Brunori, M., Antonini, E., Wyman, J., Tentori, L., Vivaldi, G., and Carta, S. (1968). *Comp. Biochem. Physiol* **24,** 519.

Bunn, H. F. (1971). *Science* **172,** 1049.

Caffin, J. P., Chauvet, J. P., and Acher, R. (1969). *FEBS Lett.* **5,** 196.

Carmena, A. O. (1971a). *Comp. Biochem. Physiol.* A **40,** 349.

Carmena, A. O. (1971b). *Comp. Biochem. Physiol. A* **40**, 517.
Chauvet, J. P., and Acher, R. (1968). *FEBS Lett.* **1**, 305.
Chauvet, J. P., and Acher R. (1970a). *FEBS Lett.* **8**, 263.
Chauvet, J. P., and Acher, R. (1970b). *FEBS Lett.* **9**, 202.
Chauvet, J. P., and Acher, R. (1970c). *FEBS Lett.* **10**, 136.
Chauvet, J. P., and Acher, R. (1971). *Int. J. Protein Res.* **3**, 261.
Chauvet, J. P., and Acher, R. (1972). *Biochemistry* **11**, 916.
Christomanos, A. A. (1967). *Enzymologia* **32**, 301.
Christomanos, A. A. (1972). *Enzymologia* **42**, 385.
Christomanos, A. A., Gardiki, V., and Dimitriades, A. (1965). *Enzymologia* **28**, 157.
Cline, M. J., and Waldmann, T. A. (1962). *Amer. J. Physiol.* **203**, 401.
Coates, M. L., and Metcalfe, J. (1971). *Resp. Physiol.* **11**, 94.
Davies, H. G. (1961). *J. Biophys. Biochem. Cytol.* **9**, 671.
Dayhoff, M. O., ed. (1972). "Atlas of Protein Sequence and Structure 1972," Vol. 5. National Biomedical Research Foundation, Silver Spring, Maryland.
Dessauer, H. C., and Nevo, E. (1969). *Biochem. Genet.* **3**, 171.
Dessauer, H. C., Fox, W., and Ramirez, J. R. (1957). *Arch. Biochem. Biophys.* **71**, 11.
DeWitt, W. (1968). *J. Mol. Biol.* **32**, 502.
DeWitt, W. (1971). *Biochem. Biophys. Res. Commun.* **42**, 266.
DeWitt, W., and Ingram, V. M. (1967). *Biochem. Biophys. Res. Commun.* **27**, 236.
DeWitt, W., Price, R. P., Vankin, G. L., and Brandt, E. M. (1972). *Develop. Biol.* **27**, 27.
Dixon, G. H. (1966). *Essays Biochem.* **2**, 147–204.
Edwards, J. A., and Justus, J. T. (1969). *Proc. Soc. Exp. Biol. Med.* **132**, 524.
Elli, R., Giuliani, A., Tentori, L., Chiancone, B., and Antonini, E. (1970). *Comp. Biochem. Physiol.* **36**, 163.
Elzinger, M. (1964). Ph.D. Dissertation, University of Illinois, Urbana.
Emilio, M. G., and Shelton, G. (1972). *J. Exp. Biol.* **56**, 67.
Flores, G., and Frieden, E. (1968). *Science* **159**, 101.
Flores, G., and Frieden, E. (1972). *Develop. Biol.* **27**, 408.
Fox, W., Dessauer, H. C., and Maumus, L. T. (1961). *Comp. Biochem. Physiol.* **3**, 52.
Foxon, G. E. H. (1964). *In* "Physiology of the Amphibia" (J. A. Moore, ed.), pp. 151–210. Academic Press, New York.
Frieden, E., and Just, J. J. (1970). *In* "Biochemical Actions of Hormones" (G. Litwack, ed.), Vol. 1, pp. 1–52. Academic Press, New York.
Friedmann, G. B., Algard, F. T., and McCurdy, H. M. (1969). *Anat. Rec.* **163**, 55.
Gahlenbeck, H., and Bartels, H. (1968). *Z. Vergl. Physiol.* **59**, 232.
Gahlenbeck, H., and Bartels, H. (1970). *Resp. Physiol.* **9**, 175.
Gatten, R. H., Jr., and Brooks, G. R. (1969). *Comp. Biochem. Physiol.* **30**, 1019.
Gillespie, J. H., and Crenshaw, J. W. (1966). *Copeia* p. 889.
Grasso, J. A., and Woodard, J. W. (1967). *J. Cell Biol.* **33**, 645.
Guimond, R. W., and Hutchinson, V. H. (1972). *Comp. Biochem. Physiol.* **42**, 367.
Guttman, S. I. (1965). *Tex. J. Sci.* **17**, 267.
Guttman, S. I. (1967). *Comp. Biochem. Physiol.* **23**, 871.
Guttman, S. I. (1969). *Copeia* p. 243.
Guttman, S. I. (1972). *In* "Evolution in the Genus *Bufo*" (W. F. Blair, ed.), pp. 265–278. Univ. of Texas Press, Austin.

Hall, F. G. (1966). *J. Cell. Physiol.* **68**, 69.
Hamada, K., and Shukuya, R. (1966). *J. Biochem. (Tokyo)* **59**, 397.
Hamada, K., Shukuya, R., and Kaziro, K. (1964a). *J. Biochem. (Tokyo)* **55**, 213.
Hamada, K., Sakai, Y., Shukuya, R., and Kaziro, K. (1964b). *J. Biochem. (Tokyo)* **55**, 636.
Hamada, K., Sakai, Y., Tsushima, K., and Shukuya, R. (1966). *J. Biochem. (Tokyo)* **60**, 37.
Harrington, D. B., and Becker, R. O. (1973). *Exptl. Cell Res.* **76**, 95.
Harris, J. A. (1972). *Comp. Biochem. Physiol. A* **43**, 975.
Hendrickson, W. A., and Love, W. E. (1971). *Nature (London)* **232**, 197.
Herner, A. E., and Frieden, E. (1961). *Arch. Biochem. Biophys.* **95**, 25.
Hollyfield, J. G. (1966a). *Develop. Biol.* **14**, 461.
Hollyfield, J. G. (1966b). *J. Morphol.* **119**, 1.
Howell, B. J., Baumgardner, F. W., Bondi, K., and Rahn, H. (1970). *Amer. J. Physiol.* **218**, 600.
Huber, R., Epp, O., Steigemann, W., and Formanek, H. (1971). *Eur. J. Biochem.* **19**, 42.
Huehns, E. R., Dance, N., Beaven, G. H., Hecht, E., and Motulsky, A. G. (1964). *Cold Spring Harbor Symp. Quant. Biol.* **29**, 327.
Huisman, T. H. J., and Schroeder, W. A. (1970). *CRC Crit. Rev. Clin. Lab. Sci.* **1**, 471–526.
Huisman, T. H. J., Schroeder, W. A., Dozy, A. M., Shelton, J. R., Shelton, J. B., Boyd, E. M., and Apell, G. (1969). *Ann. N.Y. Acad. Sci.* **165**, 320.
Ingram, V. M. (1961). *Nature (London)* **189**, 704.
Ingram, V. M. (1972). *Nature (London)* **235**, 338.
Johansen, K. (1972). *Resp. Physiol.* **14**, 193.
Johansen, K., Lenfant, C., and Hanson, D. (1970). *Fed. Proc., Fed. Amer. Soc. Exp. Biol.* **29**, 1135.
Jones, D. R. (1972). *J. Comp. Physiol.* **77**, 356.
Jordan, H. E., and Spiedel, C. C. (1923a). *Amer. J. Anat.* **32**, 155.
Jordan, H. E., and Spiedel, C. C. (1923b). *J. Exp. Med.* **38**, 529.
Jurd, R. D., and Maclean, N. (1969). *Experientia* **25**, 626.
Jurd, R. D., and Maclean, N. (1970). *J. Embryol. Exp. Morphol.* **23**, 299.
Just, J. J., and Atkinson, B. G. (1972). *J. Exp. Zool.* **182**, 271.
Kleihauer, E. F., Tang, T. E., and Betke, K. (1967). *Acta Haematol.* **38**, 264.
Kurata, Y., and Arakawa, W. (1963). *Blut* **9**, 42.
Kurata, Y., and Okada, S. (1968). *Blut* **16**, 298.
Kurata, Y., Okada, S., and Yuasa, C. (1965). *Blut* **11**, 88.
Leggio, T., and Morpurgo, G. (1968). *Nature (London)* **219**, 493.
Lenfant, C., and Johansen, K. (1967). *Resp. Physiol.* **2**, 247.
Lenfant, C., and Johansen, K. (1972). *Resp. Physiol.* **14**, 211.
Maclean, N., and Jurd, R. D. (1971a). *J. Cell Sci.* **9**, 509.
Maclean, N., and Jurd, R. D. (1971b). *Comp. Biochem. Physiol. B* **40**, 751.
Maclean, N., and Jurd, R. D. (1972). *Biol. Rev. Cambridge Phil. Soc.* **47**, 393.
Maclean, N., Brooks, G. T., and Jurd, R. D. (1969). *Comp. Biochem. Physiol.* **30**, 825.
McCutcheon, F. H. (1936). *J. Cell. Comp. Physiol.* **8**, 63.
McMahon, E. M., Jr., and DeWitt, W. (1968). *Biochem. Biophys. Res. Commun.* **31**, 176.
Maniatis, G. M., and Ingram, V. M. (1971a). *J. Cell. Biol.* **49**, 372.
Maniatis, G. M., and Ingram, V. M. (1971b). *J. Cell Biol.* **49**, 380.

Maniatis, G. M., and Ingram, V. M. (1971c). *J. Cell Biol.* **49**, 390.
Maniatis, G. M., and Ingram, V. M. (1972). *Develop. Biol.* **27**, 580.
Maniatis, G. M., Steiner, L. A., and Ingram, V. M. (1969). *Science* **165**, 67.
Manwell, C. (1966). *Comp. Biochem. Physiol.* **17**, 805.
Manwell, C., and Baker, C. M. A. (1970). "Molecular Biology and the Origin of Species." Sidgwick & Jackson, London.
Marchlewska-Koj, A. (1963). *Folia Biol. (Warsaw)* **11**, 167.
Marchlewska-Koj, A. (1964). *Folia Biol. (Warsaw)* **12**, 167.
Meints, R. H. (1969). *Comp. Biochem. Physiol.* **30**, 383.
Meints, R. H., and Carver, F. J. (1972). *Comp. Biochem. Physiol.* A. **42**, 511.
Morpurgo, G., Battaglia, P. A., and Leggio, T. (1970). *Nature (London)* **225**, 76.
Moss, B., and Ingram, V. M. (1965). *Proc. Nat. Acad. Sci. U.S.* **54**, 967.
Moss, B., and Ingram, V. M. (1968a). *J. Mol. Biol.* **32**, 481.
Moss, B., and Ingram, V. M. (1968b). *J. Mol. Biol.* **32**, 493.
Newcomer, R. J. (1967). *Amer. Natur.* **101**, 192.
Padlan, E. A., and Love, W. E. (1968). *Nature (London)* **220**, 376.
Perutz, M. F. (1970). *Nature (London)* **228**, 776.
Perutz, M. F., Kendrew, J. C., and Watson, H. C. (1965). *J. Mol. Biol.* **13**, 669.
Perutz, M. F., Muirhead, H., Cox, J. M., and Goaman, L. C. G. (1968). *Nature (London)* **219**, 131.
Perutz, M. F., Muirhead, H., Mazzarella, L., Crowther, R. A., Greer, J., and Kilmartin, J. V. (1969). *Nature (London)* **222**, 1240.
Porter, K. R. (1972). "Herpetology." Saunders, Philadelphia, Pennsylvania.
Reeves, R. B. (1972). *Resp. Physiol.* **14**, 219.
Riggs, A. (1951). *J. Gen. Physiol.* **35**, 23.
Riggs, A. (1964). *Can. J. Biochem.* **42**, 763.
Riggs, A. (1966). In *Int. Symp. Comp. Hemoglobin Struct.*, (A. Christomanos and D. J. Polychronakos, eds.) pp. 126–128, Triantafylo, Thessaloniki, Greece.
Riggs, A., and Bownds, D. (1963). *Fed. Proc., Fed. Amer. Soc. Exp. Biol.* **22**, 597.
Riggs, A., Sullivan, B., and Agee, J. R. (1964). *Proc. Nat. Acad. Sci. U.S.* **51**, 1127.
Romer, A. S. (1972). *Resp. Physiol.* **14**, 183.
Rosenberg, M. (1970). *Proc. Nat. Acad. Sci. U.S.* **67**, 32.
Rosenberg, M. (1972). *Nature (London)* **239**, 520.
Rosse, W. F., Waldmann, T., and Hull, E. (1964). *Blood* **22**, 66.
Schroeder, W. A., Huisman, T. H. J., Shelton, J. R., Shelton, J. B., Kleihauer, E. F., Dozy, A. M., and Robberson, B. (1968). *Proc. Nat. Acad. Sci. U.S.* **60**, 537.
Schroeder, W. A., Shelton, J. R., Shelton, J. B., Apell, G., Huisman, T. H. J., and Bouver, N. G. (1972). *Nature (London) New Biol.* **240**, 273.
Shontz, N. N. (1968). *Copeia* p. 683.
Sorcini, M., Orlando, M., and Tentori, L. (1970). *Comp. Biochem. Physiol.* **34**, 751.
Stratton, L. P., and Wise, R. W. (1967). *Fed. Proc., Fed. Amer. Soc. Exp. Biol.* **26**, 808.
Sullivan, B. (1967). *Biochem. Biophys. Res. Commun.* **28**, 407.
Sullivan, B. (1971). *In* "Comparative Genetics in Monkeys, Apes, and Man" (A. B. Chiarelli, ed.), pp. 213–256. Academic Press, New York.
Sullivan, B., and Nute, P. E. (1968). *Genetics* **58**, 113.
Sullivan, B., and Riggs, A. (1964). *Nature (London)* **204**, 1098.

Sullivan, B., and Riggs, A. (1967). *Comp. Biochem. Physiol.* **23**, 437.
Svedberg, T., and Hedenius, A. (1934). *Biol. Bull.* **66**, 191.
Szarski, H., and Czopek, J. (1965). *Zool. Pol.* **15**, 51.
Szarski, H., and Czopek, J. (1966). *Bull. Acad. Pol. Sci.* **14**, 433.
Szarski, H., and Kosmos, A. (1965). *Zool. Pol.* **14**, 553.
Taketa, F., and Nickerson, M. A. (1973). *Comp. Biochem. Physiol.* B. **45**, 549.
Tentori, L., Vivaldi, G., Carta, S., Salvati, A. M., Sorcini, M., and Velani, S. (1965). *Arch. Biochem. Biophys.* **109**, 404.
Tentori, L., Vivaldi, G., Carta, S., Velani, S., and Zito, R. (1967). *Biochim. Biophys. Acta* **133**, 177.
Theil, B. C. (1967). *Biochim. Biophys. Acta* **138**, 175.
Theil, B. C. (1970). *Comp. Biochem. Physiol.* **33**, 717.
Tooze, J., and Davies, H. G. (1963). *J. Cell Biol.* **16**, 501.
Trader, C. D., and Frieden, E. (1966). *J. Biol. Chem.* **241**, 357.
Trader, C. D., Wortham, J. S., and Frieden, E. (1963). *Science* **139**, 918.
Wade, M., and Rose, F. L. (1972). *Copeia 1972*, p. 889.
Watts, R. L., and Watts, D. C. (1968). *J. Theor. Biol.* **20**, 227.
Wilt, F. H. (1967). *Advan. Morphog.* **6**, 89.
Wintrobe, M. M. (1934). *Folia Haematol. (Leipzig)* **51**, 32.
Wise, R. W. (1970). *Comp. Biochem. Physiol.* **32**, 89.
Wood, S. C. (1971). *Resp. Physiol.* **12**, 53.

CHAPTER 6

Endocrinology of Amphibia

Wilfried Hanke

I. Structure and Function of Endocrine Glands 123
A. Hypothalamohypophysial System 123
B. Branchiogenic Glands 131
C. Pancreatic Islets 133
D. Mesodermal Glands 133
II. Regulation of Metabolic Processes 137
A. Energy Metabolism 137
B. Osmomineral Regulation 140
III. Regulation of Development and Metamorphosis 144
A. Influences on Metamorphosis 144
B. Regulation of Growth 146
C. Induction of Molting 147
IV. Regulation of Gonadal Function and Reproduction 148
V. Regulation of Color Change 149
References 152

During the last few years, some general reviews of the significance and function of hormones in Amphibia have been published (Gorbman, 1964; Bern and Nandi, 1964; Matty, 1966). This review will point out only some subjects of amphibian endocrinology that demonstrate the particular progress during the last 5 years.

I. Structure and Function of Endocrine Glands

A. Hypothalamohypophysial System

This system consists of a neural part, the hypothalamus and the neurohypophysis, and a glandular part, the adenohypophysis. In Amphibia as well as in all other vertebrates, there exists an interrelation between distinct parts of the brain and the glandular part of the pituitary. The chemical structure of the different hormones is similar to that known from other vertebrates. Besides, the general pattern of the regulation between the neural and the glandular part of the system resembles that of other vertebrate groups (Jørgensen, 1968).

1. Neural Part of the System

Compared to teleosts, the neurohypophysis of amphibians, especially anurans, is more distinctly divided into two parts, the eminentia mediana and the pars nervosa of the pituitary. In urodeles, the connection between the median eminence and the adenohypophysis still resembles that found in Dipnoi. Capillary connections exist between both parts, but only sometimes is a portal vein formed. In anurans, however, the neurohypophysis is stronger developed. Portal veins are always present. The capillaries fuse to bigger vessels connecting the median eminence and the pars distalis. The size of the median eminence varies considerably, especially in urodeles. Sometimes it is very small (*Necturus, Cryptobranchus*). In other cases the median eminence is more distinct (*Triturus*). In *Gymnophiona,* the area of the median eminence is a very thin tongue of tissue (Wingstrand, 1966).

Concerning the nerve terminals found in the median eminence, there are five different types near the pars distalis in *Rana esculenta:* two types of axons of peptidergic neurons containing granules of about 120–180 nm in diameter and two to three types of axons with granules of smaller size (adrenergic nerve endings). The pars nervosa contains two types of neurosecretory fibers, one with electron-dense granules (160–240 nm in diameter), the other one with lighter granules 130–200 nm) (Doerr-Schott, 1970; Doerr-Schott and Follenius, 1969, 1970b).

As in other tetrapodes, the pars nervosa of the Amphibia is isolated from the pars intermedia. In different groups, nerve fibers containing both Gomori-positive and ordinary axons run from the pars nervosa to the pars intermedia. Some of these nerves are adrenergic (Doerr-Schott and Follenius, 1970a). A special portal system that supplies the pars intermedia beside the ordinary portal system has been described by Rodriguez and Piezzi (1967).

The hypothalamus inhibits the release of melanocyte-stimulating hormone (MSH) in toads (Etkin, 1967; Jørgensen and Larsen, 1967). It has been shown by transecting the lower infundibular stem near the median eminence that the axonal pathway is responsible for the inhibitory effect. This control might be exerted through the catecholaminergic nerves that terminate closely to pars intermedia cells (Enemar and Falck, 1965; Enemar *et al.,* 1967; Iturriza, 1966, 1967). Young larva of *Xenopus laevis* are able to adapt themselves to a white background simultaneously with the appearance of aminergic centers in the caudal hypothalamus. Therefore, it is suggested that the melanophorotropin inhibiting hormone of the hypothalamus might be dopamine or another catecholamine, which is experimentally released by reserpine (Van Oordt *et al.,* 1972; Terlou *et al.,* 1974).

There are three types of vesicular organelles within the axonal terminals found in the pars intermedia of *Rana:* small vesicles (diameter 20–30 nm), larger vesicles with a dense core (60–100 nm in diameter), larger dense vesicles of 100–200 nm in diameter (similar to neurosecretory granules). These nerve endings synapse with the secretory cells (Saland, 1968; Dent and Gupta, 1967; Pehlemann, 1967a; Smoller, 1966). It is well established that both types of neurons containing the larger vesicles (60–100 nm and 100–200 nm) come into synaptic contact with secretory cells. There is some evidence that double innervation occurs. Electrophysiological studies have demonstrated that two types of spontaneously active electrical units occur in the pars intermedia. One type of neuron is inhibited by light, the receptor for which appears to be the pineal. Thus, the MSH release might be regulated (Nakai and Gorbman, 1969; Oshima and Gorbman, 1969). Besides this nervous regulation, a nucleus of bioamine-producing neurons in the caudal hypothalamus of *Xenopus* tadpoles exerts an influence on the MSH-producing cells. It is reasonable that this nucleus produces the MSH-inhibiting factor (MIF). Treatment of animals with reserpin in stage 49/50 changes the melanophores in the same way as a blockage of MSH release. That means release of MIF (Goos, 1969a). In toads, autotransplanted pars intermedia could be influenced by either dopamine or noradrenaline. Both catecholamines lightened the animals when injected near the autotransplanted pars intermedia. In the same way, transplantation of intermediate lobes to the adrenal glands did not completely darken the toads. Transplantation to the kidney, however, resulted in a high melanophore index. Stressing the toads only bleached animals that had the graft near the adrenal glands. Therefore, it seems rather evident that catecholamines inhibit the release of MSH (Iturriza, 1969).

The central nervous control of pars distalis function is mediated by neurons arising from the magnocellular nuclei in the anterior hypothalamus and other areas of the hypothalamus. Transection of the infundibular stalk, hypothalamic lesions, and transplantation experiments with the pars distalis gave evidence that a number of pars distalis functions are influenced by the central nervous system (cf Peute and Van Oordt, 1974).

The control of gonadal function by the hypothalamus was investigated in *Rana temporaria* (Dierickx, 1965, 1966, 1967a,b) and in *Bufo bufo* (Jørgensen, 1968, 1970). In *Rana,* the center that releases factors responsible for the secretion of gonadotropins is localized in the middle of the hypothalamus. The gonadotropic function is inhibited when the lobus infundibularis is dissected close to the hypophysis. Transection of the brain just posterior to the optic chiasma did not influence spermatogenesis and ovarian development.

In female *Bufo,* the area that influences the gonadotropic function of the pars distalis must be localized in a region above or anterior to the optic chiasma. Both transections in front and behind the optic chiasma reduced oocyte growth. In the male toad, on the contrary, interstitial cell activity is permitted by a group of cells in the infundibular region.

The hypothalamic control of thyroid function in Amphibia is characterized by three different patterns (Rosenkilde, 1971): (1) the general mammalian pattern (the thyroid depends on the pars distalis, which is regulated by the TRF originating from the hypothalamus), (2) the general amphibian pattern (the thyroid is controlled by TSH from the pars distalis, but this is not dependent upon the central nervous system), (3) independent thyroid activity (maintenance without thyrotropic stimulation).

The regulation of thyroid activity in adult *B. bufo* is an example for the first pattern. Hypophysectomy lowers the iodine uptake of the thyroid. Autografts of the pars distalis near to the median eminence maintain iodine uptake at the normal level. Autografts without central nervous contact do not change the level after hypophysectomy. This shows that the TSH function of the pars distalis depends on the contact with the hypothalamus.

The TRF-producing area could be localized in *Bufo* by transectioning in front of and behind the optic chiasma. The TRF-producing neurons must be located posteriorly to the optic chiasma, because both sections do not really inhibit the thyroid activity. It is only reduced to a small amount by the transection immediately behind the optic chiasma. There are evidences that both types of transections remove an inhibitory influence on the TSH secretion. It is not easy to decide whether this influence is due to neurons regulating the TRF-producing neurons or a separate regulating system which may be correlated with the influence of prolactin on the thyroid.

In other amphibian species the pars distalis has a higher degree of autonomy than in *Bufo.* Autografts of pars distalis at different places can keep the thyroid activity at the normal level. This could be demonstrated especially in urodeles (Vellano *et al.,* 1967; Vellano and Peyrot, 1970; Compher and Dent, 1970). Hypophysectomy in *Triturus cristatus* does not lower thyroid activity immediately after the operation. It needs some weeks to demonstrate reduction of activity in the thyroid. This may be due to a higher degree of autonomy of the thyroid (third pattern).

Both hypothalamic stimulation and hypothalamic inhibition of the thyroid are important for the normal course of metamorphic climax in

Amphibia. This has been discussed by Etkin (1968, 1969) and Goos (1969a) (see also Section III,A). Hypothalamic control of hypophysial corticotropic function was studied with similar methods: hypophysectomy and transplantation of pars distalis to the hypothalamus or the eye muscle. Corticotropic secretion of the pituitary in *B. bufo* was estimated from the condition of molting. It was demonstrated by Jørgensen and Larsen (1963) that normal corticotropin secretion and adrenocortical response were necessary for a normal molting process. Transection of the hypothalamus in front of and behind the optic chiasma in *B. bufo* results in abnormal molting conditions that show that the secretion of ACTH was reduced. The neurons that stimulate ACTH secretion must therefore be located anterior to the middle hypothalamus (Jørgensen, 1968).

In determining the amount of corticosterone in the blood of *Bufo*, Büchmann *et al.* (1972) confirmed this conception. They could demonstrate that removal of the pars distalis reduced the secretion rate to an undetectable amount. Autonomous secretion could not be demonstrated. The same reduction of corticosteroid production after hypophysectomy was found in *Xenopus laevis* (Chan and Edwards, 1970). These observations confirmed results on the dependency of the adrenal cortex from the hypophysis that were obtained in *R. temporaria* by indirect methods. Histological and histochemical studies on the interrenal tissue have demonstrated that hypophysectomy results in degradations of the tissue which were interpreted as impaired secretory activity (Hanke and Weber, 1964, 1965; Hanke, 1967; van Kemenade, 1968a). ACTH treatment restored the adrenocortical activity judged by means of histological and histochemical data. Histological investigations with karyometric measurements of interrenal tissue in *R. temporaria* (Dierickx and Goossens, 1970) demonstrated that the pars distalis maintained a reduced but still distinct corticotropic activity after complete infundibular transection.

From recent investigations on the pituitary–interrenal axis in *R. temporaria*, it may be concluded that other parts of the pituitary than the pars distalis secrete a substance with ACTH-like activity (Neumann, 1973). If only the pars distalis is extirpated, some immunoreactive ACTH is still present in the serum. The activity of the interrenals is higher after removal of the pars distalis than in animals totally hypophysectomized. After total hypophysectomy (removal of all parts of the pituitary), no residual immunoreactive ACTH could be detected. Furthermore, the conception described above is disturbed by the demonstration of ACTH-like activity in the hypothalamus itself (Thurmond and Hanke, 1971; Horn-Ebert *et al.*, 1974). In hypophysectomized *Xenopus* tadpoles (stage 48–52), transplantation of hypothalami of tadpoles or adult animals to a place near the kidney increased the activity of steroid dehydrogenase in the interrenal tissue judged by histochemical methods. Transplantation of hypothalamic tissue gave the same results as ACTH treatment or transplantation of the pars distalis. Although transplantation of pars distalis also stimulated development and growth of the hypophysectomized tadpoles, hypothalamic grafts did not affect these parameters. Therefore, hypothalamic tissue may exert only ACTH-like activity and no STH- or TSH-like activity. Besides, the influence is exerted only at a very short distance, as the normal hypothalamus left in these animals seems to be quite ineffective. Further analysis of these important observations must clarify whether the ACTH-like activity of the hypothalamus is due to bound ACTH secreted by the pituitary or to molecules similar to ACTH which are synthesized by the hypothalamic tissue.

The hypothalamic control of prolactin release in Amphibia was mainly studied in newts (*Triturus cristatus carnifex*). There are evidences that a prolactin-inhibiting factor secreted by the hypothalamus regulates the release of prolactin as in higher vertebrates. Prolactin-dependent characters in this newt are the increase of the height of the tail fin and the water drive. Ectopically autotransplanted pituitaries sustain water drive in summer animals for more than 2 months. Pituitaries cultured *in vitro* release a greater amount of prolactin than pituitaries left under normal control conditions. Testosterone injected in normal aestivating newts stimulates the secretion of prolactin. These experiments demonstrated that the regulatory mechanisms resemble those in mammals (Mazzi *et al.*, 1969; Peyrot *et al.*, 1969a,b).

Since the activity of the adenohypophysis is regulated by the hypothalamus, environmental influences are transmitted to endocrine activities. The activity of the pituitary follows an annual cycle and coincides with variations of light, temperature, food availability, and social environment. It is suggested that thermosensitive and other centers exist in the hypothalamus responsible for the production and release of neurohumors. The localization of these centers is still unknown, because there are a lot of sites where neurons come from, that affect the hypothalamus (Mazzi, 1970). In *R. temporaria*, long daily photoperiod and continuous darkness are known to change the secretory activity of the hypothalamus. Exposure to light results in an augmentation of aldehyde fuchsin-positive material in the preoptic nucleus and a reduction in the median eminence. The higher activity of the preoptic area could also be demonstrated by enhanced incorporation of cysteine-^{35}S and an increase of nuclear volume of the cells (Vullings, 1971).

The structure of the pars nervosa in Amphibia and the nerve supply follow the general pattern for higher vertebrates. A detailed description of the function is given in Section II,B.

There are some other endocrine functions localized in the brain and possibly independent of the hypothalamohypophyseal system. The pineal gland has found special interest. A survey is given by Wurtman *et al.* (1968). Humoral secretion of this gland may have an effect on the gonads, especially an inhibition of spermatogenesis and an induction of uterus contraction. All these effects have been but little studied in amphibia (Quay, 1969, 1970; Bagnara and Hadley, 1970).

2. Adenohypophysis

Histological and electron microscopic work has been done to clearly differentiate the cell types from each other and to study the effects of experimental conditions on these cells. In a review, Van Oordt (1968) summarized the histological results of several authors. There are two types of acidophilic cells and three types of basophils. These cell types are clearly distinguished in *Xenopus* by Kerr (1965). The acidophilic type appears identical with the α cells (Herland's nomenclature) and may produce STH. They are concentrated in the anterocentral region of the pars distalis. The second type of acidophil (η-cells of Herland's nomenclature) are suggested to secrete prolactin. They are mainly found in the dorsal caudal part of the lobe. The basophils type I are scattered throughout the lobe except the anterior part. They are quite large cells and are identical with δ cells. It is concluded from observations after thyroidectomy and inhibitors of thyroid function that they produce TSH. The second type of basophil is found in all areas of the lobe. They are by far the most numerous of the basophils (β cells). They do not arise until about 3–6 months after metamorphosis. Therefore, they are suggested to produce FSH (Kerr, 1966). The basophil type III (γ cells) are smaller. They are numerous in the area around the anterior border of the distal lobe. Kerr (1965) believed that they produced LH.

In other amphibians, these five cell types differ in shape and distribution. The staining capacity seems to be nearly the same in all species investigated (review, Van Oordt, 1968). In *Rana* species, acidophil type II could not be recognized. It may be that it is difficult to distinguish both types of acidophils in these species. The cell types of urodeles are also comparable with those of anurans (Mazzi *et al.*, 1966).

Experiments in *Rana temporaria* regarding the function of basophil type III (van Kemenade, 1969b, 1972, 1974; Larsen *et al.*, 1971) demonstrated that the rostral part of the adenohypophysis contains larger amounts of corticotropic activity than the caudal part. After treatment with metopirone, an inhibitor of interrenal activity, these basophils show signs of stimulation. Therefore, it is evident that this cell type produces ACTH. This is confirmed in *Xenopus* by studies from Evennett (1969). Investigations using immunofluorescent methods have also shown that ACTH-producing cells are located in the anterior region (Doerr-Schott and Dubois, 1972).

The different cell types show an annual cycle concerning changes of activity. In *R. temporaria,* the size of the nuclei of the basophil types II and III and acidophil type I is smallest in August. Maximal values can be found at February. During the breeding season in March, there is an extrusion of secretory granules. The seasonal changes correspond with the activity of the testis, the interrenal tissue, and the thyroid. This demonstrates that all endocrine glands of these frogs have an annual cycle with the highest activity before spawning and low activity during summer (Van Oordt *et al.,* 1968). The cytological changes in the cell types of *Rana esculenta,* especially the changes of the gonadotropin-producing cells during the reproductive cycle, after treatment with steroid hormones and gonadectomy have been described by Rastogi and Chieffi (1970a,b,c).

The ultrastructure of the different cells of the pars distalis varies with the cell type and the secretory activity of the cells. In the same stage of activity, the rough endoplasmic reticulum and the Golgi region look quite similar. However, the size and the shape of the granules are especially different, depending on the type of the cell. In some species, Doerr-Schott (1965, 1966, 1968) found spherical granules in the acidophils. These are larger in type I (200–500 nm) than in type II (100–300 nm). In basophil type II, the granules are often polymorphous.

Mira-Moser (1970, 1971) studied different cell types in the adenohypophysis of *Bufo bufo.* She could demonstrate six types; the elaboration of these depends on the fixative. The mode of synthesis and release of the secretory products may be characterized by intracytoplasmic lysis without destruction of the membrane of the granules. The glycoprotein-containing cells release granules after surgical thyroidectomy. Therefore, these cells are responsible for the production of TSH (these cells are called basophil II by the author). The cells undergo transformations into the so-called thyroidectomy cells, becoming chromophobic and hypertrophic.

In *Xenopus* larvae, Hemme (1972), Pehlemann and Hemme (1972), and Pehlemann (1974) demonstrated that the size and shape of the granules in TSH-producing cells (basophil type I) is not typical for the cell type. The diameter and the form of secretory granules vary according to the developmental stage and the activity of the cell. Treatment of the larvae (stage 51–57) with thiouracil increases the number of small, elongated granules. These granules increase in size in the older stages. Interrenalectomy changes the amount of the granules and the development of the ergastoplasma in the basophils. This was demonstrated in *R. temporaria* by Doerr-Schott and Dubois (1972) and Doerr-Schott (1970; 1974). These investigations gave evidence that the size and the shape of the granules vary and therefore cannot be used to identify the cell type.

The pars intermedia of amphibians contains one main cell type which is suggested to produce MSH. Morphological and ultrastructural changes occur in these cells in relation to different activity. Concentration or dispersion of the melanin in the melanophores are accompanied by these changes. Iturizza (1964) distinguished three cell types by electron microscopic studies of the pars intermedia. The function of these cells is not clear.

B. Branchiogenic Glands

Especially for the study of the amphibian ultimobranchial glands, a great deal of progress has been made during the last few years. In *Rana pipiens,* the glands are located laterally to the glottis. They are larger in males than in females. The glandular size changes throughout the year, increasing in winter time and decreasing to a minimum in summer. The secretory cells have been described by Robertson (1968a, b, 1971), who investigated the influences affecting secretory activity. Like the thyroids, the amphibian ultimobranchial glands have a follicular arrangement. Colloidal material appears within the cavity. But there is no incorporation of iodine. Nerve endings from the peripheral nervous system terminates on the basal portion of secretory cells. They may have an inhibitory influence on cytological activity and the release of the hormone.

Treatment of frogs with high calcium concentrations (the frogs were kept in 0.8% calcium chloride and injected with Vitamin D_2) causes glandular hypertrophy and cellular hyperplasia followed by a depletion of cytoplasmic granules. This demonstrates that the glands are the origin of a hypocalcemic factor, like the mammalian calcitonin. This has been isolated from bullfrog ultimobranchial glands by Copp *et al.* (1968). Removal of ultimobranchial glands in anuran larvae or adults results in depletion of lime sac calcium; that means reduction of the calcium stores (Robertson, 1969, 1971). A direct effect of the ultimobranchial secretory product on bone is not known. It has been suggested that calcitonin affects the kidney, because extirpation of the ultimobranchial glands caused a hypercalciuria in frogs. The knowledge of the structure of parathyroids in amphibia has been increased during the last few years (review: Cortelyou and McWhinnie, 1967; Lange and von Brehm, 1965; Rogers, 1965). In *Xenopus* (Coleman, 1969), the gland consists of closely packed epithelial cells surrounded by a capsule of connective tissue. Two cell types can be distinguished using basic dyes for staining—"dark" cells and "light" cells. During development, the number of dark and light cells is different; there are relatively more dark cells at younger stages (stage 54–59). The ultrastructure and the organelles of both cell types show no major differences. The material in the dark cells is more closely packed, and the matrix of the cytoplasma has greater electron density. These two cell types represent only different functional states of the same cell. The interpretation is controversial concerning which state is active or inactive. In *Xenopus,* the percentage of light cells increases at the end of metamorphosis. Therefore, it seems likely that light cells are more active, because calcification preferentially occurs during this period. Calcium treatment of toads increases the number of dark cells. Only small amounts of parathormone are released during a hypercalcaemic state.

The parathyroid cells in *Xenopus* do not show seasonal differences. In other anurans, however, seasonal changes have been reported (McWhinnie and Cortelyou, 1967; Boschwitz, 1967). Under cold acclimation or during winter, the metabolism of the bone and the phosphate excretion rate give evidences for a low parathormone level in the blood of *R. pipiens* and for glandular degeneration.

In *R. pipiens,* serum composition of intact, ultimobranchialectomized, parathyroidectomized frogs and frogs without both glands was examined by Robertson (1972). Frogs without parathormone were unable to mobilize calcium carbonate stores (deposits of the paravertebral lime sacs). This was investigated under normal and high carbon dioxide atmosphere.

The structure and function of the thyroid gland in amphibia is well known. The basic cellular architecture is that which is common to most vertebrates (Gorbman, 1964; Matty, 1966). Since the thyroid plays an important role during metamorphosis, histological and biochemical studies were done on the activity of the thyroid at different developmental stages (Saxén *et al.,* 1957; Etkin, 1967; Streb, 1967). Electron microscopic observations confirmed the histological results about the activity of the gland during development. The peak of follicular cell height coincides with maximal amounts of endoplasmic reticulum, maximal size of the Golgi apparatus, and a great number of mitochondria. Maximal activity judged by these parameters was found at the stage when tail resorption begins. This high activity is the response to spontaneous thyrotropic stimulation and can also be induced by injections of TSH.

Generally, the distribution of subcellular organelles is similar to that found in most other vertebrate thyroid cells (Coleman *et al.*, 1968a; report on the ultrastructure of the newt thyroid: Hearing and Eppig, 1969).

Treatment of *Xenopus* tadpoles or young toads with inhibitors of the thyroid (potassium chlorate, thiocyanate, thiourea, etc.) induces goiters. Goitrous cells accumulate "colloid" droplets and contain some membranous cytoplasmic inclusion bodies, the function of which is still unknown (Coleman *et al.*, 1968b,c).

C. Pancreatic Islets

Histological studies of the pancreatic islets demonstrated that the islets of both Anura and Urodela contain the different cell types usually found in other groups of vertebrates. It is also well known that this endocrine tissue produces two types of hormones, insulin and glucagon, which regulate carbohydrate metabolism. A survey of the older literature is given by Gorbman (1964), Bern and Nandi (1964), and Epple (1968).

Miller (1961) suggested that in urodeles, A cells, which produce glucagon are not present. Meanwhile other authors have shown that in most urodele species, A and B cells exist. The difficulties of finding A cells in islets of some urodele amphibians may be due to the histological technique used (Kern, 1962, 1966; Epple, 1966b). The islets of species of *Triturus, Ambystoma, and Salamandra* consist mainly of B cells, but by using various staining methods, a limited number of A and D cells could also be shown. The D cells or argyrophilic cells are suggested to produce a third pancreatic hormone (Epple, 1965a). For anurans the existence of A, B, and D cells is well established in *Rana catesbeiana* (Hellman and Hellerström, 1962), *Xenopus laevis* (Epple, 1965b), and *Bufo bufo* (Epple, 1966a). Seasonal changes are described in *Rana esculenta* (Poort and Genze, 1969).

Hypophysectomy did not affect islet cytology in toads despite changing carbohydrate and lipid metabolism. It is not absolutely clear whether this lack of dependancy changes with the season. But it is likely that the activity of the islet cells depends primarily on the metabolic background and not on other hormones (Epple *et al.*, 1966).

D. Mesodermal Glands

1. *Interrenal Organ*

The interrenal organ of Amphibia mainly produces corticosterone and aldosterone. 18-Hydroxycorticosterone was also detected. It is doubtful whether 17-hydroxylation is possible in interrenal tissue of adult amphibians. Therefore, cortisol might not be produced. It is not known whether the same enzymes work in larval tissue (Sandor, 1969, 1972; Chan *et al.*, 1969; Chan and Edwards, 1970; Mehdi and Carballeira, 1971a,b).

The microscopic anatomy and the histology of the interrenal tissue of Amphibia has been recently studied by some authors who tried to use histological, histochemical, and ultrastructural criteria to judge the activity or the degree of stimulation of the gland. The interrenal gland of Amphibia consists of cords of cells; groups of them are surrounded by connective tissue. Chromaffin cells are scattered between the interrenal cells. In some species of *Rana*, a special type of cells can be demonstrated, the so called Stilling (or summer) cells. Stilling cells in the typical form are present in *Rana esculenta* but absent in *Rana temporaria*. The function of these cells is still unknown. The ultrastructure resembles that of mesenchymic cells (Geyer, 1959). The Stilling cells are affected by hypophysectomy and exposure to hypertonic solutions (Scheer and Wise, 1969).

Seasonal changes have been described in the interrenal tissue of *R. temporaria* (Hanke and Weber, 1964; van Kemenade and van Dongen, 1965; van Kemenade *et al.*, 1968) and in the adrenal cortex cells of the bullfrog (Nakamura, 1967). In February, March, and April, the interrenal tissue of *R. temporaria* is characterized by large cells with large nuclei and nucleoli. The activity of steroid dehydrogenase, an enzyme that catalyzes the step from pregnenolone to progesterone in the biosynthesis of corticosteroids, is maximal in these months. The amount of lipids is rather small. This can be interpreted as high secretory activity. The characteristics of the interrenal cells in these frogs following the breeding season give evidences for a regression of activity. From July till December, the degree of stimulation increases. This cycle coincides with the activity rhythm of most other endocrine tissues.

The interrenal tissue of bullfrogs contains numerous summer cells in April and July. Especially in April, these cells show hypertrophy and contain big granules. When the summer cells are augmented, the number of lipid cells decreases. The author interpreted this correlation to mean that the lipid cells turn into summer cells. After the breeding season, regression of the corticosteroid-producing cells is correlated with an increase of the amount of summer cells. Hanke and Weber (1965) observed in *R. temporaria* mesenchymic cells, which they called migrating cells, especially in interrenal tissue during the period of regression. This favors the idea that these migrating cells are comparable to the Stilling cells and are involved in the tissue decomposition.

Mammalian ACTH injected into *R. temporaria* increases cell size and the volume of nucleus and nucleolus. The activity of steroid dehydrogenase is higher and the amount of lipids low in animals treated in that way. On the contrary, hypophysectomy reduced these criteria and increased the lipid content. This means that ACTH stimulates the interrenal tissue and removal of the pars distalis lowers the normal degree of stimulation (Hanke, 1966, 1967; van Kemenade, 1968a). The histological criteria were also used to demonstrate that metopirone or aldactone increase the ACTH output of the pituitary by inhibiting the release of corticosteroids by inducing metabolic reactions. High temperature (26°C) reduced the activity of the interrenal gland in frogs (van Kemenade, 1968b, 1969a,b).

The interrenal tissue of other amphibians (*Xenopus* and urodeles) resembles in structure and function that of *R. temporaria.* The same criteria for estimating the activity have been used in *Xenopus* tadpoles by Rapola (1962, 1963) and Leist *et al.* (1969). It could be shown that during development the highest activity can be found at the beginning of metamorphic climax.

Evidence for a pituitary–interrenal axis and a negative feedback mechanism between corticosteroids and the hypophysis has been demonstrated as well in the american bullfrog (*Rana catesbeiana*) by use of biochemical methods. Both corticosterone and aldosterone depress the amount of ACTH released by the pituitary (Piper and de Roos, 1967).

Changes in the ultrastructure of interrenal cells could also be correlated with different degrees of activity (Pehlemann and Hanke, 1968; Hanke and Pehlemann, 1969). In *R. temporaria* and *Xenopus laevis,* the most striking changes occur in the mitochondria. In unstimulated cells of hypophysectomized animals, the matrix of the mitochondria is very dense, the tubuli being packed closely together. In stimulated cells (normal or especially ACTH-treated animals) the matrix of the mitochondria is electron light and the diameter of the tubules is big. The surrounding membranes are often not visible. Concerning the degree of stimulation, variations occur in lipid droplets, amount of intercellular space, surface of cells, and the number and structure of lysosomes.

Ultrastructural studies on the interrenal cells of Amphibia regarding functional changes of cells have been reported in *R. catesbeiana* (Yoshimura and Harumiya, 1966, 1968; Volk, 1972), *Bufo arenarum* (Piezzi and Burgos, 1968), *Triturus cristatus* and *Salamandra salamandra* (Berchtold, 1969, 1970a,b), *Taricha torosa* (Bunt, 1969). There is one recent report describing, in *R. catesbeiana,* two different cell types producing different hormones, presumably aldosterone or corticosterone (Varma, 1971).

2. Endocrine Ovarian and Testicular Activity

It is well known that estrogens occur in Amphibia and are produced by the ovary (Chieffi and Lupo di Prisco, 1963; Gallien and Chalumeau-Le Foulgot, 1960). In addition, several investigators have found progesterone to be present. Work has been done about the biosynthetic pathways. There may be some—but only small—differences compared to the biosynthesis in mammals (Breuer and Ozon, 1965; Callard and Leathem, 1966; Ozon, 1969, 1972a; Eik-Nes, 1969). Testes and ovaries of adult vertebrates of all groups are capable of forming C_{19} steroids from pregnenolone or progesterone. The synthesizing activity can in most cases be demonstrated in larval tissue before metamorphosis.

The source of the two types of hormones is not absolutely clear. The follicular membranes of the growing oocyte contain enzymes that may enable this tissue to produce estrogens. But especially in *Xenopus* and *Bufo,* being best investigated in this respect, 3β-hydroxysteroid dehydrogenase could not be detected in this tissue (Chieffi and Botte, 1970; Joly, 1965). Treatment of adult female toads with pregnant mare serum gonadotropin stimulates estrogen synthesis by the ovary. The vitellogenesis is also enhanced. After treatment with PMS, the site of the hormone synthesis can be demonstrated using histochemical methods. The enzymes necessary are located in the follicular layer surrounding the oocytes and in the postovulatory follicles (Follett *et al.,* 1968; Redshaw and Nicholls, 1971). In *Salamandra,* both the corpora atretica and the corpora lutea gave a slightly positive reaction for steroid dehydrogenase. Therefore, it might be possible that preovulatory and postovulatory strucutures produce progesterone. A survey on the reaction of the ovary on steroids and the significance of hormones regulating sexual cycles was given by Barr (1968) and Redshaw (1972). The problem of luteogenesis has been discussed by Chieffi and Botte (1970).

Testosterone was identified in the plasma of the salamander *Necturus maculosus.* In testes of amphibia, the presence of C_{19} steroids is not established. There are some results that suggest the presence of progesterone. A survey about the metabolism of androgens is given by Ozon (1972b).

In Amphibia, the steroid-producing cells of testicular tissue show two different patterns of distribution; in the anuran testis, only intertubular Leydic cells produce the hormones (typical mammalian distribution), but in most urodeles, boundary cells are the source of hormones, too. These boundary cells are located in the lobule walls and are called "lobule boundary cells" (for references, see Lofts, 1968). The hormone-producing cells in the amphibian testis undergo seasonal changes of activity and variations in lipid and cholesterol content (Lofts, *et al.,*

1972). There is a correlation between the development of the secondary sex characters and the interstitial lipid cycle (Lofts, 1964).

The Sertoli cells in Amphibia also show definite cyclic changes that can be demonstrated histologically and histochemically. After spermiation these cells accumulate cholesterol-rich lipids. In *R. temporaria,* the amount of such lipoidal material is maximal at June. Afterwards, there is a release of the entire tubule generation of Sertoli cells. In other amphibians, the sudanophilic material remains for a longer time. The Sertoli cells possess the ultrastructural equipment for the production of steroids. Therefore, it is suggested that these cells take part in the production of androgens. Besides, their function and activity depend on the gonadotropin secretion by the pituitary, as can be demonstrated by histological changes. The precise mechanism of this secretion is still in question. (for review of the work done by Lofts, Van Oordt, and others, see Lofts, 1968).

II. Regulation of Metabolic Processes

A. Energy Metabolism

Carbohydrate metabolism in Amphibia is regulated by the same endocrine systems as in mammals—the hormones of the pancreatic islets, the ACTH and corticoids, and the TSH and thyroid hormones. Adrenaline and gonadal hormones are also involved in the regulation (for references, see Hanke and Neumann, 1972; Hanke, 1974).

Some investigations were done on the effects of insulin and glucagon on the blood glucose level and the amount of liver and muscle glycogen. In *Rana temporaria,* insulin lowers blood glucose to a very low level (about 5 mg/100 ml) when given in rather high doses (4 I.U./kg); 1 I.U./kg and lower doses are ineffective in this respect. It needs about 6 hours after injection to adjust blood glucose to the low level and the small amount of glucose is maintained for more than 24 hours. The same reaction occurs in other species, such as *Xenopus* (larval and postmetamorphic animals). There are three main differences between the insulin effect in mammals and amphibians: (1) relatively high doses are necessary, (2) there is a long reaction time, and (3) treatment results in very low blood glucose values despite which the animals can survive. Copeland and de Roos (1971) studying insulin effects in *Necturus maculosus* obtained similar results (for *Rana escculenta; Graul,* 1966a,b).

Glucagon treatment of *R. temporaria* (200 μg/kg) elevates blood glucose for a short time only. Maximal values (about 175%) are obtained 2–4 hours after the injection. About 6 hours after the injection, blood glucose is adjusted to the normal level. As for insulin, very high doses are necessary compared with mammals. The results suggest that insulin is more efficient in Amphibia than glucagon.

Corticosteroids induce gluconeogenesis in mammals by producing carbohydrates from proteins. Some authors conclude that corticosteroids work in the same way in Amphibia (Hunter and Johnson, 1960; Janssens, 1967; Hanke *et al.*, 1969). But there are other papers dealing with results about the activity of enzymes involved in gluconeogenesis which do not describe an increase of the enzyme activity after injection of corticosteroids (Chan and Cohen, 1964).

In *R. temporaria,* blood glucose increases after treatment with ACTH, corticosterone, and aldosterone. Hypophysectomy lowers the blood sugar level. The threshold dosage of ACTH in normal frogs is 0.04 I.U./kg; that of corticosterone and aldosterone, 0.4 μg/kg. Hypophysectomized frogs are more sensitive than normal animals. Since the level of blood glucose varies with the season and shows an annual cycle with two peaks (in March and in September), the effects to these hormones are different during the year; they depend on the metabolic background. It is important to point out that aldosterone and corticosterone are both effective regarding the elevation of blood glucose. This seems to be different from the data obtained in mammals (Hanke *et al.*, 1969). In *Rana catesbeiana,* plasma glucose also increases after injection with mammalian ACTH (Piper, 1969). In hypophysectomized *Bufo bufo,* physiological doses of ACTH cannot readjust the level of blood glucose to that found in normal animals. Only rather high doses of ACTH are effective in this respect (Hermansen and Jørgensen, 1969).

In *R. temporaria* the amounts of liver and muscle glycogen do not constantly increase after treatment with ACTH or corticosteroids during the year. In *Xenopus* tadpoles, however, and in postmetamorphic toads, augmentation of liver and tissue glycogen were demonstrated at all stages (from stage 55 onwards) after treatment with ACTH, corticosterone, or cortisol. In these animals, aldosterone is apparently less effective than the other corticosteroids. The discrimination between the different corticosteroids is very interesting. The results may indicate that aldosterone is not a natural hormone in these larvae (Hanke and Leist, 1971).

Gluconeogenesis must be accepted when an increase of blood glucose and liver and muscle glycogen is present. Summarizing the results shown in Amphibia, induction of gluconeogenesis by corticosteroids is only precisely demonstrated in a few living systems, such as tadpoles of *Xenopus.* Since the reaction depends on other hormone systems, the results are often not clear enough.

The effects of thyroxine on the metabolism of adult Amphibia are still doubtful. In *B. bufo,* Rinaudo *et al.* (1969) demonstrated, that radiothyroidectomy lowers the blood glucose level to nearly the half of what is found in normal toads. The rate of glycolysis in the liver is reduced. The activity of glucose-6-phosphatase is inhibited. Comparing the changes in tissue glycogen and oxygen consumption in adult *Rana pipiens,* McNabb (1969a) found that treatment with thyroxine decreased the amount of muscle glycogen. Changes of liver glycogen depend on the temperature. At 18°C, liver glycogen increased; there was no change at 30°C. An increase in oxygen consumption after thyroxine application was more pronounced in experiments done at 30°C than in those at 18°C. Some authors stress the opinion that thyroxine increases the oxygen consumption in amphibian tissue. Thornburn and Matty (1964, 1966). Marusic *et al.* (1966) and Taylor and Barker (1967), however, did not find changes in oxygen consumption in amphibian tissue after thyroxine treatment. Most investigations involving thyroxine have been done with tadpoles to determine the effects on development. Regarding the effects in larvae, it has been no clear distinction between primary effects on the metabolism and effects on differentiation that may secondarily affect metabolism.

Deposition of glycogen is favored by prolactin and growth hormone. The content of glycogen in the cardiac muscle of *Triturus cristatus* shows seasonal variations that correlate with fat body weight and the content of prolactin in the pituitary. Therefore, it has been suggested that in amphibians, prolactin and growth hormone induce a shift from the utilization of carbohydrates to that of lipids. This results in a decrease of the net amount of lipids and an increase of carbohydrates in animals treated with prolactin (Schalenghe *et al.,* 1968). The same correlation between the lipid content of the animals and the carbohydrates in blood and liver has been observed in *Xenopus* tadpoles. Hypophysectomy increases the amount of lipids in the animals and the size of the fat body. ACTH treatment lowers the amount of lipids and increases the carbohydrates in body fluid and organs (Gunesch, 1974; Hanke, 1974). This does not necessarily mean that lipids are converted to carbohydrates. An explanation of the results is that under the influence of ACTH or corticosteroids, lipids are metabolized preferentially.

In larval and adult *Xenopus,* cortisol and corticosterone decrease body weight. This is due to loss of water and catabolism of proteins, which can be demonstrated by a study of the excretion rate of ammonia or urea. Treatment of the animals with corticosteroids increases the amount of waste nitrogen in the urine (Janssens, 1967; Leist, 1970). Like carbohydrate metabolism, amino acid utilization depends on the thyroxine level. In thyroxine-treated *Rana pipiens,* urea excretion is reduced, but ammonia excretion does not change. This change may be due to a decrease in cellular permeability to amino acids and, therefore, reduced utilization of amino acids in the metabolic rate of the cells (McNabb, 1969b).

Estradiol induces a significant increase of lipid content of the blood. Also, a constant augmentation of the amount of plasma protein occurs for a longer period. This is correlated with the formation of yolk in the eggs. FSH diminishes the content of lipid in the blood, whereas the amount of protein remains unchanged. Progesterone and testosterone treatment do not have similar effects (Follett and Redshaw, 1968; Redshaw *et al.,* 1969; Wallace and Jared, 1969).

B. Osmomineral Regulation

The osmoregulatory system in Amphibia is very important because it represents a transitional stage between fishes and terrestrial vertebrates. Most hormonal systems of Amphibia are involved in these regulatory processes. The problems of these regulatory processes in Amphibia are discussed in some recent papers (Bentley, 1971a; Maetz, 1968, 1970; Chester Jones *et al.,* 1972; Scheer *et al.,* Chap. 3, this vol.). Special target organs that are useful for *in vitro* experiments are the skin and the bladder of amphibians, especially frogs and toads. The main hormonal systems that regulate the osmomineral content of Amphibia are neurohypophysial hormones and corticosteroids. Thyroid hormones are less effective, but some effects of these hormones are also described.

The osmoregulatory situation in Amphibia depends on the environment the animals live in. Amphibian larvae generally spend the whole time in water, usually in fresh water. Adult amphibians are adapted to very different environments which cause different degrees of dehydration. Some of them live in fresh water, like *Xenopus laevis* or the aquatic frog *Rana grylio.* Most of the Ranidae need a somewhat moist habitat. Other forms spend their adult lives in conditions that are very arid (some Bufonidae). A few species, like the crab-eating frog *Rana cancrivora,* survive very well in seawater. Larvae of some species can be adapted to salinities of about 1.2% sodium chloride (*Xenopus* larvae). Urodeles do not live in conditions as arid as those experienced by anurans. All these different habitats reflect, of course, that a different background exists regarding the reactions of hormones. The extent of hormonal effects must therefore be species specific, adapting to the needs of the animal.

The hormones of the pars nervosa of Amphibia are involved in the regulation of the water content of the animals. It is generally accepted that octapeptides of the pars nervosa exert a diuretic influence in teleosts but an antidiuretic effect in land-living tetrapods (Christensen and Jørgensen, 1972). The well known effect that amphibians kept in water and injected with neurohypophysial hormone retain water and increase body weight (Brunn effect) occurs to a different extent in different species (for different stages in *Rana catesbeiana,* see Alvarado and Johnson, 1966). There is a relation between the capacity of water retention and habitat. The terrestrial *Bufo regularis* shows maximal values of water retention, whereas the aquatic *X. laevis* has little ability to increase weight after injection of these hormones. Most urodeles do not show as striking effects as the more "terrestrial" anurans (Heller, 1965; Morel and Jard, 1968).

The released octapeptides affect the kidneys by increasing water and sodium reabsorption. Water absorption from the urine collected in the bladder is also enhanced. The water transfer across the anuran skin is also influenced by octapeptides. The skin of an aquatic species has a lower permeability than that of more terrestrial species (Schmid, 1965; Bentley, 1971a). Vasotocin, the most potent octapeptide, increases the permeability of anuran skin to water in terrestrial species. AVT is potent in this respect in different species of *Bufo* and *Rana.* It is not effective in *Xenopus, R. cancrivora,* and most urodeles, which all have a rather low degree of permeability of the skin compared with most bufonides and ranides. The octapeptides also affect the skin of tadpoles but to a lesser extent. The water retention increases after injection when metamorphosis approaches (Bentley and Greenwald, 1970). Neurohypophysial peptides also promote active sodium transport across frog skin, (Bentley, 1969; Fuhrmann and Ussing, 1951; Maetz, 1968). This effect can be demonstrated in a number of amphibian species, including *Triturus alpestris* and *X. laevis,* which do not show the hydroosmotic response. This demonstrates that the two effects on the skin mediated by octapeptides must not be linked.

The action of neurohypophysial hormones on the isolated amphibian skin, bladder, and kidney has been described by Maetz (1968). Numerous papers are concerned with the chemical nature of the octapeptides in Amphibia. It is now well established that vasotocin and mesotocin occur in a variety of anuran and urodele amphibians. Oxytocin has also been found in a few species. The amount of vasotocin present in the neurohypophysis is higher than the concentration of mesotocin. There is much more vasotocin in the neurohypophysis in terrestrial amphibians (*Bufo bufo, Triturus alpestris,* etc.) than in amphibious or aquatic animals (Heller, 1965; Bentley and Heller, 1965; Munsick, 1966; Acher *et al.,* 1968, 1969; Bentley, 1969, 1971a).

There is a special interest in the function of octapeptides in aquatic amphibians such as *Xenopus* and the crab-eating frog *R. cancrivora,* which lives in seawater. It can be expected that adaptation to water of different salinity is possible in these species because of release of different amounts of vasotocin. The neurohypophysis of *R. cancrivora* contains more hormonal activity when adapted to dilute seawater than when in fresh water (Dicker and Elliott, 1970). In *Xenopus* tadpoles kept in low sodium concentration (0.75%), most cells of the preoptic nuclei have lost their pseudoisocyanin-positive material after one or more days of hypertonic stimulation of the tadpoles (Notenboom, 1971). Preliminary results in our own laboratory have shown that *Xenopus* tadpoles gain weight (about 5%) after injection of vasotocin. This effect is more pronounced when the larvae are kept in hypertonic solution (0.9% salt solution) (K. Böke, unpublished).

Neurohypophysectomy has been done to establish the importance of the neurohypophysis for water economy in *Bufo*. But these experiments yielded conflicting results. Middler *et al.* (1967) found that in *Bufo marinus* the antidiuretic response to a salt load was reduced. Shoemaker and Waring (1968) did not find antidiuresis after destroying the preoptic nucleus. On the contrary, in *Bufo bufo*, Jørgensen *et al.* (1969) did not find a correlation between the response to dehydration and the completeness of eliminating the preoptic neurohypophysial system. Lesions in the preoptic region or neurohypophysectomy had little effect on the antidiuretic response. It could be that other systems, such as the interrenal tissue, compensate for the loss of neurohypophysial hormones in those toads.

The corticosteroids secreted by the amphibian interrenal tissue are mainly corticosterone and aldosterone. Both increase, *in vivo* and *in vitro,* the rate of sodium transport across the skin or the urinary bladder of anurans. Aldosterone is generally more potent regarding the natriferic action than other corticosteroids (Crabbé, 1964a,b; Sharp and Leaf, 1966, 1968). The aldosterone induced increase in active sodium transport by ventral skin or bladder can be reduced by inhibitors of protein synthesis (cycloheximid, actinomycin D). This demonstrates that aldosterone induces protein synthesis, and thus enhances the transport rate. A normal metabolic rate is necessary for a positive response to aldosterone (Edelman, 1969; Hutchinson and Porter, 1969; Porter, 1971).

The level of corticosteroids in the circulating blood correlates with the amount of sodium in the surroundings. As already mentioned, the secretion rate of corticosteroids is augmented by ACTH. In contrast to the correlation in mammals, this was also found for aldosterone. It may be that an extrapituitary control of interrenal function is exerted by a renin–angiotensin system. But the function and origin of this system is not precisely known (Johnston *et al.*, 1967). It is difficult, to look *in vivo* for effects due to a lack of corticosteroids. Removal of interrenal tissue is difficult because of the close contact between this gland and the kidney. Experiments done by Fowler and Chester Jones (reported by Chester Jones, 1957) have shown that the survival after cauterization of the adrenal gland depends on the season. Summer frogs die within a few days, winter frogs can survive. Interrenalectomized frogs lose sodium and accumulate potassium. The same effect has been found in mammals after adrenalectomy. These effects occur in both Anura and Urodela. Treatment of newts with aldosterone or cortisol causes retention of sodium and loss of potassium by the muscles (Ferreri *et al.*, 1967). Aldosterone is effective in *Necturus maculosus* by increasing sodium transport across the urinary bladder (Bentley, 1971b).

There are also effects of corticosteroids on the water content of the animals. *Xenopus* tadpoles lose water when treated with ACTH, corticosterone, or cortisol. Aldosterone is apparently less effective in this case. Another osmoregulatory process in *Xenopus* is the accumulation of urea in the body fluid which occurs after treatment with corticosteroids. This is important for the adaptation to higher osmolality in the environment. During the adaptation process, the interrenals are stimulated and the urea is accumulated, thus preventing water loss (Leist, 1970; Leist and Hanke, 1969).

There are other hormones involved in the regulatory system. In *Xenopus* tadpoles, the turnover rate of tritiated water decreases with the development. This could mean that the diffusional permeability to water decreases in older stages. Hypophysectomy of younger stages reduces the turnover rate. It is adjusted to the values characteristic for older larval stages. Prolactin treatment is able to restore this reduction caused by hypophysectomy (Schultheiss *et al.*, 1972). In newts, prolactin induces the water drive effect. This is the stimulation of the second metamorphosis after the terrestrial eft stage, which is accompanied by structural changes of the skin, development of mucous glands, etc. This enables the animals to return to an aquatic environment by changing the permeability of the skin.

The effect of thyroid hormones on sodium transport and water permeability of the anuran skin and bladder is controversial. Matty and Green (1962a,b, 1964) and Green and Matty (1962, 1963) were able to show that thyroid hormone increases the transfer through these tissues. Other authors could not confirm these results (Taylor and Barker, 1965, 1967). This demonstrates that the effect is variable and may depend on other parameters, such as season, food consumption, or altered hormone levels. Since the water efflux changes during development as mentioned above, thyroid hormones that promote development must also regulate the water content of the animal. Nevertheless, this effect is suggested not to be a primary one.

III. Regulation of Development and Metamorphosis

A. Influences on Metamorphosis

Endocrine studies on amphibian metamorphosis have been reviewed recently (Etkin, 1968, 1970). In particular, the effects of thyroxine to promote differentiation are well known (for references, see Frieden, 1961, 1967; Weber, 1967; Eaton and Frieden, 1969; Tata, 1969; Weber *et al.*, 1974). There are many biochemical and morphological changes during metamorphosis. All these changes can be induced by treatment of competent stages with triiodothyronine or thyroxine. The reaction mostly depends on the stage. But in all cases, induction of morphological or biochemical events is correlated with the promotion of development. This shows that the changes are induced secondarily [references for the induction of urea excretion by thyroid and other hormones are in Balinsky *et al.* (1967), Ashley *et al.* (1968), and Medda and Frieden (1970, 1972)].

Changes of activity of the thyroid during development are very important for a normal progress of metamorphosis (Saxén *et al.*, 1957; Kaye, 1961; and others). The activity, which is judged by histological methods and estimation of ^{131}I uptake, is constant at a low level during premetamorphosis. In these stages the thyroid develops from the anlage and forms larger follicles. The stages of prometamorphosis are characterized by an increasing activity of the thyroid and an increasing level of thyroid hormones in the blood. The activity reaches maximal values at the beginning of metamorphic climax or during the first stages of metamorphic climax. Then the activity decreases to the level of postmetamorphic animals.

There are two theories to explain the rhythm of activity of the thyroid observed during metamorphosis of amphibian tadpoles. In one of them (Etkin, 1963, 1966, 1968, 1970), the hypothalamic TRF mechanism becomes sensitive to thyroxine just before prometamorphosis starts. This initiates prometamorphosis, during which a positive feedback mechanism regulates the activity. This means that increasing amounts of thyroxine elevate the release of TRF and TSH. The positive feedback cycle leads to a maximal activation of the pituitary–thyroid axis. Correlated with this increase of activity of the thyroid, the production of prolactin, which may inhibit thyroid activity, is reduced to nearly zero level at the beginning of metamorphic climax. During premetamorphosis the prolactin level is rather high. The growth rate of the animals depends on this hormone. With the beginning of metamorphic climax, the positive feedback mechanism changes to a negative feedback by which the normal regulation of the thyroid during the whole life occurs.

According to the second theory (Goos, 1968, 1969b; Goos *et al.*, 1968a,b), it is not necessary to explain the facts by a shift from a positive feedback to a negative feedback during development. The thyroid hormone exerts a promotion of differentiation in all parts of the animal. The median eminence, the hypothalamus, and also the vessels that connect the central nervous system and the pituitary differentiate only in the presence of thyroid hormone (Etkin *et al.*, 1965; Etkin, 1966). The correlation between TRF, TSH, and thyroid hormone before metamorphic climax is determined by this morphogenetic effect of thyroid hormones on the hypothalamus. These hormones increase the number of cells and activate the influence the hypothalamic factors exert on the pituitary. Therefore, it now seems that thyroid hormones stimulate TRF- and TSH-release. Goos (1968, 1969a,b) and Goos and co-workers (1968a,b) removed the frontal brain, kept animals in propythiouracil, and treated animals with thyroxine after inhibition of the thyroid. Then they looked for the TSH-producing cells and pseudoisocyanine-positive cells in the hypothalamus and compared the activity of these cells with the activity of the thyroid. They drew the conclusion that, in larvae of *Xenopus*, TRF is formed in cells in the dorsal preoptic nucleus.

The antithyroid action of prolactin (Etkin and Gona, 1967; Gona, 1967; Etkin *et al.*, 1969) was demonstrated by treating tadpoles of *Rana pipiens* with mammalian prolactin. It could be shown that these animals grew faster. Metamorphic activity was inhibited. This activity has been ascribed to the goitrogenic properties of this hormone. Rather high doses of prolactin (daily doses of 0.1 mg) are necessary to stop metamorphosis at climax. Pituitary transplants placed in the tail fin during prometamorphosis reduced the rate of tail fin reduction, especially in the vicinity of the transplant. Grafting near the thyroid exerts more influence. This suggests that there are two influences, one directly on the tail fin tissue, another transmitted by the thyroid. It is concluded from other studies that prolactin interferes with the thyroidal iodine-trapping mechanism in tadpoles (Gona, 1968). This effect on the thyroid may be typical for larval anurans. In postmetamorphic *Triturus cristatus carnifex*, prolactin enhances thyroid activity by an influence on the hypothalamopituitary axis, resulting in TSH hypersecretion (Vellano *et al.*, 1967, 1969).

The direct influence of hormones on larval tissue is very important for the explanation of metamorphic events. The isolated tadpole tail is capable of surviving in a culture medium. This technique, originally designed by Shaffer (1963), has been used by Weber (1967, 1969) and Etkin and co-workers (Derby, 1968; Derby and Etkin, 1968; Etkin, 1969; Gona, 1969; Etkin and Kim, 1970, 1971). Thyroid hormone causes tissue involution when added to the culture medium. This destruction is correlated with a loss in tissue protein and an augmentation of various lysosomal enzymes. Inhibitors of protein synthesis are capable of abolishing hormone-induced tissue breakdown. Protein synthesis is, therefore involved in the local changes of tail tissue.

Tail disks change their sensitivity to thyroxine at progressive stages of development. There is a direct quantitative relationship between the concentration of thyroid hormone and tissue breakdown. Treatment of the tail disks with hormones of the anterior pituitary and implantation of pituitary grafts in the disks exposed to thyroid hormone result in a reduction of the rate of tissue involution. Mammalian prolactin and STH have similar activity as the pituitary but are less effective compared to a graft. There is an antagonism between a pituitary growth factor and thyroid hormone in the tissue cells.

B. Regulation of Growth

Prolactin and STH have been considered to promote growth in Amphibia. The somatotropic activity of both hormones may be different and depends on the ontogenetic stage. It is not known whether prolactin exists as a distinct molecular entity or whether it forms a complex with STH and other adenohypophysial hormones. It could be shown that the prolactin of amphibians is a prominent protein and cannot be distinguished from prolactins of other vertebrate groups by using polyacrylamide disk gel electrophoresis. Most prolactins have relatively high electrophoretic mobilities (Nicoll and Nichols, 1971).

In the previous section the growth promoting activity and the antithyroid action of prolactin in anuran larvae were discussed. Preparations of mammalian prolactin and STH have, in most cases, different activities concerning promotion of growth (Berman *et al.*, 1964; Nicoll *et al.*, 1965; Bern *et al.*, 1967; Enemar *et al.*, 1968; Brown and Frye, 1969a,b; Frye *et al.*, 1972). Mammalian prolactin is much more effective in tadpoles, whereas growth hormone may be the major growth factor in adult anurans. Doses of mammalian prolactin of 1–50 μg/day promote growth in *R. pipiens* tadpoles, whereas more than 50 μg/day of mammalian STH are necessary for the same effect. In postmetamorphic *R. pipiens*, 10–50 μg/day of STH promote growth given for a 2 month period: the same doses of prolactin were not effective. Growth was determined by changes in length and weight. The changes in weight were more pronounced.

Meier (1969) reported diurnal variations of metabolic responses to prolactin, especially regarding the lipid metabolism. Therefore, it is possible that experiments with prolactin and STH give different results when the daily time of hormone administration is changed. Besides, there may be an influence of these hormones on food consumption which secondarily influences the growth rate. Careful studies with toads (*Bufo boreas* and *Bufo marinus*) (Zipser *et al.*, 1969) were done observing the effects of different food consumption. In this case, it was shown that low doses of STH (0.3–0.6 μg/day) increase body weight and femur elongation and reduce lipid and glycogen storage. This increases food consumption at the same time. Prolactin does not influence the metabolism and the growth of these animals except at relatively high doses (50–100 μg/day). Injections given at different times of day did not change the results.

Treatment with prolactin and growth hormone administered to the red-spotted newt, *Notophthalmus viridescens* (Brown and Brown, 1971), gave the same results. Daily doses of 4–20 μg of growth hormone cause a significant increase in length and weight. Prolactin given in the same doses does not significantly change body weight, although a similar increase in length was observed. While prolactin increases both liver weight and the amount of liver carbohydrates, STH was without any effect on the liver. In *T. c. carnifex*, high doses of prolactin (100 I.U. per animal) increase the height of the tail (Vellano *et al.*, 1970b). This effect may be correlated with changes in the sexual behavior, like triggering water drive, which is well known as prolactin-dependent (Grant, 1966).

Summarizing all these results, it is likely that both hormones—STH and prolactin—are different entities in amphibians. Prolactin essentially promotes growth in tadpoles by exerting antithyroid influence, and STH affects growth of postmetamorphic animals. Length and weight are two parameters of growth that may be affected differently. Metabolic changes are involved that are not completely understood.

C. Induction of Molting

Molting of amphibian skin is a process which consists of different phases. It has been shown in toads that, at first, the old stratum corneum is separated and loosened from the underlying cells. This process requires 1–2 days. The second phase is the shedding of the slough. After hypophysectomy in *Bufo bufo*, a new slough is produced within 2 or 3 days, but the shedding process is inhibited. The premature formation of a slough after hypophysectomy can be further accelerated by treatment with ACTH or corticosteroids. A complete molt could be produced within about 6–9 hours after ACTH injection. This treatment is only successful done more than 17 hours after the operation. It is suggested that, in normal animals, a shedding center of the central nervous system regulates shedding behavior which is necessary for normal shedding of the slough in toads. This shedding center may be primed by the process of separation of the stratum corneum from underlying cells.

ACTH or corticosteroids are the only hormones in toads that induce molting. Corticosterone is more effective than aldosterone. Nevertheless, it is not possible to produce the normal molt rhythm by injection of ACTH or corticosteroids. This shows that natural rhythm of interrenal secretion is necessary (Jørgensen and Larsen, 1960, 1961, 1964). Also, rhythmic processes exist which are inherent in the skin. Although Clark and Kaltenbach (1961) have found in newts (*Diemictylus viridescens*) that thyroxine induces molting, Jørgensen *et al.* (1965) have demonstrated in *B. bufo* that molting is normal after complete radiothyroidectomy.

There is evidence that molting is not completely understood. In crested newts, the integument is changed after hypophysectomy, and molting is abolished. Prolactin treatment causes shedding of cornified laminae. The shedding areas excepted, the skin is smooth as in normal animals (Vellano *et al.*, 1970a).

In *Rana temporaria* molting differs from this process in toads. There is no complete slough formed, but a lot of small pieces are shed. Hypophysectomized frogs shed much more of those small peices than normal frogs. Treatment with ACTH or corticosteroids does not change the amount of the shed areas. Regarding the hormonal regulation, it can be assumed that shedding of frogs is comparable to the first phase of molting in toads, that is, the process of separation (W. Hanke and co-workers, unpublished results).

IV. Regulation of Gonadal Function and Reproduction

The pattern of ovarian activity (Barr, 1968) and testicular activity (Lofts, 1968) in Amphibia have been described recently. Therefore it is not necessary to repeat the important facts. Both these reviews deal with the hypophyseal–ovarian relationship and the endocrinology of the ovary. The pituitary control of testis function, the interstitial cell activity, and the hormonal intratesticular regulations are reported for all lower vertebrates.

The effects of progesterone and related steroids or gonadotropic hormones can be demonstrated by the meiotic division of *Xenopus* oocytes *in vitro*. The meiotic division takes place shortly before ovulation. These oocytes are responsive to steroid hormones, especially progesterone. Gonadotropic hormones, like LH which stimulate the release of progesterone or a progesterone-like principle, also induce meiosis of oocytes. This has been found in *Rana pipiens* (Masui, 1967; Schuetz, 1967a,b), *Bufo bufo* (Thornton and Evennett, 1969), and *Xenopus laevis* (Thornton, 1971). *Xenopus* oocytes are suitable for a biological assay for progesterone. Test oocytes are sensitive to about 10 ng/ml of progesterone in the incubation medium. The test is useful for detecting progesterone in plasma samples. Using this test, it was possible to determine LH activity in the pituitary of *Xenopus* between 1 and 3½ months after metamorphosis. The activity is distributed throughout the whole pituitary and may correspond with the distribution of basophil type II. Since there is no clear indication that two different gonadotropic hormones (FSH and LH) exist in Amphibia, it is likely that basophil type II is responsible for the whole gonadotropic function (Evennett and Thornton, 1971).

V. Regulation of Color Change

Color change is very important for the adaptation of amphibians. Therefore, the hormonal regulation of this process has been studied very often. There are two main problems: (1) the regulation of the release of hormones that affect the chromatophores and (2) the cellular aspects, i.e., the intracellular action that causes movement of melanin and other granules.

1. Hormone Release

MSH produces dispersion of melanin in the skin melanophores. It is released when an animal is illuminated above a dark background; then the animal darkens. The release of MSH, which is synthesized in the pars intermedia, must be discussed under the aspects described in Section I,A,1 (nervous control of the pars intermedia). It is not known whether there are hormones inducing concentration of the melanin granules and paling, which occurs on a light background. Melatonin released by the pineal and catecholamines have been discussed to be paling hormones. After injection of melatonin, dark *Rana pipiens* get pale, but it is not clear whether this effect is a physiological one (Novales and Novales, 1965a, Kastin and Schally, 1966). Though extracts of mammalian pineal glands punctate melanophores at a very low concentration (10^{-10} gm/ml), it is still doubtful whether the pineal plays an important role by releasing melatonin in Amphibia.

Melatonin is present in *Xenopus laevis* (van de Veerdonk, 1967). A fivefold increase of the level was observed just before the onset of active larval swimming (Baker, 1969). But this does not mean that melatonin is exclusively formed within the pineal. There is evidence that the compound is synthesized in the eye and is involved in the photochemical reaction of retinal cells, which is important at the time of first larval swimming. There is another opinion that melatonin forming enzyme is more broadly distributed in the tissues of lower vertebrates (Quay, 1970).

Eyeless larvae placed in darkness show the same paling reaction as normal larvae. Since the pineal has photoreceptors and endocrine function, it seems reasonable to discuss the responsibility of the pineal for this response. The effect of pinealectomy did not give significant results. Therefore, it is difficult to decide where the blanching substance is produced (Bogenschütz, 1967; for references, see Bagnara and Hadley, 1970). It can be expected that there is a direct influence on the melanophores by a blanching agent and not an indirect effect via pituitary, which causes concentration of the pigment granules.

2. *Cellular Action*

Since MSH increases the amount of cyclic AMP in dorsal frog skin, it has been suggested that the action on MSH is mediated by cyclic AMP (Robison *et al.*, 1968). If this is right, MSH must activate adenylcyclase in chromatophores. Cyclic AMP disperses melanin granules in amphibian melanophores (Bitensky and Burstein, 1965; Novales and Davis, 1967, 1969). The effect of cyclic AMP was observed for different melanophores in tissue culture, in dorsal skin of normal *Rana pipiens,* and in the foot web of hypophysectomized frogs. A positive reaction of melanophores on MSH depends on calcium (Novales and Novales, 1965b).

The effect of cyclic AMP or MSH is mediated by β-receptors. β-Adrenergic stimulation produces an increase in the content of cyclic AMP. There are some species in which β-receptors of melanophores can be stimulated by catecholamines as well as by MSH. In these cases, catecholamines disperse the granules. This is the case in *Xenopus laevis* (Graham, 1961; van de Veerdonk and Konijn, 1970) and *Scaphiopus couchi* (Goldman and Hadley, 1969). If catecholamines induce concentration of the pigment granules as in *Rana tigrina,* α-receptors mediate this effect (Gupta and Bhide, 1967). This hypothesis about different types of receptors being involved in the mechanism is supported by experiments with blocking agents (propranolol as a specific β-blocking agent, dibenamine blocks α-receptors).

The action of MSH on melanophores requires sodium, which cannot be replaced by other cations. This supports a hypothesis in which MSH is supposed to act by increasing the sodium content of the melanophore. The entry of sodium triggers the melanin dispersion. Sodium entry lowers the membrane potential of the cell and therefore produces a gradient down which the melanin granules might move (Novales *et al.*, 1962).

In addition to the action of hormones on melanophores, it is now well known that the bright-colored pigment cells of amphibians are also influenced by hormones. Iridophores, which contain purine derivatives and reflect light, concentrate their reflecting platelets after treatment with MSH. This supports the adaptation process. When the pigment of melanophores is dispersed, that of iridophores (= guanophores) is concentrated. According to studies of Hadley (1966), Taylor (1966, 1967, 1969), and Bagnara *et al.* (1969), it is not absolutely clear whether the mechanism of pigment migration in both types of pigment cells is identical. MSH treatment of larvae of *R. pipiens* is followed by a disappearance of some reflecting platelets. They seem to be replaced by droplets or vesicles. Such changes causes a "morphological" response of chromatophores. It shows that "morphological" color change is associated with a "physiological" response in so far as it might be a secondary effect.

It is accepted that xanthophores, which contain pteridines and carotinoids, are little involved in physiological color change of Amphibia. In the skin of *Hyla arenicolor*, MSH treatment causes dispersion of the granules (Bagnara *et al.*, 1968). This may be an exception in Amphibia. The content of pigment in an amphibian chromatophore is not constant. Removal of the pituitary reduces the amount of pteridines and carotinoids. Therefore, these cells take part in "morphological" color change as well (for references, see Bagnara *et al.*, 1969; Bagnara and Hadley, 1969). Melanophores, iridophores, and xanthophores are arranged in a dermal chromatophore unit (Bagnara *et al.*, 1968, 1969), which is located just below the basal membrane. Processes of the melanophores terminate in fingers between the other two chromatophores. If the melanin pigment is equally distributed within the cell, xanthophores and iridophores are covered. The color of the animal depends on the contribution which is made by each of the three types of chromatophores.

The "morphological" color change of melanophores in *Xenopus* tadpoles is affected by MSH and removal of the pituitary. The division rate of melanophores increases when the animals are kept on a black background. Grafting of pituitaries into the larvae also increases the division rate. Those grafts are uninhibited by MIF, which only works in the normally located intermediate lobe. Hypophysectomy inhibits the division of melanophores. The results demonstrate that the division rate depends on the level of MSH released from the intermediate lobe. The amount of melanoblasts which differentiate to melanophores in a definite time depends on the number of melanophores per area. Extreme augmentation of the division rate inhibits the differentiation of new melanophores. The rate of differentiation increases when the animals are kept in darkness (Pehlemann, 1967b).

REFERENCES

Acher, R., Chauvet, J., and Chauvet, M. T. (1968). *Biochim. Biophys. Acta* **154**, 255.

Acher, R., Chauvet, J., and Chauvet, M. T. (1969). *Nature (London)* **221**, 759.

Alvarado, R. H., and Johnson, S. R. (1966). *Comp. Biochem. Physiol.* **18**, 549.

Ashley, H., Katti, P., and Frieden, E. (1968). *Develop. Biol.* **17**, 293.

Bagnara, J. T., and Hadley, M. E. (1969). *Amer. Zool.* **9**, 465.

Bagnara, J. T., and Hadley, M. E. (1970). *Amer. Zool.* **10**, 201.

Bagnara, J. T., Taylor, J. D., and Hadley, M. E. (1968). *J. Cell Biol.* **38**, 67.

Bagnara, J. T., Hadley, M. E., and Taylor, J. D. (1969). *Gen. Comp. Endocrinol., Suppl.* **2**, 425.

Baker, P. C. (1969). *Comp. Biochem. Physiol.* **28**, 1387.

Balinsky, J. B., Choritz, E. L., Coe, C. G. L., and van der Schans, G. S. (1967). *Comp. Biochem. Physiol.* **22**, 53.

Barr, W. A. (1968). *In* "Perspectives in Endocrinology" (E. J. M. Barrington, and C. B. Jørgensen, eds.), pp. 163–237. Academic Press, New York.

Bentley, P. J. (1969). *Gen. Comp. Endocrinol.* **13**, 39.

Bentley, P. J. (1971a). "Endocrines and Osmoregulation." Springer-Verlag, Berlin and New York.

Bentley, P. J. (1971b). *Gen. Comp. Endocrinol.* **16**, 356.

Bentley, P. J., and Greenwald, L. (1970). *Gen. Comp. Endocrinol.* **14**, 412.

Bentley, P. J., and Heller, A. (1965). *J. Physiol. (London)* **181**, 124.

Berchtold, J.-P. (1969). *Z. Zellforsch Mikrosk. Anat.* **102**, 357.

Berchtold, J.-P. (1970a). *Z. Zellforsch. Mikrosk. Anat.* **110**, 517.

Berchtold, J.-P. (1970b). *C. R. Acad. Sci.* **270**, 626.

Berman, R. H., Bern, H. A., Nicoll, C. S., and Strohman, R. C. (1964). *J. Exp. Zool.* **156**, 353.

Bern, H. A., and Nandi, J. (1964). *In* "The Hormones" (G. Pincus, K. V. Thimann, and E. B. Astwood, eds.), Vol. 4, pp. 199–298. Academic Press, New York.

Bern, H. A., Nicoll, C. S., and Strohman, R. C. (1967). *Proc. Soc. Exp. Biol. Med.* **126**, 518.

Bitensky, M. W., and Burstein, S. R. (1965). *Nature (London)* **208**, 1281.

Bogenschütz, H. (1967). *Experientia* **23**, 967.

Boschwitz, D. (1967). *Isr. J. Zool.* **16**, 46.

Breuer, H., and Ozon, R. (1965). *Arch. Anat. Microsc. Morphol. Exp.* **54**, 17.

Brown, P. S., and Brown, S. C. (1971). *J. Exp. Zool.* **178**, 29.

Brown, P. S., and Frye, B. E. (1969a). *Gen. Comp. Endocrinol.* **13**, 126.

Brown, P. S., and Frye, B. E. (1969b). *Gen. Comp. Endocrinol.* **13**, 139.
Büchmann, N. B., Spies, I., and Vijayakumar, S. (1972). *Gen. Comp. Endocrinol.* **18**, 306.
Bunt, A. H. (1969). *Gen. Comp. Endocrinol.* **12**, 134.
Callard, I. P., and Leathem, J. H. (1966). *Gen. Comp. Endocrinol.* **7**, 80.
Chan, S. K., and Cohen, P. P. (1964). *Arch. Biochem. Biophys.* **104**, 335.
Chan, S. T. H., and Edwards, B. R. (1970). *J. Endocrinol.* **47**, 183.
Chan, S. W. C., Vinson, G. P., and Phillips, J. G. (1969). *Gen. Comp. Endocrinol.* **12**, 644.
Chester Jones, I. (1957). "The Adrenal Cortex." Cambridge Univ. Press, London and New York.
Chester Jones, I., Bellamy, D., Chan, D. K. O., Follett, B. K., Henderson, I. W., Phillips, J. G., and Snart, R. S. (1972). *In* "Steroids in Nonmammalian Vertebrates" (D. R. Idler, ed.) pp. 414–480, Academic Press, New York.
Chieffi, G., and Botte, V. (1970). *Boll. Zool.* **37**, 85.
Chieffi, G., and Lupo di Prisco, C. (1963). *Gen. Comp. Endocrinol.* **3**, 149.
Christensen, C. U., and Jørgensen, C. B. (1972). *Gen. Comp. Endocrinol.* **18**, 169.
Clark, N. B., and Kaltenbach, J. C. (1961). *Gen. Comp. Endocrinol.* **1**, 513.
Coleman, R. (1969). *Proc. Int. Symp. Calcitonin, 2nd, 1969* pp. 348–358.
Coleman, R., Evennett, P. J., and Dodd, J. M. (1968a). *Gen. Comp. Endocrinol.* **10**, 34.
Coleman, R., Evennett, P. J., and Dodd, J. M. (1968b). *Z. Zellforsch. Mikrosk. Anat.* **84**, 490.
Coleman, R., Evennett, P. J., and Dodd, J. M. (1968c). *Z. Zellforsch. Mikrosk. Anat.* **84**, 497.
Compher, M. K., and Dent, J. N. (1970). *Gen. Comp. Endocrinol.* **14**, 141.
Copeland, D. L., and deRoos, R. (1971). *J. Exp. Zool.* **178**, 35.
Copp, D. H., Low, B. S., O'Dor, R. K., and Parkes, C. O. (1968). *Calcif. Tissue Res., Suppl.* **2**, 29.
Cortelyou, J. R., and McWhinnie, D. J. (1967). *Amer. Zool.* **7**, 843.
Crabbé, J. (1964a). *In* "Hormones and the Kidney" (P. C. Williams, ed.), pp. 75–87. Academic Press, New York.
Crabbé, J. J. (1964b). *Endocrinology* **75**, 809.
Dent, J. N., and Gupta, B. J. (1967). *Gen. Comp. Endocrinol.* **8**, 273.
Derby, A. (1968). *J. Exp. Zool.* **168**, 147.
Derby, A., and Etkin, W. (1968). *J. Exp. Zool.* **169**, 1.
Dicker, S. E., and Elliott, A. B. (1970). *J. Physiol.* (*London*) **207**, 119.
Dierickx, K. (1965). *Z. Zellforsch. Mikrosk. Anat.* **66**, 504.
Dierickx, K. (1966). *Z. Zellforsch. Mikrosk. Anat.* **74**, 53.
Dierickx, K. (1967a). *Z. Zellforsch. Mikrosk. Anat.* **77**, 188.
Dierickx, K. (1967b). *Z. Zellforsch. Mikrosk. Anat.* **78**, 114.
Dierickx, K., and Goossens, N. (1970). *Z. Zellforsch. Mikrosk. Anat.* **106**, 371.
Doerr-Schott, J. (1965). *Gen. Comp. Endocrinol.* **5**, 631.
Doerr-Schott, J. (1966). *Ann. Endocrinol.* **27**, 101.
Doerr-Schott, J. (1968). *Annee. Biol.* **7**, 189.
Doerr-Schott, J. (1970). *Z. Zellforsch. Mikrosk. Anat.* **111**, 413.
Doerr-Schott, J. (1972). *Gen. Comp. Endocrinol.* **18**, 587.
Doerr-Schott; J. (1974). *Fortschr. Zool.* **22**, 245.
Doerr-Schott, J., and Dubois, M. P. (1970). *C. R. Acad. Sci.* **271**, 1534.
Doerr-Schott, J., and Follenius, E. (1969). *C. R. Acad. Sci.* **269**, 737.

Doerr-Schott, J., and Follenius, E. (1970a). Z. *Zellforsch. Mikrosk. Anat.* **106,** 99.

Doerr-Schott, J., and Follenius, E. (1970b). Z. *Zellforsch. Mikrosk. Anat.* **111,** 427.

Eaton, J. E., and Frieden, E. (1969). *Gen. Comp. Endocrinol., Suppl.* **2,** 398.

Edelman, I. S. (1969). *In* "Renal Transport and Diuretics," (K. Thurau and H. Jahrmärker, eds.), pp. 139–151. Springer-Verlag, Berlin and New York.

Eik-Nes, K. B. (1969). *Gen. Comp. Endocrinol., Suppl.* **2,** 87.

Enemar, A., and Falck, B. (1965). *Gen. Comp. Endocrinol.* **5,** 577.

Enemar, A., Falck, B., and Iturriza, F. C. (1967). Z. *Zellforsch. Mikrosk. Anat.* **77,** 325.

Enemar, A., Essvik, B., and Klang, R. (1968). *Gen. Comp. Endocrinol.* **11,** 328.

Epple, A. (1965a). *Gen. Comp. Endocrinol.* **5,** 674.

Epple, A. (1965b). *Zool. Anz., Suppl.* **29,** 459.

Epple, A. (1966a). *Gen. Comp. Endocrinol.* **7,** 191.

Epple, A. (1966b). *Gen. Comp. Endocrinol.* **7,** 207.

Epple, A. (1968). *Endocrinol. Jap.* **15,** 107.

Epple, A., Jørgensen, C. B., and Rosenkilde, P. (1966). *Gen. Comp. Endocrinol.* **7,** 197.

Etkin, W. (1963). *Science* **139,** 810.

Etkin, W. (1966). *Neuroendocrinology* **1,** 293.

Etkin, W. (1967). *In* "Neuroendocrinology" (L. Martini, and W. F. Ganong, eds.) Vol. 2, pp. 261–282, Academic Press, New York.

Etkin, W. (1968). *In* "Metamorphosis: A Problem in Developmental Biology" (W. Etkin and L. I. Gilbert eds.), pp. 314–348. Appleton, New York.

Etkin, W. (1969). *Gen. Comp. Endocrinol.* **13,** 504.

Etkin, W. (1970). *Mem. Soc. Endocrinol.* **18,** 137.

Etkin, W., and Gona, A. G. (1967). *J. Exp. Zool.* **165,** 249.

Etkin, W., and Kim, Y. S. (1970). *Amer. Zool.* **10,** 321.

Etkin, W., and Kim, Y. S. (1971). *Amer. Zool.* **11,** 654.

Etkin, W., Kikuyama, S., and Rosenbluth, J. (1965). *Neuroendocrinology* **1,** 45.

Etkin, W., Derby, A., and Gona, A. G. (1969). *Gen. Comp. Endocrinol. Suppl.* **2,** 253.

Evennett, P. (1969). *Gen. Comp. Endocrinol.* **13,** 504.

Evennett, P. J., and Thornton, V. F. (1971). *Gen. Comp. Endocrinol.* **16,** 606.

Ferreri, E., Mazzi, V., Socino, M., and Scalenghe, F. (1967). *Gen. Comp. Endocrinol.* **9,** 10.

Follett, B. K., and Redshaw, M. R. (1968). *J. Endocrinol.* **40,** 439.

Follett, B. K., Nicholls, T. J., and Redshaw, M. R. (1968). *J. Cell Comp. Physiol.* **72,** 91.

Frieden, E. (1961). *Amer. Zool.* **1,** 115.

Frieden, E. (1967). *Recent Progr. Horm. Res.* **27,** 139.

Frye, B. E., Brown, P. S., and Snyder, B. W. (1972). *Gen. Comp. Endocrinol. Suppl.* **3,** 209.

Fuhrmann, F. A., and Ussing, H. H. (1951). *J. Cell Comp. Physiol.* **38,** 109.

Gallien, L., and Chalumeau-Le Foulgot, M. T. (1960). *C. R. Acad. Sci.* **251,** 460.

Geyer, G. (1959). *Acta Histochem.* **8,** 234.

Goldman, J. M., and Hadley, M. E. (1969). *Gen. Comp. Endocrinol.* **13,** 151.

Gona, A. G. (1967). *Endocrinology* **81,** 748.

Gona, A. G. (1968). *Gen. Comp. Endocrinol.* **11,** 278.

Gona, A. G. (1969). *Z. Zellforsch. Mikrosk. Anat.* **95,** 483.

Goos, H. J. T. (1968). *Z. Zellforsch. Mikrosk. Anat.* **92**, 583.
Goos, H. J. T. (1969a). *Z. Zellforsch. Mikrosk. Anat.* **97**, 118.
Goos, H. J. T. (1969b). *Z. Zellforsch. Mikrosk. Anat.* **97**, 449.
Goos, H. J. T., de Knecht, A., and de Vries, J. (1968a). *Z. Zellforsch. Mikrosk. Anat.* **86**, 384.
Goos, H. J. T., Zwanenbeek, H. C. M., and Van Oordt, P. G. W. J. (1968b). *Arch. Anat. Histol. Embryol.* **51**, 269.
Gorbman, A. (1964). *In* "Physiology of the Amphibia" (J. A. Moore, ed.), pp. 371–425. Academic Press, New York.
Graham, J. D. P. (1961). *J. Physiol.* (*London*) **158**, 5P.
Grant, W. C. (1966). *Amer. Zool.* **6**, 354.
Graul, C. (1966a). *Acta. Biol. Med. Ger.* **17**, 404.
Graul, C. (1966b). *Biol. Rundsch.* **4**, 157.
Green, K., and Matty, A. J. (1962). *Nature* (*London*) **194**, 1190.
Green, K., and Matty, A. J. (1963). *Gen. Comp. Endocrinol.* **3**, 244.
Gunesch, K. D. (1974). *Zool. Jb. Physiol.* **78**, 108.
Gupta, I., and Bhide, N. K. (1967). *J. Pharm. Pharmacol.* **19**, 768.
Hadley, M. E. (1966). Ph.D. Thesis, Brown University, Providence, Rhode Island (University Microfilms, Ann Arbor, Michigan).
Hanke, W. (1966). *Zool. Anz., Suppl.* **30**, 132.
Hanke, W. (1967). *Proc. Int. Congr. Horm. Steroids, 2nd. 1966* Excerpta Med. Found., Int. Congr. Ser. No. 132, p. 1073.
Hanke, W. (1974). *Fortschr. Zool.* **22**, 431.
Hanke, W., and Leist, K. H. (1971). *Gen. Comp. Endocrinol.* **16**, 137.
Hanke, W., and Neumann, U. (1972). *Gen. Comp. Endocrinol., Suppl.* **3**, 198.
Hanke, W., and Pehlemann, F. W (1969). *Gen. Comp. Endocrinol.* **13**, 509.
Hanke, W., and Weber, K. M. (1964). *Gen. Comp. Endocrinol.* **4**, 662.
Hanke, W., and Weber, K. M. (1965). *Gen. Comp. Endocrinol.* **5**, 444.
Hanke, W., Bergerhoff, K., and Neumann, U. (1969). *Z. Vergl. Physiol.* **65**, 351.
Hearing, V. J., and Eppig, J. J. (1969). *J. Morpho.* **128**, 369.
Heller, H. (1965). *Arch. Anat. Microsc. Morphol. Exp.* **54**, 471.
Hellman, B., and Hellerström, C. (1962). *Acta Anat.* **48**, 149.
Hemme, L. (1972). *Z. Zellforsch. Mikrosk. Anat.* **125**, 353.
Hermansen, B., and Jørgensen, C. B. (1969). *Gen. Comp. Endocrinol.* **12**, 313.
Horn-Ebert, I., Thurmond, W., and Hanke, W. (1974). *Zool. Jb. Physiol.* (in press).
Hunter, N. W., and Johnson, C. E. (1960). *J. Cell Comp. Physiol.* **55**, 275.
Hutchinson, J. H., and Porter, G. A. (1969). *Clin. Res.* **17**, 129.
Iturriza, F. C. (1964). *Gen. Comp. Endocrinol.* **4**, 492.
Iturriza, F. C. (1966). *Gen. Comp. Endocrinol.* **6**, 19.
Iturriza, F. C. (1967). *Neuroendocrinology* **2**, 11.
Iturriza, F. C. (1969). *Gen. Comp. Endocrinol.* **12**, 417.
Janssens, P. A. (1967). *Gen. Comp. Endocrinol.* **8**, 94.
Johnston, C. I., Davis, J. O., Wright, F. S., and Howards, S. S. (1967). *Amer. J. Physiol.* **213**, 393.
Joly, J. (1965). *C. R. Acad. Sci.* **261**, 1569.
Jørgensen, C. B. (1968). *In* "Perspectives in Endocrinology" (E. J. W. Barrington and C. B. Jørgensen, eds.), pp. 469–541. Academic Press, New York.
Jørgensen, C. B. (1970). *In* "The Hypothalamus" (L. Martini, M. Motta, and F. Fraschini, eds.), pp. 649–661. Academic Press, New York.
Jørgensen, C. B., and Larsen, L. O. (1960). *Nature* (*London*) **185**, 244.
Jørgensen, C. B., and Larsen, L. O. (1961). *Gen. Comp. Endocrinol.* **1**, 145.

Jørgensen, C. B., and Larsen, L. O. (1963). *Symp. Zool. Soc. London* **9**, 59.
Jørgensen, C. B., and Larsen, L. O. (1964). *Gen. Comp. Endocrinol.* **4**, 389.
Jørgensen, C. B., and Larsen, L. O. (1967). *In* "Neuroendocrinology" (L. Martini and W. F. Ganong, eds.), Vol. 2, pp. 485–528. Academic Press, New York.
Jørgensen, C. B., Larsen, L. O., and Rosenkilde, P. (1965). *Gen. Comp. Endocrinol.* **5**, 248.
Jørgensen, C. B., Rosenkilde, P., and Wingstrand, K. G. (1969). *Gen. Comp. Endocrinol.* **12**, 91.
Kastin, A. J., and Schally, A. V. (1966). *Experientia* **22**, 389.
Kaye, N. (1961). *Gen. Comp. Endocrinol.* **1**, 1.
Kern, H. (1962). *Endokrinologie* **42**, 294.
Kern, H. (1966). *Z. Zellforsch. Mikrosk. Anat.* **70**, 499.
Kerr, T. (1965). *Gen. Comp. Endocrinol.* **5**, 232.
Kerr, T. (1966). *Gen. Comp. Endocrinol.* **6**, 303.
Lange, R., and von Brehm, H. (1965). *In* "The Parathyroid Glands, Ultrastructure, Secretion and Function" (P. J. Gaillard, R. V. Talmage, and A. M. Budy, eds.), pp. 19–26. Chicago Univ. Press, Chicago, Illinois.
Larsen, L. O., van Kemenade, J. A. M., and van Dongen, W. J. (1971). *Gen. Comp. Endocrinol.* **16**, 165.
Leist, K. H. (1970). *Zool. Jahrb., Abt. Alg. Zool. Physiol. Tiere* **75**, 375.
Leist, K. H., and Hanke, W. (1969). *Gen. Comp. Endocrinol.* **13**, 517.
Leist, K. H., Bergerhoff, K., Pehlemann, F. W., and Hanke, W. (1969). *Z. Zellforsch. Mikrosk. Anat.* **93**, 105.
Lofts, B. (1964). *Gen. Comp. Endocrinol.* **4**, 550.
Lofts, B. (1968). *In* "Perspectives in Endocrinology" (E. J. W. Barrington and C. B. Jørgensen, eds.), pp. 239–304. Academic Press, New York.
Lofts, B., Wellen, J. J., and Benraad, T. J. (1972). *Gen. Comp. Endocrinol.* **18**, 344.
McNabb, R. (1969a). *Gen. Comp. Endocrinol.* **12**, 276.
McNabb, R. (1969b). *Gen. Comp. Endocrinol.* **13**, 430.
McWhinnie, D. J., and Cortelyou, J. R. (1967). *Amer. Zool.* **7**, 857.
Maetz, J. (1968). *In* "Perspectives in Endocrinology" (E. J. W. Barrington and C. B. Jørgensen, eds.), pp. 47–162. Academic Press, New York.
Maetz, J. (1970). *Mem. Soc. Endocrinol.* **18**, 3.
Marusic, E., Martinez, R., and Torretti, J. (1966). *Proc. Soc. Exp. Biol. Med.* **122**, 164.
Masui, Y. (1967). *J. Exp. Zool.* **166**, 365.
Matty, A. J. (1966). *Int. Rev. Gen. Exp. Zool.* **2**, 44.
Matty, A. J., and Green, K. (1962a). *Life Sci.* **1**, 487.
Matty, A. J., and Green, K. (1962b). *J. Endocrinol.* **25**, 411.
Matty, A. J., and Green, K. (1964). *Gen. Comp. Endocrinol.* **4**, 331.
Mazzi, V. (1970). *In* "The Hypothalamus" (L. Martini, M. Motta, and F. Franschini, eds.), pp. 663–676. Academic Press, New York.
Mazzi, V., Peyrot, A., Anzalone, M. R., and Toscano, C. (1966). *Z. Zellforsch. Mikrosk. Anat.* **72**, 597.
Mazzi, V., Vellano, C., Peyrot, A., and Lodi, G. (1969). *Atti Accad. Sci. Torino, cl. Sci. Fis., Mot. Natur.* **104**, 771.
Medda, A. K., and Frieden, E. (1970). *Endocrinology* **87**, 2.
Medda, A. K., and Frieden, E. (1972). *Gen. Comp. Endocrinol.* **19**, 212.
Mehdi, A. Z., and Carballeira, A. (1971a). *Gen. Comp. Endocrinol.* **17**, 1.
Mehdi, A. Z., and Carballeira, A. (1971b). *Gen. Comp. Endocrinol.* **17**, 14.
Meier, A. (1969). *Gen. Comp. Endocrinol., Suppl.* **2**, 55.

Middler, S. A., Kleemann, C. R., and Edwards, E. (1967). *Gen. Comp. Endocrinol.* **9**, 38.
Miller, M. R. (1961). *In* "Comparative Physiology of Carbohydrate Metabolism in Heterothermic Animals." (A. W. Martin, ed.), pp. 125–147. Univ. of Washington Press, Seattle.
Mira-Moser, F. (1970). *Z. Zellforsch. Mikrosk. Anat.* **105**, 65.
Mira-Moser, F. (1971). *Z. Zellforsch. Mikrosk. Anat.* **112**, 266.
Morel, F., and Jard, S. (1968). *In* "Handbuch der experimentellen. Pharmakologie" (B. Berde, ed.), Vol. 23, pp. 655–716. Springer-Verlag, Berlin and New York.
Munsick, R. A. (1966). *Endocrinology* **78**, 591.
Nakai, Y., and Gorbman, A. (1969). *Gen. Comp. Endocrinol.* **13**, 108.
Nakamura, M. (1967). *Endocrinol. Jap.* **14**, 43.
Neumann, U. (1973). *Zool. Jb. Physiol.* **77**, 60.
Nicoll, C. S., and Nichols, C. W. (1971). *Gen. Comp. Endocrinol.* **17**, 300.
Nicoll, C. S., Bern, H. A., Dunlop, D., and Strohman, R. C. (1965). *Amer. Zool.* **5**, 738.
Notenboom, C. D. (1971). *Gen. Comp. Endocrinol.* **18**, 612.
Novales, R. R., and Novales, B. J. (1965a). *Progr. Brain Res.* **10**, 507–519.
Novales, R. R., and Davis, W. J. (1967). *Endocrinology* **81**, 283.
Novales, R. R., and Davis, W. J. (1969). *Amer. Zool.* **9**, 479.
Novales, R. R., and Novales, B. J. (1965b). *Gen. Comp. Endocrinol.* **5**, 568.
Novales, R. R., Novales, B. J., Zinner, S. H., and Stoner, J. A. (1962). *Gen. Comp. Endocrinol., Suppl.* **2**, 286.
Oshima, K., and Gorbman, A. (1969). *Gen. Comp. Endocrinol.* **13**, 98.
Ozon, R. (1969). *Gen. Comp. Endocrinol., Suppl.* **2**, 135.
Ozon, R. (1972a). *In* "Steroids in Nonmammalian Verterbrates" (D. R. Idler, ed.) pp. 390–413, Academic Press, New York.
Ozon, R. (1972b). *In* "Steroids in Nonmammalian Vertebrates" (D. R. Idler, ed.) pp. 328–389, Academic Press, New York.
Pehlemann, F. W. (1967a). *Gen. Comp. Endocrinol.* **9**, 48.
Pehlemann, F. W. (1967b). *Z. Zellforsch. Mikrosk. Anat.* **78**, 484.
Pehlemann, F. W. (1974). *Fortschr. Zool.* **22**, 204.
Pehlemann, F. W., and Hanke, W. (1968). *Z. Zellforsch. Mikrosk. Anat.* **89**, 281.
Pehlemann, F. W., and Hemme, L. (1972). *Gen. Comp. Endocrinol.* **18**, 615.
Peute, J., and van Oordt, P. G. W. J. (1974). *Fortschr. Zool.* **22**, 134.
Peyrot, A., Vellano, C., and Mazzi, V. (1969a). *Gen. Comp. Endocrinol.* **12**, 179.
Peyrot, A., Mazzi, V., Vellano, C., and Lodi, G. (1969b). *J. Endocrinol.* **45**, 525.
Piezzi, R. S., and Burgos, M. H. (1968). *Gen. Comp. Endocrinol.* **10**, 344.
Piper, G. D. (1969). Ph.D. Thesis, University of Missouri, Columbia.
Piper, G. D., and de Roos, R. (1967). *Gen. Comp. Endocrinol.* **8**, 135.
Poort, C., and Genze, J. J. (1969). *Z. Zellforsch. Mikrosk. Anat.* **98**, 1.
Porter, G. A. (1971). *Gen. Comp. Endocrinol.* **16**, 443.
Quay, W. B. (1969). *Gen. Comp. Endocrinol., Suppl.* **2**, 101.
Quay, W. B. (1970). *Mem. Soc. Endocrinol.* **18**, 423.
Rapola, J. (1962). *Ann. Med. Exp. Biol. Fenn., Ser. A*, **4**, 1.
Rapola, J. (1963). *Gen. Comp. Endocrinol.* **3**, 412.
Rastogi, R. K., and Chieffi, G. (1970a). *Gen. Comp. Endocrinol.* **15**, 247.
Rastogi, R. K., and Chieffi, G. (1970b). *Z. Zellforsch. Mikrosk. Anat.* **111**, 505.
Rastogi, R. K., and Chieffi, G. (1970c). *Gen. Comp. Endocrinol.* **15**, 488.
Redshaw, M. R. (1972). *Amer. Zool.* **12**, 289.
Redshaw, M. R., and Nicholls, T. J. (1971). *Gen. Comp. Endocrinol.* **16**, 85.

Redshaw, M. R., Follett, B. K., and Nicholls, T. J. (1969). *J. Endocrinol.* **43,** 47.
Rinaudo, M. T., Giunta, C., Bruno, R., Guardabassi, A., Olivero, M, and Clerici, P. (1969). *Comp. Biochem. Physiol.* **29,** 1079.
Robertson, D. R. (1968a). *Z. Zellforsch. Mikrosk. Anat.* **85,** 441.
Robertson, D. R. (1968b). *Z. Zellforsch. Mikrosk. Anat.* **85,** 453.
Robertson, D. R. (1969). *Gen. Comp. Endocrinol.* **12,** 479.
Robertson, D. R. (1971). *J. Exp. Zool.* **178,** 101.
Robertson, D. R. (1972). *Gen. Comp. Endocrinol. Suppl.* **3,** 421.
Robison, G. A., Butcher, R. W., and Sutherland, E. W. (1968). *Annu. Rev. Biochem.* **37,** 149.
Rodriguez, E. M., and Piezzi, R. S. (1967). *Z. Zellforsch. Mikrosk. Anat.* **83,** 207.
Rogers, D. C. (1965). *J. Ultrastruct. Res.* **13,** 478.
Rosenkilde, P. (1971). *Gen. Comp. Endocrinol., Suppl.* **3,** 32.
Saland, L. C. (1968). *Neuroendocrinology* **3,** 72.
Sandor, T. (1969). *Gen. Comp. Endocrinol., Suppl.* **2,** 284.
Sandor, T. (1972). *In* "Steroids in Nonmammalian Vertebrates" (D. R. Idler, ed.) pp. 253–327, Academic Press, New York.
Saxén, L., Saxén, E., Toivonen, S., and Salimaki, K. (1957). *Endocrinology,* **61,** 35.
Scalenghe, F., Andreoletti, G. E., Sampietro, P., Rotta, G. P., and Mazzi, V. (1968). *Arch. Anat., Histol. Embryol.* **51,** 627.
Scheer, B. T., and Wise, P. T. (1969). *Gen. Comp. Endocrinol.* **13,** 474.
Schmid, W. D. (1965). *Ecology* **46,** 261.
Schuetz, A. W. (1967a). *Proc. Soc. Exp. Biol. Med.* **124,** 1307.
Schuetz, A. W. (1967b). *J. Exp. Zool.* **166,** 347.
Schultheiss, H., Hanke, W., and Maetz, J. (1972). *Gen. Comp. Endocrinol.* **18,** 400.
Shaffer, B. M. (1963). *J. Embryol. Exp. Morphol.* **11,** 77.
Sharp, G. W. G., and Leaf, A. (1966). *Recent Progr. Horm. Res.* **22,** 431.
Sharp, G. W. G., and Leaf, A. (1968). *J. Gen. Physiol.* **51,** 271.
Shoemaker, V. H., and Waring, H. (1968). *Comp. Biochem. Physiol.* **24,** 47.
Smoller, C. G. (1966). *Gen. Comp. Endocrinol.* **7,** 44.
Streb, M. (1967). *Z. Zellforsch. Mikrosk. Anat.* **82,** 407.
Tata, J. R. (1969). *Gen. Comp. Endocrinol., Suppl.* **2,** 385.
Taylor, J. D. (1966). *Amer. Zool.* **6,** 587.
Taylor, J. D. (1967). Ph.D. Thesis. University of Arizona, Tucson.
Taylor, J. D. (1969). *Gen. Comp. Endocrinol.* **12,** 405.
Taylor, R. E., and Barker, S. B. (1965). *J. Endocrinol.* **31,** 175.
Taylor, R. E., and Barker, S. B. (1967). *Gen. Comp. Endocrinol.* **9,** 129.
Terlou, M., Goos, H. J. T., and van Oordt, P. G. W. J. (1974). *Fortschr. Zool.* **22,** 117.
Thornburn, C. C., and Matty, A. J. (1964). *J. Endocrinol.* **28,** 213.
Thornburn, C. C., and Matty, A. J. (1966). *J. Endocrinol.* **36,** 221.
Thornton, V. F. (1971). *Gen. Comp. Endocrinol.* **16,** 599.
Thornton, V. F., and Evennett, P. J. (1969). *Gen. Comp. Endocrinol.* **13,** 268.
Thurmond, W., and Hanke, W. (1971). *Amer. Zool.* **11,** 654.
van de Veerdonk, F. C. G. (1967). *Curr. Mod. Biol.* **1,** 175.
van de Veerdonk, F. C. G. and Konijn, T. M. (1970). *Acta Endocrinol.* (*Copenhagen*) **64,** 364.

van Kemenade, J. A. M. (1968a). *Z. Zellforsch. Mikrosk. Anat.* **92**, 549.
van Kemenade, J. A. M. (1968b). *Z. Zellforsch. Mikrosk. Anat.* **92**, 567.
van Kemenade, J. A. M. (1969a). *Z. Zellforsch. Mikrosk. Anat.* **95**, 620.
van Kemenade, J. A. M. (1969b). *Z. Zellforsch. Mikrosk. Anat.* **96**, 466.
van Kemenade, J. A. M. (1972). *Gen. Comp. Endocrinol.* **18**, 627.
van Kemenade, J. A. M. (1974). *Fortschr. Zoologie* **22**, 228.
van Kemenade, J. A. M., and van Dongen, W. J. (1965). *Nature (London)* **205**, 4967.
van Kemenade, J. A. M., van Dongen, W. J., and Van Oordt, P. G. W. J. (1968). *Z. Zellforsch. Mikrosk. Anat.* **91**, 96.
Van Oordt, P. G. W. J. (1968). *In* "Perspectives in Endocrinology" (E. J. W. Barrington and C. B. Jørgensen, eds.), pp. 405–467. Academic Press, New York.
Van Oordt, P. G. W. J., van Dongen, W. J., and Lofts, B. (1968). *Z. Zellforsch. Mikrosk. Anat.* **88**, 549.
Van Oordt, P. G. W. J., Goos, H. J. T., Peute, J., and Terlou, M. (1972). *Gen. Comp. Endocrinol. Suppl.* **3**, 41.
Varma, M. M. (1971). *Amer. Zool.* **11**, 655.
Vellano, C., and Peyrot, A. (1970). *Boll. Zool.* **37**, 409.
Vellano, C., Peyrot, A., and Mazzi, V. (1967). *Monit. Zool. Ital.* [N.S.] **1**, 207.
Vellano, C., Peyrot, A., and Mazzi, V. (1969). *Gen. Comp. Endocrinol.* **13**, 537.
Vellano, C., Lodi, G., Bani, G., Sacerdote, M., and Mazzi, V. (1970a). *Monit. Zool. Ital.* [N.S.] **4**, 115.
Vellano, C., Mazzi, V., and Sacerdote, M. (1970b). *Gen. Comp. Endocrinol.* **14**, 535.
Volk, T. L. (1972). *Z. Zellforsch. Mikrosk. Anat.* **123**, 470.
Vullings, H. G. B. (1971). *Z. Zellforsch. Mikrosk. Anat.* **113**, 174.
Wallace, R. A., and Jared, D. W. (1969). *Develop. Biol.* **19**, 498.
Weber, R. (1967). *In* "The Biochemistry of Animal Development" (R. Weber, ed.), Vol. 2, pp. 227–301. Academic Press, New York.
Weber, R. (1969). *Gen. Comp. Endocrinol., Suppl.* **2**, 408.
Weber, R., Hagenbüchle, O., and Ryffel, G. (1974). *Fortschr. Zool.* **22**, 419.
Wingstrand, K. G. (1966). *In* "The Pituitary Gland" (G. W. Harris and B. T. Donovan, eds.), Vol. 1, pp. 58–126. Butterworth, London.
Wurtman, R. J., Axelrod, J., and Kelly, D. E. (1968). "The Pineal" Academic Press, New York.
Yoshimura, F., and Harumiya, K. (1966). *Electron Microsc., Proc. Int. Congr. 6th, 1966* pp. 545–546.
Yoshimura, F., and Harumiya, K. (1968). *Endocrinol. Jap.* **15**, 94.
Zipser, R. D., Licht, P., and Bern, H. A. (1969). *Gen. Comp. Endocrinol.* **13**, 382.

CHAPTER 7

Venoms of Amphibia

Gerhard G. Habermehl

I. Introduction 161
II. Urodela 164
Salamandridae 164
III. Anura 167
A. Bufonidae 167
B. Leptodactylinae 174
C. Dendrobatidae 175
D. Atelopotidae 178
E. Hylidae 178
F. Ranidae 179
G. Discoglossidae 179
H. Pipidae 179
IV. Conclusion 180
References 180

I. Introduction

Venomous animals may be divided into two categories according to the use they make from their venom; (1) as a means for hunting or killing their prey (e.g., snakes, spiders, and scorpions) or (2) as an example for the second category, the amphibia, which make use of their venom produced in the skin glands as a protection against their enemies. Which kind of enemies are most dangerous, we shall see at the end of this chapter.

The secretions of amphibian skin glands (Fig. 1) contain an amazing variety of different substances—simple biogenic amines as well as more or less complicated peptides, steroids, and alkaloids. The pharmacological activity is still more differentiated, as there can be found cardiotoxins, neurotoxins, cholinomimetica and sympathomimetica, vasoconstrictors and hypotensive substances, local anesthetics, and even some of the most potent hallucinogens. The toxicity for mammalia differs in a broad range; an outline of the pharmacological activity and the toxicity is given in Table I.

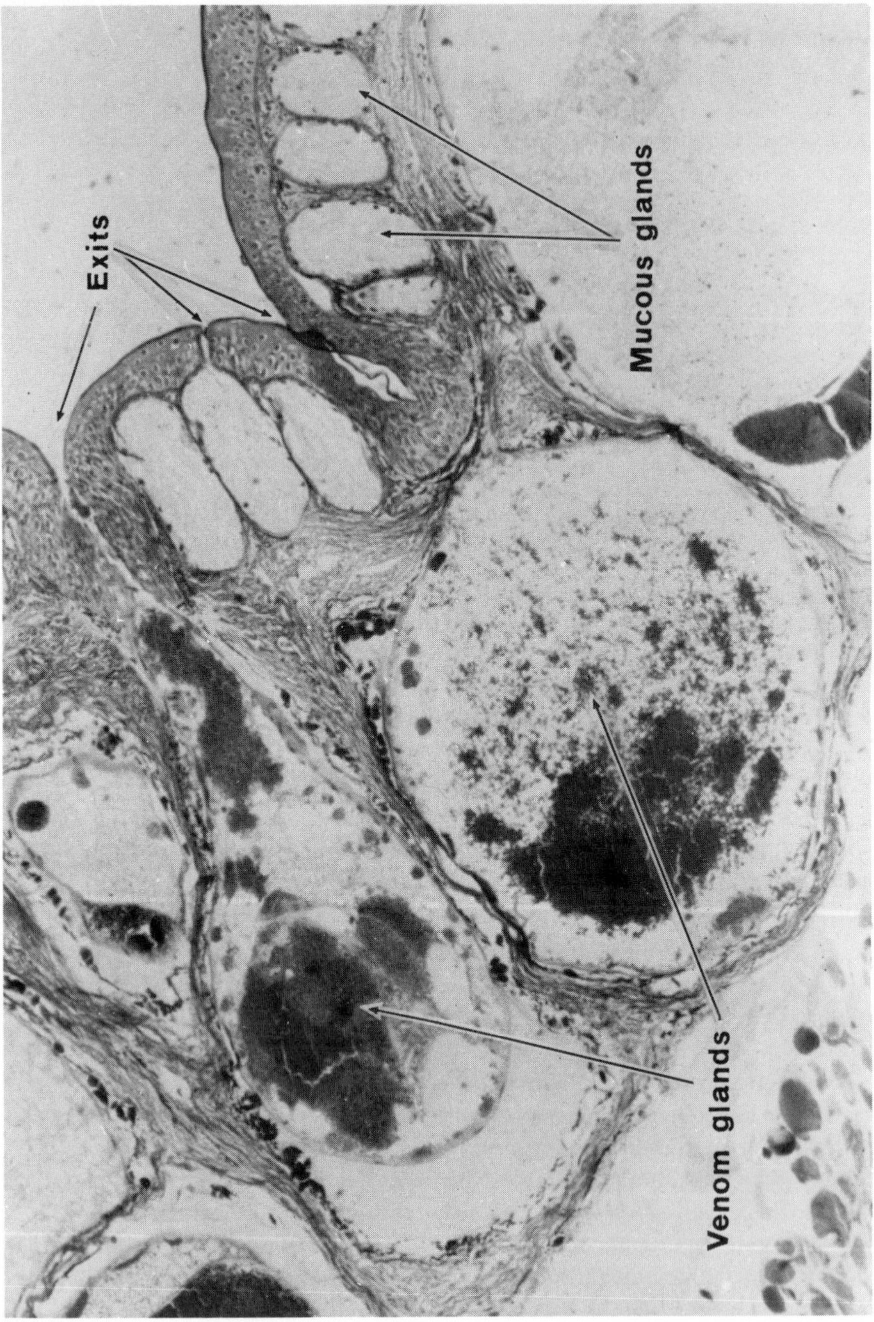

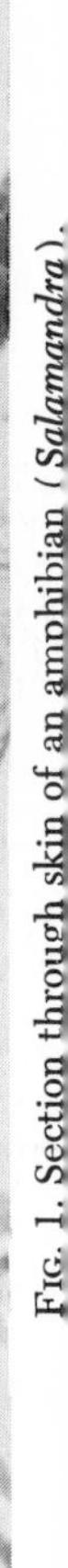

FIG. 1. Section through skin of an amphibian (*Salamandra*).

TABLE I
PHARMACOLOGICAL ACTIVITY AND TOXICITY OF VARIOUS TOXINS

Compound	Occurrence	Pharmacological activity	LD_{50}, s.c. (γ/kg mouse)
Batrachotoxin	*Phyllobates aurotaenia*	Cardio- and neurotoxin	2
Tetrodotoxin	*Taricha torosa*	Neurotoxin	8
Samandarin	*Salamandra maculosa*	Acts on central nervous system causing convulsions	1500
Bufotoxin	*Bufo vulgaris*	Cardiotoxin	400
Triturus vulgaris toxin	*Triturus vulgaris*	Hemolytic and nerve–muscle action	6700
Triturus cristatus toxin	*Triturus cristatus*		20000
Triturus marmoratus toxin	*Triturus marmoratus*		1800
Pumiliotoxin A and B	*Dendrobates pumilio, D. auratus*	Neurotoxins; nerve–muscle action	2500
			1500
Serotonin	*Leptodactylus pentadactylus*	Vasoconstrictor	300
Dehydrobufotenine	*Bufo* spp.	Acts on central nervous system causing convulsions	6000
O-Methylbufotenine	*Bufo* spp.	Hallucinogen	75000
Noradrenaline	*Bufo* spp.	Hypertensive	5000
Candicin	*Leptodactylus pentadactylus*	Cholinomimetic	10000
Leptodactylin	*Leptodactylus pentadactylus*	Cholinomimetic	10000
Atelopid toxin	*Atelopus zeteki*	Acts on central nervous system; hypotensive; cardiotoxin	16
In comparison with:			
Botulinus toxin	*Clostridium botulinum*		0.00003
Tetanus toxin	*Clostridium tetani*		0.0001
Diphtheria toxin	*Corynebacterium diphtheriae*		0.3
Cobra toxin	*Naja naja*		0.3
Crotalus toxin	*Crotalus atrox*		0.2
Curare	*Chondodendron tomentosum*		500
Strychnine	*Strychnos, nux vomica*		500
Sodium cyanide			10000
Muscarine	*Amanita muscaria*		1100

II. Urodela

Salamandridae

Among the Urodela, only salamanders and newts have been investigated. The toxicity of the black and yellow spotted salamander *Salamandra maculosa* is well known since ancient times, when Nikander of Kolophon and the younger Plinius first reported the symptoms of poisoning. Chemical investigations were started by Zalesky (1866), Faust 1898, 1900), and Gessner, (1938) and were continued by Schöpf and his co-workers after 1934 (Schöpf and Braun, 1934; Schöpf and Klein, 1954; Schöpf and Koch, 1942a,b; Schöpf and Müller, 1960; Schöpf *et al.*, 1950). He succeeded in obtaining the main alkaloid samandarine in pure and crystalline form, and from chemical experiments he obtained three possible formulae for the structure of samandarine (I, II, III). The final deter-

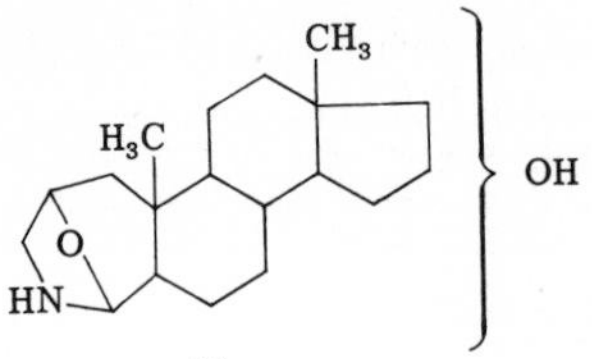

(I)

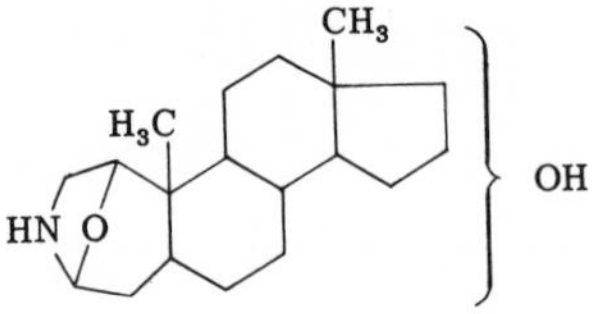

(III)

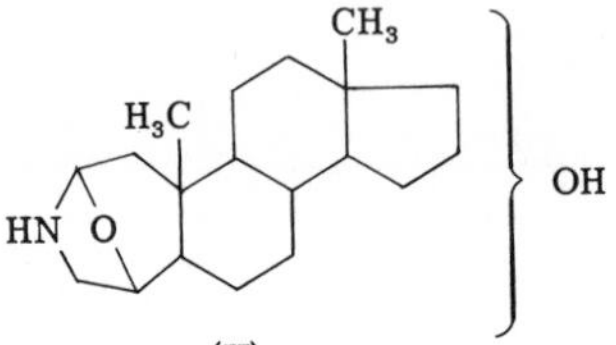

(II)

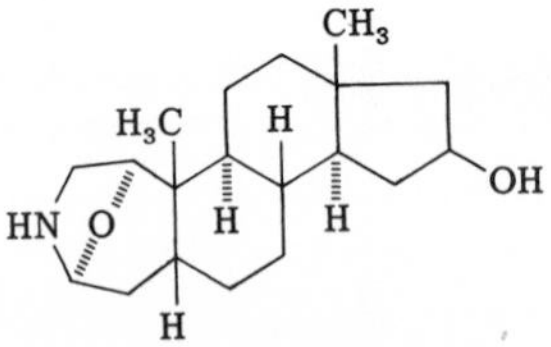

(IV)

mination of constitution and configuration of this alkaloid according to IV was accomplished by X-ray structure analysis (Wölfel *et al.*, 1961). In the same way, the structure of samandaridine has been elucidated (Habermehl, 1963a,b). An outline of all of the alkaloids, isolated from *Salamandra* venom is given in Table II.

TABLE II
ALKALOIDS OBTAINED IN SALAMANDER VENOMS

Name	Formula	Melting point (°C)	R_f[a]	Functional groups
Alkaloids with an oxazolidine system				
Samandarine	$C_{19}H_{31}NO_2$	188	0.42	—NH— —O— =CHOH
Samandarone	$C_{19}H_{29}NO_2$	190	0.52	—NH— —O— =CO
Samandaridine	$C_{21}H_{31}NO_3$	290	0.20	—NH— —O— γ-lactone
O-Acetylsamandarine	$C_{21}H_{33}NO_3$	159	0.55	—NH— —O— *O*-acetyl
Samandenone	$C_{22}H_{31}NO_2$	191	0.35	—NH— —O— C=CH—CO
Samandinine	$C_{22}H_{39}NO_3$	170	0.68	—NH— —O— *O*-acetyl —CH(CH$_3$)$_2$
Alkaloids with carbinolamine system				
Cycloneosamandione	$C_{19}H_{29}NO_2$	119	0.08	N—C—OH =CO
Cycloneosamandaridine	$C_{21}H_{31}NO_3$	282	0.75	N—C—OH γ-lactone
Alkaloids without oxazolidine and without carbinolamine system				
Samanine	$C_{19}H_{33}NO$	197		—NH— =CHOH

[a] Paper: SiO_2 paper, Schleicher & Schüll, No. 289. Solvent: cyclohexane/diethylamine 9:1.

Among the minor alkaloids, it is noteworthy to mention cycloneosamandione, as this alkaloid possesses a carbinolamine system instead of the oxazolidine group. Its structure (V) came from spectroscopic data as well as from the synthesis of this alkaloid (Habermehl and Haaf, 1968b) according to the Scheme 1.

Samanine (VI) belongs to a third type of alkaloid group in so far as it does not possess an oxygen function connected to the ring A of the steroid skeleton. The synthesis of this alkaloid according to Scheme 2 has been described by Habermehl and Haaf (1969).

Among the minor alkaloids there have been found compounds with a side chain at carbon atom 17, too. Samandaridine (VII) (Schöpf and Koch, 1942a; Habermehl, 1963a,b) and cycloneosamandaridine (VIII) (Habermehl and Haaf, 1965) possess a carboxymethylene group which is lactonized with the C-16 hydroxy group; samandinine (IX) (Habermehl and Vogel, 1969) and samandenone (X) (Habermehl, 1966) possess an isopropyl group at C-17.

Experiments with radioactive labeled cholesterol showed this substance to be the precursor of the salamandra alkaloids in the biogenetic pathway (Habermehl and Haaf, 1968a). The heterocyclic ring system arises from a fission of ring A of cholesterol between carbon atoms 2 and 3, followed by insertion of nitrogen (coming from glutamine),

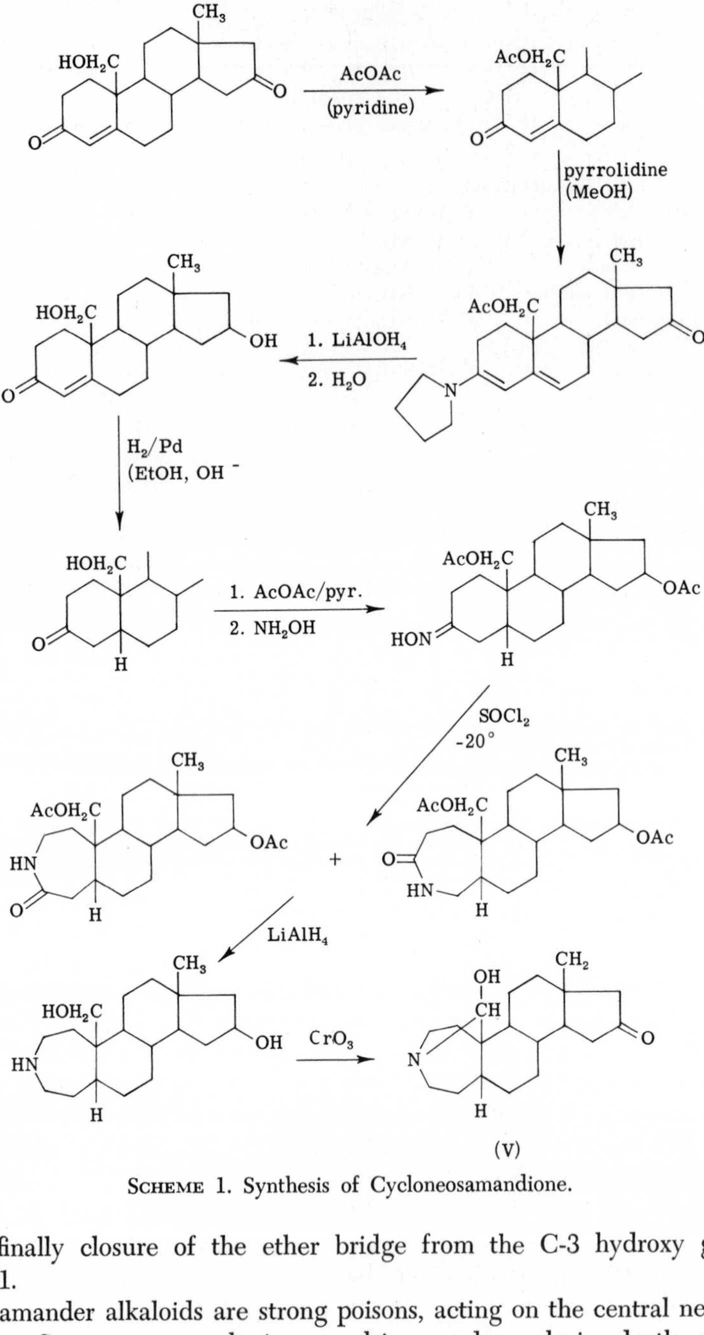

SCHEME 1. Synthesis of Cycloneosamandione.

and finally closure of the ether bridge from the C-3 hydroxy group to C-1.

Salamander alkaloids are strong poisons, acting on the central nervous system. Symptoms are clonic convulsions and paralysis; death occurs by primary respiratory paralysis. Remarkable are their hypotensive and local anaesthetic activities (Gessner, 1926; Gessner and Möllenhoff, 1932; Gessner and Esser, 1935; Gessner and Urban, 1937; E. X. Albuquerque and G. Habermehl, unpublished). For an extensive review on salamandra alkaloids, see Habermehl (1970).

From the Californian newts *Taricha torosa, Taricha rivularis,* and *Taricha granulosa,* a unique substance, tarichatoxin (XI), has been isolated. It is identical with tetrodotoxin from the buffer fish *Sphoeroides rubripes.* The structure has been elucidated by degradation reactions and finally by X-ray analysis of tedrodoic acid hydrobromide (Mosher *et al.*, 1964; Tsuda *et al.*, 1963; Tsuda, 1966). The symptoms of intoxication are a rapid and strong hypotensive effect caused by vasodilation. In the next stage muscle paralysis, convulsions, and vomiting occur. On application of larger doses (subcutaneous or oral), death occurs within a very short time.

The European newts *Triturus vulgaris, Triturus cristatus, Triturus alpestris,* and *Triturus marmoratus* possess secretions of an amazingly high hemolytical activity (Michl and Bachmayer, 1965). Besides this, phosphatases and arylamidases have been found, but no steroids or alkaloids. From the skin glands of *T. cristatus,* only tarichatoxin could be isolated (G. Habermehl, unpublished; Wakely *et al.*, 1966).

III. Anura

A. Bufonidae

Toads have been well known to be poisonous for thousands of years. The old Chinese and Japanese physicians made use of the dried secretion named *chan su* or *sen-so* as a cardiotonic. During the seventeenth and eighteenth centuries, it was imported to Europe for the same purpose until it was displaced by digitalis glycosides

1. *Biogenic Amines*

The biogenic amines are simple metabolites of amino acids, often derived by decarboxylation. Adrenaline (XIIa) and noradrenaline (XIIb) are well known as hormones of the adrenal medulla. They are very common in the skin gland secretions of many amphibia (Oestlund, 1954; Daly and Witkop, 1970; Abel and Macht, 1911). Also, dopamine and its *N*-methyl derivative epinine (XIIc) can be isolated from toad secretion (Märki *et al.*, 1962). The biosynthesis of these compounds starts from phenylalanine, which by oxidation is converted to dioxyphenylalanine. Decarboxylation yields dopamine, which is either

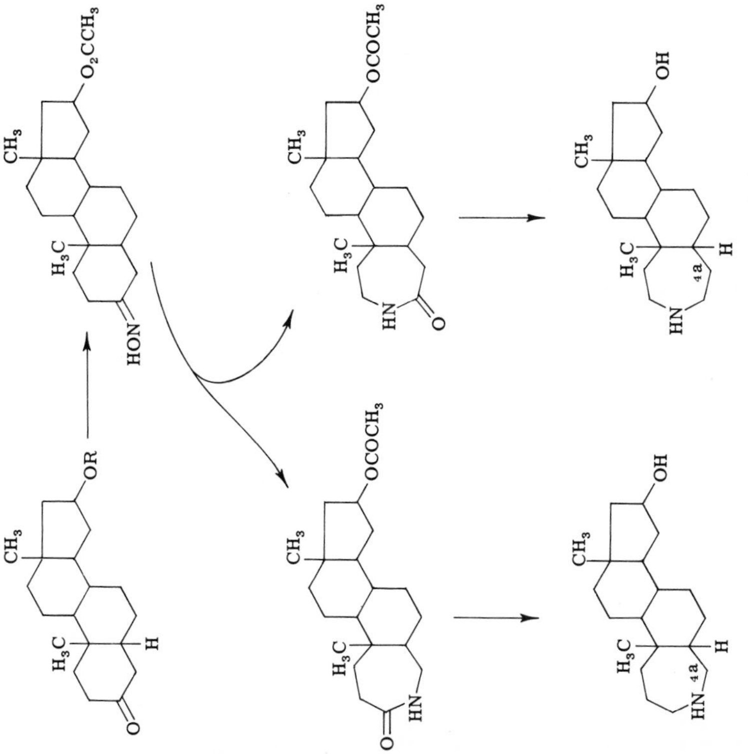

(VI)

SCHEME 2. Synthesis of Samanine.

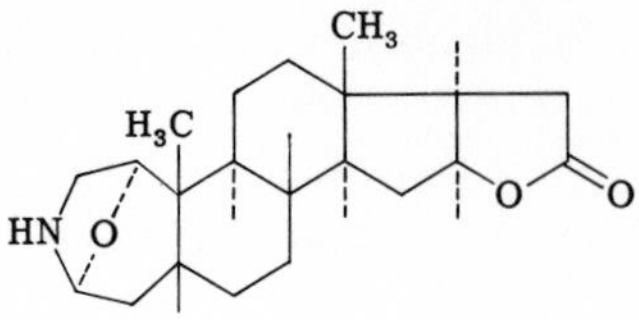

CH_3 H_3C HN O O O

(VII)

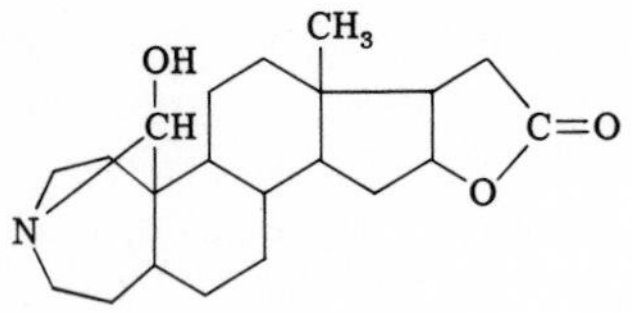

OH CH CH_3 C=O N O

(VIII)

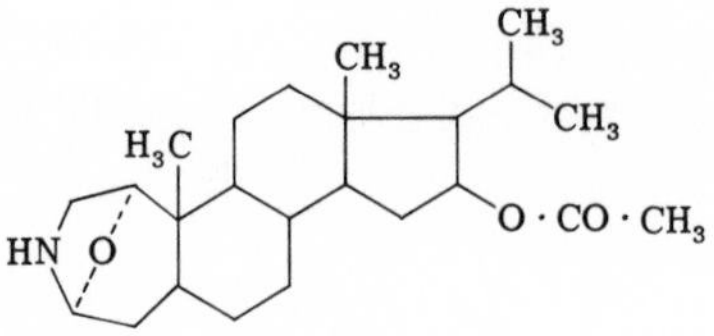

CH_3 CH_3 CH_3 H_3C HN O O · CO · CH_3

(IX)

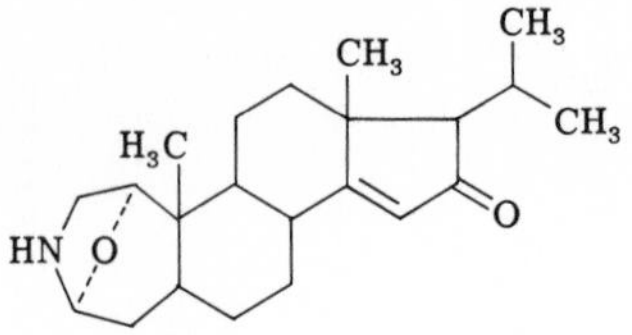

CH_3 CH_3 CH_3 H_3C HN O O

(X)

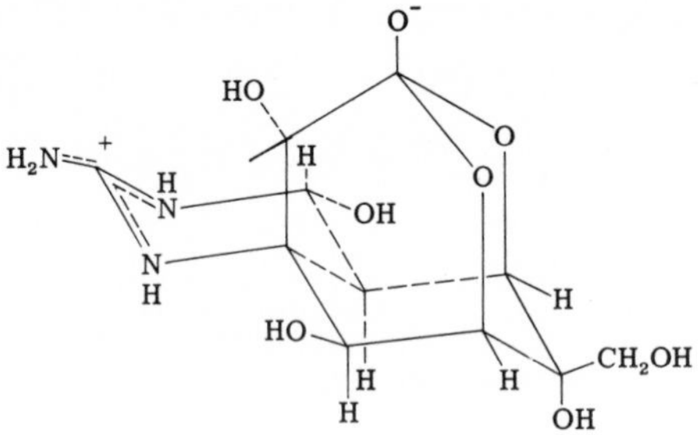

O^- HO $H_2N=^+$ H N H OH O O N H H HO H H H CH_2OH OH

(XI)

methylated to epinine or hydroxylated yielding noradrenaline, which then is methylated to adrenaline (von Euler, 1958).

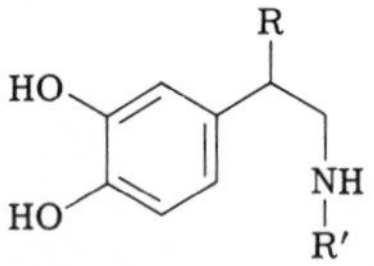

R HO HO NH R′

(XIIa) R = OH, R′ = CH_3
(XIIb) R = OH, R′ = H
(XIIc) R = H, R′ = CH_3

2. *Indolalkylamines*

Phisialix and Bertrand (1893) as well as Handovsky (1920) found basic substances with pharmacological activity among the substances from the skin glands of toads. The structure of bufotenine (XIII) has been elucidated by Wieland *et al.* (1934). Syntheses of bufotenine were performed by Hoshino and Shimodaira (1935), Harley-Mason and

Jackson (1954), Speeter and Anthony (1954), and Stoll *et al.*, (1955). *O*-Methyl-bufotenine (from *Bufo alvarius*) is one of the most potent hallucinogens (Gessner *et al.*, 1968). Investigations on its activity. which resembles that of mescaline, can be found in a paper of Fabing and Hawkins (1956). The structure of bufotenidine (XIV), which is the methylbetaine of bufotenine (XIII), was elucidated by Wieland and his co-workers (1931, 1934). Like most of the indolalkylamines, it possesses vasoconstrictor activity, which also results in hypertension. From the skin of *Bufo viridis viridis,* Erspamer and Vialli (1952) and Erspamer (1959a) isolated another indolalkylamine, bufoviridine (XV). By dilute hydrochloric acid cleavage, equivalent amounts of bufotenine and sulfuric acid are produced.

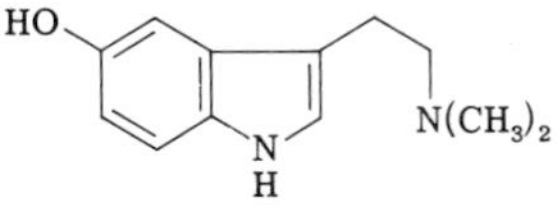

(XIII)

Two more indolalkylamines, bufothionine (XVI) and dehydrobufotenine (XVII), are related to each other in the same way as bufotenidine and bufoviridine. Bufothionine is found in the secretion of *Bufo formosus* and *Bufo arenarum* (Wieland and Vocke, 1930; Jensen, 1935). On warming of bufothionine with dilute hydrochloric acid, it yields dehydrobufotenine hydrochloride. Dehydrobufotenine has been isolated from *Bufo marinus* by Slotta and Neisser (1937); its structure was elucidated by Märki *et al.* (1961).

3. *Bufogenins*

The first investigations in this field were made by Faust (1902, 1903). He isolated bufotaline (XVIII); Wieland and Weil (1913) and Wieland

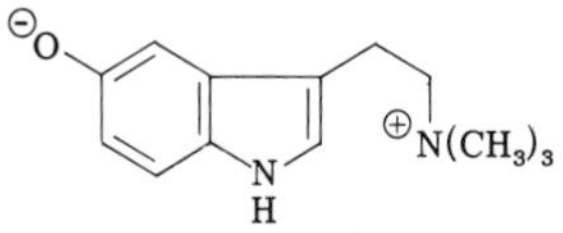

(XIV)

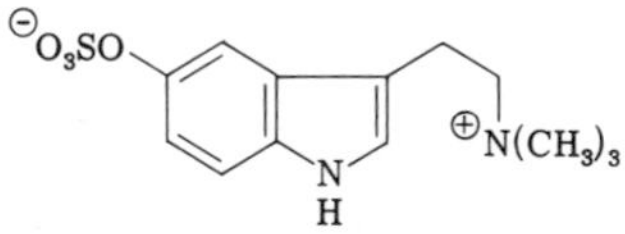

(XV)

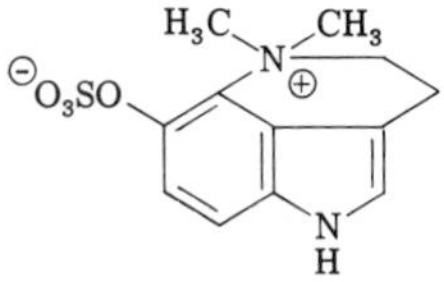

(XVI)

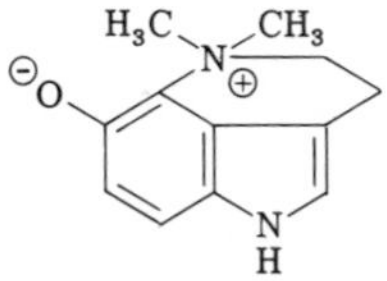

(XVII)

and Behringer (1941) then succeeded in getting important details of the structure, which was finally elucidated by Meyer (1949a,b). Characteristic for all of the bufogenins is the cis-connection of the rings A/B and C/D of the steroid skeleton, as well as the double unsaturated lactone bufadienolide ring which gives raise to the ultraviolet absorption band at 290–300 nm (log = 3.74).

A deciding step during structure elucidation of bufotalin was the oxidation by potassium permanganate to yield 3β,16β-diacetoxy-14-hydroxy-5β, 14β-etioic acid, which was already known (Meyer, 1949b). On treatment of bufotalin with hydrochloric acid in ethanol, it loses 1 mole of water and 1 mole of acetic acid. By this reaction two more double bonds are formed, and from the large increasing in the optical rotatory value at $\lambda = 584$ nm from $+5°$ to $+405°$, it was concluded

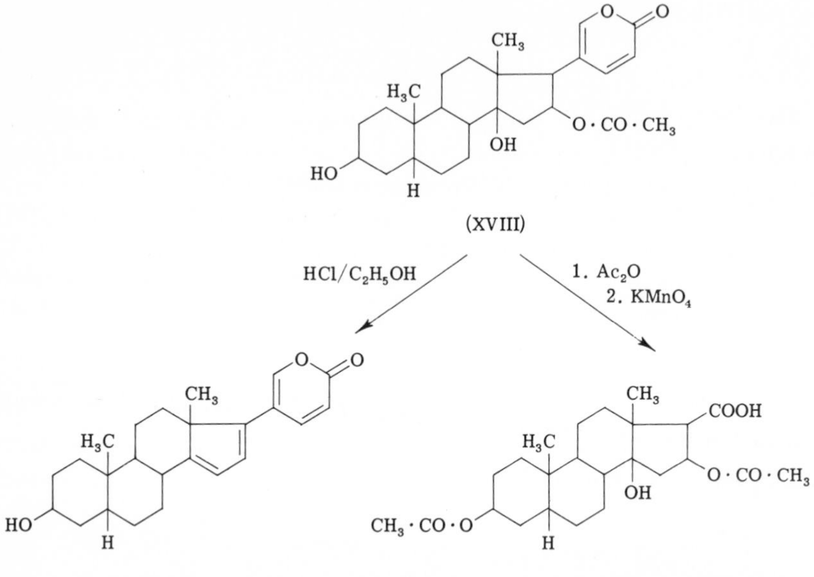

(XVIII)

HCl/C_2H_5OH

1. Ac_2O
2. $KMnO_4$

that these double bonds must be in the D ring of the steroid system. From this, the position of the tertiary hydroxy group at C-14 and the position of the *O*-acetyl group at C-16 is proven. Marinobufagin (XIX) has been obtained in cristalline form by Abel and Macht (1911). Its structure (Meyer and Pataski, 1955; Jensen and Evans, 1934; Meyer, 1951; Schröter *et al.*, 1955) was the key substance for the structure of two more bufogenins; telocinobufagin (XX) and bufotalines (XXI) (Bharucha *et al.*, 1959; Schröter *et al.*, 1958). Marinobufagin is reduced by sodium borohydride to yield telocinobufagin. Bufotalinine is converted into marinobufagin by Wolff–Kishner reduction.

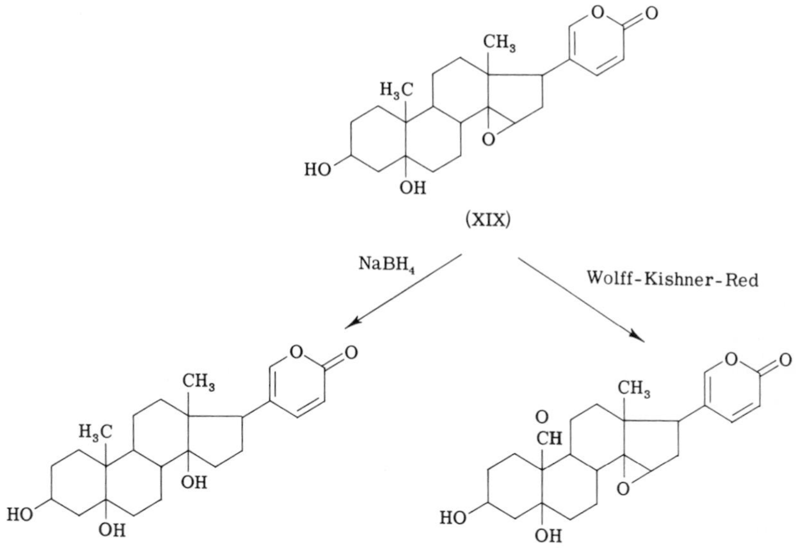

(XIX)

$NaBH_4$ Wolff-Kishner-Red

(XX) (XXI)

4. Bufotoxins

In addition to the bufogenins in the skin glands of toads, the bufotoxins are found. They are esters of suberylarginine with bufogenins. Hydrolysis is possible by enzymes only. By using acid or alkali, 1 mole of water is split off. The structure elucidation has been started by Wieland and Weil (1913) and Wieland and Allen (1922) and accomplished by Meyer (1949b). According to recent investigations of Meyer (1966), the suberylarginine residue is attached to the C-3 hydroxy group rather than to the C-14 hydroxy group as originally accepted. Bufotoxin, therefore, should possess structure XXII.

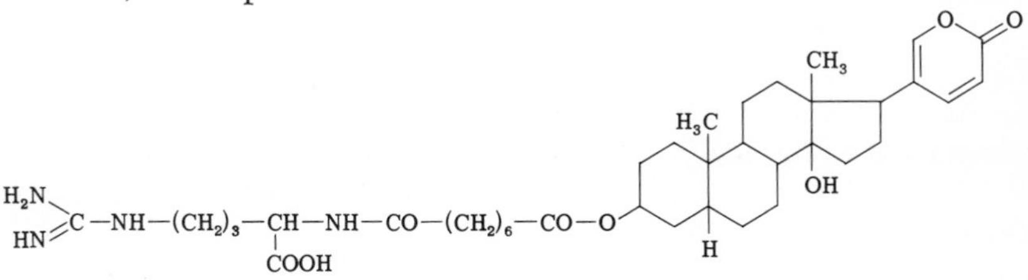

(XXII)

The biogenesis of bufogenins and bufotoxins starts with cholesterol (Siperstein *et al.*, 1957). The intermediates during this pathway are not yet known. The physiological activity of bufogenins and bufotoxins is quite similar to that of digitalis glycosides. They improve the contraction power of the sick heart, cause a raising of the tonus and a lowering of the frequency. However, they are not used therapeutically, as the plant glycosides of digitalis are more easily available. The strong local anaesthetic effect is remarkable; it is stronger than that of cocaine.

Certain relationships between structure and physiological activity have been found. Hydrogenation of the double bonds of the six-membered lactone ring (bufadienolide system) yields completely ineffective compounds; the same is true for the cis-connection of the rings C and D; elimination of the hydroxy group at C-14 or conversion into 14, 15-epoxide cause a decreasing of the heart activity and instead produces convulsion. For detailed discussion of occurrence of these compounds in toads, see Meyer and Linde (1971).

B. Leptodactylinae

The skin gland secretion of *Leptodactylus pentadactylus* and *L. ocellatus,* endemic to South and Central America, contains considerable amounts of derivatives of 5-hydroxytryptamine (10 mg/gm skin), specifically, *N*-methyl-serotonin (XXIII), bufotenine (5-hydroxy-*N*-methyltryptamine), bufotenidine, and the isomeric phenolbetaines leptodactyline (XXIV) and candicin (XXV) (Cei *et al.,* 1967; Erspamer, 1958,

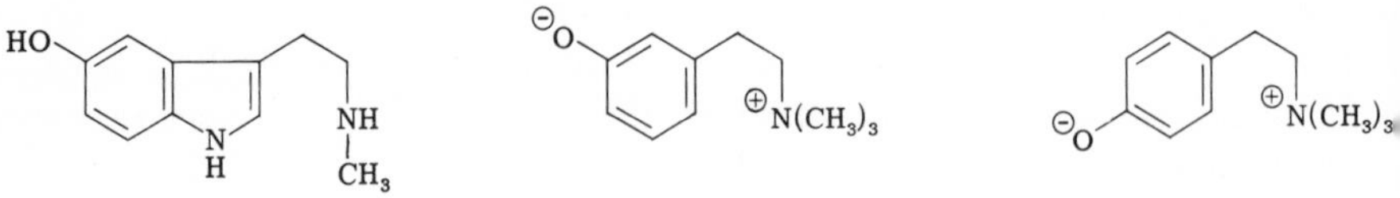

(XXIII) (XXIV) (XXV)

1959b; Erspamer *et al.,* 1964a). The biogenesis of leptodactylin starts from tyrosine (Blaschko, 1958, 1959; Nitoma *et al.,* 1957) (Eq. 1). The pharmacological activity of leptodactylin is a strong nicotinelike stimulation of the autonomous ganglia and of the neuromuscular irritation transfer as well as a neuromuscular block (Erspamer and Glaeser, 1960; Glaeser and Pasini, 1960).

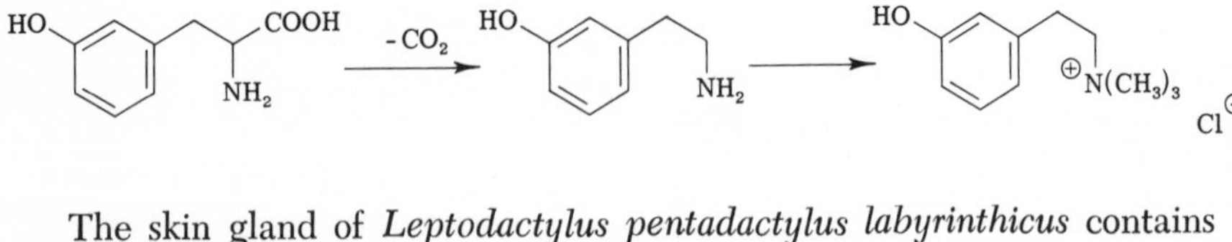

The skin gland of *Leptodactylus pentadactylus labyrinthicus* contains imidazolylalkylamines such as histamine (XXVI), *N*-methylhistamine, *N*-acetylhistamine, *N,N*-dimethylhistamine, and spinaceamine (XXVII) (Erspamer *et al.,* 1963, 1964b). Peptides of hypotensive activity have

been detected in *Physalaemus spp.* (Erspamer and Anastasi, 1965). Most of them—e.g., from *Physalaemus bresslaui,* and *Physalaemus cuvierii* as well as those of *Phyllomedusa spp.*—are not yet investigated in chemi-

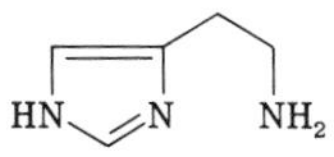

HN N NH$_2$

(XXVI)

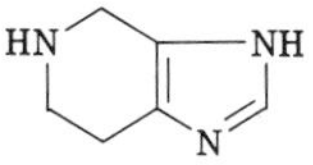

HN NH N

(XXVII)

cal respect. From the skin glands of *Physalaemus fuscumaculatus* and from *Physalaemus centralis* a peptide of strong hypotensive activity has been isolated, physalaemin:

Pyroglutamyl-Ala-Asp-Pro-Asp (NH_2)-Lys-Phe-Tyr-Gly-Leu-Met-NH_2

It possesses a strong vasodilatoric activity and is therefore strongly hypotensive (Erspamer *et al.,* 1964b).

C. Dendrobatidae

The dendrobatid frogs of South and Central America produce a number of rather interesting basic substances, which are true alkaloids. They cannot be derived from amino acids and thus are not biogenic amines. Dendrobatides have long been well known as highly toxic. Their skin gland secretions are used by the native indians as arrow poisons for hunting. Four species were investigated so far.

1. *Phyllobates aurotaenia* (previous names: *Ph. chocoenis, Ph. bicolor, Ph. latinasus, Dendrobates tinctorius*)

The structure of the toxin from *Ph. aurotaenia* has been elucidated by Witkop *et al.* (Märki and Witkop, 1963; Daly *et al.,* 1965; Tokuyama *et al.,* 1968, 1969). Altogether there are four highly toxic substances that are steroidal alkaloids. However, they do not belong to one of the known groups of steroidal alkaloids.

The most toxic of these four substances is batrachotoxin (XXVIII), which is the C-20 ester of 2,4-dimethylpyrrol-3-carboxylic acid and batrachotoxinin A, the second component. The third component is homobatrachotoxinin (formerly named as isobatrachotoxin), the C-20 ester of 2-ethyl-4-methylpyrrol-3-carboxylic acid and batrachotoxinin A. Finally, the secretion contains a very unstable substance, pseudobatrachotoxin, which on standing at room temperature yields batrachotoxinin A spontaneously. Constitution and configuration of these compounds come from an X-ray structure analysis of batrachotoxinin A-*p*-brombenzoate.

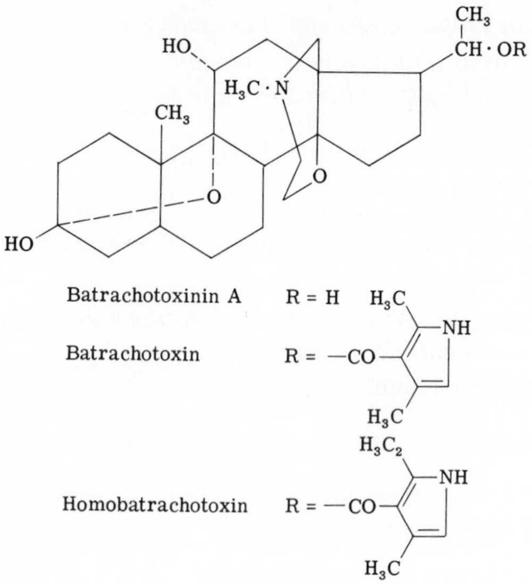

Batrachotoxinin A R = H

Batrachotoxin R = —CO—(2-ethyl... see structure: H_3C, NH, H_3C)

Homobatrachotoxin R = —CO—(H_3C_2, NH, H_3C)

(XXVIII)

Physiologically, batrachotoxin acts on the central nervous system. Batrachotoxin blocks the neuromuscular transmission irreversibly and evokes a muscle contracture in an isolated nerve–muscle preparation (Märki and Witkop, 1963). The effects of batrachotoxin in neuromuscular preparations both pre- and postsynaptically in nerve axons, in superior cervical ganglion, in heart Purkinje fibers, and in brain slices appear to be due to the selective and irreversible increase in permeability of membranes to sodium ions. The subsequent effects of this increase in Na^+ permeability evoked by batrachotoxin—such as membrane depolarization, enhanced spontaneous transmitter release, muscle contracture, and enhanced formation of cyclic AMP in brain slices—may be blocked reversibly by tetrodotoxin (Albuquerque *et al.*, 1971). Symptoms of muscle poisoning and finally respiratory paralysis occur. The activity, however, differs remarkably from that of curare. Death occurs after a few seconds; an antidote is not yet known.

The biological activity depends on—besides the seven membered heterocyclic ring system between the carbon atoms 13 and 14 of the steroid skeleton—the side chain at C-17 and the 3β-hydroxy-3α, 9α-oxido group. After reduction of this group, the activity decreases to $\frac{1}{100}$. Experiments on the biosynthesis of this interesting substance are in progress. Among other dendrobatid frogs, batrachotoxin, homobatrachotoxin, and batrachotoxinin A have been found in *Phyllobates vittatus* (Costa Rica) and *Phyllobates lugubris* (Panama).

2. *Dendrobates pumilio*

From the skin of *D. pumilio*, the following alkaloids have been isolated: pumiliotoxin A ($C_{19}H_{33}NO_2$), pumiliotoxin B ($CH_{19}H_{33}NO_3$), and pumiliotoxin C ($C_{13}H_{25}N$). Each frog contains about 200 γ of the toxins. The isolation and purification, because of the high tendency to decomposition, is very difficult. The structure of pumiliotoxin C (XXIX) has been elucidated by X-ray analysis (Daly *et al.*, 1969); its structure is, although closely related to coniine, a new type of alkaloid. Subcutaneous injection of the pumiliotoxins causes death after strong clonic convulsions.

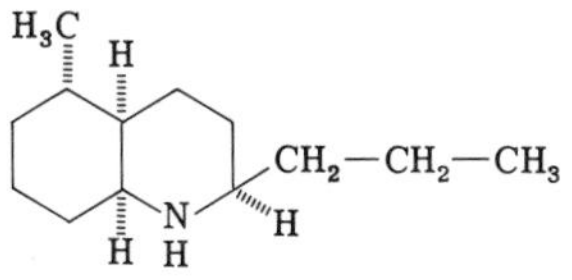

(XXIX)

3. *Dendrobates auratus*

Pumiliotoxins A, B, and C have also been isolated from *D. auratus* (Habermehl *et al.*, 1974) together with seven other closely related alkaloids.

4. *Dendrobates histrionicus*

Dendrobates histrionicus produces another type of alkaloid with a spirane system. Two of these alkaloids, histrionicotoxin ($C_{19}H_{25}NO$) (XXX) and dihydroisohistrionicotoxin ($C_{19}H_{27}NO$)(XXXI), have been

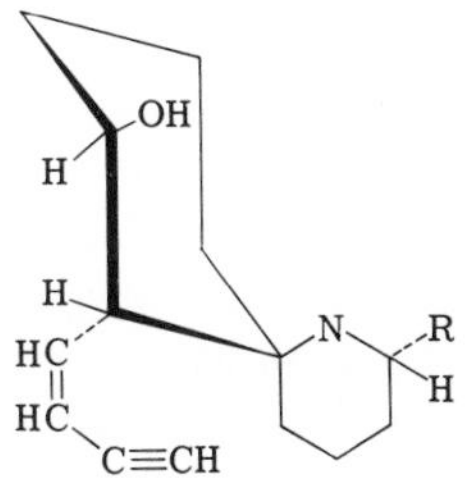

(XXX) R = $-CH_2-CH{=}CH-C{\equiv}CH$
(XXXI) R = $-CH_2-CH_2-CH{=}C{=}CH_2$

isolated; their structures have been determined by X-ray analysis (Daly *et al.*, 1971). Both of these alkaloids represent the first example of acetylenic and allenic alkaloids in animals. The histrionicotoxins are relatively nontoxic compared with the pumiliotoxins.

5. *Biosynthesis of Dendrobates Toxin*

In this field, not much is known so far. Possible radioactive precursors of the dendrobatid alkaloids, such as mevalonate, acetate, and various steroids, were not incorporated to a significant extent into the various alkaloids of these frogs. Histrionicotoxins and pumiliotoxins occur together in the same species of *Dendrobates* (Daly and Myers, 1973) and may share a common precursor, possibly a disubstituted unsaturated piperidine which could undergo cyclization to either the spiro system of the histionicotoxins or the decahydroquinoline ring system of pumiliotoxin C.

D. Atelopotidae

From the skins of *Atelopus* species, a strong, dialyzable toxin, atelopidtoxin, can be isolated (Fuhrmann *et al.*, 1969a,b; Shindelman and Mosher, 1969). It is found in *A. varius, A. ambulatorius, A. cruciger, A. planispina,* and in extremely high amounts in *A. zeteki.* The secretions of these frogs are used by the native indians of Panama, Costa Rica, and Colombia as arrow poisons. They possess a hypotensive and cardiotoxic activity and they cause convulsions.

Chemical investigations are in progress; from the results known so far, it is evident that the toxin is easily soluble in water; it may be precipitated from aqueous solution by acetone. A guanidine group has been proven, and from that some relationship with tetrodotoxin may be suspected; Atelopidtoxin however, differs in pharmacological action remarkably from tetrodotoxin, saxitoxin, and batrachotoxin.

E. Hylidae

More than 400 species of Hylidae are known; their distribution reaches all over the world. A number of them has been investigated. From *Hyla arborea,* endemic to Europe and Northern Africa, a hemolytically active peptide has been isolated, which is active in dilutions of 1:200,000; its structure, however, is not yet known.

Besides serotonin, which is found in all of the hylidae, investigated so far (Flury, 1926; Erspamer *et al.*, 1966a,b; De Caro *et al.*, 1968), *Hyla caerulea* from Australia contains histamine and caerulein (Anastasi *et al.*, 1968), a peptide of strong hypotensive activity.

Pyroglutamyl-Glu-Asp-Tyr (SO_3H)-Thr-Gly-Tryp-Met-Asp-Phe-NH_2

Hyla pearsoniana and *Hyla peroni* are known to possess bufotenine (Erspamer *et al.*, 1966a,b; De Caro *et al.*, 1968). From *Phyllomedusa rhodei,* a bradykinin-like toxin has been isolated, phyllokinin (Anastasi *et al.,* 1965, 1968):

Bradykinyl-Ileu-Tyr(SO_3H).

F. Ranidae

This family has been investigated very thoroughly. Most of these animals contain indolalkylamines known from the toads, and again serotonin (Welsh and Zipf, 1966). About twenty species have been investigated in detail. From *Rana temporaria* as well as *Rana nigromaculata* bradykinin has been isolated. Bradykinin is a peptide, the hypotensive activity of which is well known:

Arg-Pro-Pro-Gly-Phe-Ser-Pro-Phe-Arg

The skin gland secretion of the European *Rana esculenta* contains four peptides, the structures of which have not been elucidated. (Kiss and Michl, 1962). Aqueous solutions are toxic even in high dilutions. The muscles of the frog heart are paralyzed from concentrations of 1:500,000. The LD_{50} of the crude secretion is 6–12 mg/kg rabbit. Besides these peptides, a protein of strong hemolytical activity has been obtained.

G. Discoglossidae

Among this family, *Bombina bombina* and *Bombina variegata* have been investigated thoroughly by Michl *et al.* (Kiss and Michl, 1962). The dry secretion contains about 10% 5-hydroxytryptamine (serotonin). Besides this, four basic peptides have been obtained, one of which has the sequence:

Ala-Glu-His-Phe-Ala-Asp-NH_2

In the secretion of *Bombina variegata,* there were found γ-amino-butyric acid and 5-hydroxytryptamine as well as twelve free amino acids (Michl and Bachmayer, 1963, 1964). Two nonapeptides have been isolated; the amino acid sequences are (Michl and Bachmayer, 1965):

Ser-Ala-Lys-Gly-Leu-Ala-Glu-His-Phe

Gly-Ala-Lys-Gly-Leu-Ala-Glu-His-Phe

H. Pipidae

The only species investigated from the family is *Xenopus laevis,* endemic to South and Central America. Active compounds of its skin gland secretion are serotonin (Erspamer and Vialli, 1952) and bufotenidin (Jensen, 1935).

IV. Conclusion

Looking over all of the compounds from the skin glands of Amphibia isolated and investigated so far, its amazing variety in chemical respect is striking, especially if one considers that nearly each species has developed its own type of venom during its evolution. Of course, the question for the use of these compounds arises. As recently has been found (Habermehl and Preusser, 1969, 1970; Czordás and Michl, 1969), these compounds are, at least in part, a protection against microorganisms such as bacteria, fungi, and yeasts, which otherwise cause local lesions of the skin. If one considers that all of the amphibia are living in a biotope highly filled with microorganisms, and on the other hand that the skin of amphibia is a most necessary organ for respiration as well as for uptake and loss of water, the importance of these compounds as a protection is evident.

REFERENCES

Abel, J. J., and Macht, D. J. (1911). *J. Pharmacol. Exp. Ther.* **3**, 319.

Albuquerque, E. X., Daly, J. W., and Witkop, B. (1971). *Science* **172**, 995.

Anastasi, A., Erspamer, V., and Bertaccini, C. (1965). *Comp. Biochem. Physiol.* **14**, 43.

Anastasi, A., Erspamer, V., and Endean, R. (1968). *Arch. Biochem. Biophys.* **125**, 57.

Bharucha, M., Jaeger, H., Meyer, K., Reichstein, T., and Schindler, O. (1959). *Helv. Chim. Acta* **42**, 1395.

Blaschko, H. (1958). *Bull Soc. Chim. Biol.* **40**, 1817.

Blaschko, H. (1959). *Biochim. Biophys. Acta* **4**, 130.

Cei, J. M., Erspamer, V., and Roseghini, M. (1967). *Syst. Zool.* **16**, 328.

Czordás, A., and Michl, H. (1969). *Toxican* **7**, 103.

Daly, J. W., and Myers, C. W. (1974). In preparation.

Daly, J. W., and Witkop. B. (1970). *In* "Venomous Animals and Their Venoms" (W. Bucherl, E. E. Buckley, and V. Deulofeu, eds.), Vol. 2, Chapter 39, p. 492. Academic Press, New York.

Daly, J. W., Witkop, B., Bommer, P., and Biemann, K. (1965). *J. Amer. Chem. Soc.* **87**, 124.

Daly, J. W., Tokuyama, T., Habermehl, G., Karle, I. L., and Witkop, B. (1969). *Justus Liebigs Ann. Chem.* **729**, 198.

Daly, J. W., Karle, I. L., Myers, C. W. Tokuyama, T., Waters, J. A., and Witkop, B. (1971). *Proc. Nat. Acad. Sci. U.S.* (in press).

De Caro, G., Endean, R., Erspamer, V., and Roseghini, M. (1968). *Brit. J. Pharmacol. Chemother.* **33**, 38.

Erspamer, V. (1958). *Ric. Sci.* **28**, 2065.

Erspamer, V. (1959a). *Biochem. Pharmacol.* **2**, 270.

Erspamer, V. (1959b). *Arch. Biochem. Biophys.* **82**, 431.

Erspamer, V., and Anastasi, A. (1965). *In* "Hypotensive Peptides" (E. G. Erdös, H. Back, and F. Sicuteri, eds.), p. 63. Springer-Verlag, Berlin and New York.

Erspamer, V., and Glaeser, A. (1960). *Brit. J. Pharmacol. Chemother.* **15**, 14 (1960).

Erspamer, V., and Vialli, M. (1952). *Ric. Sci.* **22**, 1420.
Erspamer, V., Vitali, T., Roseghini, M., and Cei, J. M., (1963). *Experientia* **19,** 346.
Erspamer, V., Anastasi, A., Bertaccini, C., and Cei, J. M. (1964a). *Experientia* **20**, 489.
Erspamer, V., Vitali, R., Roseghini, M., and Cei, J. M. (1964b). *Arch. Biochem. Biophys.* **105**, 620.
Erspamer, V., De Caro, G., and Endean, R. (1966a). *Experientia* **22**, 738.
Erspamer, V., Roseghini, M., Endean, R., and Anastasi, A. (1966b). *Nature* (*London*) **212**, 204.
Fabing, H. D., and Hawkins, J. R. (1956). *Science* **123**, 886.
Faust, E. G. (1898). *Arch. Exp. Pathol. Pharmakol.* **41**, 229.
Faust, E. G. (1900). *Arch. Exp. Pathol. Pharmakol.* **43**, 84.
Faust, E. S. (1902). *Arch. Exp. Pathol. Pharmakol.* **47**, 278.
Faust, E. S. (1903). *Arch. Exp. Pathol. Pharmakol.* **49**, 1.
Flury, F. (1926). *Arch. Exp. Pathol. Pharmakol.* **118**, 235.
Fuhrmann, F. A., Fuhrmann, G. J., and Mosher, H. S. (1969a). *Science* **165**, 1376.
Fuhrmann, F. A., Fuhrmann, G. J., Dull, D. L., and Mosher, H. S. (1969b). *Agr. Food Chem.* **17**, 417.
Gessner, O. (1926). *Ber. Ges. Bef. Ges. Naturwiss. Marburg.* **61**, 138.
Gessner, O. (1938). *In* "Handbuch der experimentellen Pathologie," Vol. 6, Springer-Verlag, Berlin and New York.
Gessner, O., and Esser, W. (1935). *Arch. Exp. Pathol. Pharmakol.* **179**, 639.
Gessner, O., and Möllenhoff, P. (1932). *Naunyn-Schmiedebergs Arch. Exp. Pathol. Pharmakol.* **167**, 638.
Gessner, O., and Urban, G. (1937). *Naunyn-Schmiedebergs Arch. Exp. Pathol. Pharmakol.* **187**, 378.
Gessner, P. K., Godse, D. O., Krull, A. H., and McMullan, J. M. (1968). *Life Sci.* **7**, 267.
Glaeser, A., and Pasini, C. (1960). *Farmaco, Ed. Sci.* **15**, 93.
Habermehl, G. (1963a). *Chem. Ber.* **96**, 143.
Habermehl, G. (1963b). *Chem. Ber.* **96**, 840.
Habermehl, G. (1970). *In* "Venomous Animals and Their Venoms" (W. Bucherl, E. E. Buckley, and V. Deulofeu, eds.), Vol. 2, Chapter 42, p. 569, Academic Press, New York.
Habermehl, G. *et al.* (1974). In preparation.
Habermehl, G., and Haaf, G. (1965). *Chem. Ber.* **98**, 3001.
Habermehl, G. (1966). *Chem. Ber.* **99**, 1439.
Habermehl, G., and Haaf, G. (1968a). *Chem. Ber.* **101**, 198.
Habermehl, G., and Haaf, G. (1968b). *Z. Naturforsch. B* **23**, 1151.
Habermehl, G., and Haaf, G. (1969). *Justus Liebigs Ann. Chem.* **722**, 155–161.
Habermehl, G., and Preusser, H. J. (1969). *Z. Naturforsch. B* **24**, 1599.
Habermehl, G., and Preusser, H. J. (1970). *Z Naturforsch. B* **25**, 1451.
Habermehl, G., and Vogel, G. (1969). *Toxicon* **7**, 163.
Handovsky, H. (1920). *Arch. Exp. Pathol. Pharmakol.* **86**, 138.
Harley-Mason, J., and Jackson, A. H. (1954). *J. Chem. Soc., London* p. 1165.
Hoshino, T., and Shimodaira, K. (1935). *Justus Liebigs Ann. Chem.* **520**, 19.
Jensen, H. (1935). *J. Amer. Chem. Soc.* **57**, 1765.
Jensen, H., and Evans, E. A., Jr. (1934). *J. Biol. Chem.* **104**, 307.
Kiss, G., and Michl, H. (1962). *Toxicon* **1**, 33.
Märki, F., and Witkop, B. (1963). *Experientia* **19**, 329.

Märki, F., Robertson, A. V., and Witkop, B. (1961). *J. Amer. Chem. Soc.* **83**, 3341.
Märki, F., Axelrod, J., and Witkop, B. (1962). *Biochim. Biophys. Acta* **58**, 367.
Meyer, K. (1949a). *Pharm. Acta Helv.* **24**, 222.
Meyer, K. (1949b). *Helv. Chim. Acta* **32**, 1993.
Meyer, K. (1951). *Helv. Chim. Acta* **34**, 2147.
Meyer, K. (1966). *Mem. Inst. Butantan, Sao Paulo* **33**, 425.
Meyer, K., and Linde, H. (1971). *In* "Venomous Animals and Their Venoms," (W. Bucherl, E. E. Buckley, and V. Deulofeu, eds.). Vol. 2, Chapter 40, p. 521, Academic Press, New York.
Meyer, K., and Pataski, S. (1955). *Helv. Chim. Acta* **38**, 1631.
Michl, H., and Bachmayer, H. (1963). *Monatsh. Chem.* **99**, 814.
Michl, H., and Bachmayer, H. (1964). *Chem.* **95**, 480.
Michl, H., and Bachmayer, H. (1965). *Monatsh. Chem.* **95**, 1166.
Mosher, H. S., Fuhrmann, F. A., Buchwald, H. D., and Fischer, H. G. (1964). *Science* **144**, 110.
Nikander of Kolophon, "Theriaca at Alexipharmaca" (G. Schneider, ed., Leipzig, 1816).
Nitoma, C., Posner, H. S., Bogdanski, D. F., and Udenfriend, S. (1957). *J. Pharmacol. Exp. Ther.* **120**, 188.
Oestlund, E. (1954). *Acta Physiol. Scand.* **31**, Suppl. 112, 55.
Phisalix, C., and Bertrand, G. (1893). *C. R. Soc. Biol.* **45**, 477.
Plinius, C., Secundus, "Historia naturalis," Libr. XXIX and XXXVII (J. Harchinus, ed., Paris, 1741).
Schöpf, C., and Braun, W. (1934). *Justus Liebigs Ann. Chem.* **514**, 69.
Schöpf, C., and Klein, D. (1954). *Chem. Ber.* **87**, 1638.
Schöpf, C., and Koch, K. (1942a). *Justus Liebigs. Ann. Chem.* **552**, 37.
Schöpf, C., and Koch, K. (1942b). *Justus Liebigs Ann. Chem.* **552**, 62.
Schöpf, C., and Müller, O. W. (1960). *Justus Liebigs Ann. Chem.* **633**, 127.
Schöpf, C., Blödorn, H. K., Klein, D., and Seitz, G. (1950). *Chem. Ber.* **83**, 372.
Schröter, H., Tamm, C., and Reichstein, T. (1955). *Helv. Chim. Acta* **38**, 883.
Schröter, H., Tamm, C., and Reichstein, T. (1958). *Helv. Chim. Acta* **41**, 720.
Shindelman, J., and Mosher, H. S. (1969). *Toxicon* **7**, 315.
Siperstein, M. D., Murray, A. W., and Titius, E. (1957). *Arch. Biochem. Biophys.* **67**, 157.
Slotta, C. H., and Neisser, C. (1937). *Mem. Inst. Butantan, Sao Paulo* **11**, 89.
Speeter, M. E., and Anthony, W. C. (1954). *J. Amer. Chem. Soc.* **76**, 6208.
Stoll, A., Troxler, F., Peyer, J., and Hofmann, A. (1955). *Helv. Chim. Acta* **38**, 1452.
Tokuyama, T., Daly, J. W., Witkop, B., Karle, I. L., and Karle, J. (1968). *J. Amer. Chem. Soc.* **90**, 1917.
Tokuyama, T., Daly, J. W., and Witkop, B. (1969). *J. Amer. Chem. Soc.* **91**, 3931.
Tsuda, K. (1966). *Naturwissenschaften* **53**, 171.
Tsuda, K., Tamura, C., Tachikawa, R., Sakai, K., Amakasu, O., Kawamura, M., and Ikuma, S. (1963). *Chem. Pharm. Bull.* **11**, 1473.
von Euler, U. S. (1958). *Recent Progr. Horm. Res.* **14**, 483.
Wakely, J. F., Fuhrmann, G. J., Fuhrmann, F. A., Fischer, H. G., and Mosher, H. S. (1966). *Toxicon* **3**, 195.
Welsh, J. H., and Zipf, J. B. (1966). *J. Cell. Physiol.* **68**, 25.
Wieland, H., and Allen, R. (1922). *Ber. Deut. Chem. Ges.* **55**, 1789.

Wieland, H., and Behringer, H. (1941). *Justus Liebigs Ann. Chem.* **549**, 209.
Wieland, H., and Vocke, F. (1930). *Justus Liebigs Ann. Chem.* **481**, 215.
Wieland, H., and Weil, F. J. (1913). *Ber. Deut. Chem. Ges.* **46**, 3315.
Wieland, H., Hesse, G., and Mittasch, H. (1931). *Ber. Deut. Chem. Ges.* **64**, 2099.
Wieland, H., Konz, W., and Mittasch, H. (1934). *Justus Liebigs Ann. Chem.* **513**, 1.
Wölfel, E., Schöpf, C., Weitz, G., and Habermehl, G. (1961). *Chem. Ber.* **94**, 2361.
Zalesky, S. (1866). *Med. Chem. Untersuch., Hoppe-Seyler* **1**, 85.

SECTION II

REPTILIA

CHAPTER 8

Plasma Proteins of Reptilia*

Herbert C. Dessauer

I. Introduction 187
II. The Plasma Protein System 189
A. Composition and Fractionation 189
B. Comparative Aspects 192
III. Specific Proteins 193
A. Albumins 193
B. Transferrins 196
C. Immunoglobulins 200
D. Plasma Vitellin 205
E. Poorly Characterized Proteins 208
References 213

I. Introduction

Knowledge of the structure and function of the mammalian plasma proteins and of methods used in their isolation and analysis (Putnam, 1960) is often a useful guide for work on the proteins of reptiles. Mammalian physiology, however, should be considered as only a guide, for it does not offer answers to numerous questions concerning reptiles. In general, such questions relate to fundamental differences between the physiology of reptiles and mammals and to their long period of reproductive isolation.

Reptilian proteins carry out their immune, transport, and other activities in a fluid that fluctuates widely in both temperature and composition (Table I), whereas mammalian proteins function in an almost uniform environment. For example, the marine iguana's body temperature varies between 22° and 37°C while it is actively feeding in the ocean (Morgareidge and White, 1969). Preferred activity temperatures of reptiles range from as low as 14°C in the tuatara to over 42°C in certain lizards (Bogert, 1959). Blood pH extremes of 6.4 and 8.1, considered lethal for mammals, are attained by reptiles during daily events (Dessauer, 1970). When mammals feed and secrete gastric juice, the level of the chloride ion in their plasma decreases by only 2–4 meq/liter.

* Supported by the National Science Foundation of the United States (Grant GB-7294X).

TABLE I
RANGE OF ELECTROLYTE LEVELS IN PLASMA OF REPTILES AND MAMMALS

Component	Reptiles[a]	Mammals[b]
pH	6.4–8.1	7.3–7.5
Electrolytes (m*M*/liter)		
Na^+	69–242	122–187
Ca^{2+}	1–90[c]	2–4
Cl^-	7–148	83–125
HCO_3^-	7–105	16–37
lactate$^-$	trace–50	trace–18

[a] From Dessauer, 1970.
[b] From Dittmer, 1961.
[c] Largely protein bound.

In the alligator, this chloride shift is so marked, that the plasma may become virtually chloride free (Coulson *et al.*, 1950).

Mammals and the four living orders of reptiles have evolved along independent lines of descent since the Permian Period (Fig. 1) (Romer, 1966). For about 250 million years, their genomes have diverged so that the proteins of species of the five groups have acquired many structural differences. Proteins of common origin have become progressively more different. Probably new proteins have appeared in and old ones have disappeared from the blood of one or more lines.

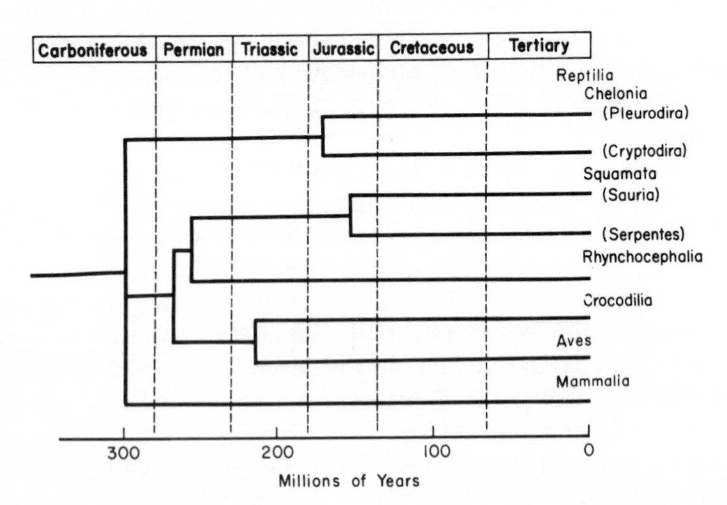

FIG. 1. Paleontological estimates of times of divergence of major evolutionary lines of the Reptilia. (From Romer, 1966.)

Evidence on the plasma proteins of reptiles is largely physicochemical and immunological. Many data have arisen as side issues from studies primarily concerned with other organisms. Though studies primarily concerned with reptiles are few, they offer new insight into problems of water balance, transport mechanisms, the nature of the immune response, and the evolutionary biology of proteins. In this chapter, I will summarize evidence concerning the plasma proteins of reptiles, pointing up their more unique properties. The evidence itself will emphasize a fact, often either ignored or not understood, that the Chelonia, Squamata, Rhynchocephalia, and Crocodilia are almost as divergent from each other as they are from class Mammalia. The monograph of Brocq-Rousseu and Roussel (1939) contains descriptions of early work. Detailed data and an extensive bibliography are given in a more recent review on blood chemistry of reptiles (Dessauer, 1970).

II. The Plasma Protein System

A. Composition and Fractionation

Proteins make up about 5% of the blood plasma (Table II). Such averages are useful only as points of reference, as levels vary markedly from individual to individual even within the same species (Dessauer, 1970). Highest values are most apt to be found in females of all groups during estrus (see Section III,D). A fall maximum and summer minimum occur in turtles (Masat and Musacchia, 1965). Plasma of a group of hybrid turtles contains less protein than that of either parental species (Crenshaw, 1965). The majority of individual plasma proteins are conjugated with carbohydrate, which composes from 3 to 7% of total protein mass (Lustig and Ernst, 1936). Sialic acids compose a significant part of this carbohydrate, especially in snakes (Seal, 1964).

Fractionation of plasma shows that it contains a complex mixture of specific proteins. Their sizes fall within molecular weight ranges comparable to the proteins of mammals. The mixture generally sediments in three or four peaks in the analytical ultracentrifuge and separates into three fractions when passed through a column packed with polydextrans such as Sephadex G-200 (Pharmacia Co.). These fractions contain proteins whose molecular weights range (1) between 40,000 and 120,000, (2) between 120,000 and 190,000, and (3) above 190,000. Proteins of the heavier fractions make up a greater percentage of the total protein than is true for mammals (Baril *et al.*, 1961; Roberts and Seal, 1965; Masat and Dessauer, 1968).

Free solution and zone electrophoresis resolve the proteins into five

TABLE II

TOTAL PROTEIN, ALBUMIN, AND TRANSFERRIN[a]

	Total protein[b]		Albumin[c]		Transferrin[d]		
	Number of species	gm/100 ml	Number of species	Percent of total	Number of species	Plasma iron (μg%)	Plasma iron-binding capacity (μg%)
Chelonia	17	4.4 (2.9–6.1)	12	17 (9–28)	6	50 (0–120)	209 (25–458)
Crocodilia	3	5.8 (5.1–6.5)	3	14 (13–15)	3	38 (18–87)	288 (230–305)
Squamata							
Sauria	13	5.1 (3.0–7.8)	16	27 (18–45)	24	68 (0–209)	218 (50–410)
Serpentes	22	4.8 (2.9–6.5)	23	22 (6–39)	40	70 (0–304)	228 (104–418)

[a] Figures in parentheses indicate range.

[b] Averages compiled from detailed table in Dessauer (1970).

[c] Averages estimated from electrophoretic patterns.

[d] Averages compiled from Barber and Sheeler, 1961; Dessauer *et al.*, 1962a; George and Dessauer, 1970.

to seven major fractions (Dessauer, 1970). Although these have mobilities comparable to the albumin, α, β, and γ fractions which typify electrophoretic patterns of human plasma proteins, the specific proteins which are present in fractions of comparable mobility are often different for species of the two classes (Fig. 2). Assumptions to the contrary can lead to incorrect conclusions. For example, an increase in the γ fraction of turtles treated with estrogens was interpreted as due to the stimulation of the immune system (Rao, 1968), whereas it probably was due to the induction of plasma vitellin (see Section III,D).

The extreme heterogeneity of the plasma proteins is revealed by high resolution electrophoresis and immunoelectrophoresis. Protein patterns of individual reptiles may exhibit as many as thirty components when subjected to starch gel electrophoresis (Dessauer and Fox, 1964); the

	Albumins			Transferrins				
	Species	α_2	Alb.	Species	γ	Sα_2	Trans. C.	Alb.
Chelonia								
Chelydridae	2			2				
Kinosternidae	6			4				
Emydidae	15			6				
Testudinidae	2			1				
Trionychidae	2			1				
Crocodilia								
Crocodylidae	2			1				
Alligatoridae	2			2				
Serpentes								
Typhlopidae	2			2				
Leptotyphlopidae	3			2				
Boidae	3			2				
Colubridae	54			44				
Elapidae	3			1				
Crotalidae	10			7				
Sauria								
Anguidae	4			4				
Anniellidae	1							
Helodermatidae	1			1				
Varanidae	2			2				
Xantusiidae	2			1				
Gekkonidae	6			5				
Lacertidae	2			1				
Cordylidae	1			1				
Scincidae	10			8				
Teiidae	8			7				
Amphisbaenidae	1							
Agamidae	2			1				
Chamaeleonidae	1			1				
Iguanidae	15			13				

FIG. 2. Ranges of electrophoretic migrations of the albumins and transferrins of species within families of the Reptilia. Mobilities of human plasma proteins form a frame of reference. Material assembled from studies listed among the references and from unpublished observations.

antigenic structure of these proteins is as complex as those of mammals (Maung, 1963; Lewis, 1964; Fox and Dessauer, 1965; Lykakis, 1971). Such techniques have also uncovered a number of intraspecific variants of specific proteins (Dessauer, 1970).

B. Comparative Aspects

Comparative studies on the plasma proteins illustrate the evolutionary divergence of the reptilian orders (Fig. 1). Antisera against plasma proteins of a species classified in one taxonomic order do not cross react with those of species from a different order (Graham-Smith, 1904; Cohen, 1955). The charge density distribution of the plasma proteins is distinct enough for species of each order that it is possible to "key out" the Chelonia, Crocodilia, and Squamata using gross features of their electrophoretic patterns as taxonomic characters (Dessauer and Fox, 1956, 1964).

Immunological cross reactions involving plasma proteins reflect degrees of divergence and suggest genetic affinities of taxa within each order. Plasma proteins of the Chinese and North American alligators and the South American caiman, which are classified in the same family of the Crocodilia, show high serological correspondence; however, those of the genera *Crocodylus* and *Alligator*, which are classified in different families, exhibit very low serological correspondence (Graham-Smith, 1904; Lewis, 1964; Gorman *et al.*, 1971b). Turtles of the suborders Cryptodira and Pleurodira, which have evolved along different lines since the Jurassic Period, show low serological correspondence. Within the Cryptodira (1) marine, (2) softshell, and (3) freshwater and land turtles are divergent groups (Frair, 1962, 1964, 1969). Squamates of the suborders Sauria (lizards) and Serpentes (snakes) apparently diverged in the Jurassic Period. Although the proteins of lizards and snakes show low immunological correspondence (Graham-Smith, 1904; Cohen, 1955), they remain similar enough that it was not possible to key out all lizards from all snakes on the basis of gross characteristics of electrophoretic patterns (Dessauer and Fox, 1964). Recent immunological studies are helping to unravel the complex taxonomy of genera of advanced snakes (Pearson, 1966, 1968; see Section III,B,3).

Numerous electrophoretic studies have demonstrated the distinctness of the protein complement of a particular species and have suggested genetic affinities of species and species groups (Deutsch and McShan, 1949; Seniów, 1963; Latifi *et al.*, 1965; Crenshaw, 1962, 1965; Kmeťová and Paulov, 1966; Maldonado and Ortiz, 1966; Newcomer and Crenshaw, 1967; Voris, 1967; Rider and Bartel, 1967; Lykakis, 1971). Earlier literature was reviewed by Dessauer and Fox (1964).

III. Specific Protein

A. Albumins

1. Isolation and Identification

Plasma of reptiles contains a protein with physicochemical properties similar to mammalian albumins. Unfortunately, albumin has no readily tested catalytic function and must be identified on the basis of its marked hydrophilic character, low molecular weight, high surface charge, and ion-binding capacity (Putnam, 1960). Studies using both molecular sieves (Masat and Dessauer, 1968; Marchalonis *et al.,* 1969) and the ultracentrifuge (Baril *et al.,* 1961; Roberts and Seal, 1965) prove that proteins are present that have a molecular weight of the proper order of magnitude. The protein that makes up the highest percentage of the fraction of low molecular weight also has the highest charge density of any plasma protein of quantitative importance. The migration rate of the protein from species of the Rhynchocephalia (Marchalonis *et al.,* 1969), Squamata, and Crocodilia equals or exceeds that of human albumin in alkaline buffers (Fig. 2). Mobilities in free solution electrophoresis of crocodilian albumins vary from -6.35 to -7.92×10^{-5} cm^2 volt^{-1} second^{-1} in 0.1 *M* barbital buffer of pH 8.6 (Rider and Bartel, 1967). The rate of migration of the protein of turtles of most families is only about that of an α_1-globulin (Fig. 2). Besides their albumin-like physical properties, these proteins are also very soluble in half-saturated ammonium sulfate and in cold alcohol–water solutions of low dielectric constant; their trichloracetic acid salts dissolve in 95% ethanol; they are insoluble in 70–100% saturated ammonium sulfate and in 2% Rivanol (Masat and Dessauer, 1968); and they do not contain enough carbohydrate to react with concanavalin A (Nakamura *et al.,* 1965; Masat and Dessauer, 1968).

2. Volume Expander Function

Albumins make a large contribution to total colloid osmotic pressure and hence to plasma volume. The contribution of the different plasma proteins to osmotic pressure was estimated from data on protein concentrations and molecular weights, appraising the contribution of the Donnan effect from electrophoretic migration rates (Masat and Dessauer, 1968). Pressures for freshwater and land turtles ranged from 48 to 160 mm of water, with albumin generally contributing 20–40% of the pressure. These low osmotic pressures were traceable to the low concentration and low charge densities of the turtle albumins (Cohen, 1954; Cohen and Stickler, 1958; Leone and Wilson, 1961; Crenshaw, 1962; Dessauer and Fox, 1964; Frair, 1964). Those with mobilities as fast as human albumin have been found only in species of the genera *Dermatemys, Kinosternon, Stenotherus, Staurotypus,* and *Chelodina* (Fig. 2). The large volume of fluid of high bicarbonate content, which is commonly present in the abdominal and pericardial cavities of freshwater turtles, probably results from the low level of albumin. This ascites may represent a physiological adaptation for buffering hydrogen ions generated by anaerobic metabolic processes during extended dives (Dessauer, 1970).

Relatively high colloid osmotic pressures characterize plasma of reptiles that are more active and adapted to dry, hot environments. For example, total colloid osmotic pressure of the lizard *Iguana iguana* was 290 mm of water, with albumin contributing over 60% of the pressure. Albumins of such species often have electrophoretic mobilities that exceed that of human albumin (Fig. 2). The concentration of albumin was lower in blood of aquatic reptiles than in semidesert or desert species (Khalil and Abdel-Messeih, 1963). Albumins apparently are present in higher concentrations in land tortoises than in freshwater turtles (Lykakis, 1971). Albumin made up an average of 32% of the total protein of land forms of advanced snakes, but only 11% of the total protein of water snakes (Dessauer, unpublished observations). The rates of evaporative water loss vary one hundredfold from desert-adapted snakes to those adapted to tropical forests or life in moist soil (Gans *et al.*, 1968). Perhaps the concentration of albumin and its charge are important factors in controlling such water loss.

3. *Binding of Anions and Other Substances*

Albumins of reptiles bind anions and a variety of potentially toxic substances. Albumin of *Alligator mississippiensis* bound tryptophane and five other indole derivatives less strongly than did albumins of a number of birds and mammals. Indole propionate, which was complexed most strongly, was used to estimate the amount of albumin present in plasma at 12% of the total protein (McMenamy and Watson, 1968), a value close to the estimate of 14% based upon analysis of electrophoretic patterns (Baril *et al.*, 1961). Albumins of reptiles bind hematin (Liang, 1957; Dessauer, personal observation) but have low affinities for dyes such as bromphenol blue (Masat and Dessauer, 1968).

Albumin is partially responsible for protecting snakes from the toxic effects of snake venom (Philpot and Smith, 1950). The protective capacity of serum is associated with the albumin rather than the immunoglobulins of the rattlesnake *Crotalus adamantes*. Neither the serum nor albumin, however formed precipitin bands during immunoelectrophoresis tests with venom (Clark and Voris, 1969). Mammalian albumins complex lysolecithins produced by the action of venom phosphatidases, thereby inhibiting the cytolytic action of these venom products (Luzzio, 1967). Perhaps reptilian albumins perform their protective function in a similar fashion.

4. Evolutionary Aspects

Albumin structure has changed relatively rapidly during evolution. The rate of change apparently has been just as rapid in the Squamata and Crocodilia as in orders of the Mammalia. Quantitative immunological data resulting from titrations using antisera against the albumins of the lizard *Iguana iguana* and *Alligator mississippiensis* were used to calculate immunological distances, a logarithmic function which reflects the extent of cross reactivity in antigen–antibody reactions. The immunological distances between the albumins of two species of reptiles were a linear function of the duration of their period of evolutionary divergence and could be used to construct a meaningful classification of iguanid lizards and the Crocodilia. Albumins have evolved so rapidly, however that cross reactions were too weak for use in studies involving species divergent since before the Tertiary Period (Gorman *et al.*, 1971b).

Albumin shows relatively little electrophoretic variability for such a rapidly evolving protein (Fig. 2). This characteristic has been useful in grouping taxa, as related organisms commonly have albumins of similar mobility. For example (Table 2), species of the Kinosternidae are almost unique among turtles in having highly charge albumins (Crenshaw, 1962; Dessauer and Fox, 1964). Four species of *Anolis* lizards of the *roquet* species group had albumins of equal and relatively fast mobility, while three had albumins of equal but relatively slow mobility. A species having a slow and one having a fast albumin were introduced into Trinidad. The albumin patterns made it possible to detect natural hybrids in the introduced population, as the plasma of the hybrids contained both the slow and fast albumins in approximately equal concentrations (Gorman and Dessauer, 1966; Gorman *et al.*, 1971a). The distribution of albumin variants in a swarm of hybrids between two species of slider turtles allowed Crenshaw (1965) to suggest a mechanism for their albumin inheritance. The albumins of many species of snakes of widespread geographic range show little variation upon electrophoresis. Western subspecies of kingsnakes from California, Nevada, and Arizona had the same albumin pattern, but it was different from that of kingsnakes from states east of the continental divide (Dessauer and Fox, 1958; F. H. Pough and H. C. Dessauer, unpublished).

B. Transferrins

1. *Isolation and Identification*

Tranferrins of reptiles and mammals have many similar properties. They are readily isolated using methods developed for human transferrin, and the reddish brown color of the protein allows one to follow its concentration during purification procedures. The transferrins are soluble in half saturated ammonium sulfate, and most that have been tested are soluble in the presence of the acridine dye Rivanol. These include the transferrins of the turtle *Terrapene carolina,* the lizard *Iguana iguana,* and snakes of the genera *Natrix, Thamnophis, Pituophis,* and *Coluber* (Masat and Dessauer, 1968; George and Dessauer, 1970; Mao and Dessauer, 1971). Transferrins of *Alligator mississippiensis* and the viperid snake *Agkistrodon acutus* are less soluble in solution containing Rivanol (Dessauer, personal observation). When plasma is subjected to molecular sieving, transferrins of all groups of reptiles separate in the low molecular weight fraction, indicating that their molecular weights fall between 70,000 and 90,000 (Masat and Dessauer, 1968). The surface charge of transferrins is highly variable in different species and even in individuals of the same species. Transferrins of reptiles may be found in any electrophoretic fraction (Fig. 2). Those of some turtles migrate as slowly as human γ-globulin, and some of iguanid and geckonid lizards migrate as rapidly as human albumin (Dessauer *et al.*, 1962a).

Transferrins are remarkably stable, a very desirable property for the comparative biochemist who often must use plasma collected years prior to its use. When saturated with iron, they resist prolonged exposure to proteolytic enzymes and to temperatures as high as 65°C (Azari and Feeney, 1958). Water snake transferrins in plasma stored frozen for 15 years were identical in their electrophoretic and immunological properties to transferrins in fresh plasma (Mao and Dessauer, 1971).

2. *Iron Transport*

The iron-binding capacity of the transferrin present in plasma generally exceeds the level of plasma iron, so that plasma normally contains both free transferrin and transferrin complexed with iron (Table II). Plasma iron is an indirect measure of the latter, as plasma iron of unhemolyzed blood is bound to transferrin. Iron-binding capacity is an indirect measure of total transferrin. Calculations based on these measurements, assuming that transferrins have a molecular weight of 90,000 and bind two atoms of iron per molecule (Putnam, 1960), show that between 100 and 300 mg% of transferrin are present in the plasma of most reptiles, as compared to 200 to 320 mg% in human plasma. Plasma of turtles and crocodiles commonly contains less iron, and a higher percentage of its transferrin is unsaturated than that of mammals and squamate reptiles. Hirschfeld and Gordon (1965) found that plasma iron was reduced in starved turtles (*Pseudemys scripta*).

Transferrins carry iron from its site of absorption or storage to the hematopoietic tissues and probably have a number of other less well understood functions. Transferrins of *Pseudemys scripta* release iron to reticulocytes, demonstrating that the reptilian protein can serve as a source of iron for hemoglobin synthesis (Fig. 3). Blood cells from normal and from anemic turtles were incubated in plasma containing transferrin labeled with iron-59, and the amount of tracer transferred to the red cells was measured at different time intervals. Maximum levels of iron incorporation were attained within 2 hours of incubation; little iron was incorporated thereafter. The higher the reticulocyte count, the greater the amount of iron transferred from plasma to cells. The tracer iron was recovered in the heme fraction of hemoglobin after a temporary period of retention in the red cell stroma (Sheeler and Barber, 1964, 1965; Sheeler, 1967).

Multiple transferrins are present in the plasma of many reptiles (Dessauer *et al.*, 1962a). The presence of two transferrins often indicates that the individual is heterozygous at the transferrin locus, and one

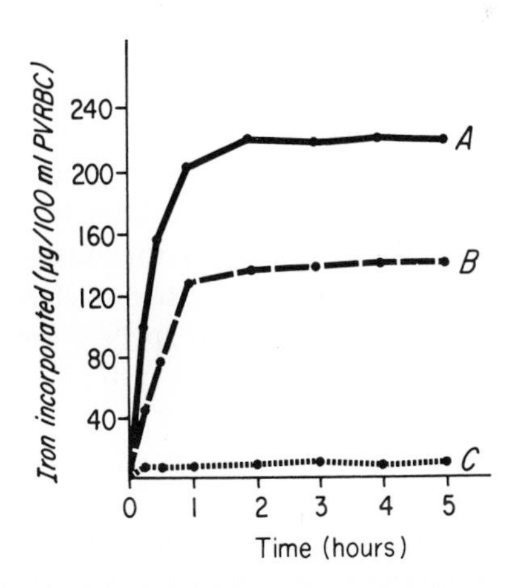

FIG. 3. Transfer of iron from transferrin to the reticulocytes of the turtle *Pseudemys scripta*. (A) Packed volume of red blood cells (PVRBC) included 73% reticulocytes; (B) PVRBC included 20% reticulocytes; (C) PVRBC included 2% reticulocytes. (From Sheeler and Barber, 1965.)

transferrin indicates that it is homozygous (Dessauer *et al.*, 1962a,b; Fox and Dessauer, 1965); however, other cases are not so easily explained. In the scincid lizard *Dasia smaragdinum* of New Guinea, three of four transferrin bands characterize electrophoretic patterns of individual animals. Do some such patterns include transferrin molecules of different physiological capabilities? Such seems the case for the two transferrins of the turtle *Pseudemys scripta.* These have different heat stabilities and unload complexed iron at different acidities. One binds iron effectively even at pH 5, an acidity at which mammalian transferrins have completely released iron (Barber and Sheeler, 1963).

Perhaps such transferrins represent adaptations to the fact that plasma of reptiles often becomes relatively acid during physiological activities. The complex between iron and transferrin maintains the level of ionic iron of plasma near zero, clearly contributing to its bacteriostatic nature (Martin *et al.*, 1963). Such heavy metal ions are also toxic to cells of higher organisms. Among other effects, they stimulate the formation of lipid peroxides. Transferrin protects the individual against this possibility (Barber and Sheeler, 1961).

3. *Evolutionary Aspects*

Transferrin is one of the most rapidly evolving proteins (Manwell and Baker, 1970). Immunological cross reactions among snakes were stronger for an albumin–antialbumin system than for the transferrin–antitransferrin system, suggesting that transferrins may have been even more mutable than albumins (George, 1969). Functional differences were evident between transferrins of closely related species. In hybrids between *Anolis trinitatus* and *Anolis aeneus,* which belong to the same species group, the transferrin inherited from *aeneus* seemed to have a greater affinity for iron than the transferrin acquired from *trinitatus* (Gorman *et al.*, 1971a). The transfer of iron from transferrin to reticulocytes was reduced fivefold when reticulocytes from the turtle *Pseudemys scripta* were incubated with transferrins of its close relative *Chrysemys picta,* suggesting that the transferrin of the two species were different enough to affect the function of this heterogenous system (Barber and Sheeler, 1963).

The structure of transferrins of numerous snakes have been compared using immunoelectrophoresis and microcomplement fixation titrations (George, 1969; George and Dessauer, 1970; Mao and Dessauer, 1971). Immunological data correlated more closely to information on the time of divergence of two species than to evidence on their external morphology, ecology or mode of reproduction. The results suggested that structural differences between transferrins largely represent selectively neutral mutations accumulated at the transferrin locus during the period of reproductive isolation of two species (Mao and Dessauer, 1971).

The evolution of certain advanced snakes was estimated by applying measures of immunological difference as an evolutionary clock, using available fossil evidence to calibrate the clock (Fig. 4). Indices of dissimilarity (ID) between transferrins of primitive and advanced snakes were very high, approaching the limit of sensitivity of the titration. Advanced snakes (ID = 70 or over) apparently diverged from the former within the Eocene Epoch of the Tertiary Period (Auffenberg, 1963). The data suggest that natricine, colubrine, xenodontine, and viperid groups of advanced snakes became independent genetic lines during the early Miocene Epoch (average ID = 11). Evolutionary lines leading to the different modern Asiatic genera of natricine snakes originated during the late Miocene Epoch (average ID = 6). By the end

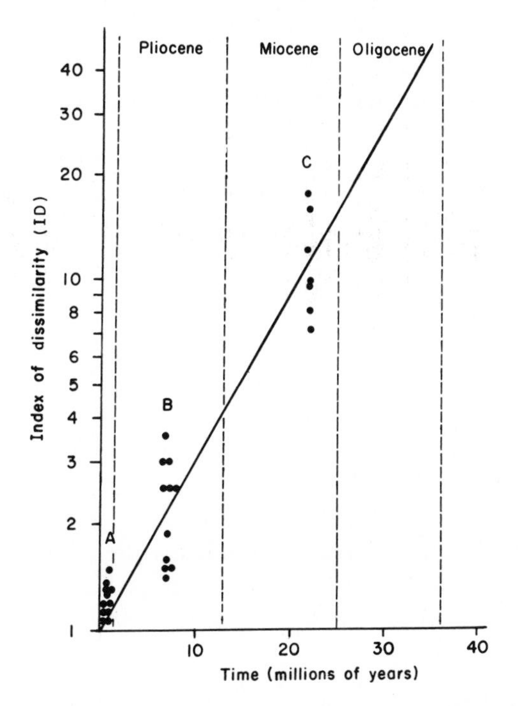

FIG. 4. Transferrin as a biological clock. Dots represent IDs of cross reactions involving transferrins and antisera against transferrins of: (A) species of North American *Natrix, Regina,* and *Storeria,* divergent since the Pleistocene Epoch; (B) species of North American *Natrix* and *Thamnophis,* divergent since the mid-Pliocene Epoch; and (C) species of different families of advanced snakes, divergent since the early Miocene Epoch. The line connects averages of transferrin IDs of the three groups (Mao and Dessauer, 1971).

of that epoch, water snakes, presently classified in genus *Natrix*, had become isolated in four continental groups. Transferrins of living species of the genus from Europe, Asia, and North America differ by an average ID of 3.8. Ancestral *Natrix* apparently underwent extensive speciation in North America since the Miocene Epoch. All living genera of North American natricine snakes are close relatives, differing on the average by an ID of 2.1.

Transferrins of many species show extensive intraspecific variation in electrophoretic mobility. Two or more transferrin patterns were found in every species of snake in which plasmas from over five individuals were examined. Ten electrophoretically distinct transferrins, arranged in eleven patterns of one or two bands were present in thirteen racers *Coluber constrictor* collected within a 25 mile radius of New Orleans, Louisiana. Six snakes of this species, captured in the sand dunes of northwest Indiana, possessed five different transferrins arranged in five distinct patterns (Dessauer *et al.*, 1962a). Although of widely different electrophoretic mobilities, these variants were immunologically similar having IDs equal to (1.0) or only slightly higher (maximum = 1.36) than that of the most common electrophoretic variant against which the antiserum was made (George, 1969).

The distribution of transferrin variants has been used to clarify a number of problems of population biology involving lizards and snakes. The distribution of two transferrin variants showed the extent of introgression of two previously isolated subspecies of the lizard *Cnemidophorus tigris* (Dessauer *et al.*, 1962b). Among two species of *Anolis* lizards of the Caribbean, the possession of a specific variant often indicated the island of origin of lizards of the same species (Gorman and Dessauer, 1965). Transferrin patterns were useful in distinguishing hybrids and parental forms in a hybridizing population of two of these species of lizards on Trinidad (Gorman *et al.*, 1971a). The distribution of two electrophoretic variant transferrins among garter snakes of the west coast of the United States showed that the aquatic and terrestrial populations were distinct, did not intergrade, and hence should be classified as different species (Dessauer *et al.*, 1962a; Fox and Dessauer, 1965).

C. Immunoglobulins

1. *Isolation and Identification*

Immunoglobulins and complement-system proteins are present in plasma of all reptiles (Hildemann, 1962; Evans *et al.*, 1965; Smith *et al.*, 1966). Many naturally occuring antibodies have been described.

Those of the carpet snake, *Morelia spilotes,* reacted with antigens of specific infecting nematodes (Timourian *et al.*, 1961). Twenty-five percent of the serum samples from a wide variety of North American snakes contained agglutinating antibodies against *Leptospira,* suggesting that epizootics of leptospirosis occur among snakes (White, 1963). Isoagglutinins were described in numerous turtles (Bond, 1940a; Frair, 1963) and heteroagglutinins in turtles (Bond, 1940a; Frair, 1963), crocodilians (Bond, 1940b), and snakes (Bond, 1939; Dujarric de la Rivière *et al.*, 1954; Timourian and Dobson, 1962). Autoantibodies, present in plasma of the Gila monster, *Heloderma suspectum* (Tyler, 1946), and the snake *Vipera aspis* (Izard *et al.*, 1961), neutralized toxins in their respective venoms.

Reptilian antibodies belong to two or more classes of proteins, having been studied in Rhynchocephalia (Marchalonis *et al.*, 1969), the Crocodilia (Saluk *et al.*, 1970), and Chelonia (Grey, 1963; Lewis, 1964; Ambrosius, 1966; Lykakis, 1968). Their physicochemical properties are similar to those of antibodies of the G and M classes of mammals. One class has a sedimentation coefficient of about 7 S and separates in the 120,000 to 190,000 molecular weight fraction of a molecular sieve column; the other group has a sedimentation coefficient of 18–19 S and a molecular weight that is greater than 190,000. During electrophoresis, immunoglobulins of both classes migrate with the fraction of slowest mobility; however, only plasmas of the Crocodilia and Chelonia have a major fraction which migrates as slowly as mammalian γ-globulin. Proteins of both classes are insoluble in half saturated ammonium sulfate (Timourian and Dobson, 1962; Tyler, 1946). Those of the 7 S class are soluble in Rivanol (Masat and Dessauer, 1968).

These immunoglobulins may behave somewhat differently in reaction with antigens than do mammalian antibodies. An agglutination reaction involving antiserum produced in a turtle gave higher titers at 7°C than at 37°C. Titers are higher at 37°C in reactions involving mammalian antisera (Maung, 1963). Simple dilution caused a decrease in the amount of precipitate obtained in a precipitin reaction involving turtle antiserum; dilution has little effect upon the magnitude of precipitin reaction involving mammalian antibodies (Grey, 1963).

Both the 7 S and 19 S immunoglobulins are fabricated from light and heavy types of polypeptide chains. To study these substructures, the intact immunoglobulins must be reduced and alkylated in strong urea, as their conformations depend upon disulfide bridges between the chains. Reducing agents such as mercaptoethanol destroyed antibody activity (Grey, 1963). Electrophoretic patterns of the dissociated 7 S and 19 S immunoglobulins in acid–urea gels were similar to patterns obtained with dissociated human immunoglobulins G and M, respectively. Light chains migrated in diffuse patterns, and the heavy chains of the 7 S and 19 S proteins had different mobilities (Marchalonis *et al.*, 1969). The heavy chains of the 7 S and 19 S immunoglobulins of the alligator were antigenically distinct. The light chains of its 7 S protein consisted of two antigenically distinct polypeptides, analogous to the κ and λ chains of human immunoglobulins (Saluk *et al.*, 1970). Immunoelectrophoresis also suggested that the 7 S immunoglobulins of the tuatara (Marchalonis *et al.*, 1969) and the turtle *Testudo hermani* (Ambrosius, 1966) consist of at least two subclasses of antibodies.

Immune hemolytic systems involving complement occur in turtles (Frair, 1963) and are common in snakes (Timourian and Dobson, 1962). The complement systems of turtles (Bond, 1940a; Frair, 1963), the lizard *Egernia* (Tait, 1969), and snakes (Bond and Sherwood, 1939) were thermolabile. Complement proteins of the carpet snake were somewhat soluble in half saturated ammonium sulfate (Timourian and Dobson, 1962).

2. *Induction*

Reptiles respond to an initial exposure to an antigen with the production of antibodies. The maximum titers obtained are comparable to those obtained in mammals, although the rate of synthesis is dependent upon temperature, season, the species involved, and its age. Figure 5 summarizes a series of experiments concerning the effect of temperature upon the antibody response of the lizard *Egernia cunninghami* to injections of sheep red blood cells. In lizards maintained at 30°C, antibodies

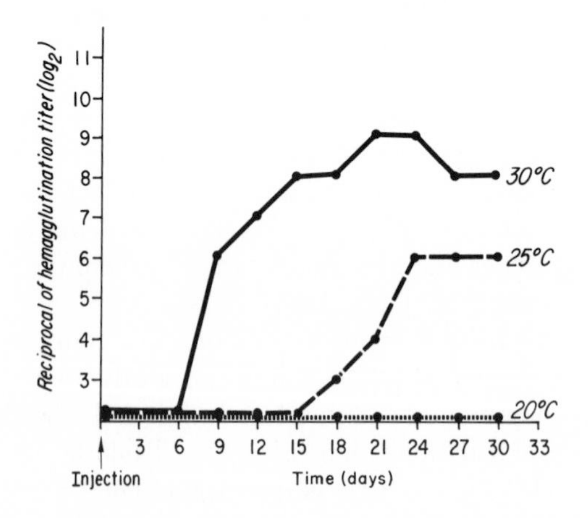

FIG. 5. Induction of immunoglobulins in the lizard *Egernia cunninghami* maintained at different temperatures (Tait, 1969).

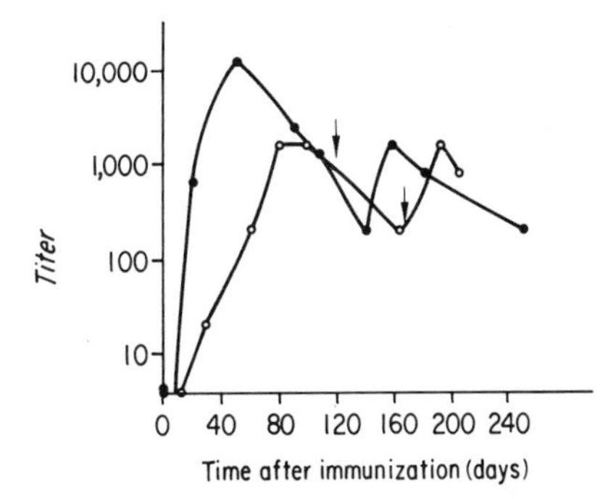

FIG. 6. Induction of immunoglobulins in two tuataras maintained at 20°C. Arrows indicate the injection of a second dose of antigen. (Marchalonis *et al.*, 1969.)

appeared in the plasma within 9 days and reached their highest titer in 3 weeks. In the group held at 25°C antibodies appeared more slowly and attained a lower maximum titer. Antibodies were not detected in the group maintained at 20°C during the first 30 days, but were present in low titers by the fifty-fourth day (Tait, 1969). The lizards *Dipsosaurus dorsalis* and *Sauromalus obesus* synthesized antibodies against the bacterium *Salmonella typhosa*, bovine serum albumin, and rabbit γ-globulin most effectively when maintained at 35°C. Synthesis was poor in lizards kept at 25°C and at 40°C (Evans, 1963a,b). The tuatara (Fig. 6) responded rapidly to an antigen when maintained at 20°C, 6° above its activity temperature. Maxium antibody titers were reached within 2 months and were comparable to levels attained in mammals (Marchalonis *et al.*, 1969). Antigens induced high antibody titers in rat snakes and turtles maintained near the upper limit of their temperature range (Frair, 1963). The turtle *Testudo*, challenged with *Brucella abortus* antigen, (Fig. 7) developed antibodies when kept at 18°–30°C, but not when kept in a cold room. When the latter were returned to the warm environment, however, they developed antibodies without the

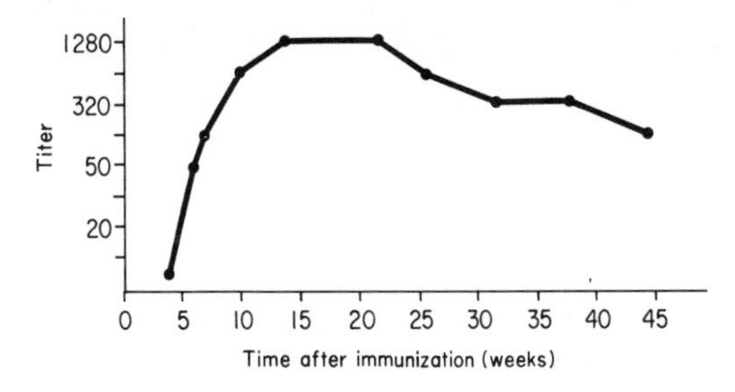

FIG. 7. Induction of immunoglobulins in the turtle *Testudo ibera* maintained between 18° and 30°C. Note that the time scale is in weeks. (Maung, 1963.)

stimulus of additional antigen. Adrenocorticoids, suggested as releasing agents for antibodies, did not affect the antibody titer of the turtles kept in the cold (Maung, 1963). Cortisone treatment caused a decrease in the protein fraction migrating in the γ-region during electrophoresis (Rao, 1968).

Reptiles respond to antigens more rapidly and effectively in spring and summer than in fall and winter. The lizard *Egernia* was not responsive during winter months of inactivity (Tait, 1969). *Testudo ibera,* maintained between 18° and 30°C, responded much more slowly to the *Brucella abortus* antigen when challenged in the winter than in the summer (Maung, 1963). Slider turtles varied in their resistance to infection during different seasons (Kaplan and Rueff, 1960).

The rate of antibody response also depends upon the species being examined. When exposed to an antigen at 20°C, the tuatara responded well (Marchalonis *et al.,* 1969), but the lizards *Dipsosaurus* and *Sauromalus* responded poorly (Evans, 1963a,b). Turtles in general responded more slowly to antigens than other reptiles. Maximum titers of antibodies in lizards were reached in 3–4 weeks (Fig. 5) (Tait, 1969), but turtles required 10–20 weeks when challenged with cellular antigens (Fig. 7). They responded even more slowly to soluble antigens such as rabbit albumin. Even the rate of clearance of such heterologous proteins was extremely slow. Rabbit albumin was detectable in the circulation of the turtle for 5 months after it had been injected. Similarly, rabbit antibodies were detectable for 3 months, indicating that passive immunization of turtles could be affective for many months (Maung, 1963).

In reptiles, as in other vertebrate animals (Abramoff and LaVia, 1970), immunoglobulins of the 19 S class appear in the circulation earlier than those of the 7 S class. The turtles *Chrysemys picta* and *Emys orbularis* gave a prolonged 19 S antibody response which only gradually shifted to immunoglobulins of 7 S class when challenged with bovine serum albumin (Grey, 1963; Lykakis, 1968). Complex antigens may induce only 19 S immunoglobulins. *Salmonella* flagellin induced only 18 S antibodies in the tuatara (Marchalonis *et al.,* 1969). Activity against *Brucella abortus* antigen in *Testudo* was largely confined to 19 S antibodies. During immunization of these turtles, the relative amount of protein increased in the electrophoretic fraction of slowest mobility (Maung, 1963).

Reptiles differ from mammals in their response to a second exposure to antigen. In mammals, this amnestic response generally occurs more rapidly and reaches a higher titer than the initial response (Abramoff and LaVia, 1970). The tuatara (Fig. 6) responded more rapidly to the second challenge with antigen, but the titer achieved in the secondary response did not exceed that of the initial response (Marchalonis *et al.*, 1969). Alligators responded in 2 days to a second exposure to an antigen (Lerch *et al.*, 1967); an amnestic response was not detected in turtles (Grey, 1963; Maung, 1963).

Snapping turtles do not become immunologically competent until 12 weeks after they hatch from the egg. Tissue cultures of the spleen of the younger turtles do not produce antibodies (Sidky and Auerbach, 1968). The capacity to reject foreign skin grafts also matures at about the same rate (Borysenko, 1969).

D. Plasma Vitellin

1. Identification and Properties

Plasma vitellin appears in the blood of vertebrates that produce yolky eggs during estrus, and its presence can be induced by estrogen injections during other seasons (Simkiss, 1967). This yolk precursor protein has been detected during the natural estrous cycle in plasma of turtles (Laskowski, 1936; Clark, 1967), lizards (Hahn, 1967; Rosenquist, 1969), and numerous snakes (Dessauer *et al.,* 1956; Dessauer and Fox, 1959; Izard *et al.*, 1961; Jenkins and Simkiss, 1968).

Plasma vitellin is a complex colloidal micelle composed of alkaline earths, lipid, and phosphoprotein. The complex is very large and readily isolated. Mere tenfold dilution of plasma with water will cause it to precipitate. Plasma vitellin of the turtle *Pseudemys scripta* sedimented in the ultracentrifuge as a single broad asymmetrical peak, having a sedimentation constant of about 17 S. Calcium sedimented with the complex and not as free colloidal calcium phosphate (Urist and Schjeide, 1960/1961; Seal, 1964). Plasma vitellin of the alligator sedimented at about 20 S and passed through a G-150 Sephadex molecular sieve with the void volume (Van Brunt and Menzies, 1971). Undenatured plasma vitellins of a number of snakes (Dessauer and Fox, 1958, 1959), the lizard *Anolis carolinensis* (Rosenquist, 1969), alligator (Van Brunt and Menzies, 1971), and turtles migrated in the β region when subjected to electrophoresis.

The appearance of plasma vitellin is accompanied by striking changes in plasma composition. Each 100 ml of plasma of the ribbon snake contained about 1 gm of the phosphoprotein and 30 mg of protein-bound calcium throughout the period of yolk production. About the time of ovulation most extreme levels occurred, e.g., a total protein as high as 8.8 gm% and calcium levels of 360 mg%. The gross composition of plasma vitellin of the garter snake *Thamnophis sauritus* was similar to that of yolk. Both contained 43% lipid and 57% protein; calcium associated with the protein in equimolar quantities to its content of phosphorus (Dessauer and Fox, 1959). The binding of Ca^{2+} to the complex was reversible and depended on the establishment of an equilibrium between ionic and bound calcium. The relative affinity of the complex for Ca^{2+} increased linearly with decreasing acidity between pH 6 and 8.6 (Van Brunt and Menzies, 1971).

Phosphoprotein isolated from the complex of *Alligator mississippiensis* had properties similar to phosvitin, the major phosphoprotein of the egg yolk of birds. Alligator phosvitin made up at least 10% of the total protein of the vitellin complex and was readily isolated using methods developed for purifying the phosvitin of birds. The phosvitin of the alligator had a sedimentation coefficient of $s_{w,20} = 2.8$, a diffusion coefficient of $D_{w,20} = 4.1 \times 10^{-7}$, a molecular weight of about 36,000, and an electrophoretic mobility that was faster than human albumin. Serine and phosphoserine accounted for at least 50% of the amino acids present in a partial acid digest of the protein (Fig. 8). The molar ratio of phosvitin phosphorus to calcium was 1:1, and that of phosphorus to serine was at least 1:1 (Van Brunt and Menzies, 1971, and a personal communication).

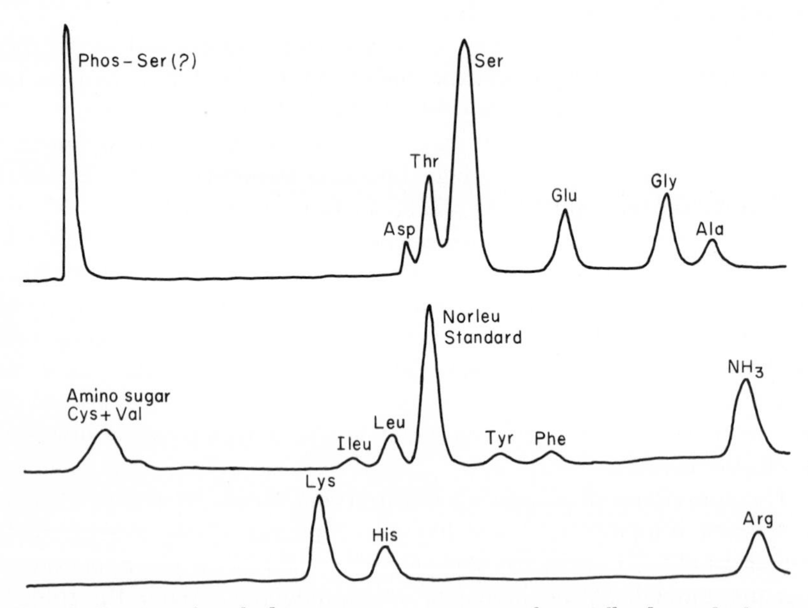

FIG. 8. Pattern of ninhydrin-positive components of partially digested phosvitin of the alligator. Protein was hydrolyzed in 6 *N* hydrochloric acid for 2 hours, and the digest was then fractionated by ion exchange chromatography. Note the predominance of hydroxyamino acids. (Courtesy of Van Brunt and Menzies.)

2. Vitellinogenesis

The physiological aspects of vitellinogenesis in reptiles are understood in broad outline. Observations on the natural process in the ribbon snake *Thamnophis sauritus* demonstrated its cyclic nature. In snakes, all follicles develop simultaneously and are ovulated within a few minutes to at most a few hours of each other, making it possible to correlate the time of appearance and disappearance of plasma vitellin with changes that take place in the ovaries and other organs. Stimulation with estrogens, presumably from the developing ovary, induces the liver to enlarge and synthesize the complex. Plasma vitellin appears in the blood shortly thereafter. The follicles enlarge taking up the vitellin from the blood. The fat bodies progressively decrease in size, probably furnishing energy for yolk synthesis. Just prior to ovulation, the liver decreases in size and ceases to synthesize phosphoprotein. Blood levels of plasma vitellin attain remarkably high levels at this time, presumably reflecting the cessation of ovarian uptake. Plasma vitellin disappears from the plasma within about 24 hours following ovulation. The protein complex is absent from the blood of females during pregnancy and anestrus, and it is not normally found in the blood of males (Dessauer and Fox, 1959).

Injection of estrogens induces the synthesis of plasma vitellin in both female and in male reptiles. The induction process has been studied in turtles (Schjeide and Urist, 1960; Urist and Schjeide, 1960/1961; Seal, 1964; Clark, 1967), lizards (Hahn, 1967; Suzuki and Prosser, 1968; Simkiss, 1967; Rosenquist, 1969), snakes (Dessauer and Fox, 1959), and crocodiles (Prosser and Suzuki, 1968; Van Brunt and Menzies, 1971). Daily injections of as little as 10^{-4} μg of 17-β-estradiol was sufficient to induce its synthesis in 5 gm males of the lizard *Anolis carolinensis* (Fig. 9). The rate and magnitude of the response was dependent upon the amount of estrogen injected in both lizards (Rosenquist, 1969) and turtles (Clark, 1967). Injections of 1 μg per gram body weight of estradiol into a lizard were followed within 24 hours by the appearance of plasma vitellin; after injections for 4–6 days, vitellin made up over 60% of the plasma protein (Hahn, 1967). Five days after treating turtles with very high levels of estrone, total lipid increased threefold, calcium fourfold, and total protein one and one-half times. There was an eightfold rise in protein-bound phosphorus. The lipid fraction contained 38% triglyceride, 35% phospholipid, 15% sterol, and 12% sterol esters (Urist and Schjeide, 1960/1961). Total plasma protein rose to 12 gm% in a turtle injected daily with 100 μg of 17-β-estradiol for 6 weeks (Seal, 1964). Rates of increase of calcium and phosphoprotein were equal

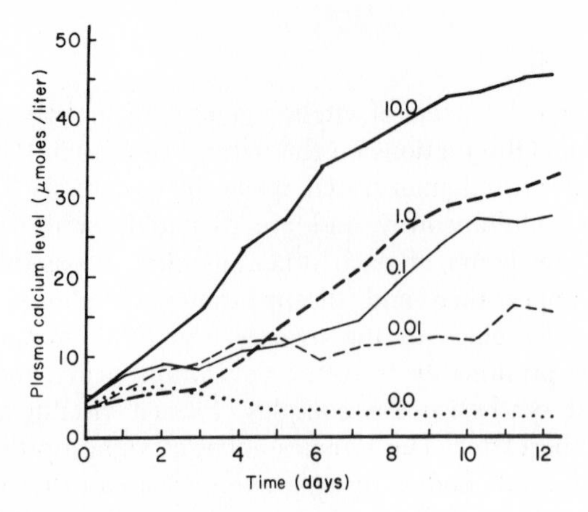

FIG. 9. Induction of plasma vitellin. Ordinate is the level of plasma calcium, an indirect measure of vitellin. Numbers represent the daily dose of β-estradiol in micrograms. (Rosenquist, 1969.)

and almost linear in male ribbon snakes (*Thamnophis sauritus*) injected with estradiol. Plasma and associated tissue changes occurred even though the snakes received no food for at least 2 weeks before and during the experiment (Dessauer and Fox, 1959).

Certain reptiles may be under calcium stress during reproduction (Simkiss, 1962, 1967; Jenkins and Simkiss, 1968). In the turtle *Stenotherus odoratus,* the loss of calcium during reproduction led to bones of decreased density (Edgren, 1960). The endolymphatic sacs of anoline and geckonid lizards store enormous deposits of calcium carbonate, which is probably available to the animal during such periods of calcium stress (Jenkins and Simkiss, 1968; Dessauer, 1970). During estrus, these sacs apparently contribute calcium to circulating plasma vitellin and hence to developing follicles. The artificial induction of plasma vitellin in male *Anolis carolinensis* was followed by a marked decrease in the X-ray opacity of the glands (Fig. 10) (Rosenquist, 1969).

E. Poorly Characterized Proteins

1. *Clotting Components*

Blood coagulation in reptiles involves plasma proteins similar to those of mammals; however, cellular factors apparently are of greater importance (Grégoire and Taynon, 1962). If carefully collected, plasma usually has a slow clotting time (Fantl, 1961; Hackett and Hann, 1967). Calcium is required for clot formation. Substances that precipitate or chelate

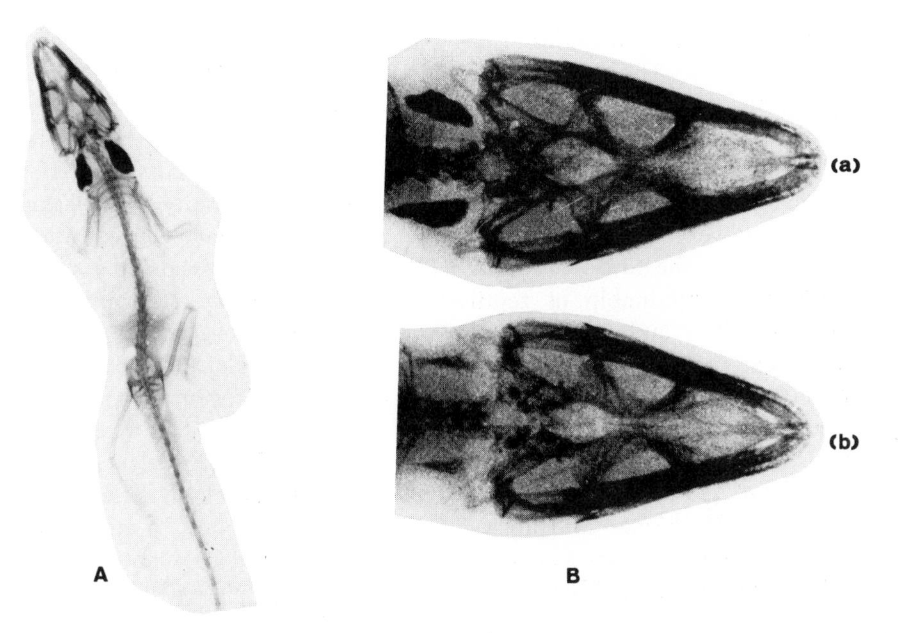

FIG. 10. The endolymphatic sacs, calcium storage organs in lizards. (A) Note the X-ray opacity of the sites of the sacs in the neck region of this male *Anolis carolinensis*. (B) X rays showing the sites in the same lizard before (a) and after (b) mobilization of calcium with ten daily injections of estradiol. (Rosenquist, 1969.)

calcium and heparin or heparin-like compounds inhibit the process (Kaplan, 1956; Jacques, 1963).

Fibrinogen of reptiles will form a fibrin clot when treated with mammalian thrombin. The rate of fibrin formation is slower than for mammalian fibrinogen, presumably because of species specificities (Hann, 1969). Fibrinogens are among the most insoluble proteins of the plasma. Those of the turtles *Chelydra* and *Pseudemys*, the lizard *Iguana*, and the snake *Natrix* are insoluble at 0°C in the 8% ethanol reagent used to isolate human fibrinogen (Dessauer, unpublished observations); however, the fibrinogens of *Chelydra* and *Pseudemys* are slightly more soluble than those of mammals (Morrison *et al.*, 1951). The fibrinogen content of plasma of a number of reptiles ranged from 115 to 970 mg% (Dessauer, 1970).

Hann (1966, 1969) examined products of the reaction of mammalian thrombin with the fibrinogen of the lizard *Tachydosaurus*. Two fibrinopeptides were released in about equal yield (about 6–7 mg per gram of fibrinogen). Both peptides migrated toward the anode during electrophoresis in a neutral buffer. The least acidic peptide had an N-terminal glutamic acid. The more acidic one lacked a free N-terminal amino group and contained a relatively high percentage of aspartic and glutamic acid and tyrosine residues. Neither peptide contained tryptophane.

Other proteins involved in the coagulation process also have properties similar to those of mammals, although their concentrations in plasma are often very different. Prothrombin is present in low concentration in plasma of the turtle *Chrysemys* (Warner *et al.*, 1939) and the snake *Notechis*. Prothrombin of reptiles can be isolated by adsorption on barium sulfate and subsequent elution with citrate (Hackett and Hann, 1967; Dessauer, personal observations). The synthesis of prothrombin by a turtle required vitamin K (Brambel, 1941). The Hageman factor was not detected in plasma of the tiger snake but was present in that of the lizard *Tiliqua*, the turtles *Chelodina* (Fantl, 1961), *Pseudemys*, and *Chelydra*, and in *Alligator* (Erdös *et al.*, 1967). Proaccelerin activity was low in turtle plasma (Murphy and Seegers, 1948). Serum of a number of colubrid and viperid snakes exhibits marked anticoagulant activity (Brazil and Vellard, 1928).

2. *Ceruloplasmin and Haptoglobin*

Ceruloplasmin and haptoglobin apparently occur in reptiles, but they have been identified indirectly on the basis of their oxidase and pseudoperoxidase activities. A copper-containing protein complexed all but 8 μg% of the 196 μg% of plasma copper of the lizard *Heloderma suspectum* (Zarafonetis and Kalas, 1960). *p*-Phenylenediamine oxidase activity was present in plasma of certain species of all major taxonomic groups. Highest values occurred in plasma of certain turtles, but the species distribution of activity was almost random. Low levels were not due to dialyzable inhibitors. Many species lacked activity, perhaps having ceruloplasmins inactive as oxidases (Seal, 1964). In species in which it was moderate to high, the activity was detectable on starch gels following electrophoresis. Migration rates of these *p*-phenylenediamine oxidases were similar in widely diverse species (Dessauer and Fox, 1964).

Haptoglobins have been found in turtles (Liang, 1957) the alligator (Coulson and Hernandez, 1964), and in numerous species of snakes. They were detected by means of the pseudoperoxidase activity of the heme moiety of the hemoglobin complexed to the haptoglobin. The level was high in the plasma of some snakes but was virtually absent in plasma of others. Very active haptoglobins occurred in the genera *Xenopeltis*, *Heterodon*, *Agkistrodon*, *Thamnophis*, and *Lampropeltis*. The affinity of a particular haptoglobin for hemoglobins of different species varied and paralleled taxonomic relationships. For example, the haptoglobin of the eastern kingsnake *Lampropeltis getulus holbrooki* complexed its own hemoglobin strongly, hemoglobins of other advanced snakes (*Coluber, Thamnophis, Arizona, Agkistrodon*) less strongly, and did not combine with hemoglobins of primitive snakes (*Lichnura, Xenopeltis, Typhlops, Leptotyphlops*), lizards, alligators, turtles, or mammals.

3. *Hydrolytic Enzymes*

a. Naphthylamidases. Enzymes that catalyze the hydrolysis of synthetic substrates such as leucine and cystine β-naphthylamide are present in plasma of a number of reptiles. The activity was absent from plasma of *Alligator mississippiensis* (Coulson and Hernandez, 1964) and snakes of the Boidae and Leptotyphlopidae, low in plasma of snakes of the Crotalidae and Elapidae, and remarkably high in the plasma of the turtle *Chelydra serpentina* and a great number of natricine and colubrine snakes (Ehrensing, 1964; Dessauer, 1967). The enzymes of many of these snakes had the same mobility when compared by means of starch gel electrophoresis (Dessauer, 1970). The naphthylamidase of the snake *Natrix rhombifera,* like that which appears in human plasma during pregnancy, catalyzed the hydrolysis of oxytocin. The oxytocinase activity in this viviparous snake, however, occurred in males as well as in females, and its level did not fluctuate during the different stages of the reproductive cycle (Ehrensing, 1964).

b. Esterases. Plasma of most of a great variety of reptiles contains enzymes capable of utilizing natural and synthetic esters as substrates. Esterase activity usually resides in a complex mixture of enzymes that have different substrate specificities and inhibitor sensitivities. One curious esterase was resolved when plasma of the turtle *Testudo graeca* was subjected to electrophoresis on a column of starch. This esterase utilized propionic acid esters of a variety of alcohols as substrates better than their corresponding acetic and butyric acid esters. It was the first example of a physostigmine-sensitive esterase that catalyzed the hydrolysis of noncholine esters more effectively than choline esters (Augustinsson, 1959).

Generally, a large number of esterases resolved when plasma was subjected to high-resolution electrophoresis in alkaline buffers. Esterase activity was highest when localized on starch gels using synthetic naphthol esters in neutral buffers. A cholinesterase had the slowest mobility of the four esterases detected in plasma of *Alligator mississippiensis.* One of the other esterases was present in plasma of only half of the individuals tested (Coulson and Hernandez, 1964). Five zones of activity appeared on electrophoretic patterns of plasma of the frilly lizard *Chlamydomonous kingii.* The enzyme of slowest mobility showed the highest activity and was of the carboxylesterase class; the four other enzymes belonged to the acetylesterase class; cholinesterases were absent (Holmes *et al.,* 1968). Variations in the electrophoretic patterns of plasma esterases were useful in distinguishing species of *Anolis* lizards (Gorman and Dessauer, 1966) and in examining the extent of introgression of two subspecies of the lizard *Cnemidophorous tigris* in southeastern Arizona (Dessauer *et al.,* 1962b).

c. Glycosidases. Weak glycosidase activities have been detected in plasma of a number of reptiles. The amylase activity of water moccasin plasma was only half that of rat serum; the amylase of the snake did not cross react with antiserum to a mammalian amylase (McGeachin and Bryan, 1964). Plasma of some turtles exhibited trehalase and maltase activity, but the magnitude of these activities varied widely from species to species (Van Handel, 1968).

4. Transport Proteins

Thyroxine-binding proteins are present in plasma of reptiles, but their capacities are lower than those of mammals. Rather than migrating in the α region as do the thyroxine-binding globulins of mammals, these reptilian proteins ranged in mobility from the post- to the prealbumin regions (Farer *et al.,* 1962; Tanabe *et al.,* 1969).

Transcortin, the corticosterone-binding protein, was studied in the alligator and garter snake. The corticosterone-binding capacity of alligator plasma, 43 μg/100 ml, was the highest of a sampling of twenty-three vertebrates, which included man and many other mammals. The binding capacity of plasma of all species was higher at 4°C than at 37°C. Transcortin of the garter snake was isolated using a hydroxyapatite column equilibrated at pH 6.8 with a 0.001 *M* phosphate buffer. Transcortin is the only plasma protein that is not adsorbed on this column. The transcortin fraction of the garter snake contained both hexose and sialic acid and made up about 2% of the total plasma protein (Seal and Doe, 1963). Transcortins of the alligator and other animals were stable for many weeks when stored in the deep freeze (Slaunwhite and Sandberg, 1959). Lizard plasma contained between 0.2 and 4.5 μg/100 ml of circulating corticosteroid (Bradshaw and Fontaine-Bertrand, 1970). Cholesterol and its esters are the predominate sterols of reptilian plasma. They precipitate with proteins that are insoluble in half saturated ammonium sulfate (Dessauer, 1970, and unpublished observations). The plasma cholesterol level of turtles increases linearly with age (Jackson *et al.,* 1970).

A number of reptilian plasma proteins are capable of binding vitamins and related substances. Plasma of iguanid lizards and certain snakes contains protein bound carotenoids which migrate as bright yellow or orange bands during electrophoresis. The carotenoid–protein of the lizard *Iguana iguana* separated in Cohn fraction IV-1 and migrated in the α_2 region when subjected to electrophoresis (H. C. Dessauer, unpublished). Plasma of a number of snakes contains remarkably high concentrations of riboflavin. Protein-bound riboflavin had L-amino acid oxidase activity and migrated during electrophoresis in the α and β regions (Villela and Prado, 1945; Villela *et al.*, 1955; Villela and Thein, 1967). Vitamin B_{12}-binding proteins were studied in *Alligator mississippiensis* and the turtle *Pseudemys scripta.* Plasma of both species contained only about 5 ng% of vitamin B_{12}, but alligator plasma was capable of binding an additional 94 ng% and turtle plasma an additional 1.8 μg% (Couch *et al.*, 1950; Rosenthal and Brown, 1954; Rosenthal and Austin, 1962).

ACKNOWLEDGMENTS

I express my sincere thanks to a number of colleagues and friends for help on this paper. Drs. P. Sheeler, A. A. Barber, N. N. Tait, R. T. Maung, J. J. Marchalonis, E. H. M. Ealey, and E. Diener gave permission to use figures from their publications. Drs. G. C. Gorman, A. C. Wilson, and M. Nakanishi furnished a copy of their paper on reptilian albumins prior to its publication. Drs. J. W. Rosenquist and R. A. Menzies and Mr. J. Van Brunt furnished unique unpublished evidence on plasma vitellin. Messrs. N. Nicosia, D. M. Alvarado, and C. W. Tate and Miss Rina Paz offered valuable technical assistance. I am indebted to the National Science Foundation (Grant GB-7294X) for financial support.

REFERENCES

Abramoff, P., and LaVia, M. F. (1970). "Biology of the Immune Response." McGraw-Hill, New York.
Ambrosius, H. (1966). *Nature (London)* **209,** 524.
Auffenberg, W. (1963). *Tulane Stud. Zool.* **10,** 131.
Augustinsson, K. (1959). *Acta. Chem. Scand.* **13,** 1081.
Azari, P. R., and Feeney, R. E. (1958). *J. Biol. Chem.* **232,** 293.
Barber, A. A., and Sheeler, P. (1961). *Comp. Biochem. Physiol.* **2,** 233.
Barber, A. A., and Sheeler, P. (1963). *Comp. Biochem. Physiol.* **8,** 115.
Baril, E. F., Palmer, J. L., and Bartel, A. H. (1961). *Science* **133,** 278.
Bogert, C. W. (1959). *Sci. Amer.* **200,** 105.
Bond, G. C. (1939). *J. Immunol* **36,** 1.
Bond, G. C. (1940a). *J. Immunol.* **39,** 125.
Bond, G. C. (1940b). *J. Immunol.* **39,** 133.
Bond, G. C., and Sherwood, N. P. (1939). *J. Immunol.* **36,** 11.
Borysenko, M. (1969). *J. Exp. Zool.* **170,** 341.
Bradshaw, S. D., and Fontaine-Bertrand, E. (1970). *Comp. Biochem. Physiol.* **36,** 37.
Brambel, C. E. (1941). *J. Cell. Comp. Physiol.* **18,** 221.
Brazil, V., and Vellard, J. (1928). *Ann. Inst. Pasteur, Paris* **42,** 907.
Brocq-Rousseu, D., and Roussel, G. (1939). "Le serum normal. Propriétés physiologiques." Masson, Paris.
Clark, N. B. (1967). *Comp. Biochem. Physiol.* **20,** 823.
Clark, W. C., and Voris, H. K. (1969). *Science,* **164,** 1402.
Cohen, E. (1954). *Science* **119,** 98.
Cohen, E. (1955). *Biol. Bull.* **109,** 394.

Cohen, E., and Stickler, G. B. (1958). *Science* **127**, 1392.
Couch, J. R., Olcese, O., Witten, P. W., and Colby, R. W. (1950). *Amer. J. Physiol.* **163**, 77.
Coulson, R. A., and Hernandez, T. (1964). "Biochemistry of the Alligator—A Study of Metabolism in Slow Motion." Louisiana State Univ. Press, Baton Rouge.
Coulson, R. A., Hernandez, T., and Dessauer, H. C. (1950). *Proc. Soc. Exp. Biol. Med.* **74**, 866.
Crenshaw, J. W., Jr. (1962). *Physiol. Zool.* **35**, 157.
Crenshaw, J. W., Jr. (1965). *Evolution* **19**, 1.
Dessauer, H. C. (1967). *Herpetologica* **23**, 148.
Dessauer, H. C. (1970). *In* "Biology of the Reptilia" (C. Gans and T. S. Parsons, eds.), Vol. 3, pp. 1–72. Academic Press, New York.
Dessauer, H. C., and Fox, W. (1956). *Science* **124**, 225.
Dessauer, H. C., and Fox, W. (1958). *Proc. Soc. Exp. Biol. Med.* **98**, 101.
Dessauer, H. C., and Fox, W. (1959). *Amer. J. Physiol.* **197**, 360.
Dessauer, H. C., and Fox, W. (1964). *In* "Taxonomic Biochemistry and Serology" (C. A. Leone, ed.), p. 625, Ronald Press, New York.
Dessauer, H. C., Fox, W., and Gilbert, N. L. (1956). *Proc. Soc. Exp. Biol. Med.* **92**, 299.
Dessauer, H. C., Fox, W., and Hartwig, Q. L. (1962a). *Comp. Biochem. Physiol.* **5**, 17.
Dessauer, H. C., Fox, W., and Pough, F. H. (1962b). *Copeia,* p. 767.
Deutsch, H. F., and McShan, W. H. (1949). *J. Biol. Chem.* **180**, 219.
Dittmer, D. S., ed. (1961). "Blood and Other Body Fluids," Biological Handbooks. Fed. Amer. Soc. Exp. Biol., Washington, D.C.
Dujarric de la Rivière, R., Eyquem, A., and Fine, J. (1954). *Experientia* **10**, 159.
Edgren, R. A. (1960). *Comp. Biochem. Physiol.* **1**, 213.
Ehrensing, R. H. (1964). *Proc. Soc. Exp. Biol. Med.* **117**, 370.
Erdös, E. G., Miwa, I., and Graham, W. J. (1967). *Life Sci.* **6**, 2433.
Evans, E. E. (1963a). *Fed. Proc., Fed. Amer. Soc. Exp. Biol.* **22**, 1132.
Evans, E. E. (1963b). *Proc. Soc. Exp. Biol. Med.* **112**, 531.
Evans, E. E., Kent, S. P., Attleberger, M. H., Sieberg, C., Bryant, R. E., and Booth, B. (1965). *Ann. N.Y. Acad. Sci.* **126**, 629.
Fantl, P. (1961). *Aust. J. Exp. Biol. Med. Sci.* **39**, 403.
Farer, L. S., Robbins, J., Blumberg, B. S., and Rall, J. E. (1962). *Endocrinology* **70**, 686.
Fox, W., and Dessauer, H. C. (1965). *Yearb. Amer. Phil. Soc. p.* 263.
Frair, W. (1962). *Bull. Serol. Mus. New Brunsw.* **27**, 7.
Frair, W. (1963). *Science* **140**, 1412.
Frair, W. (1964). *In* "Taxonomic Biochemistry and Serology" (C. A. Leone, ed.), p. 535. Ronald Press, New York.
Frair, W. (1969). *Bull. Serol. Mus. New Brunsw.* **42**, 1.
Gans, C., Krakauer, T., and Paganelli, C. V. (1968). *Comp. Biochem. Physiol.* **27**, 747.
George, D. W. (1969). Master's Thesis, Louisiana State Univ. Medical Center, New Orleans
George, D. W., and Dessauer, H. C. (1970). *Comp. Biochem. Physiol.* **33**, 617.
Gorman, G. C., and Dessauer, H. C. (1965). *Science* **150**, 1454.
Gorman, G. C., and Dessauer, H. C. (1966). *Comp. Biochem. Physiol.* **19**, 845.
Gorman, G. C., Licht, P., Dessauer, H. C., and Boos, J. O. (1971a). *Syst. Zool.* **20**, 1.
Gorman, G. C., Wilson, A. C., and Nakanishi, M. (1971b). *Syst. Zool.* **20**, 167.

Graham-Smith, G. S. (1904). *In* "Blood Immunity and Blood Relationship" (G. H. F. Nuttall ed.), p. 336. Cambridge Univ. Press, London and New York.
Grégoire, C., and Taynon, H. J. (1962). *Comp. Biochem.* **4**, 435.
Grey, H. M. (1963). *J. Immunol.* **91**, 819.
Hackett, E., and Hann, C. S. (1967). *J. Comp. Pathol.* **77**, 175.
Hahn, W. E. (1967). *Comp. Biochem. Physiol.* **23**, 83.
Hann, C. S. (1966). *Biochim. Biophy. Acta* **124**, 398.
Hann, C. S. (1969). *Biochim. Biophy. Acta* **181**, 342.
Hildemann, W. H. (1962). *Ann. N.Y. Acad. Sci.* **97**, 139.
Hirschfeld, W. J., and Gordon, A. S. (1965). *Anat. Rec.* **153**, 317.
Holmes, R. S., Masters, C. J., and Webb, E. C. (1968). *Comp. Biochem. Physiol.* **26**, 837.
Izard, Y., Detrait, J., and Boquet, P. (1961). *Ann. Inst. Pasteur, Paris* **100**, 539.
Jackson, C. G., Jr., Holcomb, C. M., and Jackson, M. M. (1970). *Comp. Biochem. Physiol.* **35**, 491.
Jacques, F. A. (1963). *Comp. Biochem. Physiol.* **9**, 241.
Jenkins, N. K., and Simkiss, K. (1968). *Comp. Biochem. Physiol.* **26**, 865.
Kaplan, H. M. (1956). *Herpetologica* **12**, 269.
Kaplan, H. M., and Rueff, W. (1960). *Proc. Anim. Care Panel* **10**, 63.
Khalil, F., and Abdel-Messeih, G. (1963). *Comp. Biochem. Physiol.* **9**, 75.
Kmeťová, S., and Paulov, Š. (1966). *Acta Fac. Rerum. Natur. Univ. Comenianae, Zool.* **23**, 251.
Laskowski, M. (1936). *Biochem. Z.* **284**, 318.
Latifi, M., Shamloo, K. D., and Amin, A. (1965). *Can. J. Biochem.* **43**, 459.
Leone, C. A., and Wilson, F. E. (1961). *Physiol. Zool.* **34**, 297.
Lerch, E. G., Huggins, S. E., and Bartel, A. H. (1967). *Proc. Soc. Exp. Biol. Med.* **124**, 448.
Lewis, J. H. (1964). *Protides Biol. Fluids, Proc. Colloq.* **12**, 149.
Liang, C. (1957). *Biochem. J.* **66**, 552.
Lustig, B., and Ernst, T. (1936). *Biochem. Z.* **289**, 365.
Luzzio, A. J. (1967). *Toxicon* **5**, 97.
Lykakis, J. J. (1968). *Immunology* **14**, 799.
Lykakis, J. J. (1971). *Comp. Biochem. Physiol.* **39**, 83.
McGeachin, R. L., and Bryan, J. A. (1964). *Comp. Biochem. Physiol.* **13**, 473.
McMenamy, R. H., and Watson, F. (1968). *Comp. Biochem. Physiol.* **26**, 392.
Maldonado, A. A., and Ortiz, E. (1966). *Copeia*, p. 179.
Manwell, C., and Baker, C. M. A. (1970). "Molecular Biology and the Origin of the Species." Univ. of Washington Press, Seattle.
Mao, S. H., and Dessauer, H. C. (1971). *Comp. Biochem. Physiol.* **40**, 669.
Marchalonis, J. J., Ealey, E. H. M., and Diener, E. (1969). *Aust. J. Biol. Med. Sci.* **47**, 367.
Martin, C. M., Jandl, J. H., and Finland, M. (1963). *J. Infec. Dis.* **112**, 158.
Masat, R. J., and Dessauer, H. C. (1968). *Comp. Biochem. Physiol.* **25**, 119.
Masat, R. J., and Musacchia, X. J. (1965). *Comp. Biochem. Physiol.* **16**, 215.
Maung, R. T. (1963). *J. Pathol. Bacteriol.* **85**, 51.
Morgareidge, K. R., and White, F. N. (1969). *Nature (London)* **223**, 587.
Morrison, P. R., Scudder, C., and Blatt, W. (1951). *Biol. Bull.* **101**, 171.
Murphy, R. C., and Seegers, W. H. (1948). *Amer. J. Physiol.* **154**, 134.
Nakamura, S., Tominaga, S., Katsuno, A., and Murakawa, S. (1965). *Comp. Biochem. Physiol.* **15**, 435.
Newcomer, R. J., and Crenshaw, J. W. (1967). *Copeia*, p. 481.
Pearson, D. D. (1966). *Bull. Serol. Mus. New Brunsw.* **36**, 8.

Pearson, D. D. (1968). *Proc. Pa. Acad. Sci.* **42**, 49.
Philpot, V. B., and Smith, R. G. (1950). *Proc. Soc. Exp. Biol. Med.* **74**, 521.
Prosser, R. L., III, and Suzuki, H. K. (1968). *Comp. Biochem. Physiol.* **25**, 529.
Putnam, F. W., ed. (1960). "The Plasma Proteins," 2 vols. Academic Press, New York.
Rao, C. A. P. (1968). *Comp. Biochem. Physiol.* **26**, 1119.
Rider, J., and Bartel, A. H. (1967). *Comp. Biochem. Physiol.* **20**, 1005.
Roberts, R. C., and Seal, U.S. (1965). *Comp. Biochem. Physiol.* **16**, 327.
Romer, A. S. (1966). "Vertebrate Paleontology," 3rd. ed. Univ. of Chicago Press, Chicago, Illinois.
Rosenquist, J. W. (1969). Ph.D. Dissertation, Tulane University, New Orleans, Louisiana.
Rosenthal, H. L., and Austin, S. (1962). *Proc. Soc. Exp. Biol. Med.* **109**, 179.
Rosenthal, H. L., and Brown, C. R., Jr. (1954). *Proc. Soc. Exp. Biol. Med.* **86**, 117.
Saluk, P. H., Krauss, J., and Clem, L. W. (1970). *Proc. Soc. Exp. Biol. Med.* **133**, 365.
Schjeide, A. O., and Urist, M. R. (1960). *Nature (London)* **188**, 291.
Seal, U. S. (1964). *Comp. Biochem. Physiol.* **13**, 143.
Seal, U. S., and Doe, R. P. (1963). *Endocrinology* **73**, 371.
Seniów, A. (1963). *Comp. Biochem. Physiol.* **9**, 137.
Sheeler, P. (1967). *Amer. Zool.* **7**, 203.
Sheeler, P., and Barber, A. A. (1964). *Comp. Biochem. Physiol.* **11**, 139.
Sheeler, P., and Barber, A. A. (1965). *Comp. Biochem. Physiol.* **16**, 63.
Sidky, Y. A., and Auerbach, R. (1968). *J. Exp. Zool.* **167**, 187.
Simkiss, K. (1962). *Comp. Biochem. Physiol.* **7**, 71.
Simkiss, K. (1967). "Calcium in Reproductive Physiology." Van Nostrand-Reinhold, Princeton, New Jersey.
Slaunwhite, W. R., Jr., and Sandberg, A. A. (1959). *J. Clin. Invest.* **38**, 384.
Smith, R. T., Meischer, P. A., and Good, R. A. (1966). "Phylogeny of Immunity." Univ. of Florida Press, Gainesville.
Suzuki, H. K., and Prosser, R. L., III. (1968). *Proc. Soc. Exp. Biol. Med.* **127**, 4.
Tait, N. N. (1969). *Physiol. Zool.* **42**, 29.
Tanabe, Y., Ishii, T., and Tamaki, Y. (1969). *Gen. Comp. Endocrinol.* **13**, 14.
Timourian, H., and Dobson, C. (1962). *J. Exp. Zool.* **150**, 27.
Timourian, H., Dobson, C., and Sprent, J. F. A. (1961). *Nature (London)* **192**, 996.
Tyler, A. (1946). *Proc. Nat. Acad. Sci. U.S.* **32**, 195.
Urist, M. R., and Schjeide, A. O. (1960/1961). *J. Gen. Physiol.* **44**, 743.
Van Brunt, J., III, and Menzies, R. A. (1971). *Fed. Proc., Fed. Amer. Soc. Exp. Biol.* **30**, 1160.
Van Handel, E. (1968). *Comp. Biochem. Physiol.* **26**, 561.
Villela, G. G., and Prado, J. L. (1945). *J. Biol. Chem.* **157**, 693.
Villela, G. G., and Thein, M. (1967). *Experientia* **23**, 722.
Villela, G. G., Mitidieri, E., and Ribeiro, L. P. (1955). *Arch. Biochem. Biophys.* **56**, 270.
Voris, H. K. (1967). *Physiol. Zool.* **40**, 238.
Warner, E. D., Brinkhous, K. M., and Smith, H. P. (1939). *Amer. J. Physiol.* **125**, 296.
White, F. H. (1963). *Amer. J. Vet. Res.* **24**, 179.
Zarafonetis, C. J. D., and Kalas, J. P. (1960). *Copeia*, p. 240.

CHAPTER 9

Intermediary Metabolism of Reptiles

R. A. Coulson and Thomas Hernandez

I. Introduction 217
II. Metabolism of Carbohydrates 218
A. Glycogenolysis and Glycolysis 218
B. Hyperglycemic Agents 221
C. Metabolism of Various Sugars and the Effect of Insulin 222
III. Metabolism of Amino Acids 224
A. Digestion of Proteins and Fate of the Products 224
B. Rate of Deamination of Amino Acids 225
C. Effect of Insulin on Amino Acid Deamination 226
D. Nitrogen Transport 227
E. Metabolism of α-Keto Derivatives of Amino Acids 228
F. Amino Acid Interconversions in Crocodilia 228
G. Ammonium Bicarbonate Synthesis in Crocodilia 231
H. Amino Acid Metabolism in the Chameleon 234
I. Summary of the Metabolism of Amino Acids in the Alligator, Caiman, Turtle, Chameleon, and *Egernia cunninghami* 237
IV. Intermediary Metabolism and Metabolic Rate 241
A. Body Temperature 241
B. Substrate Concentrations in Vertebrates 242
C. Enzyme Concentration by Measurements *in Vitro* 242
D. Enzyme Concentration by Measurements *in Vivo* 243
E. Substrate Delivery as a Function of Blood Flow 244
F. Body size 246
References 246

I. Introduction

Biochemical transformations have not been investigated at all in scores of species of reptiles, and information gained from most of the others is superficial. Two species of crocodilia (*Alligator mississipiensis* and *Caiman crocodilus crocodilus*) were studied in detail, and there is also information on intermediary metabolism in one species of turtle (*Pseudemys scripti elegans*) and two lizards (*Anolis carolinensis* and *Egernia cunninghami*).

Hundreds of individual reactions have been discovered in fragmented tissues of animals over the past 50 years. Where information is available, most of the pathways appear common to a large number of vertebrates and to many of the plants as well. However, it is possible to overemphasize the importance of the existence of similar pathways in excised tissues of various animals, as each may differ in the degree of emphasis accorded any given reaction. There may also be a problem in applying some of the results of the *in vitro* experiments toward an understanding of reactions *in vivo*. The discovery of a particular enzyme in a tissue of an animal is often considered sufficient proof that the reaction catalyzed by that enzyme occurs in the intact animal. It is also commonly believed that if a single compound may be converted by two different enzymes into either of two derivatives, then if one of the enzymes in the tissue homogenate has several times the activity of the other, the product of the reaction catalyzed by that enzyme will be formed at several times the rate of the other product. However, the intact animal is complex, and the physiological reaction rates may not resemble those predicted from results of estimation of the enzymes *in vitro*, since membranes are often interposed between the reactants.

In comparing intermediary metabolism in the reptile *in vivo* with that of the more familiar mammal *in vitro*, one is comparing two different systems, and when the results diverge, the reason may be more in the difference in the systems than in the animals. In the following discussion, pathways proved to exist in the living animal will receive the major emphasis.

II. Metabolism of Carbohydrates

Crocodilia have plasma glucose concentrations of about 50 mg/100 ml in the winter and about 120 mg/100 ml in the summer. The glycogen content of the liver is about 10% of its total wet weight, and that of the muscle (and several other tissues) is about 0.5%. Of the total body glycogen, about half is in the liver and half in the muscle (Coulson and Hernandez, 1953). Neither liver nor muscle glycogen appear to be influenced much by variations in food intake, since the glycogen content of an alligator that has been deprived of food for several months is about the same as that of an alligator fed recently. A constant level of liver glycogen is not universal in reptiles. Both the Australian lizard *Egernia cunninghami* (Barwick and Bryant, 1966) and *Anolis carolinensis* (Dessauer, 1970) showed marked seasonal variations in liver glycogen, and in the case of *Egernia*, considerable increases were also observed after feeding.

A. Glycogenolysis and Glycolysis

A captive alligator in a pen appears phlegmatic. When approached slowly there is no visible sign of nervousness, and the animal is motionless. If blood is removed within a minute or so of entering the room, blood lactate is about 5 mmoles/liter, a value as high as the total plasma organic acids of an average man. Blood removed at intervals for the next 12 hours declines in lactate until by the twelfth hour, none can be demonstrated (Stevenson *et al.*, 1957). When one considers that time is required for lactic acid to be produced from muscle glycogen, and that after it is formed, more time will be required for it to appear in the plasma, it is apparent that glycolysis is almost instantaneous. When it was observed that lactate rose rapidly and that glucose did not, it was concluded that muscle glycolysis was a rapid reaction and that liver glycogenolysis was slow.

Epinephrine was injected to study the rates of breakdown of the two glycogens and the rates of their resynthesis. Lactate, already high from handling the animal, went much higher immediately and peaks above 25 mmoles/liter of plasma were seen often. Pyruvic acid concentrations paralleled those of lactate at a ratio of about 40 to 1 in favor of lactate. Movement of glucose from liver glycogen into the plasma was so slow that it was not possible to observe a significant rise for several hours after epinephrine injection, leading to the conclusion that blood glucose derived from liver glycogen was of little energy value in times of immediate stress (Stevenson *et al.*, 1957).

If we assume that the values derived from mammalian experiments apply to the alligator, we can calculate the energy derived from muscle glycolysis. A 1 kg alligator has a muscle mass equal to 50% of the body weight. The glycogen content is 0.5%, giving about 2.5 gm of glycogen per kilogram of alligator. If all the glycogen were converted into lactate (molecular weight 90), one would get about 2.5 gm of lactic acid or 0.028 moles of lactic acid distributed in 700 ml of water (since lactic acid is dissolved in body water). This would give a maximum value of about 0.040 moles per liter of body water or 40 mmoles/liter. Since values almost this high were observed after injection of epinephrine, the alligator must be capable of converting most of the muscle glycogen into lactic acid on sufficient provocation. The energy released could be calculated on the assumption that the conversion of 1 mole of glycogen (glucose) to lactic acid gives rise to the formation of 2 moles of ATP, or 56,000 cal at 100% efficiency, or about 16,000 cal at 30% efficiency (the number usually taken as a reasonable estimate). If we convert the 2.5 gm of glucose (0.014 moles) derived from muscle glycogen into lactate we should get an instantaneous release of $0.014 \times 16{,}000$ or 224 cal, or 5% of the daily energy requirement.

Figure 1 shows the sequence of events where the glycogen stores in the muscle, first depleted by epinephrine, are replenished largely

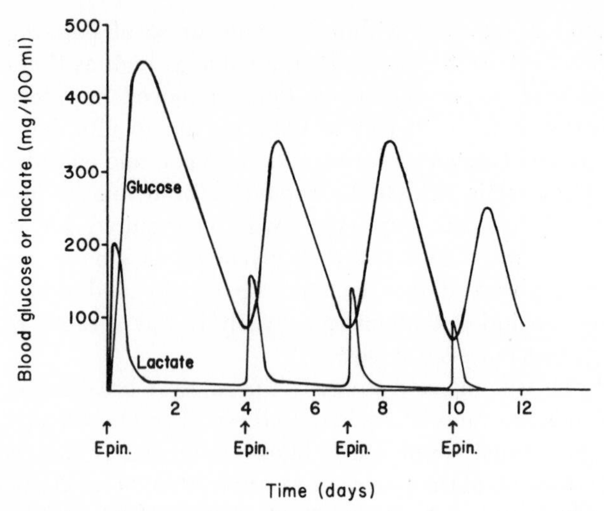

FIG. 1. The effect of repeated injections of epinephrine (1 mg/kg) into 1 kg alligators. Derived from Stevenson *et al.* (1957).

by glucose from liver glycogen, and liver glycogen is probably replenished by gluconeogenesis. After repeated injections of epinephrine, the rise in both glucose and lactate was only slightly smaller than in the initial experiment, a finding suggesting that liver and muscle glycogen resynthesis is a high priority reaction in the alligator (Stevenson *et al.*, 1957).

Glycolysis in reptiles must be very important. Energy is released in the absence of oxygen, energy not dependent on respiration, metabolic rate, or blood flow. In essence, an animal with a very low metabolic rate may be converted into one capable of a brief burst of energy sufficient to make him more than a match for warm blooded animals of equivalent size. Since an excited alligator produces more lactate than an excited mammal, it is possible that the alligator is able to release proportionately more energy anaerobically than a mammal. The power source is depleted rather quickly and in a prolonged contest the animal with the low metabolic rate would be at a disadvantage.

Experiments to determine the exact role of epinephrine in activating muscle and liver phosphorylases in reptiles have not been reported. In the absence of information to the contrary, it may be assumed that the system is similar to that in the mammals and that the synthesis of cyclic AMP is the most important effect of epinephrine in the glycolytic sequence.

The alligator is moderately sensitive to the action of glucagon (Fig. 2), and as in the mammal, glucagon depletes the liver glycogen and

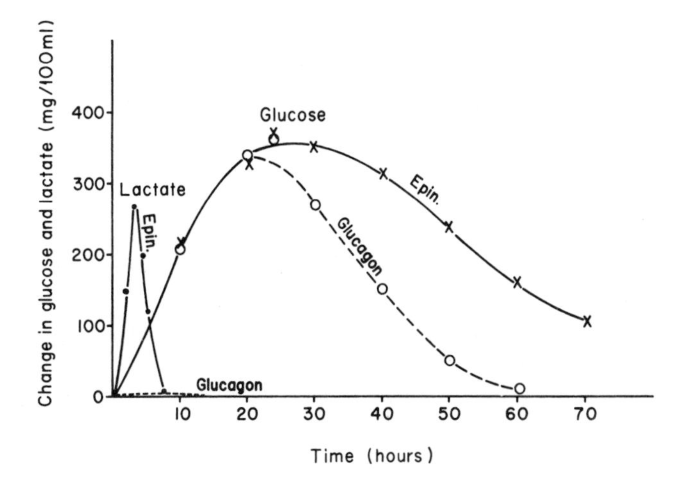

FIG. 2. A comparison of the effect of epinephrine (1 mg/kg) and glucagon (10 mg/kg) on blood glucose and lactate of the alligator. The hormones were injected at time 0. Derived from Stevenson *et al.* (1957).

raises blood glucose, but does not affect blood lactate (Stevenson *et al.*, 1957). The alligator then must possess at least two phosphorylases, one in the liver activated by glucagon and epinephrine, and one in the muscle activated by epinephrine only. The source of glycogen and glucose is probably protein, since digestion is a slow process and any glucose or glycogen in the animals eaten would be converted to lactic acid by the enzymes in the food itself or by bacterial action in the stomach.

B. HYPERGLYCEMIC AGENTS

In addition to epinephrine and glucagon, several other agents have been found to elevate blood glucose. Crude acetone-dried pituitary powder injected for a period of several days increased blood glucose by about 100 mg/ml (Hernandez and Coulson, 1952). Growth hormone (somatotrophin) and hydrocortisone also caused a mild hyperglycemia, and when the two were combined, a condition resembling diabetes resulted (Stevenson *et al.*, 1957). The purines caffeine and theophylline were sympathomimetic, as they activated both liver and muscle phosphorylase. Theobromine appeared to be inactive (Hernandez and Coulson 1956).

Seasonal changes in blood glucose in many reptiles are probably the result of changes in the production of various pituitary hormones. Reptiles grow more in the summer than in the winter even when the temperature is kept at a constant high level (Coulson and Hernandez, 1964).

C. Metabolism of Various Sugars and Effect of Insulin

Experiments were designed to compare the rates of metabolism of eighteen sugars in the hope that this might provide information on the catabolic routes employed in their disposal. Each was injected at a dose of 2 gm/kg, and the blood was analyzed at intervals until it had returned to normal. The rates were expressed in terms of the percent utilized and/or excreted per hour (Fig. 3). The fates of the sugars were subject to great variations, with a few apparently not metabolized (D-glucosamine, 6-*O*-methylglucose, and sucrose), and others metabolized quite rapidly; with a few excreted rapidly, and others completely reabsorbed by the renal tubules; with most distributed evenly in body water and a few restricted to the extracellular fluids (Coulson and Hernandez, 1964).

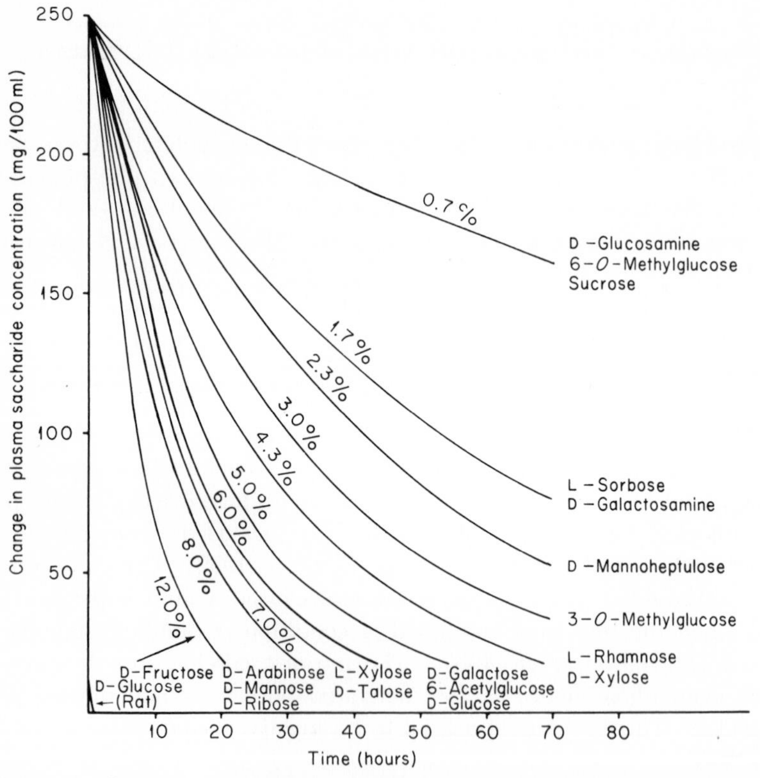

Fig. 3. A schematic representation of the rates of disappearance of various saccharides (1 gm/kg) injected intraperitoneally into 1 kg alligators. The numbers in percents refer to the decay rate as measured in percent per hour. In the extreme inside corner of the graph is shown the rate of disappearance of an equivalent amount of glucose injected into a 250 gm rat intraperitoneally. (From Coulson and Hernandez, 1964.)

In experiments on the fasted rat, one can give a sugar, wait an interval, and determine whether the glycogen content of the liver was increased. If it increased, it is proof that the sugar may be converted into glycogen. Unlike the rat, the fasting alligator has a liver so rich in glycogen that addition of sugars in reasonable amounts has no detectable effect on its content and therefore changes in glycogen cannot be used to determine which sugars may serve as precursors.

A single dose of insulin (100 units/kg) is effective for about a month in a 1 kg alligator, and although severe hypoglycemia finally occurs, the animal can survive for the first 2 or 3 days without need for glucose. To test the relationship of carbohydrate structure and the action of insulin on the fate of the carbohydrate, in a second series of experiments, each of the eighteen sugars was injected into alligators primed with insulin. Table I shows the distribution of the sugars in body fluids and

TABLE I
EFFECT OF INSULIN ON THE RATE OF DISAPPEARANCE OF VARIOUS SACCHARIDES IN ALLIGATOR[a]

Saccharide	Fluid distribution[b]	Rate of disappearance after insulin ÷ rate of disappearance without insulin	Metabolized (+) or not metabolized (−)
L-Xylose	ECF	1.26	+
D-Glucose	ECF	1.55	+
6-*O*-Acetylglucose	ECF	—[c]	−
D-Galactose	ECF	1.94	+
D-Xylose	ECF	1.44	+
L-Rhamnose	ECF	1.92	+
3-*O*-Methylglucose	ECF	1.63	−
D-Mannoheptulose	ECF	1.10	−
L-Sorbose	ECF	1.70	−
Sucrose	ECF	1.10	−
6-*O*-Methylglucose	ECF	—[c]	−
D-Glucosamine	ECF	3.23	−
D-Fructose	BW	1.52	+
D-Ribose	BW	1.08	+
D-Mannose	BW	2.00	+
D-Arabinose	BW	1.22	+
D-Talose	BW	1.23	+
D-Galactosamine	BW	1.94	−

[a] 100 units/kg insulin; saccharides injected intraperitoneally at a dose of 2 gm/kg. Data from Coulson and Hernandez (1964).

[b] ECF = extracellular fluid (40% of the body weight); BW = body water (70% of the body weight).

[c] Not determined.

the effect of insulin on the rates of their disappearance. Considering the differences in chemical structure and the fact that the cell membrane permits some of the sugars to pass while excluding others, it seems strange that insulin should have increased the rate of disappearance of them all. One might expect that insulin, a hormone known for its ability to facilitate movement of glucose across a cell membrane, would affect chiefly those sugars normally restricted to the extracellular spaces while having no affect on those to which the cell membrane was no barrier. This was not the case (Coulson and Hernandez, 1964).

III. Metabolism of Amino Acids

A. Digestion of Proteins and Fate of the Products

In one of the few reptile nutrition experiments, caimans were force fed and both the rate of digestion and the fate of the amino acids released were determined (Coulson and Hernandez, 1970a). Over 90% of the food disappeared from the gastrointestinal tract in 48 hours, and the intestinal contents remained almost devoid of free amino acids at all times during digestion. The high rate of absorption of the mixed amino acids released by digestion of protein was in contrast to the very slow rate of absorption of crystalline amino acids injested singly, even though the initial gut concentration of single amino acids was hundreds of times as great. Protein, which is synthesized rapidly from a balanced mixture of amino acids, is a reservoir of almost unlimited capacity in which to put free amino acids. Although great differences existed in the rate of absorption of different amino acids given singly, all amino acids liberated in protein hydrolysis were absorbed at the same percentage rate.

When fasted crocodilians were permitted unlimited food, the total free amino acid concentration rose in the plasma from the fasting value of about 3 mmoles/liter to about 15 mmoles/liter at the peak of intestinal absorption. Essential amino acids were increased in tissues and body fluids but slightly, while three nonessentials (glycine, alanine, and glutamine) were increased out of all proportion (Fig. 4). At the peak, these three compounds had incorporated in their structure over 80% of the free α-amino nitrogen in the entire body (Coulson and Hernandez, 1967). It was evident that some special significance should be attached to glycine, alanine, and glutamine, and that these compounds were probably responsible for the transfer of almost all nitrogen not required for protein synthesis. Great as the rise in glycine, alanine, and glutamine was, most of the nitrogen consumed was converted efficiently into protein. When a growing crocodilian eats, nearly 50% of the weight of food

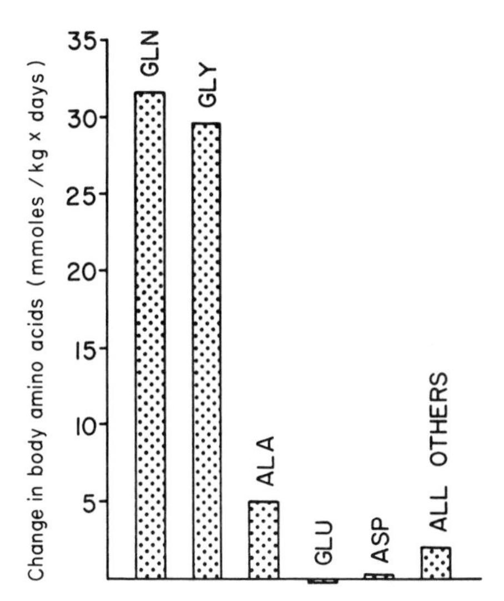

FIG. 4. Changes in free amino acids in small caimans force fed fish. The bars represent the areas of the curves where the increase in the amino acids in the body fluids was multiplied by the length of time each amino acid exceeded the fasting control concentration. (From Coulson and Hernandez, 1967.)

is used for growth. Were this not true, the animals would have so much excess nitrogen in the body that the kidneys would need weeks to excrete it.

B. Rate of Deamination of Amino Acids

Amino acids absorbed from the gut have two possible metabolic pathways—to protein or to carbon dioxide, water, and nitrogen compounds such as uric acid, etc. An amino acid in excess will be catabolized at a rate dependent on its concentration at the proper site and the availability of enzymes and cofactors to degrade it. It is generally agreed that for most amino acids, deamination combined with oxidation is the first step in catabolism.

It was possible to determine the rate of removal of the α-amino group from each free amino acid occurring in proteins, and from a few others as well (Hernandez and Coulson, 1961). The rationale of the experiments was as follows: If a single amino acid is injected, and if it is still plentiful in the body long after the injection, then that amino acid is deaminated slowly. Conversely, if it disappears rapidly, it is deaminated rapidly. Only the first of a complex series of reactions is tested and that is the primary one of deamination. The rates of deamination of the different amino acids shown in Fig. 5 varied from 0.5% per hour for methionine to thirty times as great for serine. The declining segments

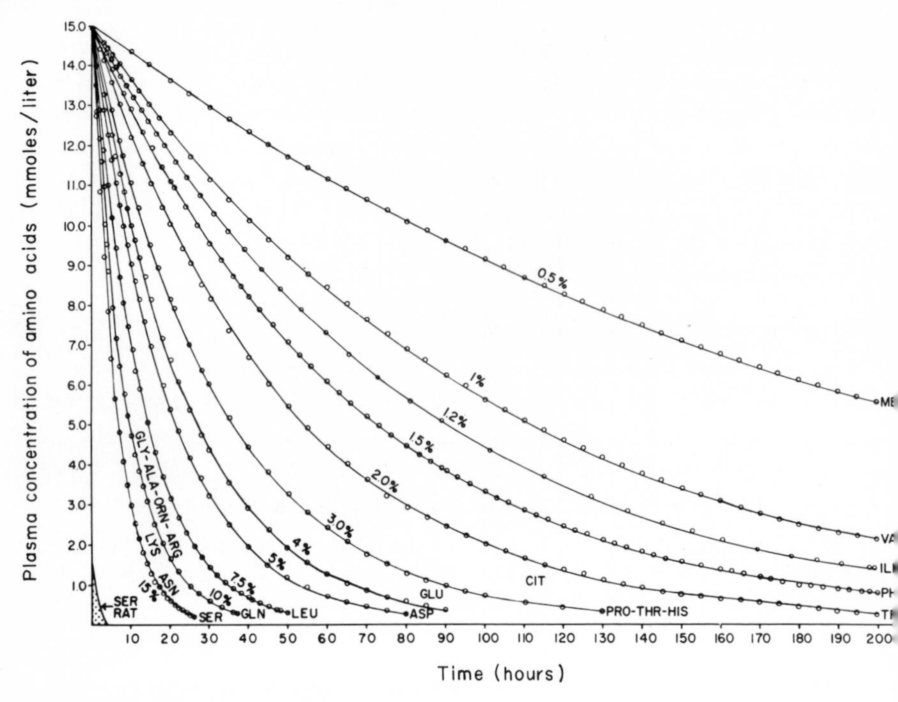

FIG. 5. Schematic representation of the rates of disappearance of amino acids (10 mmoles/kg) injected intraperitoneally into 1 kg alligators. The numbers in percents refer to the decay rate in percent removed per hour. Derived from Hernandez and Coulson (1961). The shaded curve (lower left corner) shows the rate of disappearance of serine in a rat that received 10 mmoles/kg.

of the curves were all of the same shape, slopes that could be described mathematically as log curves indicative of first order reactions. Neither glutamic nor aspartic acid were deaminated rapidly, and amino acids considered essential in mammals were deaminated much more slowly than the nonessentials. Considering the possibility that these unexpected results in the crocodilian experiments were unique, the experiments were replicated in rats, dogs, turtles, and chameleons (Coulson and Hernandez, 1965). In all of the animals, aspartate, glutamate, and most of the essential amino acids were deaminated slowly, and the remaining nonessential amino acids were deaminated rapidly. The absolute rates at which the different species deaminated amino acids were dependent on the relative metabolic rates.

C. Effect of Insulin on Amino Acid Deamination

In the crocodilia at least, insulin decreases the concentration of every amino acid in the plasma as effectively as it does glucose (Hernandez and Coulson, 1961). In the tissues, all of the essential amino acids are decreased by insulin, but glycine, alanine, and glutamine are either unaffected or actually increased (Hernandez and Coulson, 1968a, 1969). When insulin is injected along with an amino acid, the rate of deamination of that amino acid is increased greatly (Hernandez and Coulson, 1961). No single explanation explains adequately the means by which insulin hastens the disappearance from the body of scores of compounds of unrelated structure and function.

D. Nitrogen Transport

Only those amino acids that may be deaminated rapidly could possibly serve in a nitrogen transport system. If methionine, for example is injected into any animal, there can be no rapid increase in urea, uric acid, or urinary ammonia, the compound is deaminated too slowly. One can compute an index that should indicate the amino acids available as efficient nitrogen carriers by multiplying the decay rate (rate of deamination) by the plasma concentration for each of the animals for which we have the necessary information. From the data in Table II,

TABLE II

Apparent Ability of Amino Acids to Carry and Transfer Nitrogen[a]

Amino acid	Alligator fasting	Alligator fed	Turtle fasting	Chameleon fasting	Rat fasting	Dog fasting
Gly	26.5	44.1	26.0	49.0	14.1	20.4
Ala	17.5	15.1	4.2	10.9	22.9	19.9
Ser	7.6	3.0	15.0	6.8	13.9	11.1
Thr	1.2	0.4	3.5	4.2	2.5	3.0
Asp	0.3	0.1	0.2	2.2	0.2	0.2
Glu	2.0	0.7	2.5	2.3	1.0	0.5
Gln	15.2	26.8	8.0	7.2	28.5	32.6
Val	0.7	0.4	1.0	1.5	1.4	0.6
Leu	6.3	2.8	6.0	6.1	3.2	4.3
Ile	0.2	0.1	0.5	0.5	0.8	0.7
Met	0.1	0.1	0.7	1.1	0.2	0.1
Orn	3.4	0.9	17.5	2.2	1.3	0.5
Arg	4.5	0.9	3.5	0.8	4.0	3.7
Cit	0.1	0.0	—	—	—	—
His	3.9	1.3	3.5	4.2	0.4	0.7
Lys	7.9	2.2	6.0	0.9	4.9	0.9
Trp	0.5	0.2	3.0	—	—	0.1
Phe	0.6	0.1	0.8	0.3	0.6	0.5

[a] The plasma concentration was multiplied by the relative rate of deamination—the higher the product, the higher the probable rate of nitrogen turnover. Values are in terms of percent of the total turnover. (Coulson and Hernandez, 1970b.)

it is evident that glycine, alanine, and glutamine transport most of the excess nitrogen to the liver and kidney of the alligator, caiman, chameleon, rat, dog, and turtle.

E. Metabolism of α-Keto Derivatives of Amino Acids

The rate of catabolism of an amino acid appears to be a function of the rate at which it is deaminated, and this could be controlled in part by the ease with which the product of deamination is removed. The rates of disappearance of six α-keto compounds derived from amino acids were determined (Coulson and Hernandez, 1965). Of the group, only α-ketoglutarate was metabolized as slowly as its amino acid precursor (Fig. 6). If these α keto derivatives are representative, then the rate-controlling step in amino acid degradation is the one associated with deamination.

F. Amino Acid Interconversions in Crocodilia

When one examines a modern textbook of biochemistry or a recent review on intermediary metabolism, different reactions appear to receive the same emphasis, and the yields of the products are rarely mentioned since they are exceedingly difficult to determine accurately. In experiments on living animals, it is possible to estimate the yield of amino acid derivatives, and the results are often not those that might have been predicted from the observations of others on tissue fragments.

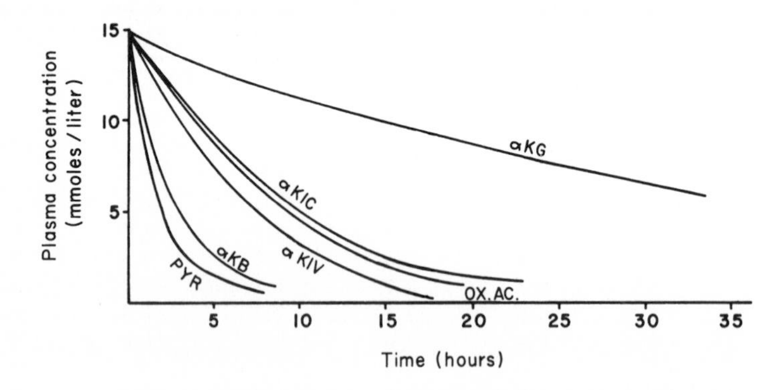

Fig. 6. Schematic representation of the rates of disappearance of α-keto derivatives (10 mmoles/kg) injected intraperitoneally into 1 kg alligators. The decay rates in percent per hour were: α-ketoglutarate (αKG), 3%; α-ketoisocaproate (αKIC), 12%; oxaloacetate (OX.AC.), 13%; α-ketoisovalerate (αKIV), 14%; α-ketobutyrate (αKB), 35%; and pyruvate (PYR), 40%. Compare this figure with Fig. 5, which illustrates the much slower rate of removal of all the parent amino acids except glutamic acid. Derived from Coulson and Hernandez (1965).

1. Transamination

In theory, all but one or two amino acids are normally deaminated by means of specific transaminases, and α-keto derivatives are formed in the process. In the living animal, this theory is not entirely satisfactory. Only eight compounds participate in transamination with any enthusiasm, five are amino acids (aspartate, glutamate, alanine, glutamine, and asparagine), and three are α-keto acceptors (pyruvate, oxaloacetate, and α-ketoglutarate). None of the other amino acids transfer amino groups toward synthesis of aspartate or glutamate in amounts greater than traces so far as could be determined (Coulson and Hernandez, 1970b). If transamination is not the principal reaction for removal of amino groups from most amino acids, then a simple deamination (probably oxidative) must be.

2. Synthesis of Glycine, Alanine, Glutamine, and Glutamic Acid

In the process of catabolism, either the carbons or the nitrogens of various amino acids are incorporated in other amino acids. Figure 7 shows the known routes of amino acid interconversions *in vivo* and also syntheses from several nonnitrogenous compounds (Coulson and Hernandez, 1970b). The relative yield is indicated by a number next to the arrows; the lower the number, the higher the yield. Those reactions enclosed in rectangles or brackets are theoretical intermediates in the catabolic routes. Most of the free amino acids are utilized in whole or in part for the synthesis of large amounts of glycine, alanine, and glutamine, and these in turn donate both nitrogen and carbon toward the synthesis of hepatic uric acid or renal ammonia. Limitations in space do not permit a detailed discussion of the organs involved, and for more information, reference to an earlier review is suggested (Coulson and Hernandez, 1970b).

Inhibition of protein synthesis *in vivo* by cycloheximide increased the body fluid concentration of essential amino acids slightly and glycine, alanine, and glutamine greatly (Coulson and Hernandez, 1971). Apparently, failure to synthesize protein leads to the destruction of protein, and glycine, alanine, and glutamine transport the excess nitrogen to the liver and kidney.

3. Glutamine and Glutamic Acid Synthesis from Nonnitrogenous Precursors

Crocodilia synthesize large amounts of glutamine, primarily in the liver, but also in lesser amounts in the brain and other tissues (Coulson and Hernandez, 1967). In the classic theory, glutamic acid combines

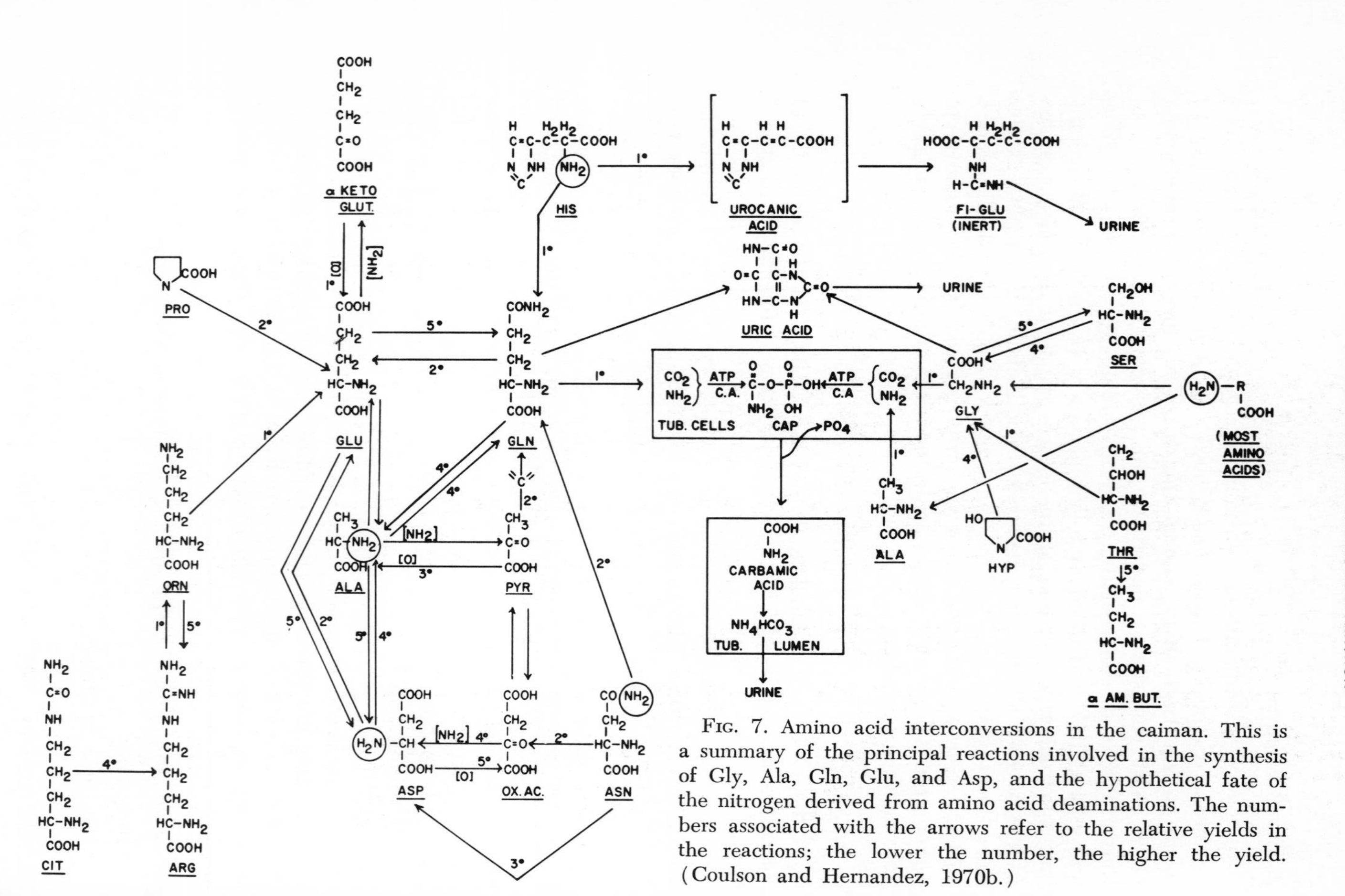

FIG. 7. Amino acid interconversions in the caiman. This is a summary of the principal reactions involved in the synthesis of Gly, Ala, Gln, Glu, and Asp, and the hypothetical fate of the nitrogen derived from amino acid deaminations. The numbers associated with the arrows refer to the relative yields in the reactions; the lower the number, the higher the yield. (Coulson and Hernandez, 1970b.)

with ammonia in the presence of glutamine synthetase, various cofactors, etc. to give glutamine. There are objections to the theory. For example, although glutamine is an excellent precursor of urinary ammonia in the mammal or reptile, glutamic acid injections do not affect ammonia excretion, and they should if glutamine is derived from glutamate.

Some forty-five different compounds were injected into caimans and alligators. Of all the compounds injected, pyruvic acid gave the highest yield of glutamine, and it was one of the best precursors of glutamate (Coulson and Hernandez, 1970b). Pyruvate injected as a ^{14}C uniformly labeled compound gave about the same labeled yield of glutamine and glutamate as when it was injected labeled in the one, two, or three positions, and it was concluded that all three carbon atoms were utilized for synthesis of the carbon skeleton of both glutamyl derivatives (Herbert and Coulson, 1971). Since neither glutamate nor α-ketoglutarate had any appreciable activity in the glutamine synthesizing system, the immediate precursor of glutamine and glutamic acid is unknown. Those compounds capable of yielding pyruvic acid were all instrumental in increasing glutamine, and when they were given as ^{14}C isotopes, glutamine was labeled, and to a somewhat lesser extent, so was glutamate. Whether pyruvate first goes to oxaloacetate in the well known carbon dioxide fixation reaction is not known. If this is the reaction, then a second reaction must occur whereby the four-carbon oxaloacetate adds yet another carbon to form the five-carbon chain of the unknown glutamine precursor.

For several reasons, the Krebs' cycle does not appear to be a major pathway in the conversion of the three-carbon chain of pyruvic acid to the five-carbon chain of glutamine or glutamic acid. Other components of the cycle (succinate, citrate, fumarate, malate) were injected without producing any rise in glutamine or glutamate, and acetate, the usual starting point for carbon degradation, was also without activity (Herbert and Coulson, 1971). These results suggest the likelihood that in this system glutamine and glutamate are synthesized outside the mitochondria.

G. Ammonium Bicarbonate Synthesis in Crocodilia

In the urine of a hydrated alligator the principal ions are ammonium and bicarbonate with the ammonia concentration exceeding that of bicarbonate by about 30%. Although the measured pH is about 7.80, the composition of the urine resembles the acid urine of a mammal to which an excess of ammonium bicarbonate has been added (Coulson and Hernandez, 1957).

There are physical factors at work in the tubules that regulate the movement of ammonia and carbon dioxide. If we assume that the alligator kidney is subject to the same laws that apply to frogs and mammals (Pitts, 1963), neither ammonium nor bicarbonate can diffuse into the lumen of the renal tubules, but the nonionic ammonia and carbon dioxide can. If the tubular lumen is acid, ammonia should diffuse rapidly into it; if the lumen is alkaline, carbon dioxide would be encouraged to diffuse in the same manner. In the Pitts' theory (1963) on the function of the mammalian kidney, one could have a urine containing the ammonium ion or bicarbonate, but certainly not both, yet both appear in almost equal quantity in the urine of an alligator. One needs postulate the existence of a system whereby ammonium bicarbonate could be formed regardless of the pH of the glomerular filtrate. A possible mechanism is outlined below along with the evidence for the theory.

1. Ammonia Synthesis from Glycine, Alanine, and Glutamine

All three amino acids increased ammonia excretion greatly, while other amino acids were inactive or nearly so (Coulson and Hernandez, 1959). When insulin was given, amino acids decreased in the plasma to values as low as one-fourth the normal level, ammonia production was almost eliminated, and considerable amounts of Na^+ appeared in the urine (Coulson and Hernandez, 1962). Ammonia synthesis, almost abolished by insulin, was restored immediately by injection of any of the three amino acids (Coulson and Hernandez, 1962). In fed animals, high ammonia production continued only as long as glycine, glutamine, or alanine remained elevated in the blood (Coulson and Hernandez, 1959).

2. Bicarbonate Synthesis from Glycine, Alanine, and Glutamine

In a 1 kg alligator injected with glycine, alanine, or glutamine (10 mmoles), as much as 7 mmoles of bicarbonate appear in the urine in a day. In the form of a gas, this quantity would represent over 150 ml, or 15% of the total output of carbon dioxide by the entire alligator in a day (Hernandez and Coulson, 1952). Since the oxygen consumption of the combined kidneys at an RQ of 1.00 would then need to be 15% of the total of the body, the carbon dioxide in the urine must come from a reaction (such as decarboxylation) that does not require a large amount of oxygen.

After injection of ^{14}C-labeled glycine, alanine, or glutamine large amounts of labeled bicarbonate appeared in the urine immediately. When other ^{14}C-labeled amino acids which did not produce urinary ammonia were injected, almost no labeled bicarbonate appeared in the urine (Hernandez and Coulson, 1971b).

3. *Carbonic Anhydrase Requirement for Renal Bicarbonate Synthesis*

Injection of a carbonic anhydrase inhibitor (acetazoleamide), almost eliminated bicarbonate from the urine without affecting ammonia excretion (Hernandez and Coulson, 1954). Injection of acetazoleamide and ^{14}C-labeled glycine, alanine, or glutamine resulted in excretion of the ammonium ion (as the chloride) and the small amount of bicarbonate of the urine contained only traces of ^{14}C labeling (Hernandez and Coulson, 1968b).

4. *Renal Synthesis of Carbamyl Phosphate from Ammonia and Carbon Dioxide*

Carbamyl phosphate is formed whenever free ammonia is produced if the proper enzyme is present and if the cell has sufficient ATP (Cohen and Brown, 1960). The sequence of the reactions is usually considered to be:

$$NH_3 + CO_2 + HOH \rightarrow NH_4HCO_3 \text{ (nonenzymatic)} \tag{1}$$

$$NH_4HCO_3 + (2?)ATP \xrightarrow[\text{carbamyl } PO_4 \text{ kinase?}]{\text{carbonic anhydrase}} H_2N\text{—}C(=O)\text{—}O\text{—}P(=O)(OH)_2 \tag{2}$$

In the second reaction, the equilibrium is displaced toward the right in the presence of carbonic anhydrase (Jones and Lipmann, 1960) and a specific enzyme (carbamyl phosphate kinase?), and to the left in its absence. Other cofactors appear to be necessary.

5. *Conversion of Renal Carbamyl Phosphate into Urinary Ammonium Bicarbonate*

Carbamyl phosphate is an energy-rich package of carbon dioxide and ammonia. The high-energy phosphate bond could supply energy for the secretion of carbamic acid into the lumen with the retention of inorganic phosphate in the peritubular cells (Hernandez and Coulson, 1971b). The secreted carbamic acid inside the tubular lumen would now be in a medium free of both ATP and carbonic anhydrase, and the reaction would be shifted toward the breakdown of carbamic acid and synthesis of ammonium bicarbonate. The initial secretion of carbamic acid into the lumen would be accomplished if the contents of the lumen were neutral, acid, or alkaline. If the contents of the lumen were acid, a reaction with the newly formed ammonium bicarbonate would occur, free carbon dioxide would be evolved and reabsorbed into the blood, and the ammonium ion in the urine would exceed the concentration of the bicarbonate ion greatly. Neither the ammonium ion nor free ammonia are reabsorbed from the lumen of the tubule (apparently an ancient adaptation to prevent ammonia intoxication), and whatever ammonia enters the tubule reaches the cloaca.

Although inhibition of carbonic anhydrase by acetazoleamide does not prevent the formation of ammonium bicarbonate in the peritubular cells, carbonic anhydrase is essential for the displacement of the ammonium bicarbonate–carbamyl phosphate reaction in favor of carbamyl phosphate (Jones and Lipmann, 1960). In the virtual absence of carbonic anhydrase, ammonia liberated by amino acid deamination is subject to the usual NH_3–NH_4^+ equilibrium and the ammonia present as NH_3 would be free to diffuse into the lumen where it would replace Na^+. On the other hand, carbon dioxide would not be free to diffuse, since it would not go into an acid lumen. Therefore, in carbonic anhydrase inhibition in crocodilia, the urine becomes quite acid, rich in ammonia, and almost devoid of bicarbonate.

H. Amino Acid Metabolism in the Chameleon

The chameleon *Anolis carolinensis* is found in countless numbers in southeastern United States, and it has been a subject of study for a considerable time (Dessauer, 1970). However, until recently, no attempts had been made to investigate metabolic pathways. The small size of the animal suggested that experiments designed to follow the metabolism of a compound in any one organ would be difficult, and therefore the first series of experiments involved the injection of a single amino acid followed by analyses of the entire carcass for free amino acids (Coulson and Hernandez, 1968). As was to be expected, yields of the various derivatives were small, since inclusion of tissues that did not participate in the catabolic reactions had the effect of diluting the contribution of those that did.

Later, a shortage of caimans prompted a renewed attempt to study the fate of amino acids in a single organ in the chameleon, and the liver was selected as the organ of choice. A grown chameleon weighing 6–8 gm has a liver weight of about 100 mg, a size sufficient for accurate amino acid analysis by modern methods.

The results of the experiments were gratifying. When only the liver was studied, the yields of various derivatives of the amino acids were often high, even higher than in the crocodilian liver (Herbert, 1973). The principal ninhydrin-reactive compounds synthesized from other amino acids in the liver were glutamine, glutamic acid, serine, and glycine. Some of the reactions involved appear complex, and a few seem

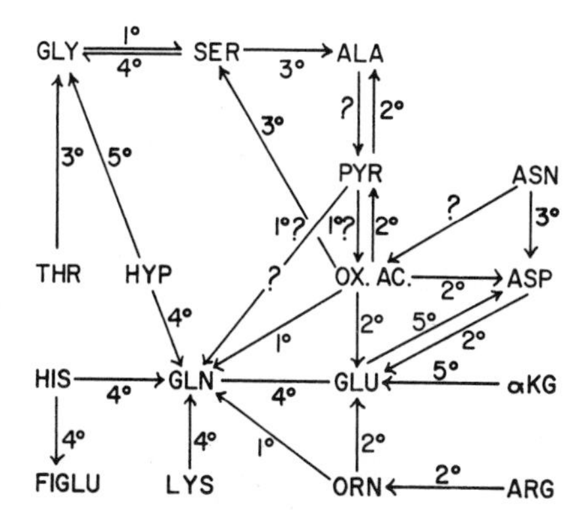

FIG. 8. Amino acid interconversions in chameleon liver. The number in degrees associated with each arrow indicates the relative yield of a product, with 1 representing the highest yield and 5 the lowest. In this figure, no distinction is made between the synthesis of a derivative through donation of nitrogen or carbon or both. For interconversions involving carbon incorporation only, see Fig. 9. (J. D. Herbert, R. A. Coulson, and T. Hernandez, unpublished observations, 1971.)

to be unprecedented (Fig. 8). In this figure, as in the one showing interconversions in the caiman and alligator (Fig. 7), the lower the number next to the arrow, the higher the yield.

1. *Synthesis of Glutamine and Glutamic Acid*

The four compounds effective in increasing liver glutamine greatly were pyruvic and oxaloacetic acids, arginine, and ornithine. Although no ^{14}C-labeled arginine or ornithine was injected into the chameleon, arginine and ornithine probably supplied the carbon chain of both glutamic acid and glutamine in the classic pathways. The great yield of glutamine after ornithine seems to have no precedent, even in crocodilia. The rise in glutamine and glutamate following pyruvate or oxaloacetate resembles that seen in the crocodilian experiments. Repeated experiments with ^{14}C-labeled pyruvic acid gave high yields of labeled glutamine and glutamate, and once again, it was necessary to conclude that the three-carbon pyruvate gains two more carbons to form a five-membered precursor of glutamine or glutamate. The distribution of ^{14}C after the administration of several labeled compounds is shown in Fig. 9. Glutamine was not synthesized from α-ketoglutarate, and only traces were synthesized from glutamate; therefore, glutamine synthetase (if it exists) is of little practical importance in the liver of the chameleon. Since glutamate was synthesized from α-ketoglutarate, α-ketoglutarate must have entered the cells; and since injection of glutamate itself also led to its intracellular accumulation, failure of the dicarboxylic acids to enter

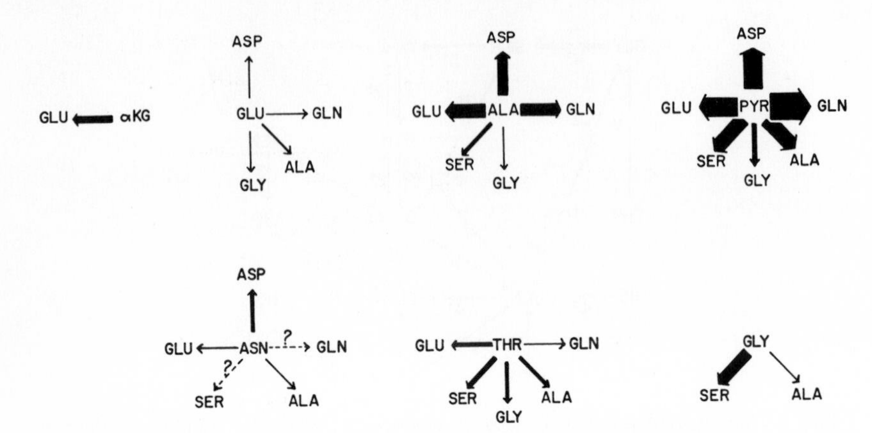

FIG. 9. Distribution of ^{14}C in chameleon liver following injection of uniformly labeled amino acids. The labeled compound injected was the one at the source of the arrows. The thickness of the arrows is proportional to the relative yield of labeled carbon in the derivatives. (Herbert and Coulson, 1972).

the hepatic cells cannot have been responsible for their inability to serve as precursors of glutamine.

2. *Synthesis of Serine and Glycine*

Unlike the crocodilians, the chameleon favors serine over glycine in any equilibrium involving the two. Threonine increases both glycine and serine, possibly first through an aldolase cleavage to glycine followed by conversion to serine.

3. *Synthesis of Alanine and Aspartic Acid*

Only pyruvate and oxaloacetate were major precursors of alanine in the liver of the chameleon. The fact that oxaloacetate was apparently as good as pyruvate in alanine synthesis suggests that the conversion of oxaloacetate to pyruvate is rapid and efficient. Other compounds that might have been expected to increase liver aspartate were able to do so, but the total yield was small.

4. *Other Amino Acid Interconversions*

Injection of several amino acids (valine, isoleucine, leucine, and α-aminobutyric acid) did not result in increases in any other ninhydrin reactants in the chameleon under the conditions of these experiments. While in the caiman, proline gave rise to considerable amounts of glutamic acid, the increase in glutamic acid after injecting proline into a chameleon was very small. Unlike in the turtle or caiman, citrulline proved inert or nearly so in the chameleon, and there was no sign of its conversion to either arginine or ornithine. Phenylalanine was converted in part to tyrosine, and methionine to very small amounts of cystine, α-aminobutyric acid, and larger amounts of methionine sulfoxide.

Histidine was not a good precursor of glutamic acid, although it did increase glutamine significantly. The fate of the small amounts of formiminoglutamic acid that appeared in the liver following histidine administration was not determined.

I. Summary of Metabolism of Amino Acids in the Alligator, Caiman, Turtle, Chameleon, and *Egernia cunninghami*

1. *Glycine*

In the free state, this is probably the most plentiful amino acid in vertebrates, since it concentration in the muscle of many animals exceeds that of all the others combined (Hernandez and Coulson, 1971a). It is also first or second in concentration in the extracellular fluids, and partially as a consequence, very important in the delivery of excess nitrogen to the liver and kidney for disposal. From the available evidence, in the reptiles the usual fate of glycine is destruction in the liver with the formation of uric acid and in the kidney with the formation of ammonia. When glycine is injected, serine is the only amino acid formed in quantity in the chameleon, and even this reaction is barely detectable in the turtle and crocodilia.

The major pathways involved in the synthesis of glycine are still unknown. Synthesis from serine and threonine, although they may be shown to occur, cannot account for the massive amounts of glycine formed in fed crocodilia or fed chameleons, and other major sources must exist.

2. *Alanine*

In the living reptiles, alanine is important in transamination. Since alanine may form pyruvic acid, its carbon chain is also incorporated into aspartic acid, glutamic acid, and glutamine. An important constituent of the cell fluids, it is also a good precursor of ammonia in crocodilia and is important in nitrogen transport.

3. *Serine*

Serine is a minor amino acid in the caiman, alligator, and turtle and a major one in the chameleon, in which it is synthesized in great amounts from glycine and in fair amounts from pyruvic acid, oxaloacetic acid, alanine, and even threonine. The reaction in which injected serine is converted to glycine is a minor one in all the reptiles studied, and since serine is metabolically the most active amino acid in many animals, other routes of destruction must exist. In the chameleon, serine is a good source of alanine.

4. *Threonine*

Threonine is an essential amino acid for all species tested. When injected, it is converted almost quantitatively into glycine in crocodilia and into small amounts of glycine in the other reptiles. In common with other essential amino acids, it is degraded slowly. In the chameleon, it serves as a source of serine and alanine.

5. *Aspartic and Glutamic Acids*

These two dicarboxylic acids resemble each other in some respects but not in others. Both function in transamination systems, and neither is rapidly catabolized. Aspartic acid is never present in the body in large amounts, while glutamic acid is a major constituent of several tissues. Neither compound is synthesized in appreciable amounts in the muscle, and although both are formed in the liver and kidney from their respective amido analogs, neither is efficient in the reverse direction, i.e., toward the synthesis of glutamine or asparagine. Dicarboxylic amino acids are not increased in the body in the free state after feeding. Since both are major constituents of the protein that is being synthesized rapidly, there would appear to be a mechanism by which the rate of their synthesis parallels exactly the rate of protein synthesis.

In some respects, the body appears to consider these acidic compounds as foreign substances. A crocodilian will often eat 10% of its weight in food at one time. If the food is 15% protein, and if aspartic and glutamic acids account for 30% of the amino acids in the protein, then a digesting 1 kg crocodilian would release about 4.5 gm of aspartic and glutamic acids in the process of hydrolysis. This amount is about 32 mmoles/kg, and at a blood pH of 7.4, it would require nearly 60 meq of sodium bicarbonate per kilogram to neutralize the dicarboxylic acids. A fatal acidosis could follow rapid absorption of these compounds in the free state, since neither is catabolized rapidly.

6. *Valine, Leucine, and Isoleucine*

All three are essential in the reptiles. Leucine is catabolized the most rapidly but at only a fraction of the rate of several nonessential amino acids. The branched-chain amino acids do not produce detectable amounts of any other amino acid.

7. *Cystine and Methionine*

Although the insolubility of cystine prevented any accurate assessment of its catabolic pathways, if it resembles methionine it is catabolized slowly. A single injection of methionine elevated the blood level for several days in chameleons and for weeks in turtles and crocodilia. Very small amounts of cystine and large amounts of methionine sulfoxide were derived from methionine by all reptiles tested, but no amino acid caused any increase in body methionine, an indication of its essentiality. Curiously, no α-aminobutyrate is formed from methionine by caimans, but synthesis begins in the alligator 4 days after methionine injection. The reason for the delay in α-aminobutyrate production is unknown (Coulson and Hernandez, 1965).

8. *Ornithine, Arginine, and Citrulline*

With respect to nitrogen excretion, there are two (or perhaps three) systems employed by the reptiles. Aquatic turtles favor urea over uric acid, while lizards and crocodilians rely on uric acid synthesis or uric acid and ammonia. Long ago, animals were classified as "ureotelic" or "uricotelic" partly on the basis of the presence or absence of the enzyme arginase in the liver. By definition, arginase is an enzyme that converts arginine into ornithine and urea, and any other enzyme that converted arginine into ornithine without urea formation would not be called arginase. The reptiles studied converted arginine into ornithine but only the turtle (*Pseudemys scripta*) formed urea in the process. Once ornithine was formed, it, too, was metabolized rapidly and glutamic acid was one of the derivatives. In the chameleon liver, glutamine was synthesized from ornithine in greater amounts than from any other amino acid. The injection of ornithine led to synthesis of small amounts of arginine in crocodilia and turtles but not in the chameleon (Coulson and Hernandez, 1970b). Conversion of arginine to ornithine and urea in the turtle liver was so rapid, that injections of arginine did not increase the liver content (Hernandez and Coulson, 1966). All species catabolized arginine at about the same relative rate, regardless of the pathway employed.

Crocodilia, turtles, and chameleons catabolize citrulline slowly. In the alligator, the compound is catabolized to arginine almost quantitatively; in the turtle, ornithine was the only recognizable product; and in the chameleon, no arginine or ornithine was formed. In none of the animals was there any visible citrulline synthesis from ornithine or arginine. Elucidation of the metabolism of citrulline and arginine in reptiles will require considerably more study than has been devoted to the problem to date.

9. *Proline and Hydroxyproline*

The classic pathways seem to describe the fate of these amino acids fairly well. Both are metabolized slowly by catabolic routes that are distinctly different. Hydroxyproline forms a small amount of glycine in the chameleon liver, and perhaps in the other reptiles as well, but no glutamic acid is derived from this compound. Proline, on the other hand, increases the level of glutamic acid in the reptiles without affecting the glycine concentration.

10. *Lysine*

Lysine is an essential amino acid in the reptiles but one that differs from the others in being catabolized slowly by mammals and rapidly by reptiles (Coulson and Hernandez, 1965). In the reptiles studied, small amounts of glutamic acid appear to be synthesized from lysine, but since the apparent synthesis was not checked by isotope experiments, it would be better to reserve judgment. In the chameleon liver, lysine injection also led to an increase in glutamine.

11. *Tryptophan*

Tryptophan is an essential amino acid and one that is fairly toxic. The increase in 5-hydroxytryptophan observed in all the reptile experiments may partially explain the toxicity if 5-hydroxytryptophan is a precursor of serotonin in the reptiles. No other detectable ninhydrin-reactive compounds were observed (Coulson and Hernandez, 1970b).

12. *Histidine*

Histidine is an essential amino acid in the reptiles with a curious catabolic fate. It increased glutamine in the turtle and caiman, but when histidine-^{14}C was injected, the absence of labeled glutamine indicated that histidine served as a donor of nitrogen only (Herbert, 1968). The carbon skeleton found its way almost quantitatively into formiminoglutamic acid, an inert compound that was excreted unchanged. Whether urocanic acid is involved in the normal conversion of histidine to formiminoglutamic acid has not been solved. The turtle and chameleon synthesized a small amount of a compound that resembled citrulline in its position on the chromatographic column and in its chemical nature (Herbert, 1968).

3. *Phenylalanine and Tyrosine*

Phenylalanine is an essential amino acid in reptiles, and small amounts of tyrosine are synthesized from it. Next to methionine, phenylalanine is the most difficult amino acid to catabolize in reptiles or mammals. Tyrosine catabolism was not studied directly, as it was so insoluble that it could not be administered in sufficient concentration.

4. *Amino Acid Synthesis in Egernia*

Radioactive glucose, acetate and succinate were incubated with homogenates of *Egernia cunninghami* liver and several components of the Krebs' cycle were isolated. The major components were labeled, as were aspartate, glutamine, asparagine, glutamate, α-aminobutyrate, and alanine. In this experiment *in vitro,* the Krebs' cycle intermediates appeared to furnish carbon for amino acid synthesis (Barwick and Bryant, 1966).

IV. Intermediary Metabolism and Metabolic Rate

In cell-free systems, the rate at which a catalyzed reaction proceeds is dependent upon the temperature, the nature and concentration of the enzyme, and the concentration of the substrate. Within the vertebrates, metabolic rates vary widely, and on a weight basis, hummingbirds and shrews consume at least 2000 times as much oxygen as some of the larger reptiles. Since the total metabolic rate is a summation of the rates of hundreds of individual chemical reactions, theoretically the great differences in metabolic rates of different species should then be related to equally great differences in enzyme contents or substrate concentrations or both. The factors that could be responsible for differences in reaction rates in living animals are discussed below.

A. Body Temperature

In comparing the metabolic rates of cold-blooded animals with those that are warm-blooded, body temperature should be considered. Most reptiles in warm climates seem to prefer temperatures around 32°C, a figure enough below the body temperature of the mammal to affect the metabolic rate. The importance of temperature in interspecies comparison may be exaggerated, however. At the same temperature, the chameleon has several times the metabolic rate of an alligator, and the body temperature of a shrew is almost the same as that of an elephant whose metabolic rate is only a fraction as great.

B. Substrate Concentrations in Vertebrates

Perhaps the most important limiting factor in the regulation of substrate concentration is the one of osmotic pressure. In all vertebrate species from which we have information, the osmotic pressure of the extracellular and intracellular fluids is about 290 mosm/liter. Crowded within the totals are the principal ions of sodium, potassium, chlorine and bicarbonate and some protein, phosphates, etc. There is no room for high concentrations of necessary substrates such as free amino acids, acetoacetic acid, components of the Krebs' cycle, and pyruvate and lactate. Differences in metabolic rates cannot be ascribed to differences in substrate concentration, since there is no evidence that species with high metabolic rates have higher substrate concentrations than those requiring less oxygen.

C. Enzyme Concentration by Measurements *in Vitro*

To estimate the "concentration" of a soluble enzyme, a tissue is homogenized and centrifuged. Substrate and necessary cofactors are added to portions of the clear supernatent, and the rate of the disappearance of the substrate (or appearance of a derivative) is determined. Within limits, the greater the enzyme concentration the faster the substrate will be utilized. On the theory that enzyme concentrations were the principal determinant of metabolic rate, concentrations of several soluble enzymes (lactic dehydrogenase, isocitric dehydrogenase, and glutamic–oxaloacetic transaminase) were estimated in homogenates of entire caimans, turtles, mice, chameleons, and rats (Coulson, unpublished observations). In all cases the amount in the homogenate seemed unreasonably large, exceeding the quantity that could possibly be utilized in the intact animal by a thousand times. Although the rat had about a 30% higher apparent enzyme content as measured *in vitro* than caimans or turtles, variations between species were still small and none had as much as three times the concentration of any other.

In vitro measurements of enzyme concentrations are subject to considerable error, and we doubted the validity of our measurements. The most thorough study of enzyme concentration seems to have been that of Schmidt and Schmidt (1960). They estimated seventeen different enzymes in each of sixteen tissues or organs of man. Some of the tissues were obtained by biopsy and some from fresh cadavers awaiting autopsy. Their results on man were similar to ours on rats and reptiles, and since theirs are quoted often, it is probably safe to assume their values have been verified by others in the years since their figures were reported. In international units, the computed approximate average for a kilogram of "average" tissue from man was: malic dehydrogenase 100,000, lactic dehydrogenase 100,000, pyruvic kinase, 50,000, and glutamic–oxaloacetic transaminase 20,000. All four of the above are of some interest in principal metabolic pathways, and they should serve to illustrate the general enzyme levels encountered. Since 1 I.U. is defined as that quantity of enzyme needed to catalyze the conversion of 1 μmole of substrate in 1 minute, then 1 average kg of human tissue could catabolize (maximally) 100,000 μmoles of malate, or 100,000 μmoles of lactate, or 50,000 μmoles of pyruvate, or 20,000 μmoles of glutamate in 1 minute. In millimoles per kilogram per minute, the values would be 100, 100, 50, and 20, respectively, and on the basis of a 70 kg man, they would be increased seventyfold to 7000, 7000, 3500, and 2000 mmoles/minute. The total would be from 2 to 7 moles/minute, a considerable quantity. If the man ate only malate (molecular weight 117), and each gram provided 3 kcal of energy, he could utilize only about 1000 gm (8.6 moles) per day or 0.006 moles/minute. The apparent enzyme "concentration" appears to be in excess by a factor of about 1200.

G. Ehrensvaard estimated the enzyme content in percent of the total tissue protein from the data of Schmidt and Schmidt (1960). In the different tissues, the totals varied from over 5% to about one-tenth this amount, and he concluded that enzyme contents exceeded requirements for catalysis by a great margin (Ehrensvaard, personal communication).

D. Enzyme Concentration by Measurements *in Vivo*

If we assume that the physical rules of kinetics apply in the living animal as well as in the flask, it should be possible to estimate the functional enzyme content of the entire body by giving substrates and then determining their rate of disappearance by frequent analyses. If, for example, 1 μmole of glycine is deaminated in 1 minute in an animal weighing 1 kg, then the enzyme responsible for removing the amino group must be at a concentration of 1 I.U./kg. Such a unit would represent a measure of the enzyme actually in use under physiological conditions. Table III shows the rates of disappearance of eighteen amino acids in five different species. In all cases the enzyme "concentration" was estimated when the substrate concentration in the plasma was 10 mmoles/liter, a concentration not too much in excess of that found under physiological conditions.

From Table III it is possible to deduce that 1 kg of alligator in 1 minute deaminates 17 μmoles of glycine, compared with the deamination of 25 μmoles in the turtle, 95 in the chameleon, 232 in the dog, and 680 in the rat. With respect to the apparent physiological "international units" of all the different deaminating enzymes in the five species, the

TABLE III
APPARENT "CONCENTRATION" OF FUNCTIONAL DEAMINATING ENZYMES IN VARIOUS INTACT ANIMALS[a]

	Alligator	Turtle	Chameleon	Dog	Rat
Gly	17	25	95	232	680
Ala	14	17	102	150	1292
Ser	26	33	66	214	1292
Thr	5	6	39	41	238
Asp	5	12	48	44	357
Asn	20	15	100	160	1020
Glu	6	7	26	51	153
Gln	21	16	54	250	1054
Val	1.6	2.7	14	20	153
Leu	13	10	53	119	442
Ile	2	2.2	13	61	289
Phe	3	1.4	8	41	170
His	8	1.9	12	37	102
Lys	21	14	22	31	238
Orn	18	23	105	54	578
Arg	17	16	105	292	578
Met	0.7	1.5	10	7	51
Trp	4	1.4	16	10	119
Average	11.8	11.4	49.7	100.8	489.8
Oxygen used (mmoles/gm/hour	40	45	220	580	2000
Oxygen consumed per enzyme unit	3.4	4.0	4.4	5.8	4.1

[a] All were injected with 10 mmoles of each amino acid per kilogram body weight. Values in I.U./kg (μmoles of substrate utilized/minute/kg at a time when the concentration of the substrate in the plasma was 10 m*M*).

alligator and turtle were about equal, the chameleon was 4.5 times as great, the dog 9 times, and the rat 40 times. The rates of oxygen consumption in the five species are proportional to the numerical physiological enzyme contents, and dividing the apparent enzyme content by the oxygen consumption for each species gives about the same figure. In these experiments, the substrate concentration was about the same in all cases, and from what is known, the enzyme concentrations were similar, yet a fortyfold difference in reaction rates was observed.

E. SUBSTRATE DELIVERY AS A FUNCTION OF BLOOD FLOW

The apparent lack of association between enzyme concentration as determined *in vitro*, substrate concentration in the extracellular fluids, and metabolic rate seems paradoxical. If enzymes are there and substrates are present, why are the reaction rates subject to such great nterspecies variations? If neither enzyme nor substrate concentration are responsible for the metabolic rate, some other variable must be nvolved. The actual ability of an animal to deliver substrate (and oxygen) to the enzyme site may be that variable.

In 1 minute, the entire body of a resting man receives 5 liters of blood containing 1 liter of oxygen at standard temperature and pressure (S.T.P.). At rest, less than half the oxygen is removed from the 5 liters, and therefore the tissues receive about 500 ml of oxygen (22 mmoles/minute). If the body utilized only glucose, then there is enough oxygen to burn 3.7 mmoles of glucose each minute, since six molecules of oxygen react with one of glucose. Regardless of intracellular enzyme concentration, there is sufficient oxygen to burn 5.3 moles of glucose (0.0037 × 1440), enough to furnish 3800 kcal/day and no more. To increase the production of energy, either more blood must be delivered, or it must be delivered at a higher concentration. Arterial blood is saturated with oxygen, therefore an increased rate of blood flow is the only possible way to increase the rate of delivery of both substrate and the oxygen to burn it.

A 1 kg alligator uses 1000 ml of oxygen per kilogram per day or 44.6 mmoles/day. He could then burn one-sixth this molar quantity of glucose, or about 7.4 mmoles of glucose per day or 518 mmoles/70 kg, a quantity equal to about 93 gm and supplying about 373 kcal/day. His metabolic rate is then one-tenth that of man. To deliver the 70 liters of oxygen required each day (70 kg × 1 liter), at least 700 liters of blood would be needed if each liter delivered 100 ml oxygen at S.T.P. The actual blood flow required is greater, since the blood oxygen capacity is only half that of man. For a resting alligator to receive the same amount of oxygen as a man, the blood flow would have to be increased at least ten times and more probably twenty times, since not all of the oxygen is removed from the arterial blood at each pass through the tissues.

A resting gray shrew uses 13,700 mm^3 of oxygen per gram per hour (Pearson, 1948), which is equivalent to 16 liters of oxygen per minute in a theoretical shrew the size of a man (70 kg). With a blood oxygen capacity of 20 vol%, 80 liters of blood must be supplied per minute if all the oxygen is removed from the arterial blood, and 160 liters/minute if the venous blood returns to the heart half saturated. This is about 50 gallons/minute in the animal at rest and several times that if he is active. This phenomenal flow approximates that of a fire hose. To consume 16 liters of oxygen each minute, an absolute minimum of 160 liters of blood must be delivered to the tissues of the man-sized shrew.

Although information on large numbers of species is scanty, the rela tionship between blood flow and oxygen consumption has been deter mined for six species of reptiles, seven of amphibia, ten of fish, and five o mammals (Lenfant *et al.*, 1970). The blood flow/oxygen consumptio ratios for animals with widely divergent metabolic rates were very close indicating that throughout the animal kingdom the blood flow is prob ably proportional to the metabolic rate.

Different animals with the same substrate and enzyme concentration have different metabolic rates if they have different rates of blood flow In final analysis, it appears that it is not the concentration of substrat in moles per liter that is important, but the total quantity of substrat passing an enzyme site in a given length of time. True "concentration can only be expressed in terms of the likelihood of collision betwee a molecule of substrate and an enzyme, and the likelihood is directl proportional to blood flow.

F. Body Size

Not enough is known of the influence of body size on metabolic rat in any one species of reptile. From what is known (Altman and Dittmer 1968), metabolic rate decreases with increase in size. If metabolic rat is a function of blood flow, and that in turn is related to body surfac area, then it is probable that the larger the animal the smaller the bloo flow per unit mass.

REFERENCES

Altman, P. L., and Dittmer, D. S., eds. (1968). "Metabolism." Fed. Amer. Soc Exp. Biol., Washington, D.C.

Barwick, R. E., and Bryant, C. (1966). *Physiol. Zool.* **39**, 1.

Cohen, P. P., and Brown, G. S. (1960). In "Comparative Biochemistry" (M. Florki and H. S. Mason, eds.), Vol. 2, pp. 161–244. Academic Press, New York.

Coulson, R. A., and Hernandez, T. (1953). *Endocrinology* **53**, 311.

Coulson, R. A., and Hernandez, T. (1957). *Amer. J. Physiol.* **188**, 121.

Coulson, R. A., and Hernandez, T. (1959). *Amer. J. Physiol.* **197**, 873.

Coulson, R. A., and Hernandez, T. (1962). *Amer. J. Physiol.* **202**, 83.

Coulson, R. A., and Hernandez, T. (1964). "Biochemistry of the Alligator—A Stud of Metabolism in Slow Motion." Louisiana State Univ. Press, Baton Rouge.

Coulson, R. A., and Hernandez, T. (1965). *Fed. Procs, Fed. Amer. Soc. Exp Biol.* **24**, 927.

Coulson, R. A., and Hernandez, T. (1967). *Amer. J. Physiol.* **212**, 1308.

Coulson, R. A., and Hernandez, T. (1968). *Comp. Biochem. Physiol.* **25**, 861.

Coulson, R. A., and Hernandez, T. (1970a). *J. Nutr.* **100**, 810.

Coulson, R. A., and Hernandez, T. (1970b). *In* "Comparative Biochemistry o Nitrogen Metabolism" (J. W. Campbell, ed.), Vol. 2, p. 640. Academic Press New York.

Coulson, R. A., and Hernandez, T. (1971). *Comp. Biochem. Physiol.* **40 B**, 741.

Dessauer, H. C. (1970). *In* "Biology of the Reptilia" (C. Gans, ed.), Vol. 3, pp. 1–54. Academic Press, New York.
Herbert, J. D. (1968). *Comp. Biochem. Physiol.* **24**, 229.
Herbert, J. D., and Coulson, R. A. (1971). *Fed. Proc., Fed. Amer. Soc. Exp. Biol.* **30**, 641.
Herbert, J. D., and Coulson, R. A. (1972). *Comp. Biochem. Physiol.* **42B**, 463.
Herbert, J. D. (1973). *Comp. Biochem. Physiol.* 46(2B), 229.
Hernandez, T., and Coulson, R. A. (1952). *Proc. Soc. Exp. Biol. Med.* **79**, 145.
Hernandez, T., and Coulson, R. A. (1954). *Science* **119**, 291.
Hernandez, T., and Coulson, R. A. (1956). *Amer. J. Physiol.* **185**, 201.
Hernandez, T., and Coulson, R. A. (1961). *Biochem. J.* **79**, 596.
Hernandez, T., and Coulson, R. A. (1966). *Comp. Biochem. Physiol.* **20**, 291.
Hernandez, T., and Coulson, R. A. (1968a). *Comp. Biochem. Physiol.* **26**, 991.
Hernandez, T., and Coulson, R. A. (1968b). *Fed. Proc., Fed. Amer. Soc. Exp. Biol.* **27**, 741.
Hernandez, T., and Coulson, R. A. (1969). *Amer. J. Physiol.* **217**, 1846.
Hernandez, T., and Coulson, R. A. (1971a). *Comp. Biochem. Physiol.* **38 B**, 679.
Hernandez, T., and Coulson, R. A. (1971b). *Proc. Int. Congr. Physiol. Sci.*, **9**, 244.
Jones, M. E., and Lipmann, F. (1960). *Proc. Nat. Acad. Sci. U.S.* **46**, 1194.
Lenfant, C., Johansen, K., and Hanson, D. (1970). *Fed. Proc., Fed. Amer. Soc. Exp. Biol.* **29**, 1124.
Pearson, O. P. (1948). *Science* **108**, 44.
Pitts, R. F. (1963). "Physiology of the Kidney and Body Fluids." Yearbook Publ., Chicago, Illinois.
Schmidt, E., and Schmidt, F. W. (1960). *Klin. Wochensch.* **38**, 957.
Stevenson, O. R., Coulson, R. A., and Hernandez, T. (1957). *Amer. J. Physiol.* **191**, 95.

CHAPTER 10

Digestion in Reptiles[1]

G. Dandrifosse

I. Feeding Habits 249
II. Anatomical Considerations 251
A. Gross Anatomy 251
B. Microscopic Anatomy 251
III. Digestive Juice 252
A. Bile 253
B. pH of Digestive Juice 253
C. Digestive Enzymes 253
IV. Mechanism of Digestion 260
A. Gastric Secretion 261
B. Intestinal Absorption 267
References 270

Little attention has been paid to digestion in reptiles, although popular interest has traditionally been directed to aspects of the snake feeding phenomenon, namely to the use of venom and the ability to ingest spectacularly large prey. In 1932, Benedict and Planck partly reviewed the problem. In 1941, Vonck reported the data available in the literature at this date. Since that time, many observations have been published in direct or indirect relation to the digestion phenomenon, particularly in the fields of feeding habits or mechanisms, of the histology of the salivary glands, gastric mucosa, or pancreas, and of the movement of some molecules across the digestive tract. However these reports have generally been limited to a small number of animal species. Many problems have almost entirely been neglected, as for example the study of the hormonal control of the digestive secretions. The present review will therefore appear very incomplete.

I. Feeding Habits

The purpose of this review is not to describe the feeding habits of reptiles. It will suffice here to point out that the feeding habits of these

[1] This work was partly realized at the Duke University Marine Laboratory during a stay supported by a grant, No. 12157, from NIMH.

animals under normal conditions of life, in captivity or in accidental circumstances, have been reported by many authors following direct observations or analyses of stomach contents (for example, Schmidt, 1932; Knowlton and Thomas, 1936; Pell, 1940; Schonberger, 1945; Fitch and Twining, 1946; Fowler, 1947; Carr, 1952; Kennedy, 1956; Barton and Allen, 1961; Johnson, 1966; Smith and Milstead, 1971). It appears, as with all the large groups of mammals, that reptiles can be divided into herbivorous, carnivorous, and omnivorous (Bellairs, 1969; Pope, 1955, 1957). The carnivores strongly predominate and include all snakes (except perhaps some species) (Irvine, 1953; Oliver, 1954; Rose, 1954; Salmon, 1954) and crocodiles and Tuatara (Klauber, 1956; Wright and Wright, 1957; Logier, 1958; Fitzsimons, 1962). True herbivores occur only in a few genera of lizards and tortoises, while those with a tendency towards omnivorous habits are more numerous (Pope, 1939; Mertens, 1960; Szarski, 1962; Ostrom, 1963; Sokol, 1967).

The herbivorousness of lizards in relation to evolution has been a matter of discussion. Szarski (1962) is surprised that lacertilians have not made more use of the herbivorous niche. Ostrom (1963) believes that, because of their inability to develop dental grinding mechanisms, it is odd that there are any herbivorous lizards at all. Sokol (1967) suggests that herbivorousness is directly related to the physical strength of the species.

Nonfood items, sometimes ingested voluntarily (geophagy, lithophagy) are observed in the stomach of reptiles (Glaser, 1955; Kennedy, 1956; Villers, 1958; Rick and Bowman, 1961; Cott, 1961; Kennedy and Brockman, 1965; Peaker, 1969; Brazaitis, 1969; Sokol, 1971). Their function, if any, is problematical, and a number of possibilities have been suggested—digestion (maceration of food) (Johnson, 1966), ballast (enabling the animal to lie submerged in streams with strong currents of water) (Cott, 1961), reculturing mechanism maintaining a beneficial intestinal flora (Sokol, 1971). The understanding of their role will be only realized after accumulation of much additional data.

However, our knowledge of the feeding of many reptiles, namely of young animals, is always scanty and perhaps, in certain cases, inexact (Bellairs, 1969). Indeed, it is possible that the stomach contains prey secondarily or accidentally ingested (Neill and Allen, 1956). A captive animal may accept food that it would not encounter in the wild. Most of the literature presents lists of food items consumed without relation to size, sex, or age of the predator, seasonal needs, or food availibility. The food habits of reptiles have been studied briefly through examination of their fecal material. The drinking needs of these animals are even less well known (Breder, 1924; Auffenberg, 1963; Meyer, 1966;

Henderson, 1970; Myer and Kowell, 1971a,b). Many animals drink very little water and may sometimes be able to subsist on water derived from their food (Aldoph, 1943; Klauber, 1956; Schmidt-Nielsen and Bentley, 1966).

II. Anatomical Considerations

A. Gross Anatomy

All reptiles have a relatively simple digestive tract—an anterior part with the buccal cavity, the esophagus, and the stomach, and a posterior part corresponding to the intestine, the cloaca, and the glands annexed to the digestive tract (Guibé, 1970). In this field, there are no important differences between the species of reptiles, except in the case of the buccal cavity, which is sometimes adapted to kill, to capture, or swallow large prey without masticating.

A series of studies attempting to correlate morphological and behavioral observations, eventually in relation with evolution, has been published (for references, see Gans, 1961). The detailed anatomical disposition of different portions of the digestive tract is partly well documented in excellent reviews (see Plenck, 1932; Vonck, 1941; Gans, 1961; Anthony, 1970; Guibé, 1970). Herbivorous animals do not seem to specialize in the ability to culture cellulolytic gut symbionts or to possess grinding mechanisms. However, the length of the colon in relation to the intestine total length is generally smaller in the carnivorous species than in the vegetarian animals (Guibé, 1970).

B. Microscopic Anatomy

The microscopic anatomy of the digestive tract of the reptiles has been examined in few species. Observations concerning the histology of the salivary glands (Gabe and Saint Girons, 1969), the gastric mucosa (Gabe and Saint Girons, 1972), the pancreas (Miller and Lagios, 1970), and the cloaca (Gabe and Saint Girons, 1965) have been the subject of different reports. However, few studies of the submicroscopic structures of these organs have been undertaken (Kobayashi, 1968; Miller and Lagios, 1970; Ferri, 1971; Hansen, 1972). Many more detailed studies are still required and this is also true for the other visceral organs.

It appears that the feeding habits of reptilia are not related to the structure of their salivary glands. As suggested by Gabe and Saint Girons (1969), it seems that little fragmentation of the food is not favorable to an important action of the salivary glands in the digestion mechanism.

The cytoplasmic organization of the exocrine pancreas is very similar to that classically reported for mammals (Miller and Lagios, 1970). The process of protein secretion is basically identical to that of the other acinar cells of vertebrates.

The esophagus of turtles and lizards is lined with ciliated columnar epithelium interspersed with goblet cells (Ballmer, 1949; Gorbman and Bern, 1962; Thiruvathukal, 1965). The anterior part of this organ is provided with simple tubular glands, whereas the posterior two-thirds of the organ shows an admixture of simple and compound tubular glands which seem sometimes to be mucus-secreting in nature (Ballmer, 1949; Thiruvathukal, 1965). Yet, the presence of esophageal glands has been questioned in the case of turtles (Ballmer, 1949). These glands have not been discovered throughout the length of alligator esophagus (Reese, 1913, 1915). The mucosa of the stomach generally consists of two parts—the anterior cardiac region and the posterior pyloric region. In the case of turtles (Thiruvathukal, 1965), of the rattle snake (Allen and Lhotka, 1961), and of the alligator (Reese, 1913, 1915), the mucosa seems similar in many respects to that of mammals. However, Staley (1925), Plenck (1932), Wright *et al.* (1957), and Luppa (1961) maintain that the stomach of reptiles lacks parietal cells and contains, as in amphibians and birds, cells undoubtedly secreting digestive enzymes and hydrochloric acid (oxynticopeptic cells). Following Gabe and Saint Girons (1972), three anatomical types of fundic glands, in relation with the systematic position of the reptile species considered, could be observed. There was no clear-cut relation between these histological structures and the alimentary habits of the animals. The intestinal mucosa of *Sceloporus occidentalis* and *Chrysemys picta* is lined with simple columnar cells and scattered goblet cells (Thiruvathukal, 1965; Musacchia and Wurth, 1964; Wurth and Musacchia, 1964; Johnson *et al.*, 1967). Some entities that seem to be subepithelial buds have been observed. There are no important morphological differences in the entire length of the reptile intestine (Reese, 1913; Ballmer, 1949; Thiruvathukal, 1965).

III. Digestive Juice

It is generally admitted that the efficiency of the digestive juice of reptiles, particularly ophidians, is very high (Klauber, 1956; Neill and Allen, 1956; Fitch and Twining, 1946; Skoczylas, 1970a; Blain and Campbell, 1942). This belief seems to arise from the common observation that some snakes are able to ingest entire large prey, such as mammals, eggs, or birds, without ejecting anything through the mouth and to reject feces without traces of bones or teeth, but containing the hairs unchanged and the feathers, bills, or shells of the eggs, and perhaps claws only somewhat disintegrated. However, systematic studies of this problem are scanty. Detailed biochemical analyses of food compared to feces have not been published. The pH of the digestive juice, the chemical composition of the bile, or the nature and the properties of the digestive enzymes do not allow any conclusion to be drawn concerning the problem of the digestion efficiency. The nature of the bile salts is well known for several species (see Chapter 12), but these substances have evidently no lytic activity.

A. Bile

The chemical composition and the role of the bile have been reported in detail in different reviews (Sobotka, 1937; Josephson, 1941; Sobotka and Bloch, 1943; Shimizu, 1948; Tayeau, 1949; Haslewood, 1955, 1962, 1964, 1967).

B. pH of Digestive Juice

The pH of the gastric or intestinal juices have been measured in few species of reptilia. The results obtained are summarized in Table I. From these scanty studies one may suggest that the pH of the stomach juice is exceedingly variable among the animals of the same species. This apparently depends upon the functional state of the organ. At rest, the pH is generally close to neutrality, at temperatures higher than 15°C, it becomes acidic following stimulus. In the intestine, a neutral or slightly alkaline pH is rapidly attained as the gastric juice mixes with the pancreatic and bile secretions. The acidity of the stomach content results from a secretion of hydrochloric acid (Staley, 1925; Anderson and Wilbur, 1948; Wright *et al.*, 1957). The reason for the alkalinity of the pancreatic juice is not known; but by analogy with the data obtained in the case of mammals, it may be presumed that this arises from a bicarbonate secretion by the pancreas. The presence of secretin, a hormone specifically increasing this secretion in mammals, has been demonstrated at the level of the reptiles intestine (Koschtojanz and Iwanoff, cited by Gorbman and Bern, 1962).

C. Digestive Enzymes

A list of distribution of the digestive enzymes observed among the reptiles is reported in Table II. The data obtained in the case of reptile venom enzymes are not included here. They are reported elsewhere in this volume (see Chapter 16) and in recent reviews (Condrea and de Vries, 1962; Russell and Scharffenberg, 1964; Grassé, 1970; Boquet, 1970). Thus, it suffices to point out that the venom, a salivary secretion that facilitates the acquisition of food by the poisonous species of reptiles

TABLE I
DIGESTIVE JUICE pH IN REPTILES

Species	Gastric juice pH	Intestinal juice pH	Experimental conditions	Reference
Chrysemys belli	±4.0	±7.0	—	Kenyon, 1925
Chrysemys cinera	±4.0	±7.0	—	Kenyon, 1925
Chrysemys picta	2.0–6.0	6.6–8.3	Fed animals	Fox and Musacchia, 1959
	1.8–7.5	6.3–8.9	Fasting animals	Fox and Musacchia, 1959
	7.0–8.0	7.0–8.0	Cold and fasting animals	Fox and Musacchia, 1959
	5.8–8.0	6.4–8.0	Cold force-fed animals	Fox and Musacchia, 1959
Chelydra serpentina	±4.0	±7.0	—	Kenyon, 1925
Pseudemys scripta	1.8–6.7	—	Fasting animal, histamine injected subcutaneously	Anderson and Wilbur, 1948
Testudo graeca	2.0–3.0	—	Fed animals	Wright *et al.*, 1957
	7.5–8.0	—	Fasting animals	Wright *et al.*, 1957
	1.0–2.0	—	Histamine injected intramuscularly	Wright *et al.*, 1957
	7.0–7.5	—	Pilocarpine injected intraarterially	Wright *et al.*, 1957
	1.0	—	Vagal stimulation	Wright *et al.*, 1957
Tachysaurus rugosus	2.0	—	Fed animals	Wright *et al.*, 1957
	1.0	—	Vagal stimulation	Wright *et al.*, 1957
	2.0–3.0	—	Fasting animals	Wright *et al.*, 1957
Tiliqua nigro-lutea	1.5–2.1	—	Fed animals	Wright *et al.*, 1957
	1.0–4.0	—	Histamine injected intramuscularly	Wright *et al.*, 1957
	1.0–8.0	—	Carbachol injected intramuscularly	Wright *et al.*, 1957
Pituophis sayi	±3.0	—	Fed animals	Kenyon, 1925
	±7.0	—	Fasting animals	Kenyon, 1925
	—	±7.0		Kenyon, 1925
Garter snake	±3.0	—	Fed animals	Kenyon, 1925
Grass snake	2.1–5.6	5.7–7.5	Fasting animal (24 hours)	Mennega, 1938
Natrix natrix	7.0–7.7	—	Fasting animals	Skoczylas, 1970b
	6.8–7.6	—	Cold, fed animals	Skoczylas, 1970b
	3.9–7.2	—	Fed animals	Skoczylas, 1970b

by enabling these animals to immobilize their prey without danger to themselves, probably allows the digestive processes to start immediately after biting (Pope, 1957; Zeller, 1948). The reptile venoms may contain, among other substances, several enzymes such as L-amino-acidoxidase, phosphodiesterase, phosphomonoesterase, hyaluronidase, and peptidase. The composition of the venom is often characteristic of a given species or family (Boquet, 1970). For example, the elapic venom contains cholinesterase, whereas those of Crotalidae, Viperidae, and Hydrophiidae do not (Ghosh *et al.*, 1939; Ghosh, 1940; Chang and Lee, 1955); the venoms of Crotalidae and Viperidae possess a trypsin-like enzyme, but none of the Elapidae venoms examined show the proteolytic activity (Tu *et al.*, 1965). Substances contained in the venom are often able to modify the activity of some enzymes and physiological properties of the prey.

Table II shows that the presence of few digestive enzymes has been investigated among reptiles and that data have been obtained for a small number of species. Certain adaptive features have been described in the relation of digestive enzymes to food habits. The animals that include plants in their diet show a striking difference in the amount of carbohydrate-splitting enzymes (except for chitinase) (Jeuniaux, 1963b), from those that are exclusively carnivorous (Kenyon, 1925; Vonck, 1941; MacGeachin and Bryan, 1964; Abrahamson and Maher, 1967). For example, the snakes *Pituophis sayi*, *Agkistrodon piscivorus*, and *Crotalus atrox* possess little amylase as compared with that present in the turtles *Chrysemys belli* and *Chrysemys cinera* (Kenyon, 1925). In the case of chitinase, it seems that this enzyme is only present in the digestive tracts of animals that ingest chitin (Jeuniaux, 1961b, 1963b). Apart from defining the gross lytic abilities of various digestive enzymes, little has generally been accomplished in the context of general enzymology. In 1941, Vonck reviewed the scanty literature available up to that time. The recent observations are summarized below.

1. *Amylase*

The pancreas of reptiles usually secretes an amylase, the pH optimum of which is about 7 (Vonck, 1941; MacGeachin and Bryan, 1964; Wolvekamp, 1928). The enzyme is most probably an α-amylase as in the case of the pancreatic amylase of mammals, but this fact needs confirmation. The effect of the temperature on the enzyme extracted from the pancreases of *Uromastix aegyptia* (Hussein, 1960), *Terrepene carolina*, *Chrysemys* sp. (Chesley, 1934), *Eumeces obsoletus*, *Dipsosaurus dorsalis*, and *Crotalus atrox* (Abrahamson and Maher, 1967) has been studied. It has been suggested that between the species of reptiles, there

TABLE II
DISTRIBUTION OF SOME DIGESTIVE ENZYMES IN REPTILES

	Organ	Digestive enzyme	Reference
Chelydra serpentina	Stomach	Amylase; pepsin	Kenyon, 1925
	Pancreas	Amylase	Kenyon, 1925
		Ribonuclease	Zendzian and Barnard, 1967a,b; Barnard, 1969
		Trypsin	Kenyon, 1925; Zendzian and Barnard, 1967a,b; Möckel and Barnard 1969a
		Chymotrypsin	Zendzian and Barnard, 1967a,b; Möckel and Barnard, 1969a,b; Derechin *et al.*, 1969; Barnard, 1969
		"Caseinolytic activity"	Zendzian and Barnard, 1967a,b;
		Carboxypeptidase A	Zendzian and Barnard, 1967a
	Intestine	Proteinase; invertase; amylase; maltase	Kenyon, 1925
Chrysemys belli	Stomach	Pepsin; trypsin	Kenyon, 1925
Chrysemys picta	Pancreas	Ribonuclease; trypsin; chymotrypsin; "caseinolytic activity"	Zendzian and Barnard, 1967a
Chrysemys sp.	Pancreas	Amylase	Chesley, 1934
Clemmys caspice rivulata	Stomach; pancreas	Chitinase	Jeuniaux, 1963b
Clemmys leprosa	Stomach	Chitinase	Jeuniaux, 1963b; Dandrifosse *et al.*, 1965
	Pancreas	Chitinase	Jeuniaux, 1963b
Emys europaea	Pancreas	Amylase	Wolvekamp, 1928
	Intestine	Amylase	Wolvekamp, 1928
Emys orbicularis	Stomach	Chitinase	Jeuniaux, 1963b; Dandrifosse *et al.*, 1965
	Pancreas	Chitinase	Jeuniaux, 1963b
Malaclemys centrata centrata	Stomach	Amylase; chitinase	G. Dandrifosse (unpublished results)
	Pancreas	Chitinase; amylase	G. Dandrifosse (unpublished results)
		Trypsins; chymotrypsin	S. Bricteux (personal communication)
Podocnemis unifillis	Pancreas	Ribonuclease; trypsin; chymotrypsin; "caseinolytic activity"	Zendzian and Barnard, 1967a
Pseudemys elegans	Pancreas	Ribonuclease	Zendzian and Barnard, 1967a,b
		Trypsin	Zendzian and Barnard, 1967a,b; Möckel and Barnard, 1969a
		Chymotrypsin	Zendzian and Barnard, 1967a,b; Möckel and Barnard, 1969a,b; Barnard and Hope, 1969; Derechin *et al.*, 1969

		"Caseinolytic activity	Zendzian and Barnard, 1967a,b
		Carboxypeptidase A	Zendzian and Barnard, 1967a
Terrepene carolina	Pancreas	Amylase	Chesley, 1934
Testudo graeca	Stomach	Pepsin	Vonck, 1927, 1941; Wright *et al.*, 1957
	Pancreas	Amylase	Wolvekamp, 1928
Testudo hermanni	Stomach	Chitobiase	Jeuniaux, 1961a,b, 1963b
	Intestine	Chitobiase	Jeuniaux, 1961a,b, 1963b
		Maltase; trehalase; isomaltase; sucrase	Zoppi and Shmerling, 1969
Anguis fragilis	Stomach	Chitinase	Jeuniaux, 1963b
Anolis carolinensis	Stomach; pancreas	Chitinase	Jeuniaux, 1963b
Caiman crocodilus	Pancreas	Chymotrypsin	Möckel and Barnard, 1969a
Chamaeleo vulgaris	Stomach; pancreas	Chitinase	Jeuniaux, 1963b
Dipsosaurus dorsalis	Pancreas	Amylase	Abrahamson and Maher, 1967
Eumeces obsoletus	Pancreas	Amylase	Abrahamson and Maher, 1967
Iguana iguana	Pancreas	Chymotrypsin	Möckel and Barnard, 1969a
Lacerta viridis	Stomach	Chitobiase	Jeuniaux, 1961a,b, 1963b
		Chitinase	Jeuniaux, 1961a,b, 1963b; Dandrifosse *et al.*, 1965; Micha, 1966
	Pancreas	Chitobiase; chitinase	Jeuniaux, 1961a,b, 1963b
	Intestine	Chitobiase	Jeuniaux, 1963b
Tachysaurus rugosus	Stomach	Pepsin	Wright *et al.*, 1957
Tiliqua nigrolutea	Stomach	Pepsin	Wright *et al.*, 1957
Uromastix aegyptia	Pancreas	Amylase	Hussein, 1960
Uromastix acanthinurus	Stomach	Chitinase	Jeuniaux, 1963b
Agkistrodon piscivorus	Esophagus	Amylase	MacGeachin and Bryan, 1964
	Stomach	Amylase	MacGeachin and Bryan, 1964
	Pancreas	Amylase	MacGeachin and Bryan, 1964
	Intestine	Amylase	MacGeachin and Bryan, 1964
Bungarus fasciatus	Pancreas	Chymotrypsin	Möckel and Barnard, 1969a
Crotalus atrox	Pancreas	Amylase	Abrahamson and Maher, 1967
Natrix natrix	Esophagus	Pepsin	Langley, 1881
	Stomach	Pepsin	Langley, 1881
Pituophis sayi	Stomach	Amylase; pepsin	Kenyon, 1925
	Pancreas	Trypsin	Kenyon, 1925
	Intestine	Protease; amylase	Kenyon, 1925

are thermal differences at the subcellular level that are similar to the differences in the preferred temperature of the animals (Abrahamson and Maher, 1967). This hypothesis is in agreement with the conclusions of Licht (1964), who shows similar thermal relations for the myosin–ATPase in several species of lizards. Chesley (1934) concludes from other similar experiments that the optimum temperature of the pancreatic amylases decreases as the period of digestion increases.

The amylase of the *Agkistrodon piscivorus* pancreas differs from the mammalian enzyme. It is not inhibited by the antisera to hog pancreatic amylase, which suppresses the activity of this enzyme arising from different organs of mammals (MacGeachin and Bryan, 1964). The freezing of the pancreas appears to denature the amylase, in direct contrast to the findings in mammals.

2. *Chitinase*

The gastric chitinase of reptilia has a pH optimum close to 3.5, an important chitinolytic activity being always measured at pH 1.5 (Jeuniaux, 1963a,b; Micha, 1966). These observations are different from those obtained with the enzyme of bacterial or pancreatic origin (pH optimum 5, low activity at pH 1.5). The activity of the chitinase secreted (experimental conditions described below, see p. 263), when it is estimated by the method of Jeuniaux (1961b, 1962, 1963b), is not influenced by many substances (For example, see Table III, p. 264).

The gastric enzymes of *Lacerta viridis* (Dandrifosse and Schoffeniels, 1967) and *Malaclemys centrata centrata* have been partly purified by following the first part of the method described for the *Streptomyces antobioticus* chitinase purification (Jeuniaux, 1963b). The specific activity obtained in both cases is close to 250 units/mg proteins (G. Dandrifosse, unpublished results).

3. *Ribonuclease*

Ribonuclease has been detected in the pancreas of different turtles (see Table II). This enzyme does not seem to be concentrated in this organ (Zendzian and Barnard, 1967a). However, in the case of *Chelydra serpentina,* a wide range of ribonuclease levels between individual pancreases has been observed. When the homogenization is performed at pH 5, the ribonuclease extracted from *Chrysemys picta* or *Chelydra serpentina* pancreas is acid stable but is rapidly destroyed in the same experimental conditions when it originates from *Podocnemis unifillis.* When the occurrence of the acid-stable ribonuclease, which is probably the secreted enzyme, is considered in diverse vertebrate groups, it may be considered a relatively ancient separation of a homologous series of this protein (Zendzian and Barnard, 1967a). This hypothesis is supported by results showing the similar active center reactivities of this group of ribonucleases, the latter being inactivated by the same specific reagent (Zendzian and Barnard, 1967b).

4. *Trypsin-Like Enzymes*

High levels of trypsin-like enzymes have been measured in the pancreases of several reptiles (see Table II). As with bovine trypsin, the enzymes show an activity on *p*-toluene sulfonyl-L-arginine methyl ester comparable with that on *N*-benzoyl-L-arginine ethyl ester and are present entirely as zymogens that can be activated by bovine trypsin (Zendzian and Barnard, 1967a,b). In the case of *Chelydra serpentina* and *Pseudemys elegans,* these trypsin-like enzymes are inhibited by α-*N*-tosyl-L-lysine chloromethyl ketone and diisopropyl fluorophosphate. The zymogens of *Pseudemys elegans* trypsin-like enzymes may be easily separated from chymotrypsinogens by chromatography (Möckel and Barnard, 1969a). The trypsinogen of *Chelydra serpentina* always shows very similar behavior to chymotrypsinogen in chromatography and electrophoresis. The pancreas of *Malaclemys centrata centrata* seems to contain at least two trypsin-like enzymes and one chymotrypsin-like enzyme (S. Bricteux-Grégoire, personal communication). One of these trypsin-like enzymes seems unstable, the second possesses two activation peptides, the sequences of which have not yet been determined.

5. *Chymotrypsins*

Two chymotrypsinogens, I and II, originating from the pancreas of *Pseudemys elegans* and *Chelydra serpentina* have been isolated in highly purified forms by chromatographic procedures (Möckel and Barnard, 1969a,b; Barnard and Hope, 1969). After activation with bovine trypsin, *Chelydra serpentina* chymotrypsinogen II and a mixture of the two *Pseudemys elegans* chymotrypsinogens hydrolyze the substrate *N*-benzoyl-L-tyrosine ethyl ether about three times as rapidly as bovine chymotrypsin (Möckel and Barnard, 1969a). The relative activities of these enzyme preparations on tyrosine and leucine substrates and the inactivation rate of these proteins by chloromethyl ketone derivatives of L-phenylalanine and L-leucine have been measured. From these data, a partial similarity to porcine chymotrypsin C has been discerned.

Different results obtained, particularly after alkylation of the *Pseudemys elegans* chymotrypsinogens by [carbonyl-^{14}C]tosyl-L-phenylalanine chloromethyl ketone, suggest that the active center of these enzymes is basically identical in structure throughout the vertebrates from fishes to mammals (Barnard and Hope, 1969). The chymotrypsins I and II of *Pseudemys elegans* are each inactivated by *N*-tosyl-L-phenylalanine chloromethyl ketone (Möckel and Barnard, 1969b).

The zymogen I from *Pseudemys elegans* and II from *Chelydra serpentina* release no free peptide or amino acid in forming the active enzyme, so that probably one or more internal cleavages occur. The zymogen II from *Pseudemys elegans* releases a fourteen-residue peptide, a tripeptide (Arg_2, Phe), and tyrosine after complete activation by bovine trypsin (Möckel and Barnard, 1969b). The chymotrypsinogens have a monomeric molecular weight of approximately 25,000, but the accuracy of this data varies with the protein analyzed, due to reversible dimerizations which are observed both in the enzymes and their zymogens (Derechin *et al.*, 1969).

6. Other Proteases

The caseinolytic activity of pancreatic extracts of different turtle species have been examined (see Table II). The results obtained (Zendzian and Barnard, 1967a,b) show that this activity is much higher than what would be expected from the combined trypsin and chymotrypsin-like enzyme activity, as estimated from their action on synthetic substrates (*N*-benzoyl-L-arginine ethyl ester and benzoyl-L-tyrosine ethyl ester). The uncharacterized enzymes are of very limited specificity; they only degrade casein to a strictly limited extent. They appear to exist in the pancreas in a zymogen form and are activated by bovine trypsin. In the case of *Chelydra serpentina* and *Pseudemys elegans*, when the trypsin and chymotrypsin activities are suppressed by specific inhibitors such as diisopropyl fluorophosphate, a considerable caseinolytic is always observed. This activity may partly be attributed to carboxypeptidase A. A strong activity is measured with the specific substrate of this enzyme (hippuryl-L-phenylalanine), and in the case of *Pseudemys elegans*, the caseinolytic activity is reduced in the presence of ethylenediaminetetraacetate, which is an inhibitor of bovine carboxypeptidase (Vallee, 1955).

IV. Mechanism of Digestion

The results published up to 1941 concerning the mechanisms of the digestion in reptiles have been reviewed by Vonck (1941). It seems that there has not been much interest in the matter since that time, except in the case of the gastric secretion of hydrochloric acid and chitinase or in the case of the intestinal absorption of some organic molecules. However, generally speaking, the elucidation of the biochemical processes involved still falls far short of that which has been achieved for mammals. Few species have been examined and the studies have rarely been accomplished *in vivo*. From a general point of view, different authors have analyzed the problem of adaptation to a long period of fasting (survival in a desert environment, hibernation). It has been noted that starvation of aroused animals can involve a reduction of metabolism (Belkin, 1965) or, in the case of reptiles living in a desert environment, the drawing from important reserves, especially from fat deposits (Bustard, 1967). The problem of hibernation is reviewed in this volume (Chapter 13).

The duration of digestion, which can be very long, has been studied (Reichert, 1936; Blain and Campbell, 1942; Salmon, 1954; Pope, 1955; Klauber, 1956; Root, 1961; Fox and Musacchia, 1959; Barton and Allen, 1961; Algauhari, 1967; Bellairs, 1969; Skoczylas, 1970a,b; Myer and Kowell, 1971a,b). It appears that age, amount of food consumed, and especially temperature affect this phenomenon. The studies concerning this problem are too scanty to allow a generalization, but it may be noted that, usually, the digestion is completely inhibited at a temperature lower than 10°C, proceeds only very slowly at 15°C, and is normal around 25°C. However, the digestive process is always slower than in higher vertebrates (Benedict, 1932; Lueth, 1941; Fox and Musacchia, 1959; Henderson, 1970; Guibé, 1970; Skoczylas, 1970a).

The question of the regulation of food intake by the American chameleon *Anolis carolinensis* has been examined (Fox and Dessauer, 1958; Dessauer, 1955). The appetite in this animal is found to be subject to photoperiodic regulation. Toh and Mohiuddin (1958) and Kim *et al.* (1965) have studied the peristaltic movement of the tortoise intestine under different experimental conditions.

A. Gastric Secretion

1. *Hydrochloric Acid Production*

At temperatures up to 17°C, food ingestion and injections, subcutaneous or intramuscular, of histamine increase the hydrochloric acid secretion by the gastric mucosa of *Testudo graeca, Pseudemys scripta troosti, Terrapene carolina, Tiliqua nigro lutea,* and *Tachysaurus rugosus* (Anderson and Wilbur, 1948; Wright *et al.*, 1957). A marked difference within species with respect to the effective dose of histamine, the presence of which is demonstrated at the level of the reptile stomach (Wright and Trethewie, 1956), has been observed. In *Testudo graeca,* gastric acid secretion is increased by the combined action of histamine and pilocarpine and slightly by vagal stimulation, in *Tiliqua nigro lutea* by carbachol, in *Tachysaurus rugosus* by vagal stimulation (Wright *et al.*, 1957), and lastly, in *Pseudemys scripta troosti* and *Terrapene carolina* by sulfanilamide (Anderson and Wilbur, 1948).

Anderson and Wilbur (1948) have studied the effect of this last substance in combination with histamine on acid secretion with the aim of specifying the role of carbonic anhydrase in the phenomenon. It has been proposed that this enzyme might be important in the secretion of hydrochloric acid (Davenport, 1939), and it has been shown that sulfanilamide is an inhibitor of carbonic anhydrase activity (Anderson and Wilbur, 1948; Maren, 1967). Anderson and Wilbur (1948) have suggested that the stimulation of hydrochloric acid production by sulfanilamide may be the result of an inhibition of the carbonic anhydrase activity of the blood. Indeed, in this case, blood carbon dioxide would increase in sulfanilamide treated animals. A greater amount of gastric hydrochloric acid might be secreted, provided gastric carbonic anhydrase remains active enough. Similarly, a high blood carbon dioxide concentration would tend to offset the inhibition of gastric anhydrase activity by sulfanilamide.

In the case of *Lacerta viridis*, it has been demonstrated that *in vitro* the acid secretion is nearly completely inhibited by ouabain added at a concentration of 3×10^{-7} *M* or higher to the serosal saline (Hansen, 1972; Hansen *et al.*, 1972). This substance does not seem to act directly on the acid secretion mechanism but seems to affect it by modifying the cellular sodium–potassium concentration by way of an inhibition of a sodium–potassium-activated MgATPase activity. The acid secretion could also be indirectly related to a gastric anion-sensitive MgATPase, which could act as a bicarbonate transport system (Hansen, 1972; De Pont *et al.*, 1972). The acid secretion rate is reduced when chloride or sodium is omitted from the serosal bathing medium, chloride omission being the most effective (Hansen, 1972). In the case of the acid-secreting mucosa, the net chloride flux equals the sum of acid secretion rate and the current used to short circuit the electrical transmucosal potential. This has been defined as composed of two potential steps: (1) between the interior of the cell and the serosal side of the gastric mucosa, described as the sum of ion diffusion potential for potassium and chloride; (2) between the interior of the cell and the mucosal side of the epithelium not affected by variations of the potassium or chloride concentration in the mucosal saline, but lowered by decreasing the sodium concentration in this solution. A net chloride flux from the serosal to the mucosal side of the gastric mucosa has been reported.

2. Pepsin Secretion

The pepsin secretion by the gastric mucosa of *Testudo graeca* and of *Tachysaurus rugosus* is increased by vagal stimulation (Wright *et al.*, 1957). Histamine also seems to act on this phenomenon, but this observation may be questioned if, as suggested in the case of mammals, the effect is due to a washing out of the glandular ducts.

3. *Chitinase Secretion*

The mechanism of gastric secretion has been studied *in vitro*. An isolated gastric mucosa, stripped from the adjacent muscle layers, is mounted between two perspex half-chambers as described by Schoffeniels (1960) and Baillien and Schoffeniels (1961a,c). In this experimental condition, the epithelium separates two pools of physiological saline, and the spontaneous electrical potential difference existing between the two solutions may be measured together with the secretion of the enzymes. The obtained results show that gastric chitinase of *Lacerta viridis, Clemmys leprosa, Malaclemys centrata centrata,* and *Emys orbicularis* is excreted in the solution bathing the mucosal side of the epithelium during at least 6 hours of incubation (Dandrifosse and Schoffeniels, 1965; Dandrifosse *et al.*, 1965; Dandrifosse, 1972b). During this time, the electrical potential difference existing between the two salines remains practically constant, indicating that the mucosa is in good metabolic condition.

The curve relating the amount of enzyme excreted to time is generally hyperbolic, indicating a decrease with time in chitinase secretion. This observation does not result from an intracellular diminution of enzyme concentration (Dandrifosse, 1965; Dandrifosse and Schoffeniels, 1965). The important secretion measured during the first hour of incubation does not seem to depend on a diffusion of chitinase molecules from the glandular lumen to the mucosal saline or of parasympathic, sympathic, or mechanical stimulations (Dandrifosse and Schoffeniels, 1967, and unpublished results). Thus, it appears that the decrease in enzyme secretion originates rather from a modification of the chitinase transfer rate across the cell membrane (Dandrifosse and Schoffeniels, 1965). A quantitative treatment of the experimental data obtained has been proposed.

The mechanism of enzyme extrusion by the gastric mucosa of *Lacerta viridis* has been analyzed after incubation of the epithelium in a uniformly ^{14}C-labeled glutamic acid solution for different periods of time (Dandrifosse and Schoffeniels, 1967). The results show that the chitinase excreted has a lower specific radioactivity than that in the subcellular particules but is always very close to that found in the soluble fraction of the cell. It has been found that the agranular cellular fraction cannot contain chitinase as artifact. This enzyme does not arise from the glandular ducts from microgranules or from damaged zymogen granules. These results suggest that chitinase present in the glandular lumen originates

TABLE III

PROPERTIES OF THE MECHANISM OF CHITINASE SECRETION BY THE ISOLATED GASTRIC MUCOSA OF *Lacerta viridis*[a]

Experimental conditions (I)		10^4 K_1 chitinase		Electrical potential difference		Water net flux		Conclusions		
		Control (II)	Exper. (III)	Before (IV)	During (V)	Before (VI)	During (VII)	K_1 chit. (VIII)	PD (IX)	Water net flux
External electromotive power		26	28	22	0	—	—	*	signif	—
		31	27	18	40	—	—	*	signif	—
		33	36	22	−20	—	—	*	signif	—
S + maltose 110 mM	i/o	12	14	14	12.5	+3.1	0	*	*	*
S + maltose 110 mM	o	33	57	21	16.5	0	+6.2	signif	*	←signif
S + maltose 110 mM	i	4	2	20.5	20.5	0	−6.2	signif	*	→signif
S + glucose 83.7 mM	i/o	47	51	—	—	—	—	*	—	—
S + cellobiose 87.6 mM	i/o	13	18	—	—	—	—	*	—	—
S + sucrose 58.4 mM	i/o	2	15	—	—	—	—	signif	—	—
S + lactose 58.4 mM	i/o	25	47	—	—	—	—	signif	—	—
S + chitobiose 0.18 mM	i/o	2	110	—	—	—	—	signif	—	—
S + trehalose 52.8 mM	i/o	7	17	—	—	—	—	signif	—	—
S + turanose 58 mM	i/o	34	85	—	—	—	—	signif	—	—
S + acetylcholine 4.4 mM	i/o	40	112	26	22.5	0	−3.1	signif	signif	*
S + adrenaline 0.02 mM	i/o	12	12	27.5	27	—	—	*	*	—
S + carbamylcholine 0.7 mM	i/o	33	54	—	—	—	—	signif	—	—
S + oxytocin 0.1 IU/ml	i/o	12	13	14	14.5	—	—	*	*	—
S + maltose 110 mM + oxytocin 0.1 IU/ml	i/o	11	37	12.5	12.5	+3.1	+15.9	signif	*	←signif
S sulfate	i/o	8	3	—	—	—	—	signif	—	—
S $Cl^- = I^-$	i/o	30	1	—	—	—	—	signif	—	—
S + $NaHCO_3$ 10 mM	i/o	2	15	—	—	—	—	signif	—	—

S $Na^+ = K^+$	i/o	1	59	—	—	—	—	signif	—	—
S $Na^+ = Cs^+$	i/o	65	105	—	—	—	—	signif	—	—
S $Na^+ = NH_4$	i/o	32	85	—	—	—	—	signif	—	—
S + NaI 50 mM	o	27	22	11	10	+3.1	+9.4	*	*	←signif
S + NaI 40 mM	o	6	7	7.5	8.5	+4.8	+12.7	*	*	←signif
+ oxytocin 0.1 IU/ml	i									
S + $(NH_4)_2SO_4$ 100 mM	i/o	8	90	20	6.5	0	−3.1	signif	signif	—
S + NH_4Cl 50 mM	o									
+ NH_4Cl 100 mM	i	2	20	15	7	+3.1	−4.8	signif	signif	→signif
S + NaF 10 mM	i	7	35	20	3	—	—	signif	signif	—
S + monoiodoacetate 10 mM	i/o	9	0	20	0	—	—	signif	signif	—
Anaerobiosis		8	6	15	5	—	—	*	signif	—
CO (darkness)		8	8	15	10	—	—	*	signif	—
S + KCN 1 mM	i	4	10	15	5	—	—	signif	signif	—
S + arsenate 3 mM	i/o	9	7	—	—	—	—	*	—	—
S + arsenate 1 mM	i/o	10	11	17	9.5	—	—	*	signif	—
S + 2, 4-dinitrophenol 5 mM	i/o	25	29	16	−3	—	—	*	signif	—
S + dicoumarol 0.1 mM	i	25	27	—	—	—	—	*	—	—
S + ouabain 1 mM	i/o	33	36	23.5	4	—	—	*	signif	—
S + chloromercuribenzoate 1 mg/ml	i/o	11	3	11	0	—	—	signif	signif	—
S + iodosobenzoate 1 mg/ml	i/o	18	4	15	0	—	—	signif	signif	—

[a] *Explanations:* S + X (column I): the substance X is added in the saline bathing the mucosal (o) or the serosal (i) side of the gastric epithelium. S sulfate: saline the cations of which are at the same concentration as in the basic saline and the anions Cl^- of which are replaced by SO_4^{2-} ions. SA = B: Basic saline the ions A of which are replaced by the ions B. Control (column II) and experimental (column III) values of the cellular permeability coefficient to chitinase (K_1). Value (mV) of the electrical difference of potential (PD) existing spontaneously between the solutions bathing the epithelium just before (column IV) and during (after 1 hour, column V) the experiment. The value ($\mu l/h/cm^2$) of the water net flux crossing the epithelium is measured during the hour preceding (column VI) or following (column VII) the beginning of the experiment. (Columns VIII, IX and X): → The water net flux is directed from the mucosal to the serosal face of the epithelium. ← : The water net flux is directed from the serosal to the mucosal face of the epithelium. *: no significant variation; signif: significant variation; —: no data.

in a cellular pool of soluble enzymes. Thus, the chitinase would be excreted molecule by molecule across the membrane of the cell at the level of specific sites. This suggestion contradicts the theories implying a mode of enzyme excretion in bulk through the formation of vesicules (for example, see Palade, 1959; Ekholm, *et al.* 1962; Helander, 1962; Parks, 1962; Ito and Winchester, 1963; Amsterdam *et al.*, 1969; Miller and Lagios, 1970; Vandermeers-Piret *et al.*, 1971).

Similar conclusions have been expressed in the case of the secretion of other digestive enzymes (Kramer and Poort, 1968; Rothman, 1967, 1970, 1972; Dandrifosse, 1970b, 1972a). They do not imply that exocytosis does not exist, but indicate that this phenomenon only obtains in particular cases or is not an essential step in the process of digestive enzyme secretions. By considering the details of the results obtained in the same experiments. it has been concluded that chitinase is synthesized in the microsomal fraction of the cell, then accumulated in zymogen granules as observed for other digestive enzymes in vertebrates (for example, see Siekevitz and Palade, 1960; Caro and Palade, 1964; Jamieson and Palade, 1967a,b). After this step the chitinase is released into the cytoplasm.

The properties of the mechanism of chitinase secretion have been analyzed *in vitro*. Some results obtained are reported in Table III. From these and others detailed in specific publications (Dandrifosse, 1961–1962, 1964, 1969a,b, 1970a; Dandrifosse and Schoffeniels, 1966), it has been suggested that the secretion involves the formation of a complex between the chitinase and an ultrastructural element of the cell membrane. There is no direct relationship between the enzyme excretion, the electrical potential difference existing spontaneously between the two salines bathing the isolated gastric mucosa, and the water net influx obtained from osmotic gradient crossing this epithelium. It is possible to specifically modify the chitinase excretion by many substances, for example, lactose, sucrose, turanose, chitobiose, K^+, NH^+, Cs^+, Cl^-, HCO_3^-, I^-, acetylcholine, carbamylcholine, potassium cyanide, cysteine, chloromercuribenzoate.

Some hypotheses concerning the mechanism of chitinase excretion have been presented. The enzyme might cross the cell membrane as a glycoprotein, the carbohydrate moiety of which would act as a chemical label, which upon interaction with the membrane receptor would promote the transport of the enzyme into the extracellular space (Dandrifosse, 1969a). Some inorganic ions could modify the excretion of chitinase by acting on a cholinergic mechanism (Dandrifosse, 1970a). The extrusion of this enzyme would be in relation to the movement of some ions across the cellular membrane. The resting excretion of chitinase is a metabolically dependent process, the energy of which is not derived from the transfer of electrons in the respiratory chain (Dandrifosse, 1969b).

B. Intestinal Absorption

The intestinal absorption of different fatty acids, sugars, and amino acids has been studied in several species of turtles.

1. *Fatty Acids*

Fox (1965b) has analyzed the net uptake of fatty acids by the painted turtle *Chrysemys picta* by using sac preparations. Thirteen of these substances have been tested. Absorption increases with chain lengths up to twelve carbons, then slightly declines as the chain lengthens to twenty-two carbons. Stearic (18:0), oleic (18:1), and linoleic (18:2) acids are absorbed to the same extent. The net uptake of behenic acid (22:0) is greater than that of erucic acid (22:1). These results are in agreement with data on fatty acid absorption in mammals (Bloom *et al.*, 1951; Gelb and Kessler, 1963). Observations on the accumulation of fatty acids on the serosal side of the intestine or detailed biochemical analyses on the absorption mechanism have not yet been reported.

2. *Sugars*

The movement of D-glucose across the intestinal mucosa of *Chrysemys picta* has been studied following the method described by Crane and Wilson (1958), which makes use of everted sacs. The sugar is taken up at the mucosal side of the epithelium and moves into the serosal saline against a glucose concentration difference. This uptake of glucose seems directly related to the concentration of this sugar in the mucosal saline (Fox and Musacchia, 1960; Fox, 1961a,b). Segments taken from the upper portion of the intestine transport D-glucose at a faster rate than segments from the lower part of this mucosa. The absorption is least at 2°C and greatest at 20–30°C. The source of the sugar entering the serosal saline appears to be in part glucose absorbed from the mucosal saline and in part intracellular carbohydrate. Thus, the glucose absorbed is either metabolized or translocated.

Phlorizin, iodoacetate, and ouabain are potent inhibitors of the movement of D-glucose across the intestinal epithelium, 2,4-dinitrophenol, malonate, cyanide, and azide are less effective (Fox, 1963, 1965a). No significant inhibition is observed when nitrogen replaces aeration of the mucosa. From this study of the action of inhibitors, it is concluded that D-glucose transport depends on the energy supplied by the glycolysis. Transport of D-galactose and 3-*O*-methylglucose against a concentration gradient of these sugars is also observed from the mucosal to the serosal side of an isolated intestinal segment of the same turtle (Fox, 1962). Uptake of each of these sugars by the epithelium appears directly dependent on the concentration of sugar in the mucosal saline. It is concluded that neither the substitution of the hydroxyl group at the third carbon of the glucose ring by a methyl group nor reversal of hydroxyl and hydrogen groups at the fourth position of this molecule block the active absorption of this substance by the *Chrysemys picta* intestine.

The uptake of glucose, fructose, arabinose, and xylose has been studied in aroused and hibernating desert lizards. (*Uromastix hardwickii*) (Latif *et al.*, 1967). In the active animal, the relative absorption rate of these sugars are, in decreasing order: glucose, fructose, xylose, and arabinose. Glucose alone moves from the mucosal to the serosal side of the intestinal mucosa against a concentration gradient of this substrate. The presence of an active transport of glucose, galactose, α-methylglucoside, and 3-*O*-methylglucopyranose is suggested at the level of the *Testudo graeca* intestine.

The relations between this phenomenon and the electrical potential difference existing across the epithelium have been analyzed *in vitro* (Wright, 1966). It is proposed that the increase in the difference of potential observed when one of the sugars mentioned above is added to the mucosal saline, is due to the presence of an electrogenic ion pump at the serosal face of the epithelial cell.

3. *Amino Acids*

The mechanism of the amino acid absorption by the intestinal epithelium of *Testudo hermanni hermanni* has been carefully examined by Gilles-Baillien and Schoffeniels. These analyses have been realized *in vitro* using the apparatus described above (see p. 263). In aroused turtles, glycine, L-alanine, and L-histidine are actively transported through the small intestine, while L-glutamic acid moves passively. These amino acids passively cross the colon epithelium (Baillien, 1961; Baillien and Schoffeniels, 1961a). When the specific activity of L-alanine is calculated in the mucosal saline, in the intracellular medium, and in the serosal saline after a 3 hours' incubation of the small intestine in the presence of L-alanine-U-^{14}C, it is found that the specific radioactivity of this amino acid in the intracellular medium is equal to that of the mucosal solution and higher than that in the serosal saline (Gilles-Baillien and Schoffeniels, 1966; 1968). It has been suggested that this variation in the specific radioactivity of alanine would result from a synthesis of this amino acid in or near the serosal border of the epithelium. The synthesis of alanine and active transport mechanism would be intimately related if not identical.

The properties of this mechanism have been studied. Schoffeniels and Gilles-Baillien have analyzed the relations existing between this phenomenon, the electrical potential difference observed between the two salines bathing the epithelium isolated, the intracellular ion content, and the sodium concentration in the mucosal saline (Gilles-Baillien, 1970a,b,c; Baillien and Schoffeniels, 1962; Gilles-Baillien and Schoffeniels, 1965, 1966, 1968). They have concluded that the increase in the electrical potential difference observed when alanine is added to the mucosal saline can be explained, at least partially, by a modification in the permeability of potassium from the serosal membrane of the cell. In this case, the chloride movement across the epithelium is modified, leading to a complete reversal of the net flux of this ion. It has been suggested that a more direct relation exists between the transport of alanine across the intestinal mucosa and the intracellular potassium concentration than between the movement of this amino acid and the sodium concentration in the mucosal saline, as proposed for other intestinal epithelium.

4. *Other Substances*

The permeability of the intestinal epithelium of *Testudo hermanni* to urea, thiourea, lactic acid, and phenylacetic acid has been analyzed (Lippe *et al.*, 1963a,b, 1964, 1965, 1966a,b). An asymmetric distribution of these substances is observed in the mucosal and serosal spaces. The mechanism of this phenomenon seems to be different for urea and thiourea, for lactic acid and phenylacetic acid. The action of amphotericin B on the permeability of thiourea and phenylacetic acid has been tested (Lippe and Giordana, 1967a,b,c). The obtained results show that amphotericin B affects differently the movement of these substances across the intestinal epithelium. They suggest differences in structure or composition of the mucosal membranes of both small and large intestines.

Different authors have investigated the permeability characteristics to ions of the turtle intestinal epithelium. In short, according to Baillien (1961), Baillien and Schoffeniels (1961a,b, 1962), Gilles-Baillien and Schoffeniels (1967a,b), Giordana *et al.* (1969, 1970), and Gilles-Baillien (1970b), in the case of aroused animals, active transport of Na^+ and Cl^- exists in the mucosal and serosal sides of the small intestine. Active transport of Cl^- is located at the level of both sides of the colon, while the same mechanism of Na^+ transport is only observed at the level of the serosal face of this mucosa. The sodium and chloride passive movements across the sides of the small intestine are more important than across those of the colon. K^+ passively crosses the serosal face of the last organ and both sides of the jejunum. Pituitrin and arginine vasotocin do not seem to influence the permeability characteristics of the large intestine (Bentley, 1962). The secretion of bicarbonate is higher in the distal half of the turtle colon than in the proximal part of this organ (Hajjar, 1971). Junqueira *et al.* (1966) have estimated the water reabsorption in the cloaca and large intestine of *Xenodon, Phylodria,* and *Crotalus.* They have suggested that active Na^+ transport takes place in these organs. This could be followed by passive movement of water. However, the plasma colloidal osmotic pressure could also provide a force sufficient to withdraw water from the cloaca, as demonstrated by Murrish and Schmidt-Nielsen (1970) in the case of *Dipsosaurus dorsalis.* The cloaca and the large intestine could have a supplementary function to the kidney in maintaining homeostasis in these reptiles.

REFERENCES

Abrahamson, Y., and Maher, M. (1967). *Can. J. Zool.* **45**, 227.

Adolph, E. F. (1943). (1943). "Physiological Regulations." Jacques Cattell Press, Lancaster, Pennsylvania.

Algauhari, A. E. I. (1967). *Z. Vergl. Physiol.* **54**, 395.

Allen, R. F., and Lhotka, J. F., Jr. (1961). *Anat. Rec.* **139**, 202.

Amsterdam, A., Ohad, I., and Schramm, M. (1969). *J. Cell Biol.* **41**, 753.

Anderson, N. G., and Wilbur, K. M. (1948). *J. Cell Comp. Physiol.* **31**, 293.

Anthony, J. (1970). *In* "Traité de Zoologie" (P. P. Grassé ed.), Vol. 14, Part 2, p. 549. Masson, Paris.

Auffenberg, W. (1963). *Anim. Behav.* **11**, 72.

Baillien, M. (1961). *Arch. Int. Physiol. Biochim.* **69**, 260.

Baillien, M., and Schoffeniels, E. (1961a). *Biochim. Biophys. Acta* **53**, 521.

Baillien, M., and Schoffeniels, E. (1961b). *Biochim. Biophys. Acta* **53**, 537.

Baillien, M., and Schoffeniels, E. (1961c). *Nature (London)* **190**, 1107.

Baillien, M., and Schoffeniels, E. (1962). *Arch. Int. Physiol. Biochim.* **70**, 286.

Ballmer, G. W. (1949). *Mic. Acad. Sci. Arts Lett.* **35**, 91.

Barnard, E. A. (1969). *Annu. Rev. Biochem.* **38**, 677.

Barnard, E. A., and Hope, W. C. (1969). *Biochim. Biophys. Acta* **178**, 364.

Barton, A. J., and Allen, W. B. (1961). *Zoologica (New York)* **46**, 83.

Belkin, D. A. (1965). *Copeia,* 367.

Bellairs, A. (1969). "The Life of Reptile," Weidenfel & Nicolson, London.

Benedict, F. G. (1932). "The Physiology of Large Reptiles," Carnegie Institution, Washington, D.C.

Bentley, P. J. (1962). *Gen. Comp. Endocrinol.* **2**, 323.

Blain, A. W., and Campbell, K. N. (1942). *Amer. J. Roentgenol. Radium Thes.* **48**, 229.

Bloom, B., Chaikoff, I. L., and Reinhardt, W. O. (1951). *Amer. J. Physiol.* **166**, 451.

Boquet, P. (1970). *In* "Traité de Zoologie" (P. P. Grassé, ed.), Vol. 14, Part 2, p. 599. Masson, Paris.

Brazaitis, P. (1969). *Herpetologica* **25**, 63.

Breder, R. B. (1924). *Copeia,* 62.

Bustard, H. R. (1967). *Science* **158**, 1197.

Caro, L. G., and Palade, G. E. (1964). *J. Cell Biol.* **20**, 473.
Carr, A. (1952). "Handbook of Turtles. The Turtles of United States, Canada and Baja California," Cornell Univ. Press. (Comstock), Ithaca, New York.
Chang, C. C., and Lee, C. Y. (1955). *J. Formosan Med. Ass.* **54**, 103.
Chesley, L. C. (1934). *Biol. Bull.* **66**, 330.
Condrea, E., and de Vries, A. (1962). *Toxicon* **2**, 261.
Cott, H. B. (1961). *Trans. Zool. Soc. London* **29**, 211.
Crane, R. K., and Wilson, T. H. (1958). *J. Appl. Physiol.* **12**, 145.
Dandrifosse, G. (1961–1962). *Ann. Soc. Roy. Zool. Belg.* **92**, 199.
Dandrifosse, G. (1964). *Arch. Int. Physiol. Biochim.* **72**, 517.
Dandrifosse, G. (1965). *Ann. Soc. Roy. Zool. Belg.* **95**, 81.
Dandrifosse, G. (1969a). *Arch. Int. Physiol. Biochim.* **77**, 639.
Dandrifosse, G. (1969b). *Bull. Cl. Sci., Acad. Roy. Belg.* [S] **55**, 701.
Dandrifosse, G. (1970a). *Arch. Int. Physiol. Biochim.* **78**, 339.
Dandrifosse, G. (1970b). *Comp. Biochem. Physiol.* **34**, 229.
Dandrifosse, G. (1972a). *Comp. Biochem. Physiol.* **B41**, 559.
Dandrifosse, G. (1972b). *Amer. Zool.* **12**, XX, Abst. 12.
Dandrifosse, G., and Schoffeniels, E. (1965). *Biochim. Biophys. Acta* **94**, 165.
Dandrifosse, G., and Schoffeniels, E. (1966). *Ann. Endocrinol.* **27**, 517.
Dandrifosse, G., and Schoffeniels, E. (1967). *Biochim. Biophys. Acta* **148**, 741.
Dandrifosse, G., Schoffeniels, E., and Jeuniaux, C. (1965). *Biochim. Biophys. Acta* **94**, 153.
Davenport, H. W. (1939). *J. Physiol. (London)* **97**, 32.
De Pont, J. J. H. H. M., Hansen, T. D., and Bonting, S. L. (1972). *Biochim. Biophys. Acta* **274**, 189.
Derechin, M., Möckel, W., and Barnard, E. A. (1969). *Biochim. Biophys. Acta* **191**, 379.
Dessauer, H. C. (1955). *Proc. Soc. Exp. Biol. Med.* **90**, 524.
Ekholm, R., Zelander, T., and Edlund, Y. (1962). *J. Ultrastruct. Res.* **7**, 61.
Ferri, S. (1971). Ph.D. Thesis, São Paulo University.
Fitch, H. S., and Twining, H. (1946). *Copeia,* 64.
Fitzsimons, V. F. M. (1962). "Snakes of Southern Africa," Purnell, Capetown, Johannesburg.
Fowler, J. A. (1947). *Copeia,* 210.
Fox, A. M. (1961a). *Amer. J. Physiol.* **201**, 295.
Fox, A. M. (1961b). *Comp. Biochem. Physiol.* **3**, 285.
Fox, A. M. (1962). *Proc. Iowa Acad. Sci.* **69**, 600.
Fox, A. M. (1963). *Prox. Iowa Acad. Sci.* **70**, 423.
Fox, A. M. (1965a). *Biol. Bull.* **129**, 490.
Fox, A. M. (1965b). *Comp. Biochem. Physiol.* **14**, 553.
Fox, A. M., and Musacchia, X. J. (1959). *Copeia,* 337.
Fox, A. M., and Musacchia, X. J. (1960). *Fed. Proc., Fed. Amer. Soc. Exp. Biol.* **19**, 182.
Fox, W., and Dessauer, H. C. (1958). *Z. Exp. Zool.* **134**, 557.
Gabe, M., and Saint Girons, H. (1965). *Mem. Mus. Hist. Nat., Paris, Ser. A* **53**, 149.
Gabe, M., and Saint Girons, H. (1969). *Mem. Mus. Hist. Nat., Paris, Ser. A* **58**, 1.
Gabe, M., and Saint Girons, H. (1972). *Zool. yahrb. Abt. Anat. Ontog. Tiere* **89**, 579.

Gans, C. (1961). *Am. Zool.* **1**, 217.
Gelb, A. M., and Kessler, J. I. (1963). *Amer. J. Physiol.* **204**, 821.
Ghosh, B. N. (1940). *Oesterr. Chem.-Ztg.* **43**, 158.
Ghosh, B. N., Dutt, P. K., and Chowdhurry, D. K. (1939). *J. Indian Chem. Soc.* **16**, 75.
Gilles-Baillien, M. (1970a). *Arch. Physiol. Biochim.* **78**, 119.
Gilles-Baillien, M. (1970b). *Arch. Int. Physiol. Biochim.* **78**, 327.
Gilles-Baillien, M. (1970c). *Life Sci.* **9**, 585.
Gilles-Baillien, M., and Schoffeniels, E. (1965). *Arch. Int. Physiol. Biochim.* **73**, 355.
Gilles-Baillien, M., and Schoffeniels, E. (1966). *Life Sci.* **5**, 2253.
Gilles-Baillien, M., and Schoffeniels, E. (1967a). *Arch. Int. Physiol. Biochim.* **75**, 754.
Gilles-Baillien, M., and Schoffeniels, E. (1967b). *Comp. Biochem. Physiol.* **23**, 95.
Gilles-Baillien, M., and Schoffeniels, E. (1968). *Life Sci.* **7**, 53.
Giordana, B., Bianchi, A., and Lippe, C. (1969). *Ist. Lomb. Rend. Sci. B* **103**, 115.
Giordana, B., Bianchi, A., and Lippe, C. (1970). *Comp. Biochem. Physiol.* **36**, 395.
Glaser, R. (1955). *Copeia*, 248.
Gorbman, A., and Bern, H. A. (1962). "A Textbook of Comparative Endocrinology." Wiley, New york.
Grassé, P. P. (1970). *In* "Traité de Zoologie" (P. P. Grassé, ed.), Vol. 14, Part 2, p. 676. Masson, Paris.
Guibé, J. (1970). *In* "Traité de Zoologie" (P. P. Grassé, ed.), Vol. 14, Part 2, p. 521. Masson, Paris.
Hajjar, J. J. (1971). *Comp. Biochem. Physiol.* A **40**, 39.
Hansen, T. D. (1972). Ph.D. Thesis, Nijmegen University.
Hansen, T. D., Bonting, S. L., Slegers, J. F. G., and De Pont, J. J. H. H. M. (1972). *Pfluegers Arch.* **334**, 141.
Haslewood, G. A. D. (1955). *Physiol. Rev.* **35**, 178.
Haslewood, G. A. D. (1962). *In* "Comparative Biochemistry" (M. Florkin and H. S. Mason, eds.), Vol. 3, Part A, p. 205. Academic Press, New York.
Haslewood, G. A. D. (1964). *Biol. Rev. Cambridge Phil. Soc.* **39**, 537.
Haslewood, G. A. D. (1967). *J. Lipid Res.* **8**, 535.
Helander, H. J. (1962). *J. Ultrastruct. Res.* **4**, 1.
Henderson, R. W. (1970). *Herpetologica* **26**, 520.
Hussein, M. F. (1960). *Proc. Egypt. Acad. Sci.* **15**, 53.
Irvine, F. R. (1953). *Brit. J. Herpetol.* **1**, 173.
Ito, S., and Winchester, R. J. (1963). *J. Biophys. Biochem. Cytol.* **16**, 541.
Jamieson, J. D., and Palade, G. E. (1967a). *J. Cell Biol.* **34**, 577.
Jamieson, J. D., and Palade, G. E. (1967b). *J. Cell Biol.* **34**, 597.
Jeuniaux, C. (1961a). *Arch. Int. Physiol. Biochim.* **69**, 384.
Jeuniaux, C. (1961b). *Nature (London)* **192**, 135.
Jeuniaux, C. (1962). *Ann. Soc. Roy. Zool. Belg.* **92**, 27.
Jeuniaux, C. (1963a). *Arch. Int. Physiol. Biochim.* **71**, 307.
Jeuniaux, C. (1963b). "Chitine et chitinolyse." Masson, Paris.
Johnson, D. R. (1966). *Amer. Midl. Natur.* **76**, 504.
Johnson, T. S., Dornfeld, E. J., and Conte, F. P. (1967). *Can. J. Zool.* **45**, 63.
Josephson, B. (1941). *Physiol. Rev.* **21**, 463.
Junqueira, L. C. U., Malnic, G., and Monge, C. (1966). *Physiol. Zool.* **39**, 151.
Kennedy, J. P. (1956). *Tex. J. Sci.* **8**, 328.

Kennedy, J. P., and Brockman, H. L. (1965). *Brit. J. Herpetol.* **3**, 201.
Kenyon, W. A. (1925). *Bull. U.S. Fish. Bur.* **41**, 181.
Kim, I. Y., Cha, W. K., and Kim, D. W. (1965). *Experientia,* **21**, 540.
Klauber, L. M. (1956). "Rattlesnakes," Univ. of California Press, Berkeley.
Knowlton, G. F., and Thomas, W. L. (1936). *Copeia,* 64.
Kobayashi, S. (1968). *Arch. Histol.* **28**, 525.
Kramer, M. F., and Poort, C. (1968). *Z. Zellforsch. Mikrosk. Anat.* **86**, 475.
Langley, J. N. (1881). *Phil. Trans. Roy. Soc. London* **172**, 663.
Latif, S. A., Zain, B. K., and Zain-Ul-Abedin, M. (1967). *Comp. Biochem. Physiol.* **23**, 121.
Licht, P. (1964). *Comp. Biochem. Physiol.* **12**, 331.
Lippe, C., and Giordana, B. (1967a). *Biochim. Biophys. Acta* **135**, 966.
Lippe, C., and Giordana, B. (1967b). *Boll. Soc. Ital. Biol. Sper.* **43**, 435.
Lippe, C., and Giordana, B. (1967c). *Boll. Soc. Ital. Biol. Sper.* **43**, 437.
Lippe, C., Bianchi, A., and Capraro, V. (1963a). *Arch. Int. Physiol. Biochim.* **71**, 768.
Lippe, C., Bianchi, A., and Capraro, V. (1963b). *Boll. Soc. Ital. Biol. Sper.* **39**, 669.
Lippe, C., Cremaschi, D., and Capraro, V. (1964). *Boll, Soc. Ital. Biol. Sper.* **40**, 1004.
Lippe, C., Bianchi, A., Cremaschi, D., and Capraro, V. (1965). *Arch. Int. Physiol. Biochim.* **73**, 43.
Lippe, C., Cremaschi, D., and Capraro, V. (1966a). *Rev. Roum. Biol., Ser. Zool.* **11**, 129.
Lippe, C., Cremaschi, D., and Capraro, V. (1966b). *Comp. Biochem. Physiol.* **19**, 179.
Logier, E. B. S. (1958). "The Snakes of Ontario," Univ. of Toronto Press, Toronto.
Lueth, F. X. (1941). *Copeia,* 125.
Luppa, H. (1961). *Acta Histochem.* **12**, 137.
MacGeachin, R. L., and Bryan, J. A. (1964). *Comp. Biochem. Physiol.* **13**, 473.
Maren, T. H. (1967). *Physiol. Rev.* **47**, 595.
Mennega, A. M. W. (1938). Ph.D. Dissertation, Utrecht (cited by Vonck, 1941).
Mertens, R. (1960). "The World of Amphibians and Reptiles" (transl. by H. W. Parker). McGraw-Hill, New York.
Meyer, D. E. (1966). *Copeia,* 126.
Micha, J. C. (1966). M.S. Thesis. Liège University.
Miller, M. R., and Lagios, M. D. (1970). *In* "Biology of the Reptilia (C. Gans, ed.), Vol. 3, p. 319. Academic Press, New York.
Möckel, W., and Barnard, E. A. (1969a). *Biochim. Biophys. Acta* **178**, 354.
Möckel, W., and Barnard, E. A. (1969b). *Biochim. Biophys. Acta* **191**, 370.
Murrish, D. E., and Schmidt-Nielsen, K. (1970). *Science* **170**, 324.
Musacchia, X. J., and Wurth, M. A. (1964). *Copeia,* 220.
Myer, J. S., and Kowell, A. P. (1971a). *J. Comp. Physiol. Psychol.* **75**, 5.
Myer, J. S., and Kowell, A. P. (1971b). *Physiol. & Behav.* **6**, 71.
Neill, W. T., and Allen, E. R. (1956). *Herpetologica* **12**, 172.
Oliver, J. A. (1954). *Brit. J. Herpetol.* **1**, 192.
Ostrom, J. H. (1963). *Evolution* **17**, 368.
Palade, G. E. (1959). *In* "Subcellular Particules" (Hayashi, T., ed.), p. 64. Ronald Press, New York.
Parks, H. F. (1962). *J. Ultrastruct. Res.* **6**, 449.

Peaker, M. (1969). *Brit. J. Herpetol.* **4**, 103.
Pell, S. M. (1940). *Copeia,* 131.
Plenck, H. (1932). *In* "Handbuch der mikroskopischen Anatomie des Menschen" (W. von Möllendorff, ed.), Vol. 2, p. 1. Springer-Verlag, Berlin and New York.
Pope, C. H. (1939). "Turtles of the United States and Canada." Knopf, New York.
Pope, C. H. (1955). "The Reptiles World." Knopf, New York.
Pope, C. H. (1957). "Reptiles Round the World." Knopf, New York.
Reese, A. M. (1913). *Anat. Rec.* **7**, 105.
Reese, A. M. (1915). "Alligator and its Allies." Knickerbocker Press, New York.
Reichert, E. (1936). *Bl. Aquar.-Terrarienfreunde* **47**, 228.
Rick, C. M., and Bowman, R. I. (1961). *Evolution* **15**, 407.
Root, H. D. (1961). Thesis, Minneapolis (cited by Skoczylas (1970a,b).
Rose, W. (1954). *Brit. J. Herpetol.* **1**, 225.
Rothman, S. S. (1967). *Nature (London)* **213**, 460.
Rothman, S. S. (1970). *Amer. J. Physiol.* **218**, 372.
Rothman, S. S. (1972). *Nature (London)* **240**, 176.
Russell, F. E., and Scharffenberg, R. S. (1964). "Bibliography of Snake Venoms and Venomous Snakes." Bibliographic Associate, Inc. West Covina, California.
Salmon, F. (1954). *Brit. J. Herpetol.* **1**, 193.
Schmidt, K. P. (1932). *Copeia,* 6.
Schmidt-Nielsen, K., and Bentley, P. J. (1966). *Science* **151**, 1547.
Schoffeniels, E. (1960). *Arch. Int. Physiol. Biochim.* **68**, 1.
Schonberger, C. F. (1945). *Copeia,* 120.
Shimizu, T. (1948). *J. Jap. Biochem.* Soc. **20**, 118.
Siekevitz, P., and Palade, G. E. (1960). *J. Biophys. Biochem. Cytol.* **7**, 619.
Skoczylas, R. (1970a). *Comp. Biochem. Physiol.* **33**, 793.
Skoczylas, R. (1970b). *Comp. Biochem. Physiol.* **35**, 885.
Smith, D. D., and Milstead, W. W. (1971). *Herpetologica* **26**, 147.
Sobotka, H. (1937). "Physiological Chemistry of the Bile." Williams & Wilkins, Baltimore, Maryland.
Sobotka, H., and Bloch, E. (1943). *Annu. Rev. Biochem.* **12**, 45.
Sokol, O. M. (1967). *Evolution* **21**, 192.
Sokol, O. M. (1971). *J. Herpetol.* **5**, 69.
Staley, H. F. (1925). *J. Morphol.* **40**, 169.
Szarski, H. (1962). *Evolution* **16**, 529.
Tayeau, F. (1949). *Exposes Annu. Biochim. Med.* **10**, 251.
Thiruvathukal, K. V. (1965). *J. Anim. Morphol. Physiol.* **12**, 220.
Toh, C. C., and Mohiuddin, A. (1958). *Brit. J. Pharmacol. Chemother.* **13**, 113.
Tu, A. T., James, G. P., and Chua, A. (1965). *Toxicon* **3**, 5.
Vallee, B. L. (1955). *Advan. Protein Chem.* **10**, 317.
Vandermeers-Piret, M. C., Camus, J., Rathe, J., Vandermeers, A., and Christophe, J. (1971). *Amer. J. Physiol.* **220**, 1037.
Villers, A. (1958). "Tortues et Crocodiles de l'Afrique noire." Institut Français d'Afrique Noire, Dakar.
Vonck, H. J. (1927). *Z. Vergl. Physiol.* **5**, 445.
Vonck, H. J. (1941). *Advan. Enzymol.* **1**, 371.
Wolvekamp, H. P. (1928). *Z. Vergl. Physiol.* **7**, 454.
Wright, A. H., and Wright, A. A. (1957). "Handbook of Snakes." Cornell Univ. Press (Comstock), Ithaca, New York.

Wright, E. M. (1966). *J. Physiol.* (*London*) **185**, 486.
Wright, R. D., and Trethewie, E. R. (1956). *Nature* (*London*) **178**, 546.
Wright, R. D., Florey, H. W., and Sanders, A. S. (1957). *Quart J. Exp. Physiol. Cog. Med. Sci.* **42**, 1.
Wurth, M. A., and Musacchia, X. J. (1964). *Anat. Rec.* **148**, 427.
Zeller, E. A. (1948). *Advan. Enzymol.* **8**, 459.
Zendzian, E. N., and Barnard, E. A. (1967a). *Arch. Biochem. Biophys.* **122**, 699.
Zendzian, E. N., and Barnard, E. A. (1967b). *Arch. Biochem. Biophys.* **122**, 714.
Zoppi, G., and Shmerling, D. H. (1969). *Comp. Biochem. Physiol.* **29**, 289.

CHAPTER 11

Water and Mineral Metabolism in Reptilia

William H. Dantzler and W. N. Holmes

I. Regulation of Urinary Excretion of Water and Ions 277
A. Introduction 277
B. Glomerular Filtration Rate 278
C. Renal Tubular Function 288
D. Bladder and Cloacal Function 306
E. Interrelations among Glomerular, Tubular, and Bladder or Cloacal Functions 309
II. The Extrarenal Excretory System 312
A. Introduction 312
B. The Marine Environment 318
C. The Arid Terrestrial Environment 330
References 334

I. Regulation of Urinary Excretion of Water and Ions

A. Introduction

Among the reptiles, which cannot produce urine hyperosmotic to plasma, the regulation of the urinary excretion of water and ions involves not only the regulation of glomerular filtration rate (GFR) and of tubular reabsorption and secretion, but also the modification of the ureteral urine by the bladder or cloaca. The reptilian bladder or cloaca appears to perform a number of functions performed by the distal tubule and collecting ducts of the mammalian nephron. In reptiles, the regulation of these apparently extrarenal processes should be considered together with the regulation of the more strictly renal glomerular and tubular processes. The regulation at any one of these three levels—glomerular, tubular, and bladder or cloaca—may be related to regulation at the other two, to the habitat of the animal, to the excretory end products of nitrogen metabolism, and to the tolerance of changes in plasma osmolality, as well as the function of clearly nonurinary routes for excretion of ions. The function and regulation of nonurinary routes for ion excretion will be considered in detail in a later section. In the present section, the integration of glomerular, tubular, and cloacal or bladder roles in the regulation of the urinary excretion of ions and water will be considered.

Although it is important to consider the role of the bladder or cloaca along with that of the kidney in understanding the overall regulation of ion and water excretion in the urine, the very fact that the bladder or cloaca may modify the ureteral urine makes it necessary to separate ureteral urine from bladder urine in studies evaluating the separate components of the excretory system. In many studies this has not been done, and it is impossible to define clearly the individual roles of glomerular filtration rate, tubular transport, and cloacal or bladder reabsorption in the regulation of urinary excretion or to understand how these functions are related. As nearly as possible, the data presented in this discussion have been limited to those obtained in studies where glomerular and tubular function were studied separately from bladder or cloacal function. Where it has been necessary to present data from studies that did not discriminate among glomerular and tubular and bladder or cloacal function, this has been clearly stated.

B. Glomerular Filtration Rate

It is possible to compare the effects of various states of hydration on the glomerular filtration rate (GFR) in reptiles from different environments representing three of the four living orders of reptiles—the chelonia, the squamata, and the crocodilia. These data are shown in Table I. Unfortunately, in some animals from arid regions, such as the desert tortoise *Gopherus agassizii,* the effects of dehydration on renal function could not be easily studied over a period of a few hours as they could in some reptiles from moist terrestrial or aquatic habitats, such as the freshwater turtle *Pseudemys scripta.* Under natural conditions, the desert tortoise can live for long periods of time without eating or drinking, and dehydration in the laboratory caused inconsistent changes in plasma osmolality (Dantzler and Schmidt-Nielsen, 1966). The differing magnitude of the effects of dehydration in these two species probably resulted from the fact that respiratory and cutaneous water losses in desert tortoises are about one-eighth of those in freshwater turtles (Schmidt-Nielsen and Bentley, 1966). Similar differences have been shown for water loss across the skin of other chelonian, saurian, and ophidian reptiles from arid and moist environments (Bentley and Schmidt-Nielsen, 1965; Prange and Schmidt-Nielsen, 1969) (Table II). The structural and physiological mechanisms involved in these differences have yet to be examined.

It was found, however, that the effects of dehydration on renal function could be mimicked in freshwater turtles by the intravenous administration of a hyperosmotic (1 *M*) sodium chloride solution (Dantzler and Schmidt-Nielsen, 1966). The changes in renal function with a given increase in plasma osmolality were similar whether the increase was produced by dehydration or salt-loading. For this reason, the effect of increased plasma osmolality induced by a salt load was studied and compared in these and other reptilian species shown in Table I. This method has the advantage that the osmolality of the blood can be controlled but the disadvantage that the salt load also expands the extracellular fluid volume. This may then affect volume regulation in the animal. It should be noted that among the species in Table I, only the marine turtle *Chelonia mydas* and the sea snake *Laticauda colubrina* have an extraurinary route for sodium chloride excretion of known importance.

The glomerular filtration rate of reptiles tends to decrease with dehydration or salt loading and to increase with water loading (Table I). However, there are differences in the sensitivity of the glomerular responses, which appear related to habitat as well as to tubular and bladder or cloacal function (see below). Among the chelonians, freshwater turtles (*Ps. scripta*) ceased producing urine when the plasma osmolality was increased 20 mOsm by dehydration or salt loading (Table I) (Dantzler and Schmidt-Nielsen, 1966). With increases in plasma osmolality of between 10 and 20 mOsm, GFR tended to decrease (Fig. 1). On the other hand, desert tortoises (*Gopherus agassizii*) of the southwestern United States continued to produce urine until the plasma osmolality was increased 100 mOsm (Table I) (Dantzler and Schmidt-Nielsen, 1966). Increases in plasma osmolality of less than 50 mOsm had no consistent effect on GFR (Fig. 1). In both desert tortoises and freshwater turtles, however, an intravenous water load (administered as 5% dextrose solution) produced a marked increase in urine flow and GFR (Table I). The only data on ureteral urine of a marine turtle (*Ch. mydas*) indicate that the normal GFR is nearly three times that of the wholly terrestrial desert tortoise and the semiaquatic freshwater turtle (Table I) (Schmidt-Nielsen and Davis, 1968).

A somewhat similar pattern of glomerular function is found in ophidian reptiles (Table I). *Pituophis melanoleucus* obtained from the deserts of Arizona and New Mexico had a lower GFR during a control diuresis (produced by the intravenous infusion of 1.25% mannitol) (Komadina and Solomon, 1970) than freshwater *Natrix sipedon* (Dantzler, 1967a, 1968). The GFR of desert snakes, like that of desert tortoises, did not decrease during the intravenous administration of a 5% sodium chloride solution. Instead the filtration rate increased markedly (Table I) (Komadina and Solomon, 1970). On the other hand, freshwater snakes responded to an acute 1 *M* salt load like freshwater turtles with a decrease in GFR (Dantzler, 1968). Renal function appeared

TABLE I

Renal Function in Some Reptiles during Normal Hydration, Dehydration, Salt Loading, and Water Loading[a]

Species	Environment and mode of existence	Condition	GFR ml/kg/hour	Osmolal U/P	Osmolal U/P (range for all conditions)
A. Chelonians					
Desert tortoise, *Gopherus agassizii*	Arid, terrestrial	Normal	4.74 ± 0.60 (9)	0.36 ± 0.02 (9)	0.3–0.7
		Salt load	2.94 ± 0.91 (20)	0.57 ± 0.05 (20)	
		No urine flow when plasma osmolality increased 100 mOsm			
		Water load	15.12 ± 6.64 (17)	0.61 ± 0.03 (17)	
Fresh-water turtle, *Pseudemys scripta*	Fresh-water, semiaquatic	Normal	4.73 ± 0.69 (40)	0.62 ± 0.03 (40)	0.3–1.0
		Salt load	2.77 ± 0.90 (10)	0.84 ± 0.06 (10)	
		No urine flow when plasma osmolality increased 20 mOsm			
		Water load	10.27 ± 2.00 (25)	0.60 ± 0.03 (25)	
Marine turtle, *Chelonia mydas*	Salt water, aquatic	Normal	14.3	0.95	—
B. Ophidians					
Bull snake, *Pituophis melanoleucus*	Arid, terrestrial	Salt load	16.08 ± 1.06 (5)	0.72 ± 0.05 (5)	0.5–1.0
		Water load	10.96 ± 1.07 (5)	0.73 ± 0.09 (5)	
Fresh-water snake, *Natrix sipedon*	Freshwater, semiaquatic	Salt load	13.12 ± 1.26 (11)	0.58 (3)	0.1–1.0
		Water load	22.84 ± 1.75 (12)	0.27 (3)	
		No urine flow when plasma osmolality increased 50 mOsm			
Banded sea snake, *Laticauda colubrina*	Marine, aquatic	Normal	0.5	0.85	
C. Saurians					
Blue-tongued lizard, *Tiliqua scincoides*	Terrestrial	Normal	15.9 ± 1.0	0.50 ± 0.07	
		Dehydration	0.65	0.79	
		Salt load	14.5 ± 0.5	0.66 ± 0.03	
		Water load	24.5 ± 2.0	0.43 ± 0.04	
Horned lizard,	Arid, terrestrial	Normal	3.54 ± 0.32 (23)	0.93 ± 0.04 (22)	

Species	Habitat	Condition			Range
Phrynosoma cornutum		Dehydration	2.14 ± 0.20 (14)	0.97 ± 0.02 (14)	
		Salt load	1.73 ± 0.40 (10)	1.00 ± 0.01 (10)	
		Water load	5.52 ± 0.54 (19)	0.90 ± 0.04 (19)	
Galapagos lizard, *Tropidurus* sp.	Arid, terrestrial	Normal	3.62 ± 0.36 (17)	0.96 ± 0.02 (17)	
		Dehydration	1.23 ± 0.23 (5)	0.97 ± 0.02 (5)	
		Salt load	2.42 (1)	1.01 (1)	
		Water load	4.53 (3)	0.99 (3)	
Puerto Rican gecko, *Hemidactylus* sp.	Moist, terrestrial	Normal	10.40 ± 0.77 (14)	0.64 ± 0.02 (12)	
		Dehydration	3.33 ± 0.37 (6)	0.74 ± 0.02 (6)	
		Salt load	11.01 ± 2.18 (4)	0.80 ± 0.02 (4)	
		Water load	24.30 ± 1.67 (12)	0.74 ± 0.02 (12)	
D. Crocodilians					
Crcoodylus acutus	Freshwater, semiaquatic	Normal	9.6 ± 1.0 (13)	0.80 ± 0.01 (13)	0.55–0.95
		Dehydration	6.1 ± 0.6 (13)	0.84 ± 0.01 (13)	
		Salt load	7.3 ± 0.8 (13)	0.82 ± 0.02 (13)	
		Water load	15.2 ± 2.0 (18)	0.67 ± 0.02 (18)	
Crocodylus johnstoni	Freshwater, semiaquatic	Normal	6.0 ± 1.5	0.71 ± 0.06	
		Dehydration	1.9 ± 0.2	0.94 ± 0.01	
		Water load	3.3 ± 1.1	0.83 ± 0.04	
Crocodylus porosus	Saltwater, semiaquatic	Normal	1.5 ± 0.2	0.95 ± 0.04	
		Salt load	2.8 ± 0.9	0.78 ± 0.15	
		Water load	18.8 ± 2.3	0.45 ± 0.05	

[a] Values are means ±S.E. Numbers in parentheses indicate number of determinations. No standard errors are given when there were fewer than four determinations. Approximate range for osmolal U/P ratio observed under all conditions is given for each species when it was available. Data on chelonians were compiled from Dantzler and Schmidt-Nielsen (1966) and Schmidt Nielsen and Davis (1968); on ophidians, from Dantzler (1967a, 1968), Komadina and Solomon (1970), and Schmidt-Nielsen and Davis (1968); on saurians, from Roberts and Schmidt-Nielsen (1966) and Schmidt-Nielsen and Davis (1968); on crocodilians, from Schmidt-Nielsen and Skadhauge (1967) and Schmidt-Nielsen and Davis (1968). Where numbers of determinations are not given in parentheses, values were obtained from four to six determinations (Schmidt-Nielsen and Davis, 1968).

TABLE II

RESPIRATORY AND CUTANEOUS WATER LOSS IN VARIOUS REPTILES IN DRY AIR[a]

Species	Environment and mode of existence	Temperature (°C)	Water Loss: Total weight (mg/cm²/day)	Water Loss: Respiratory (mg/gm/day)	Water Loss: Respiratory (mg/ml O_2)	Water Loss: Cutaneous (mg/cm²/day)	*N*
A. Chelonians							
Desert tortoise, *Gopherus agassizii*	Arid, terrestrial	23	2.0 ± 0.22	0.4 ± 0.12	1.5 ± 0.42	1.5 ± 0.21	6
Box turtle, *Terrapene carolina*	Terrestrial	23	7.2 ± 0.31	2.6 ± 0.29	4.2	5.3 ± 0.41	6
Freshwater turtle, *Pseudemys scripta*	Freshwater, semiaquatic	23	15.8 ± 1.70	4.3 ± 0.48	4.2	12.2 ± 1.44	6
B. Ophidians							
Gopher snake, *Pituophis catenifer affinis*	Arid, terrestrial	25	5.6 ± 0.48		1.3 ± 0.18	3.7 ± 0.48	9
Brown water snake, *Natrix taxispilota*	Freshwater, semiaquatic	25	19.1 ± 2.14		1.6 ± 0.18	16.7 ± 2.15	7
C. Saurians							
Chuckwalla, *Sauromalus obesus*	Arid, terrestrial	23	1.7 ± 0.17	0.6 ± 0.08	0.5	1.3 ± 0.10	6
Iguana, *Iguana iguana*	Moist, terrestria	23	6.7 ± 0.41	3.4 ± 0.48	0.9	4.8 ± 0.50	8
D. Crocodilians							
Caiman, *Caiman sclerops*	Freshwater, semiaquatic	23	3.7 ± 2.11	9.6 ± 1.2	4.9	32.9 ± 2.45	8

[a] Values are means ±SE. Data on chelonians are from Bentley and Schmidt-Nielsen (1966) and Schmidt-Nielsen and Bentley (1966); on ophidians, from Prange and Schmidt-Nielsen (1969); on saurians, from Bentley and Schmidt-Nielsen (1966); and on crocodilians, from Bentley and Schmidt-Nielsen (1966).

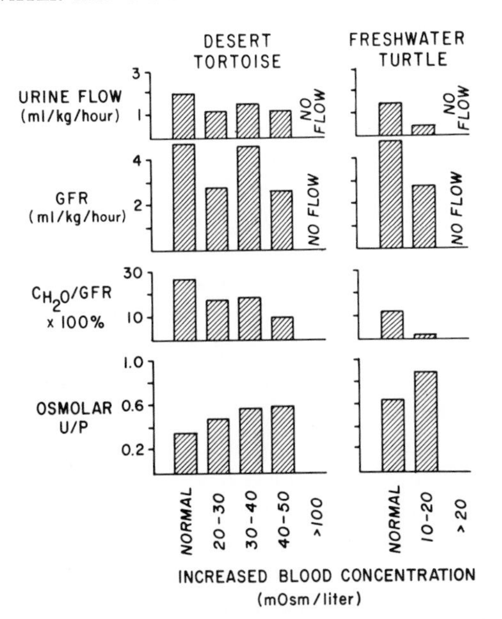

FIG. 1. Effect of salt loading on renal function in the desert tortoise and the freshwater turtle (modified from Dantzler and Schmidt-Nielsen, 1966). Urine flow ceased in the desert tortoise when the osmolality of the blood was increased more than 100 mOsm. Urine flow ceased in the freshwater turtle when the osmolality of the blood was raised more than 20 mOsm. The results represent a total of twenty-nine clearance periods in the desert tortoise and fifty clearance periods in the freshwater turtle.

to cease when plasma osmolality had been raised 50 mOsm (Dantzler, 1968). In contrast to data on marine turtles, however, the only data on filtration rate in sea snakes (*Laticauda colubrina*) indicate that the normal filtration rate in these animals is very low (Table I) (Schmidt-Nielsen and Davis, 1968). Since these data were not obtained under the same conditions of controlled diuresis as those in the freshwater and desert snakes, they cannot be compared readily with the data obtained for the latter species.

The glomerular function of saurian reptiles from habitats varying from moist to arid follows a pattern similar to that for chelonian and ophidian reptiles (Table I). The data for the two arid-living species—*Phrynosoma cornutum*, a horned lizard from the western United States, and *Tropiduras* sp., from the Galapagos Islands—are similar (Table I) (Roberts and Schmidt-Nielsen, 1966). The control GFR in these animals was only about one-third to one-fourth that found in lizards from a temperate region of moderate rainfall (*Tiliqua scincoides,* from Australia) and a moist tropical region (*Hemidactylus* sp. from Puerto Rico) (Table I) (Schmidt-Nielsen and Davis, 1968; Roberts and Schmidt-Nielsen, 1966). In contrast to the filtration rate of arid-living chelonian and ophidian species, the GFR of these arid-living lizards decreased markedly during dehydration. However, the value never fell so low as that seen in *Tiliqua.* Also, the percentage decrease in GFR was greater in those lizards from a more moist environment. A water load produced a marked increase in GFR in the species from a moist environment but only a modest increase in the horned lizard and Galapagos lizard. It is more difficult to interpret the effects of a salt load in lizards than in turtles and snakes, since it could only be administered safely following hydration. Nevertheless, a similar load produced a decrease in GFR from the control level in the two arid living forms but no change in the tropical gecko or blue-tongued lizard. On the whole, however, the filtration rate in animals from a moist environment was greater and more variable than that in animals from arid environments.

In studies on the Australian lizard *Tiliqua* (formerly *Trachysaurus*) *rugosa,* which occupies a rather arid habitat, Shoemaker *et al.* (1966) found that urine flow varied with body temperature, increasing tenfold between 14° and 24°C. This appeared to result from changes in GFR. The authors suggested that these were due to changes in hydrostatic pressure within the glomerular arterioles with temperature, since it had been shown (Templeton, 1964a) that arterial pressure in the lizard *Sauromalus obesus* exhibited a Q_{10} of about 2.4 between 5° and 25°C. In addition, the viscosity of the plasma being filtered should increase at low temperatures, reducing the filtration coefficient and the filtration rate. Shoemaker *et al.* (1966) emphasized the importance of studying renal function in poikilotherms at temperatures appropriate for that species. However, variations in temperature should not have influenced the renal response to different states of hydration in the above animals (Table I), since all studies were carried out at temperatures between 24° and 30°C. Over this temperature range, urine flow and arterial pressure appear to be virtually independent of temperature (Templeton, 1964a; Shoemaker *et al.*, 1966).

The GFR of crocodilians also varies with states of hydration, but this variation differs in animals from different habitats (Table I) (Schmidt-Nielsen and Skadhauge, 1967; Schmidt-Nielsen and Davis, 1968). In salt-water crocodiles (*Crocodylus porosus*) a water load caused a marked increase in filtration rate. Less variation in filtration rate was seen in freshwater crocodiles (*Crocodylus acutus* and *Crocodylus johnstoni*) as a result of dehydration or water loading. In the one species of crocodile extensively studied (the fresh-water crocodile *C. acutus*) (Schmidt-Nielsen and Skadhauge, 1967), the decrease of GFR to about 64% of normal during dehydration and the increase to about 158% of normal with a water load was somewhat less than that seen in semiaquatic turtles (*Ps. scripta*) and snakes (*N. sipedon*) (Table I).

Variations in GFR in reptiles appear to reflect changes in the number of functioning glomeruli (LeBrie and Sutherland, 1962; Dantzler and Schmidt-Nielsen, 1966; Dantzler, 1967a). This possibility is most clearly suggested by studies on chelonian and ophidian reptiles showing that the tubular maximum (T_m) for the transport of *p*-aminohippurate (PAH) varied directly with GFR (see example of this relationship for a freshwater turtle in Fig. 2) (Dantzler and Schmidt-Nielsen, 1966; Dantzler, 1967a). If changes in GFR had resulted from changes in the amount filtered by each glomerulus, but all had continued to function, the T_m for PAH would not have been expected to change (Ranges *et al.*, 1939; Forster, 1942). Although some controversy exists about the use of the T_m for PAH transport as a measure of tubular mass (Deetjen and Sonnenberg, 1965; Tanner and Isenberg, 1970), the correlation between $T_{m\ PAH}$ and GFR agrees rather well with direct observations on glomerular function in frogs (Sawyer, 1951) and thus appears to give

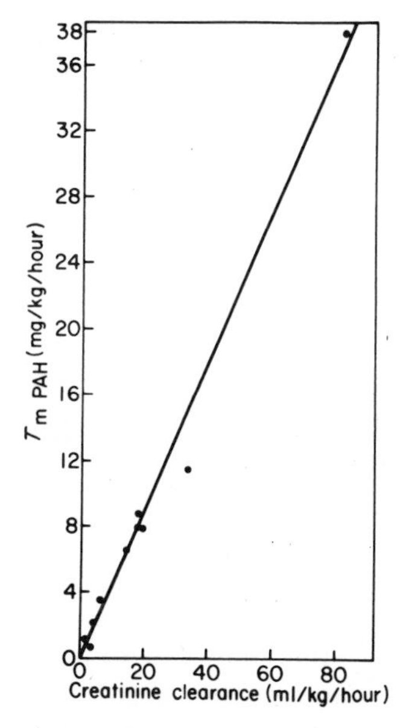

FIG. 2. Relationship between tubular maximum for secretion of *p*-aminohippurate ($T_{m_{PAH}}$) and GFR for a single freshwater turtle (Dantzler and Schmidt-Nielsen, 1966).

a reasonable indication of glomerular activity. Moreover, recent microscopic studies on freshwater crocodiles (*C. johnstoni*), saltwater crocodiles (*C. porosus*), nonarid terrestrial lizards (*Tiliqua scincoides*), sea snakes (*L. colubrina*), and marine turtles *Ch. mydes*) (Schmidt-Nielsen and Davis, 1968) indicate that the ratio of the number of open to closed tubular lumina correlates roughly with the filtration rate. If it is assumed that a tubule would tend to collapse when its glomerulus stopped filtering, then these observations also support the concept that changes in filtration result from changes in the number of functioning nephrons. This type of regulation appears reasonable in these nonmammalian vertebrates where nephrons do not function together in the production of a concentrated urine.

Although the mechanisms involved in the regulation of glomerular filtration rate in reptiles are not well understood, the neurohypophysial hormones appear to play an important regulatory role. Arginine vasotocin (AVT) has been identified in the pituitaries of all reptiles studied (Sawyer, *et al.*, 1959; Munsick, 1966) and is generally considered to be the natural antidiuretic hormone. However, this has never been completely documented by the production of hormone deficiencies by ablation of the hypothalamic neurosecretory cells and correction of the defect by hormone replacement. The renal effects of this hormone have been most extensively studied in freshwater snakes (*N. sipedon*) (Dantzler, 1967a). Arginine vasotocin in doses as low as 0.7 vasopressor mU/kg produced a clear depression of GFR (Fig. 3) without any apparent effect on systemic circulation. As can be seen in Fig. 3, the magnitude and, to a lesser extent, the duration of the decrease in GFR increased with increasing dose levels of AVT. The maximum decrease in GFR varied with log dose of AVT administered (Dantzler, 1967a). If the neurohypophysis of *Natrix* contains a quantity of AVT similar to that found in the rattlesnake (*Crotalus atrox*) neurohypophysis (about 500 vasopressor mU according to Munsick, 1966), then all the dose levels used in this study were well within the range of physiological release by the pituitary. A few experiments with turtle pituitary extract on freshwater turtles (*Ps. scripta*) indicated that large doses of extract (about 72 vasopressor mU and 135 oxytocic mU/kg) also produce a depression of GFR in these reptiles (Dantzler and Schmidt-Nielsen, 1966). However, the vasodepressor effects of these doses were not studied, and it is possible that the glomerular effect was secondary to a depression of the systemic arterial pressure. A decrease in glomerular filtration rate with the administration of vasopressin (Pitressin) has also been observed in crocodilians (Burgess *et al.*, 1933; Coulson and Hermandez, 1964; Sawyer and Sawyer, 1952). Although no definite informa-

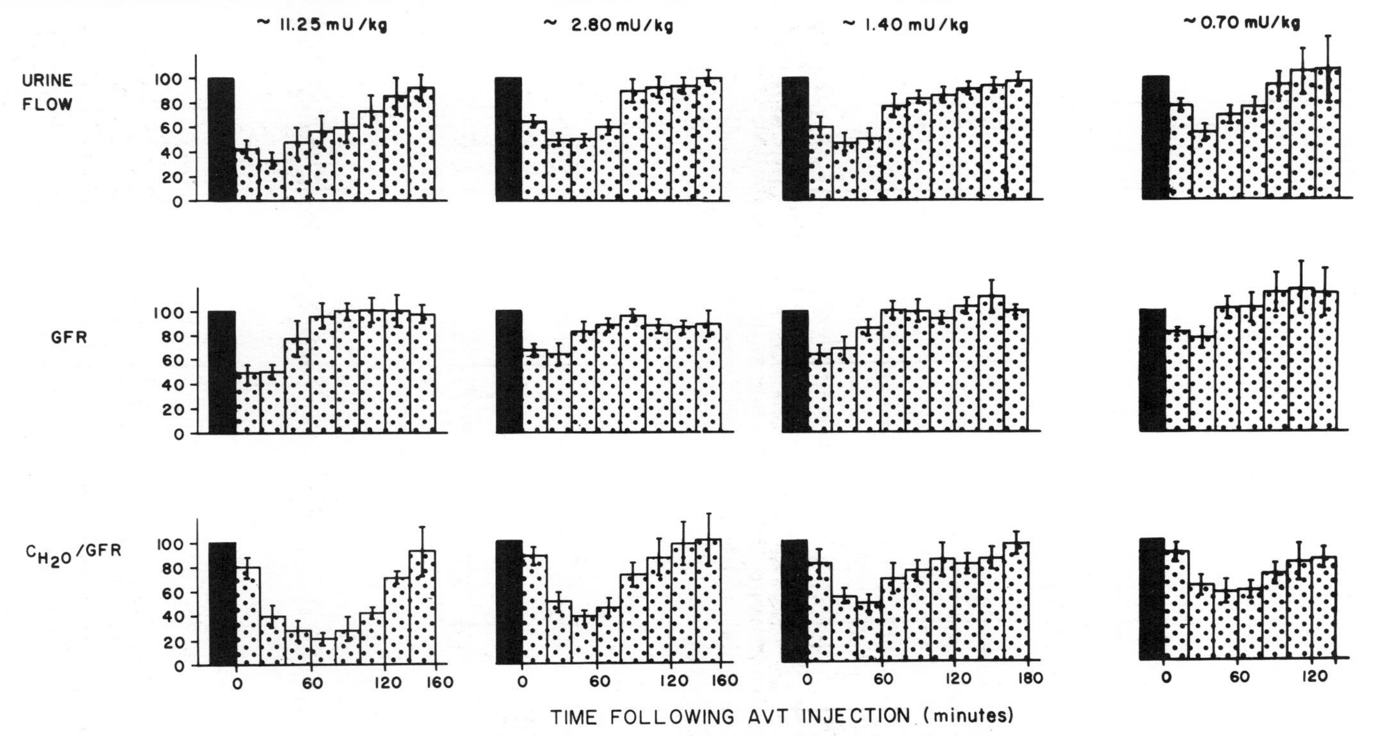

FIG. 3. Urine flow, glomerular filtration rate (GFR), and relative free-water clearance (C_{H_2O}/GFR) following injection of AVT (Dantzler, 1967a). Values are expressed as percents of the average value for three to four control periods in the same animal before injection of AVT. Black bars represent control values expressed as 100%. Each stippled bar represents the mean value for that clearance period. Numbers in parentheses indicate the number of determinations at each dose level. Vertical lines represent standard error.

tion is available on the effect of antidiuretic hormone on glomerular function in saurian reptiles, Bentley (1959) has demonstrated that Pitressin can cause an antidiuresis in a lizard *Tiliqua* (formerly *Trachysaurus*) *rugosa.* However, the doses he used (300–600 vasopressor mU/kg) could have caused a marked systemic vasodepressor response in this animal (Woolley, 1959). Thus, the nature of the antidiuretic response to vasopressin in these saurian reptiles remains obscure.

The reptilian neurohypophysis contains oxytocic as well as vasopressor activity. This may result from the presence of oxytocin, 8-isoleucine oxytocin, or both (Sawyer, *et al.,* 1959; Munsick, 1966). In high doses (20 oxytocic mU/kg in the case of oxytocin; 40 mU/kg or more in the case of 8-isoleucine oxytocin), these peptides had an antidiuretic effect on water snakes which was wholly glomerular in nature but which appeared to be secondary to a depression of systemic arterial blood pressure (Dantzler, 1967a). In alligators, however, a dose of oxytocin of only 10 oxytocic mU/kg ($\frac{1}{500}$ of that necessary to produce a systemic vasodepressor response in these animals) can cause a slight depression of GFR (Sawyer and Sawyer, 1952). Bentley also showed an antidiuretic effect of oxytocin in *T. rugosa,* but again, the filtration rate was not measured and the dose used was 15–30 times that necessary to produce systemic vasodepression (Woolley, 1959). Thus, it is not possible to be sure if there was an antidiuretic effect separate from the effect on the systemic circulation.

It appears most likely from the studies on freshwater snakes, in which a very low level of AVT caused glomerular depression without an effect on the systemic circulation (Dantzler, 1967a), that AVT is the major hormone regulating GFR. Presumably, this hormone acts to reduce the filtration rate by a vasoconstrictor effect on the afferent glomerular arteriole. However, as noted above, changes in GFR appear to reflect changes in the number of functioning nephrons, and yet all the nephrons appear to have the same general structure. The way in which the hormone can interact with the afferent arterioles of some of these apparently identical nephrons and not others to produce intermittent nephron function remains to be explained.

C. Renal Tubular Function

1. *Regulation of Tubular Permeability to Water*

Reptiles can produce a urine varying from about one-tenth the osmolality of the plasma to isosmotic with the plasma. It can reasonably be assumed that, as in other vertebrates (Walker *et al.,* 1937; Wirz,

1956), reabsorption of filtrate is isosmotic in the proximal tubule. Thus, dilution, when it occurs, must take place in the distal tubule by the hyperosmotic reabsorption of solute. Variations in dilution in any species appear to result from variations in the permeability of the luminal membrane of the distal tubular cells to water. However, the capacity for altering the permeability to water and thereby varying the osmolality of tubular urine differs in reptiles of different species. This may be related in part to habitat, but also to the excretory end products of nitrogen metabolism and to bladder or cloacal function.

The differences in ability to vary tubular permeability to water are apparent in Table I. Among the chelonians, freshwater turtles (*Pseudemys scripta*) can produce a urine varying from about three-tenths the osmolality of the plasma to isosmotic with the plasma (Dantzler and Schmidt-Nielsen, 1966). With a rise in plasma osmolality between 10 and 20 mOsm produced either by dehydration or by salt loading, the urine osmolality rose toward that of the plasma, and the free-water clearance decreased from about 11% of the filtration rate to only 2% (Fig. 1). Generally, the increase in tubular permeability to water preceded any decrease in GFR during dehydration and persisted after GFR had returned to normal with rehydration. In this respect the tubular and glomerular responses of freshwater turtles resembled those of frogs (Schmidt-Nielsen and Forster, 1954). In contrast, desert tortoises (*Gopherus agassizii*) showed little variation in tubular permeability to water as the plasma osmolality increased (Dantzler and Schmidt-Nielsen, 1966) (Fig. 1 and Table I). The urine remained hypoosmotic to the plasma and the osmolal urine-to-plasma (U/P) ratio varied only from about 0.3 to 0.6. In no experiments with desert tortoises did the osmolal U/P ratio rise above 0.7. The free-water clearance did not fall below 10% of the filtration rate and averaged about 15% of the filtration rate (Fig. 1). Thus, even with marked increases in plasma osmolality, these animals were never able to conserve additional water by producing urine isosmotic with the plasma.

A somewhat similar pattern for tubular permeability is found for semiaquatic freshwater snakes and desert-dwelling terrestrial snakes (Table I). The osmolal U/P ratio for freshwater snakes (*Natrix sipedon*) can vary from 0.1 during a marked water diuresis to 1.0 during a maximal antidiuresis (Dantzler, 1967a). During a salt load, the distal tubular permeability to water increased and the urine osmolality approached that of the plasma (Komadina and Solomon, 1970). In desert snakes (*Pituophis melanoleucus*), as in desert tortoises, there was little change in distal tubular reabsorption of water, the average value remaining remarkably constant during both a salt load and a water load (Table I)

(Komadina and Solomon, 1970). However, the values from individual animals studied by Komadina and Solomon (1970) show that the osmolal U/P ratio can vary at least from 0.5 to 1.0. Thus, these animals are capable of producing a hypoosmotic or isosmotic urine, even though the average osmolal U/P ratios did not change during salt loading or water loading.

Distal tubular permeability to water appears to be even less variable in saurian reptiles than in desert chelonian and ophidian reptiles. Horned lizards (*Phrynosoma cornutum*) and Galapagos lizards (*Tropidurus* sp.) from arid habitats produced a urine nearly isosmotic with the plasma in every state of hydration (Table I) (Roberts and Schmidt-Nielsen, 1966). In contrast, tropical geckos (*Hemidactylus* sp.) and blue-tongued lizards (*Tiliqua scincoides*) from more moist environments always produced ureteral urine hypoosmotic to the plasma (Table I) (Roberts and Schmidt-Nielsen, 1966; Schmidt-Nielsen and Davis, 1968). Moreover, the distal tubules in *Hemidactylus* did not appear even to regulate the degree of dilution very well, since the osmolal U/P ratio averaged 0.7 during both dehydration and water loading (Table I). This failure of the distal tubule of many lizards to regulate urine osmolality with changes in hydration has been supported by the recent micropuncture studies of H. Stolte and B. Schmidt-Nielsen (personal communication) on *Sceloporus cyanogenys*, a terrestrial lizard from a nonarid environment. They found that the osmolal tubular fluid-to-plasma ratio (osmolal TF/P ratio) was consistently 0.5 to 0.6.

The crocodilians appear to show relatively little variability in distal tubular permeability to water during different states of hydration regardless of their natural habitat (Table I) (Schmidt-Nielsen and Skadhauge, 1967; Schmidt-Nielsen and Davis, 1968). In general, the ureteral urine tends to be somewhat hypoosmotic. In this characteristic, the crocodilians tend to resemble desert chelonian and ophidian reptiles. However, the individual values for the crocodilian most thoroughly studied from this point of view (the freshwater crocodile *Crocodylus acutus*) show that the osmolal U/P ratio can vary from about 0.55 to 0.95 (Table I) (Schmidt-Nielsen and Skadhauge, 1967). In this respect, the crocodilians closely resemble the desert-dwelling ophidians.

Changes in permeability of the distal tubule permitting the production of a hypoosmotic or isosmotic urine are, like changes in GFR, presumably under the control of arginine vasotocin. This regulation has been most extensively studied in freshwater snakes (*N. sipedon*) (Dantzler, 1967a). As noted above, AVT in doses as low as 0.7 vasopressor mU/Kg produced a marked antidiuresis (Fig. 3). Although a portion of this antidiuretic effect resulted from a decrease in GFR, a larger portion resulted from a reduction in relative free-water clearance (C_{H_2O}/GFR). Moreover, at each dose level, the filtration rate returned to the control level before the urine flow, and the continued depression of urine flow corresponded to the depression in C_{H_2O}/GFR (Fig. 3). Only a small portion of the difference between duration of depression of C_{H_2O}/GFR. and duration of depression of GFR could be explained by dead-space errors. The greater magnitude and duration of the depression of C_{H_2O}/GFR compared with the depression of GFR at each dose level suggests that tubular permeability to water is affected more than glomerular filtration rate by the presence of AVT. These results, if applicable to freshwater turtles, would explain the persistence of increased tubular permeability to water when GFR returned to normal following a period of dehydration and rehydration. The mechanism underlying this effect, however, remains obscure. Furthermore, these data do not fully explain the appearance of increased tubular permeability to water prior to a decrease in filtration rate during the early stages of dehydration in freshwater turtles. A few preliminary experiments on freshwater turtles with turtle pituitary extract indicated that a relatively small dose of extract (2–9 vasopressor mU/kg) had a tubular but not a glomerular effect (Dantzler and Schmidt-Nielsen, 1966). Larger doses, as noted above, had a glomerular as well as a tubular effect. Thus, small amounts of AVT released during early stages of dehydration would cause increased tubular permeability to water, and larger amounts released during later stages would cause a decrease in GFR as well. However, in more detailed studies with freshwater snakes, the smallest dose of AVT capable of producing an antidiuresis (0.1 vasopressor mU) caused a depression of both GFR and C_{H_2O}/GFR (Dantzler, 1967a).

As noted above (Table I), in some reptiles, variation in the permeability of the distal tubule to water is limited. Thus, the distal tubule in these animals appears poorly responsive to AVT. H. Stolte and B. Schmidt-Nielsen (personal communication) examined this by micropuncture of the distal tubule in *S. cyanogenus* and found no effect of AVT on the osmolal TF/P ratio. Finally, the other neurohypophysial peptides which might affect distal tubular permeability, oxytocin and 8-isoleucine oxytocin, do not appear to have a tubular effect in freshwater snakes even in very high doses (Dantzler, 1967a).

2. *Regulation of Sodium Transport*

Few data on tubular regulation of sodium excretion are available for reptiles. Moreover, the regulation of sodium transport appears to be related to the excretion of the end products of nitrogen metabolism and to cloacal and bladder function to be discussed below. Among the chelonians, freshwater turtles (*Pseudemys scripta*) showed increased tubular reabsorption of sodium, as indicated by a decrease in the fraction of filtered sodium excreted (C_{Na}/GFR) during dehydration (Dantzler and Schmidt-Nielsen, 1966). A large salt load enhanced C_{Na}/GFR in desert tortoises (*Gopherus agassizii*), indicating a reduction in tubular reabsorption of sodium.

Somewhat more information is available about the renal tubular regulation of sodium excretion in snakes. During a control diuresis produced by an infusion of 1.25% mannitol in freshwater snakes, about 10% of the filtered sodium was excreted in the ureteral urine (Dantzler, 1967a). Stop-flow studies under the same conditions (Dantzler, 1967b) indicated that, as in mammals, peak reabsorption of sodium and water occurs in the distal portions of the nephrons.

Arginine vasotocin may also influence sodium reabsorption in freshwater snakes (Dantzler, 1967a). Following injections of 2.8 vasopressor mU or more of AVT/kg, C_{Na}/GFR decreased significantly. The mean value during the period of maximum decrease following this dose of AVT was about 25% below the control level. This suggests that AVT enhances tubular reabsorption of sodium. It must be noted, however, that this increased tubular reabsorption of sodium may not result from a direct effect of AVT on the sodium transport mechanism. Instead, with a concurrent decrease in GFR, it may result from a decreased filtered load of sodium delivered to the site of sodium reabsorption. The effects of adrenalectomy on renal function in freshwater snakes (*Natrix cyclopion cyclopion* and *Natrix cyclopion floridana*) have also been studied (Elizondo and LeBrie, 1969). Plasma sodium, chloride, and osmolality decreased following adrenalectomy. After a peritoneal water load, neither the urine flow nor the GFR of adrenalectomized animals, determined on ureteral urine from a single kidney, differed significantly from that of control water-loaded animals. However, the rate of sodium excretion nearly doubled and the percentage of filtered sodium reabsorbed fell from about 90% to about 70%. Fractional water reabsorption by the renal tubules also decreased. The authors suggest that the animals were in a sufficient diuresis that all water reabsorption was occurring proximally and that a depression of such reabsorption indicates a proximal action for the mineralocorticoids. However, it is not clear from the authors' data that the animals were in this state of diuresis. Also, even in the absence of antidiuretic hormone, some water may accompany solute reabsorption in the distal tubule. Thus, the site of action, whether in the proximal or distal portions of the nephron, remains speculative.

These workers also studied the effect of aldosterone administration during the intraperitoneal administration of 20 ml/kg of 1.5% sodium chloride in *N. cyclopion* (LeBrie and Elizondo, 1969). The large intraperitoneal salt load did raise the plasma osmolality significantly, but, in contrast to the acute administration of a much more concentrated load (Dantzler, 1968; Komadina and Solomon, 1970), did not depress GFR. Apparently volume expansion suppressed ADH release and helped maintain the filtration rate at control levels. The saline load did decrease the fraction of filtered sodium reabsorbed by about 20%. Administration of aldosterone (200 μg per animal over 2 days) with the sodium chloride load increased the fraction of filtered sodium reabsorbed by about 24% and the fraction of filtered water reabsorbed by about 21%. The authors again speculate that the primary site of action is in the proximal tubule, but it appears difficult to decide, on the basis of the data, in what portion or portions of the nephron the hormone has its effect.

Those studies available on saurian reptiles indicate that tubular regulation of sodium reabsorption is distinctly limited in some species. Among those species studied by Roberts and Schmidt-Nielsen (1966), the horned lizard (*Phrynosoma cornutum*) and the Galapagos lizard (*Tropidurus* sp.) excreted about 45% of the filtered sodium in the ureteral urine. Tubular reabsorption of sodium increased somewhat during water diuresis, but a very high proportion of the filtered sodium was still excreted. Ultrastructural studies of the tubule cells in these desert-dwelling lizards (Roberts and Schmidt-Nielsen, 1966) reveal very flat basal cell membranes without infoldings and relatively few, randomly scattered mitochondria. The few, randomly arranged mitochondria, which contrast with the numerous, well organized mitochondria in the basal region of mammalian tubule cells, may be related to the small fraction of filtered sodium reabsorbed. This structure in the distal tubule cells may also be related to the inability of these cells to dilute the urine by reabsorbing sodium and leaving water behind. However, as already noted, the permeability of the luminal cell membrane is equally important in the production of a hypoosmotic or isosmotic urine.

In the same study, geckos from Puerto Rico (*Hemidactylus* sp.) excreted only about 15% of the sodium filtered in the ureteral urine during normal hydration (Roberts and Schmidt-Nielsen, 1966). Although this is a high percentage of the filtered sodium that is not reabsorbed, it is much lower than that observed in the two desert-dwelling saurians studied. These geckos showed an increased excretion of sodium during both dehydration and water loading. The proximal and distal tubule cells of these geckos have more mitochondria located near the basal and lateral cell surfaces than do those of the horned and Galapagos lizards (Roberts and Schmidt-Nielsen, 1966). These may be related to the greater percentage of filtered sodium reabsorbed by the tubules of the geckos. Moreover, the distal tubule cells of these animals show basal infoldings of the plasma membranes with elongated parallel mitochondria. This structure resembles that of mammalian cells, but is less well developed with fewer mitochondria. It may be related to the ability of these animals to dilute the urine by reabsorbing sodium without water. However, as emphasized above, the relative permeability of the apical cell membrane to water is equally important in the production of a dilute urine.

Temperature also appears important in the regulation of tubular sodium reabsorption of saurian reptiles. Although the percentage of filtered sodium excreted in the ureteral urine of the Australian skink (*Tiliqua rugosa*) increased with increasing urine flow, it decreased with increasing temperature at any specific flow rate (Shoemaker *et al.*, 1966). Thus, tubular reabsorption of sodium is temperature dependent, increasing with increasing temperature. Some of the changes in sodium reabsorption at different urine flows during salt loading and water loading in the study by Roberts and Schmidt-Nielsen (1966) might also have been related to temperature, since that study was performed at temperatures ranging from 24° to 30°C.

Little information is available on the renal tubular regulation of sodium excretion in crocodilians. However, estimates derived from the data obtained by Schmidt-Nielsen and Skadhauge (1967) on *Crocodylus acutus* indicate that the sodium excreted in the ureteral urine varies from about 2 to 8% of the filtered load during all states of hydration and salt loading. This is a much higher percentage than is found normally in mammals, but it does indicate greater tubular reabsorption of sodium than in many saurian and ophidian reptiles. Moreover, it has been estimated from clearance data and anatomical data that 50% of the filtered sodium is reabsorbed in the proximal tubule at a rate of 196×10^{-10} Eq/cm^2/second (Davis and Schmidt-Nielsen, 1967). This is nearly three times the rate of reabsorption in a rat and appears to be very high for a poikilothermic animal.

3. *Coupling of Sodium and Water Transport*

As noted above, the reabsorption of sodium and water by the renal tubular epithelium is isosmotic in the proximal tubule and may vary from isosmotic to hyperosmotic in the distal tubule. The coupling of water movement to active solute transport in epithelium is now generally considered to be due to local osmotic gradients set up within the epithelium (Diamond, 1964, 1971). Ultrastructural studies of a variety of transporting epithelia (see summary by Schmidt-Neilsen, 1971) have suggested that these local osmotic gradients are produced within lateral intercellular spaces. According to Diamond's (1964, 1971) hypothesis, the active transport of solute (usually sodium) into the closed luminal ends of these spaces produces the local osmotic gradient. Water flows continually into these spaces because of the osmotic gradient and sweeps solute toward the open ends of the spaces. In the steady state, a standing osmotic gradient is maintained in the spaces by active solute transport, with the osmolality decreasing from the closed to the open ends. Fluid emerging from the open ends of these spaces will be either isosmotic or hyperosmotic, depending on such factors as the length and radius of the space and the permeability of the membranes. Diamond and Bossert's (1967) mathematical treatment of this model predicted that the transported fluid would be isosmotic to the luminal fluid if the intercellular spaces were long and narrow, since osmotic equilibrium could be attained between the cells and transported solution. However, it predicted that the transported fluid would be hyperosmotic if the spaces were shorter and wider.

Ultrastructural studies of the kidneys of chelonian, ophidian, saurian, and crocodilian reptiles during various states of hydration and salt loading (Davis and Schmidt-Nielsen, 1967; Schmidt-Nielsen and Davis, 1968) indicated that lateral intercellular spaces between proximal tubular cells were always open when fluid transport was occurring. They were closed when sodium transport was inhibited by ouabain or the tubules were closed during salt loading (see earlier discussion of this phenomena). These data agree with the concept of Diamond. The situation differed, however, in the distal tubules where the transported fluid varied from isosmotic to hyperosmotic. With isosmotic transport, the intercellular spaces were open; with hyperosmotic transport (i.e., when the tubular fluid was hypoosmotic to the blood), the intercellular spaces were essentially closed. These data deviate from the prediction of Diamond and Bossert (1967) that with hyperosmotic transport, short, wide spaces should be found. Thus, it is not yet clear whether hyperosmotic transport in reptilian distal tubules occurs via the intercellular channels (Schmidt-Nielsen, 1971). It is possible that when sodium is transported with little water, the reabsorbate can move directly across the cells.

4. *Regulation of Potassium and Hydrogen Ion Transport*

The regulation of renal tubular potassium and hydrogen ion transport in reptiles is not well understood and, as in the case of sodium and water transport, is related to the excretion of nitrogen and the function of the bladder or cloaca. The concentration of potassium increased and the concentration of sodium markedly decreased in the ureteral urine of desert tortoises (*Gopherus agassizii*) following a potassium load (Dantzler, 1964). When potassium excretion is necessary, the renal tubules can apparently increase the secretion of potassium as more sodium is reabsorbed in exchange for it. The ureteral urine of freshwater turtles (*Pseudemys scripta*) maintained on a high protein, acid ash diet is slightly acid (pH 6.7) (Dantzler and Schmidt-Nielsen, 1966). However, following a bicarbonate load, the urine can become alkaline (pH 8.0), presumably with an increase in bicarbonate excretion and a decrease in tubular hydrogen ion secretion. Although carbonic anhydrase has been found in the kidneys of several species of turtles (Maren, 1967), nothing is known about its effect on hydrogen ion secretion or bicarbonate reabsorption.

Among the ophidian reptiles, freshwater snakes of the genus *Natrix* maintained on a high protein, acid ash diet secreted about 25–75% of the filtered potassium in the ureteral urine during a control diuresis produced by the infusion of 1.25% mannitol (Dantzler, 1967a,b, 1968). At this time, the pH of the ureteral urine averaged 6.1. Stop-flow studies under these conditions (Dantzler, 1967b) did not show any potassium secretion in the most distal portions of the nephron. There was, however, a slight potassium peak in more proximal regions. Precise localization by stop-flow is difficult. The continued reabsorption in the most distal samples may reflect the situation in collecting ducts where reabsorption continues in mammalian nephrons. The more proximal peak, suggesting secretion, may be in the region of the distal tubules, where secretion has been demonstrated by micropuncture in mammalian nephrons (Malnic *et al.*, 1964, 1966a,b). The secretory peak may also be related to uric acid transport (see below). In any case, potassium secretion by the renal tubules can occur, for, with bicarbonate loading, the urine pH rose to 7.0 and the potassium excretion exceeded the filtered load (Dantzler, 1968). If, as in mammalian nephrons, filtered potassium is essentially completely reabsorbed proximally and a variable amount secreted distally by passive movement out of the tubule cells, then presumably more potassium is secreted as sodium is reabsorbed when hydrogen ion secretion is reduced. This mammalian pattern of potassium and hydrogen ion secretion also can be seen in the effect of carbonic anhydrase inhibitors. The administration of acetazolamide led to the production of an alkaline urine with a marked increase in the excretion of sodium and potassium but no change in the excretion of chloride (Dantzler, 1968). In the absence of measurements of urine bicarbonate or renal carbonic anhydrase activity, this pattern can be taken as presumptive evidence of the presence of carbonic anhydrase in the renal tubule cells and of its role in hydrogen ion secretion and bicarbonate reabsorption.

There is some evidence of possible hormonal control of tubular potassium secretion. Following the administration of 12.5 ng (2.8 vasopressor mU) or more of AVT/kg, the percentage of filtered potassium decreased significantly (Dantzler, 1967a). The mean value for the period of maximum decrease was about 75% of the control level. Decreases were also seen with lower dose levels of AVT. Although these may have been significant, there was great variation. These data indicate a decrease in tubular secretion of potassium, but this may not be a direct effect of AVT on the potassium transport system. A reduction in the amount of sodium reaching the potassium secretory region of the nephron may, as in mammals (Giebisch, 1969), reduce the intratubular electronegativity, thereby reducing the electrical gradient for potassium diffusion out of the cells. The reduced flow rate through this portion of the nephron may also reduce potassium excretion.

In contrast to mammals, adrenalectomy caused a significant decrease in serum potassium in snakes of the genus *Natrix*, but no significant change in urinary potassium excretion (Elizondo and LeBrie, 1969). These observations remain to be explained adequately.

Snakes of the genus *Natrix* also appeared to be unable to excrete an acid load one-third to one-tenth of that handled by mammals or birds, and the pH of the ureteral urine never fell below 5.8 even when the blood pH was below 7.0 (Dantzler, 1968). This suggests an inability of the tubules to secrete hydrogen ions against a high gradient. Although it might be possible to demonstrate further tubular acidification in the presence of poorly reabsorbed anions such as sulfate, it also seems possible that further acidification might occur in the cloaca (see below).

Little is known about tubular transport of potassium in saurian reptiles, but net tubular secretion can occur in the Australian skink (*Tiliqua rugosa*) (Shoemaker *et al.*, 1966). The data suggest that net tubular secretion of potassium may be more frequent at lower temperatures. If, as in mammals, filtered potassium is reabsorbed actively and secreted by passive diffusion, the effect of temperature might be explicable. At low temperatures, the active reabsorption of filtered potassium would probably be reduced, but there need not be any effect on the permeability of the luminal membranes. Moreover, the greater load of sodium delivered to distal regions because of reduced sodium reabsorption (see above) might increase the electronegativity of the tubular lumen and further enhance potassium secretion.

In the crocodilian *Crocodylus acutus*, the excretion of potassium in the ureteral urine remained low under most conditions of hydration or salt loading, ranging from about 10 to 30% of the filtered load (Schmidt-Nielsen and Skadhauge, 1967). Net secretion was not demonstrated. However, in one experiment by Schmidt-Nielsen and Skadhauge (1967), 85% of the filtered potassium was excreted during a sodium load. It seems likely that potassium can be secreted by the tubules and that this secretion is enhanced by increased sodium excretion.

Hydrogen ion excretion in crocodilians appears to be related to the pattern of ammonium and bicarbonate excretion, which is different from that of mammals. Ammonium and bicarbonate appear to be the principal ions in either ureteral or cloacal urine (Coulson and Hernandez, 1964; Schmidt-Nielsen and Skadhauge, 1967). In mammals, ammonium ions are excreted extensively only when urine is acid and bicarbonate ions only when urine is alkaline. In the crocodilians, the urine pH is normally alkaline (about 7.8) despite a high protein, acid ash diet (Coulson and Hernandez, 1964). Apparently, marked ammonia production by the renal tubule cells permits adequate buffering of hydrogen ions and the continued reabsorption of sodium and potassium. It appears that bicarbonate exchanges for chloride. This is suggested by the effects of carbonic anhydrase inhibitors on the cloacal urine of the alligator (*Alligator mississippiensis*) studied by Coulson and Hernandez (1964). Following acetazolamide administration, urinary bicarbonate excretion fell and urinary chloride excretion increased. This effect of the inhibition of carbonic anhydrase suggests that carbon dioxide is hydrated in the tubule cells to produce bicarbonate which is exchanged for luminal chloride.

5. *Regulation of Calcium and Phosphate Transport*

Very little is known about the renal regulation of calcium and phosphate in reptiles. The excretion of calcium and phosphate has been studied by clearance techniques with ureteral urine in control and parathyroidectomized, unanesthetized water snakes of the genus *Natrix* (Clark and Dantzler, 1972). In intact animals, the clearance of phosphate was about 2.6 times the filtered load, indicating that there was net tubular secretion of phosphate. However, in these same animals, only about one-tenth of the filtered calcium was excreted. Since the concentration of calcium in the urine was less than that in an ultrafiltrate of the plasma, it appeared probable that calcium was actively reabsorbed by the renal tubules. In parathyroidectomized animals, only about 30% of the filtered phosphate was excreted, indicating that there now was net tubular reabsorption of phosphate. Since the concentration of phosphate in the urine was still greater than that in the plasma, there was no indication of active tubular reabsorption of phosphate. Parathyroidectomy did not alter the tubular reabsorption of calcium. Ninety percent of the filtered calcium was still reabsorbed. The intravenous administration of mammalian parathyroid extract restored net tubular secretion of phosphate in parathyroidectomized animals and elevated phosphate secretion in intact animals within 1 hour. The hormone did not produce any effect on glomerular filtration rate in contrast to its effect in mammals (Hiatt and Thompson, 1957). The hormone also had no significant effect on the tubular transport of calcium in intact or parathyroidectomized animals.

Studies with bladder or cloacal urine have indicated that the administration of mammalian parathyroid extract causes hyperphosphaturia in intact and parathyroidectomized turtles (*Chrysemys picta*) (Clark, 1965) and both hyperphosphaturia and hypercalciuria in intact lizards (McWhinnie and Cortelyou, 1968). Thus, the hormone may affect tubular transport of phosphate and calcium in these animals, but this remains to be studied.

Although Coulson and Hernandez (1964) collected cloacal samples from alligators (*Alligator mississippiensis*), they found evidence of apparent net tubular secretion of phosphate. Almost no calcium appeared in the urine even with calcium loading. Instead it was excreted in the feces. Thus, tubular reabsorption of calcium appeared to be quite effective. As these investigators noted, significant amounts of calcium in an alkaline urine with a high ammonia concentration would be conducive to the formation of renal calculi.

6. *Regulation of Nitrogen Excretion*

A summary of the percentage of urinary nitrogen appearing in the forms of ammonia, urea, and uric acid for the four living orders of reptiles is given in Table III. A more specific breakdown for some individual species can be found in the references from which this table was compiled. Two individual chelonian species were included because these are the only ones for which the distribution of total urinary nitrogen was obtained with ureteral urine. Only factors in nitrogen excretion relating clearly to renal function will be considered here.

a. Ammonia. Only the aquatic or semiaquatic chelonians and the crocodilians excrete a significant portion of ureteral urinary nitrogen as ammonia regardless of the pH of the urine. It has already been noted that the urine pH in crocodilians is neutral or alkaline (Coulson and Hernandez, 1964). In freshwater turtles (*Psuedemys scripta*) on an acid ash diet, the ureteral urine tended to be slightly acid. However, even when the urine became alkaline (pH 8) following a bicarbonate infusion, a significant portion (about 15%) of the nitrogen in the ureteral

TABLE III
APPROXIMATE PERCENTAGE OF TOTAL URINARY NITROGEN IN THE FORM OF AMMONIA, UREA, AND URIC ACID IN REPTILES[a]

Reptile	Percent of total urinary nitrogen		
	Ammonia	Urea	Uric acid
Chelonia			
Wholly aquatic	20–25	20–25	5
Semiaquatic	6–15	40–60	5
Wholly terrestrial			
Moist environment	6	30	7
Dry environment	5	10–20	50–60
Gopherus agassizii (desert tortoise)	3–8	15–50	20–50
Pseudemys scripta (freshwater turtle)	4–44	45–95	1–24
Squamata			
Sauria	Insignificant–highly significant(?)	0–8	90
Ophidia	Insignificant–highly significant(?)	0–2	98
Crocodilia	25	0–5	70
Rhyncocephalia			
Sphenodon punctatus	3–4	10–28	65–80

[a] Data on chelonia are from Moyle (1949) and Baze and Horne (1970) with values on *Gopherus agassizii* and *Pseudemys scripta* from Dantzler and Schmidt-Nielsen (1966); on sauria, from Khalil (1951), Dessauer (1952), Perschmann (1956), and J. E. Minnich (personal communication); on ophidia, from Khalil (1948a,b) and J. E. Minnich (personal communication); on crocodilia, from Khalil and Haggag (1958); on rhynchocephalia, from Hill and Dawbin (1969).

urine was in the form of ammonia (Dantzler and Schmidt-Nielsen, 1966). The excretion of significant amounts of ammonia is possible in these animals with abundant access to water. Moreover, extensive production of ammonia by the renal tubule cells could buffer hydrogen ions in these animals on an acid ash diet and free sodium ions from salts of strong acids for continued reabsorption. In mammals, ammonia transport out of the renal tubule cells is considered to occur by passive nonionic diffusion (Balagura and Pitts, 1962; Pitts, 1968). This is enhanced by an acid urine, for the ammonia is converted to ammonium ions, which cannot readily diffuse back into the cells. In the presence of an alkaline urine, ammonia excretion decreases markedly. The high level of ammonia excretion in an alkaline urine in crocodilian and some chelonian reptiles suggests that an additional mechanism of tubular secretion may be present in these animals.

Ammonia forms only a small fraction of ureteral urinary nitrogen in ophidian reptiles. In unanesthetized freshwater snakes of the genus *Natrix*, the ammonia concentration in the ureteral urine averaged about 9.8 mM/liter and the excretion rate about 19 μM/kg/hour during a control diuresis, although the urine pH was about 6.2 (W. H. Dantzler, unpublished observations). This is one-fifth or less of that seen in crocodilians (Schmidt-Nielsen and Skadhauge, 1967). A few preliminary experiments suggest that ammonia excretion in snakes does not increase during a metabolic acidosis (W. H. Dantzler, unpublished observations). However, as noted earlier, the pH of the ureteral urine never fell below 5.8 (Dantzler, 1968). Thus, if ammonia is secreted by nonionic diffusion in ophidian reptiles, this small decrease in pH might not have led to a significant increase in its excretion. The possibility that nonionic diffusion, as in mammals, is the mechanism for tubular ammonia secretion in ophidians is strengthened by changes observed in ammonia excretion following carbonic anhydrase inhibition. As urine pH rose from about 6.0 to about 7.2 in a typical mammalian response, the excretion of ammonia decreased by about 75% (W. H. Dantzler, unpublished observations). Thus, in terms of ammonia excretion in ureteral urine, ophidian reptiles appear to resemble mammals more closely than chelonian or crocodilian reptiles.

Recently, J. E. Minnich (personal communication) has found significant amounts of ammonium excreted with precipitated urates from the cloaca of monitor lizards and nonconstictor snakes. Possibly some additional ammonia secretion takes place in the cloaca.

b. Urea. Only the chelonia and rhynchocephalia excrete urea as a consistently significant fraction of urinary nitrogen (Table III). However, in some saurian reptiles—*Anolis carolinensis* (Dessauer, 1952) and *Lacerta viridis* (Perschmann, 1956)—significant quantities of urea have been found in bladder urine. In those chelonian reptiles in which urea excretion was studied in ureteral urine (*Pseudemys scripta* and *Gopherus agassizii*), there was no convincing evidence that urea was actively secreted or reabsorbed by the renal tubules (Dantzler and Schmidt-Nielsen, 1966). It seems most likely that urea is handled by filtration and passive back diffusion across the tubule cells (Dantzler and Schmidt-Nielsen, 1966). However, urea is actively secreted by the renal tubules of the only living rhynchocephalian (*Sphenodon punctatus*) (B. Schmidt-Nielsen, personal communication).

The studies of *Testudo leithii* and *Testudo sulcata* by Khalil and Haggag (1955) and of *Chelodina longicollis* by Rogers (1966) indicated that with dehydration, the percentage of nitrogen excreted as uric acid increased and that excreted as urea decreased. However, Khalil and

Haggag (1955) studied spontaneous excretions. Since crystalline urates are stored in the bladder of tortoises and may be voided spontaneously while water and urea may be reabsorbed from the bladder (Dantzler and Schmidt-Nielsen, 1966), spontaneous urine samples cannot be used to indicate a change from ureotelism to uricotelism. Similarly, Rogers (1966) collected bladder samples. Because urea diffuses rather well across epithelial membranes and because chelonian bladders may be highly permeable (see below), urea would not be expected to accumulate in the bladder. Therefore, it is necessary to collect ureteral urine in order to determine if there is a true change in the renal excretion of urea and uric acid with dehydration. Collections of ureteral urine in the desert tortoise (*Gopherus agassizii*) showed no consistent change in the proportions of urea and uric acid excreted as the animals were dehydrated (Dantzler, 1964). Finally, Baze and Horne (1970) have found an increase in the ornithine–urea cycle enzymes in the livers of dehydrated *Gopherus berlandieri* and have suggested that increased urea synthesis during dehydration might, in part, account for higher plasma urea concentrations.

c. Uric Acid. Uric acid is the major excretory end product of nitrogen metabolism in the Squamata, Crocodilia, Rhynchocephalia, and many of the Chelonia (Table III). Since it has a low solubility in water (0.384 m*M*/liter), it contributes very little to the osmotic pressure of the urine and can be excreted with very little water. However, most of the uric acid in the ureteral urine probably does not exist in the acid form. The pK_a for the monobasic salt of uric acid is 5.67. Since the pH of the normal ureteral urine of reptiles is only slightly acid (for example, chelonians, Dantzler and Schmidt-Nielsen, 1966; ophidians, Dantzler, 1968) or even alkaline (for example, crocodilians, Coulson and Hernandez, 1964; saurians, Minnich, 1970), most of the uric acid must exist in the form of monobasic salt. The cation associated with urate will probably be primarily sodium or potassium, the predominant one depending upon whether the animal is carnivorous or herbivorous. Minnich (1970) found that in the herbivorous desert iguana (*Dipsosaurus dorsalis*) uric acid was excreted primarily as monopotassium urate. In crocodilians, of course, ammonium urate may be the important excretory salt, although this is less soluble than the sodium or potassium salt (Porter, 1963a). As noted above, J. E. Minnich (personal communication) also found large quantities of ammonium urate in pellets excreted from the cloaca of monitor lizards and nonboid snakes.

Although the solubilities of the monobasic urate salts in water are not very great (6.76 m*M*/liter for sodium urate; 12.06 m*M*/liter for potassium urate), they are much greater than that for uric acid. In chelonian reptiles (Dantzler and Schmidt-Nielsen, 1966), the concentration of urate in ureteral urine was well below the level of saturation. In other reptiles, the concentration of urate may be above the level of saturation (Dantzler, 1968, and unpublished observations). In these cases, a portion of the urate in the tubular and ureteral urine may be in the form of a colloidal suspension. Evidençe for this possibility has been presented for birds (McNabb and Poulson, 1970) and dalmatian dogs (Porter, 1963b), which also excrete large amounts of urate.

Precipitated uric acid and urates are found in the distal large intestine of snakes (Dantzler, 1968), in the cloaca of lizards (Minnich, 1969, 1970) and crocodiles (Khalil and Haggag, 1958; Coulson and Hernandez, 1964), and in the bladder of a number of turtles (Dantzler and Schmidt-Nielsen, 1966; Baze and Horne, 1970). The precipitation of uric acid could occur through the acidification of the urine in the bladder or cloaca by the secretion of hydrogen ions or the reabsorption of bicarbonate (Dantzler, 1968, 1970b). This acidification would convert urate to less soluble uric acid, permitting further reabsorption of sodium and excretion of acid. However, obligatory reabsorption of sodium in exchange for hydrogen ions might also require an extraurinary route for excretion of ions (Schmidt-Nielsen *et al.*, 1963). Precipitation of uric acid and urates could also occur through the cloacal or bladder reabsorption of water or through the dispersion of colloidal urates in the claoca (Minnich, 1970) (see below for more detailed discussion of transport by reptilian bladder and cloaca). The precipitation and excretion of sodium or potassium urate rather than urid acid would permit the excretion of inorganic ions by the urinary system without water (Minnich, 1970).

In those reptilian species on which clearance data have been obtained, urate is secreted by the renal tubules (Dantzler and Schmidt-Nielsen, 1966; Dantzler, 1967b, 1968). Tubular urate transport may be bidirectional in all these animals, but stop-flow studies in conscious water snakes of the genus *Natrix* gave no evidence of tubular reabsorption of urate (Dantzler, 1967b).

Acid–base balance may affect tubular urate transport. Tubular secretion of urate in conscious water snakes increased during an acute alkalosis (produced by intravenous infusion of sodium bicarbonate) but did not change during an acute acidosis (produced by intravenous infusion of hydrochloric acid) (Dantzler, 1968). The increase in urate secretion with alkalosis appeared to be correlated with an increase in blood pH and not with urine pH, since the administration of sufficient acetazolamide to produce an alkaline urine without a change in blood pH did not affect urate secretion. Since almost all the uric acid is in the form of the sodium or potassium salt, an increase in tubular secretion during metabolic alkalosis would eliminate base in the same fashion as an increased excretion of sodium bicarbonate.

An increased excretion of uric acid during acidosis, had it occurred, could also have been advantageous. But the secretion of uric acid did not change. Since the urine pH in this study did not fall to the level of the pK_a for uric acid, most of the uric acid would still have been in the form of the sodium or potassium salt. Therefore, the failure to increase urate secretion during acidosis seems quite appropriate. However, with a relatively high urine pH, a decrease in urate secretion would seem to have been an even more appropriate response. The failure of urate secretion to decrease during acidosis may have been related to further acidification of the urine in the cloaca by the secretion of hydrogen ions in exchange for sodium ions, as noted above. If hydrogen ions normally exchanged for sodium ions of sodium urate in the cloaca during acidosis, this would permit the continued excretion of nitrogenous waste during acidosis while sodium and water were being conserved and acid eliminated.

Other factors that might influence the transport of urate have been evaluated with kidney slices and isolated perfused renal tubules *in vitro*. The uptake of uric acid by kidney slices from desert-dwelling saurian reptiles exceeded that by slices from nonarid ophidian reptiles (Dantzler, 1969, 1970a,b, 1971c). The steady-state urate slice-to-medium ratio (S/M ratio) for slices from the desert spiny lizard (*Sceloporus magister*) was more than twice that observed with slices from garter snakes (*Thamnophis* sp.) or freshwater snakes (*Natrix* sp.) with the same concentration of urate (3×10^{-7} *M*) in the incubation medium (Table IV). Whether this difference is related to a greater need for tubular secretion of urate in the arid-living forms is not yet clear.

Other differences exist in the requirement for potassium for transport by these slices. Potassium depletion reduced the urate uptake by garter snake and desert spiny lizard slices (Table IV), but inhibition of transport was not complete. This contrasts with the effects seen with mammalian and avian kidney slices in which comparable potassium depletion produced complete inhibition of active transport (Berndt and Beechwood, 1965; Dantzler, 1969). Also, the inhibition of transport with *Sceloporus* slices was less than with *Thamnophis* slices. Since plasma potassium levels are more variable in reptiles than in mammals and birds, it is tempting to consider that these differences in the sensitivity of urate transport to potassium depletion are related to the differences in variability in the plasma potassium concentration. Since urate is the major excretory end product of nitrogen metabolism in ophidian and

TABLE IV
EFFECTS OF VARIATIONS IN MEDIUM POTASSIUM ON URATE ACCUMULATION BY REPTILIAN KIDNEY SLICES[a]

Species	3 mMK Ringer	0 mMK Ringer	40 mMK Ringer
Freshwater snakes (*Natrix* sp.)	1.69 ± 0.03(21)	—	—
Terrestrial snakes, nonarid environment (*Thamnophis* sp.)	1.67 ± 0.07(50)	1.35 ± 0.05(16)	1.60 ± 0.04(18)
Terrestrial lizards, arid environment (*Sceloporus magister*)	3.49 ± 0.20(42)	2.13 ± 0.18(6)	5.70 ± 0.28(18)

[a] Values are mean slice-to-medium ratios ±SE. Figures in parentheses indicate number of experiments. Values for garter snakes are from Dantzler (1969). Values for freshwater snakes are from Dantzler (1970a). Values for terrestrial lizards are from Dantzler (1971c)

saurian reptiles, it appears important that its tubular secretion not be greatly reduced by natural decreases in plasma potassium.

However, the latter consideration does not explain why a high medium potassium concentration, which led to similar increases in tissue potassium concentration in snakes and lizards, resulted in a marked increase in urate uptake by *Sceloporus* slices but none by *Thamnophis* slices. It would seem advantageous for any uricotelic vertebrate to show an increase in tubular urate secretion.

There are also differences between the responses of potassium-depleted snake and lizard kidney slices to increasing medium potassium concentrations. With potassium-depleted slices from both *Sceloporus* and *Thamnophis,* urate transport only tends to be restored with incubation in a high (40 m*M*) potassium concentration. However, tissue potassium content of *Thamnophis* slices is restored following incubation in 3 m*M* potassium medium, while that of *Sceloporus* slices is restored only with incubation in 40 m*M* potassium. In both cases it appears that severe depletion of tissue potassium impairs some portion of the urate transport system which can only be restored by incubation in high potassium medium. In the case of *Sceloporus,* this may also be related to a parallel inhibition of potassium transport. It does not appear that differences in potassium and urate transport are related to changes in sodium and potassium-activated adenosine triphosphatase (Na-K-ATPase) (Dantzler, 1971a). A detailed understanding of the relationship of potassium to urate transport in reptiles will require further knowledge of the steps in transport at the membrane level.

The site of urate transport in the reptilian nephron is not yet clear. However, stop-flow studies on conscious water snakes (Dantzler, 1967b) have suggested that such transport may occur throughout the nephron. In these studies, a slight peak did occur in what may have been either the late proximal or early distal tubular regions, and this peak corresponded to a peak for potassium secretion (see above). It is possible that potassium and urate secretion are linked at this point, but such a relationship awaits further study.

Preliminary studies with isolated perfused proximal renal tubules from garter snakes (*Thamnophis* sp.) indicate that urate is transported in this region and that the rate of transport increases with increasing rates of perfusion (Dantzler, 1971b). Variation of the rate of transport with the rate of flow through the tubules indicate that urate transport ceases in nephrons that stop filtering during dehydration. This agrees with the data obtained on PAH secretion in intact animals (Dantzler, 1967a) and indicates that, even in animals with a renal portal system, tubular transport cannot continue when filtration stops. The nature of the regulation of these processes awaits further study.

D. Bladder and Cloacal Function

In recent years, the function of the chelonian bladder in the regulation of ions and water has been evaluated *in vivo* and the transport processes have been studied extensively *in vitro*. Studies on intact desert tortoises (*Gopherus agassizii*) indicated that the bladder was quite permeable to water, small ions, and large molecules (Dantzler and Schmidt-Nielsen, 1966). The state of hydration of the animal had no effect on the rate of water or ion movement. Although the bladder had a tight sphincter, and ureteral urine could bypass it, ureteral urine that entered the bladder attained a composition similar to that of the body fluids except for urates, which precipitated out in the bladder. When desert tortoise bladders were studied *in vitro*, the net water movement was only about one-twentieth of that in the intact animal with a similar osmotic gradient, and larger molecules did not penetrate the bladder. The differences between the *in vivo* and *in vitro* studies can possibly be explained by the vascularization of the intact bladder. Inadequate mixing in the *in vitro* preparation may also have played a role. In any case, the differences between the *in vivo* and *in vitro* preparations suggest that the application of *in vitro* results to intact animals must be made with caution.

Nevertheless, the lack of a distinct bladder sphincter in other chelonian species, such as freshwater turtles (*Pseudemys scripta*), makes it extremely difficult to prevent ureteral urine from entering the bladder and to study the bladder *in vivo* (Dantzler and Schmidt-Nielsen, 1966). Extensive studies have been performed with *Pseudemys* bladders *in vitro*. These studies, employing both open-circuit bladders mounted in the form of sacs and short-circuited bladders, have demonstrated the active transport of both sodium and chloride from the mucosal to the serosal surface (Brodsky and Schilb, 1966; Gonzalez *et al.*, 1967a,b). Water movement across the bladder from mucosal to serosal surface is passive and is secondary to both the existing osmotic gradient and the active transport of sodium and chloride (Brodsky and Schilb, 1965). Although the bladders are less permeable to water than amphibian bladders and are unresponsive to antidiuretic hormone (Brodsky and Schilb, 1960), water reabsorption from bladder urine would continue in the intact animal with continued reabsorption of sodium and chloride. The freshwater turtle is equipped to tolerate prolonged oxygen deprivation during diving (Robin *et al.*, 1964), and sodium transport by the bladder, although reduced, can continue under anaerobic conditions (Klahr and Bricker, 1964, 1965; Gonzalez *et al.*, 1969).

The isolated turtle bladder also acidifies the solution bathing its mucosal surface, reducing the pH from 7.4 to below 4.5 (Schilb and Brodsky, 1966). Two mechanisms have been suggested to account for this acidifying effect. One involves the transport of bicarbonate ions from the mucosal to the serosal bathing solution (Schilb and Brodsky, 1966; Brodsky and Schilb, 1967). The other involves the secretion of hydrogen ions from the epithelium into the mucosal solution (Steinmetz, 1967; Green *et al.*, 1970). The evidence suggests that bicarbonate reabsorption is the primary process, but does not rule out some hydrogen ion secretion (Brodsky and Schilb, 1967). Carbonic anhydrase has now been demonstrated in the isolated freshwater turtle bladder (Scott *et al.*, 1970), and this may play a role in bicarbonate, chloride, and hydrogen ion transport. It may catalyze the hydration of carbon dioxide to carbonic acid in both the bladder epithelium and the musocal fluid, thereby helping to provide hydrogen ions in the cells for secretion into mucosal fluid as well as bicarbonate ions in the mucosal fluid for transport to the serosal side.

Whatever the method of acidification of the mucosal fluid, it would aid these animals in eliminating acid and conserving base. As noted earlier (see section on tubular regulation of hydrogen and potassium secretion), the pH of the ureteral urine of freshwater turtles is not low even on a high protein acid ash diet. In addition, acidification of urine in the bladder would cause precipitation of uric acid in the region where it could be stored and easily excreted later. Urate deposits have been found in bladders of dehydrated freshwater turtles (Baze and

Horne, 1970) and consistently in the bladders of desert tortoises (Dantzler and Schmidt-Nielsen, 1966).

Far less is known about the properties of the cloaca of other reptiles than is known about the chelonian bladder. Junqueira *et al.* (1966) have studied cloacal and distal intestinal reabsorption in snakes of the genera *Xenodon, Phylodria,* and *Crotalus.* Weight loss during dehydration was much greater in animals in which ureteral urine was not allowed to enter the cloaca than in normal controls, suggesting that water reabsorption occurred in the cloaca or distal intestines. Potential measurements were made across the intestine and cloaca both *in vivo* and *in vitro.* Short-circuit current measurements were also made *in vitro.* Potentials of about 20–60 mV, mucosal side negative to serosal side, and a short-circuit current of about 19 $\mu A/cm^2$ were found. The potential difference was rapidly abolished by the substitution of choline chloride Ringer for sodium chloride Ringer and restored when sodium chloride Ringer was again used. These data suggest that sodium is actively transported from the mucosal to serosal surface.

Some studies have also been performed on the cloacal function of saurian reptiles. Braysher and Green (1970) studied reabsorption *in vivo* from an isosmotic Ringer solution placed in the cloaca of an Australian lizard (*Varanus gouldii*). Both sodium and water were reabsorbed. Since the sodium concentration decreased, the rate of reabsorption of sodium appeared to exceed that of water, and the fluid reabsorbed was hyperosmotic to the plasma. These data suggested that water reabsorption was secondary to active sodium reabsorption. Following the administration of 100 ng of arginine vasotocin per kilogram (approximately 25 mU/kg), the rates of reabsorption of both sodium and water more than doubled and reabsorption appeared to be isosmotic. This appears to be the only demonstration thus far of an effect of a neurohypophysial hormone on sodium and water movement in the cloaca or bladder of a reptile. Although no direct study of acidification of cloacal fluid was made, Braysher and Green (1970) noted that the pH of the cloacal fluid in these lizards ranged from 5.2 to 5.8. Acidification in the cloaca as well as water reabsorption may have contributed to urate precipitation.

In contrast to these studies, Murrish and Schmidt-Nielsen (1970), using the wick method devised by Scholander *et al.* (1968), demonstrated that the colloidal osmotic pressure of the plasma proteins was sufficient to account for water reabsorption from urine in the cloaca of the desert iguana (*Dipsosaurus dorsalis*). As these authors noted, water movement due to colloid osmotic pressure does not exclude the active transport of ions with water accompanying it. And preliminary data on potentials and short-circuit current across isolated *Dipsosaurus* cloacas (J. E. Minnich, personal communication) suggests that sodium may be transported from mucosal to serosal surface. It is possible that both active sodium reabsorption and the colloid osmotic pressure of the plasma proteins play a role in determining fluid reabsorption in the cloaca. Possibly the active reabsorption of sodium is the driving force for water reabsorption, while the colloid osmotic pressure of the plasma helps regulate the rate of such reabsorption as postulated by Lewy and Windhager (1968) for the mammalian proximal renal tubule.

By collecting alternate cloacal and ureteral urine samples from crocodilians (*Crocodylus acutus*) Schmidt-Nielsen and Skadhauge (1967) demonstrated sodium, chloride, and water reabsorption from ureteral urine in the cloaca. The sodium reabsorption averaged about 7–20 μM/kg/hour regardless of the state of hydration or salt loading. Sodium reabsorption generally exceeded water reabsorption, so that the osmolality of the ureteral urine fell in the cloaca. Ammonia did not appear to be reabsorbed in the cloaca. Somewhat similar data were obtained by Bentley and Schmidt-Nielsen (1965) in the caiman *Caiman sclerops*. In addition, *in vitro* measurements of the potential across the isolated cloaca gave a value of about 10–15 mV (mucosal side negative). This potential was essentially abolished by the substitution of choline for sodium at the mucosal surface. These data are all compatible with the active reabsorption of sodium by the cloaca of crocodilians.

E. Interrelations among Glomerular, Tubular, and Bladder or Cloacal Functions

For a few chelonian, ophidian, and crocodilian species about which enough data are available, it is possible to make some generalizations about the glomerular, tubular, and bladder or cloacal roles in the regulation of the urinary excretion of ions and water. These roles appear to be related to the habitat of the animals, the tolerance of changes in plasma osmolality, and the end products of nitrogen metabolism. It should be noted that the few species considered here do not have extra-urinary routes for ion excretion. In those reptiles in which such routes exist, different modifications of renal and bladder or cloacal function may occur.

1. *Chelonian Reptiles*

Freshwater turtles, *Pseudemys scripta,* live a semiaquatic amphibious life and are primarily aminoureotelic. They lose water rapidly from the skin and respiratory tract compared with terrestrial turtles (Table II) (Schmidt-Nielsen and Bentley, 1966) and cannot tolerate great increases in plasma osmolality during dehydration (Dantzler and Schmidt-Nielsen, 1966). They can easily excrete excess water through an increased filtration rate and production of a dilute urine. Dehydration leads first to a decrease in free-water clearance, second to reduction in filtration rate, and finally to complete renal shut down when plasma osmolality has increased 20 mOsm. Although the permeability of the turtle bladder is not changed by antidiuretic hormone, reabsorption of sodium and chloride will cause a slow reabsorption of bladder urine which becomes important when urine flow becomes low or ceases. Since most of the excretory nitrogen is in the form of soluble urea and ammonia, continued production of urine at low rates would accomplish little, since the urinary constituents would be reabsorbed anyway.

Desert tortoises, *Gopherus agassizii,* which live an arid, terrestrial life, respond in a different manner which appears to be related in part to the fact that a large fraction of their excretory nitrogen is in the form of urates. They lose water slowly from the skin and respiratory tract (Table II) (Schmidt-Nielsen and Bentley, 1966) and can tolerate increases in plasma osmolality of 200–300 mOsm during dehydration (Dantzler and Schmidt-Nielsen, 1966; J. E. Minnich, personal communication). The ability of desert reptiles to survive and, indeed, function well with marked increases in plasma osmolality was first clearly demonstrated by Bentley (1959). It should be noted that most of the increase in plasma osmolality in the desert tortoise resulted from an increase in the concentration of urea and that vertebrate tissues are less sensitive to high concentrations of urea than to high concentrations of electrolytes (Thesleff and Schmidt-Nielsen, 1962; McClanahan, 1964). Although freshwater turtles are predominantly ureotelic, a marked increase in plasma urea concentration occurred only shortly before death from dehydration.

Renal function does not cease or even change significantly as a result of a moderate increase in plasma osmolality. Tubular reabsorption of water does not increase, and urine remains hypoosmotic to the plasma. This reduces the problem of precipitation of urates in the renal tubules. In the bladder, urates precipitate out and the urine equilibrates with the blood. This type of excretion leads to temporary storage of urates in the bladder. Thus, in the predominantly uricotelic desert tortoise, the continued kidney function during moderate dehydration helps to get rid of excretory nitrogen.

The high permeability of the desert tortoise bladder indicates that all urine entering the bladder would equilibrate with the blood and only urate would be excreted. However, in well hydrated animals, the urine by-passed the bladder and passed directly out the cloaca. This apparently resulted from the closing of the bladder sphincter. How this sphincter is regulated is unknown.

2. *Ophidian Reptiles*

Although all ophidian reptiles are uricotelic, the glomerular and tubular functions of freshwater snakes, *Natrix sipedon,* resemble those of freshwater turtles, while the glomerular and tubular functions of desert snakes, *Pituophis melanoleucus,* resemble those of desert tortoises. The desert dwellers lose water less rapidly from the skin and respiratory tract during dehydration than the freshwater forms (Table II) (Prange and Schmidt-Nielsen, 1969), but no comparable data are available on tolerance of increases in plasma osmolality. The differences in renal function may be related to as yet undetermined differences in cloacal reabsorption of sodium and water. If reabsorption from the distal intestine or cloaca in freshwater snakes were only hyperosmotic even during dehydration, it would create a problem in solute excretion and water conservation in animals without any extraurinary route for solute excretion. Under these circumstances, storage of urate in the distal intestine might not be of sufficient importance for filtration to continue.

It is possible that the function of the cloaca or distal intestine in desert-dwelling snakes more nearly resembles bladder function in the desert tortoise than cloacal function in freshwater snakes. If water is reabsorbed passively from a dilute urine in this region as urate is precipitated, it may be of benefit to these animals to continue producing hypoosmotic urine during dehydration.

3. *Crocodilian Reptiles*

The crocodile (*Crocodylus acutus*) has a semiaquatic, amphibious mode of existence similar to that of freshwater turtles and snakes. However, it is ammonouricotelic. The water losses across the skin and respiratory tract have not been measured but are probably similar to those of the caiman (Table II) (Bentley and Schmidt-Nielsen, 1966). These losses are higher, on both a weight and surface area basis, than those for other reptiles. Although the filtration rate was high, it did not show the variation with water loading and salt loading found in freshwater turtles and snakes. Tubular regulation of water tended to resemble that of desert ophidians. The osmolar U/P ratio could be as low as 0.5 (less dilute than freshwater turtles and snakes) and as high as 0.95, but it averaged 0.8 under all conditions of water loading and salt loading. As suggested by Schmidt-Nielsen and Skadhauge (1967), however, the adaptations of this reptile to fresh water may be related to its excretion of ammonia and its cloacal function. The excretion of ammonia and bicarbonate as the principle ions in ureteral urine may allow sodium, potassium, and chloride to be conserved by a tubule that cannot reduce the urine osmolality to a low level. Furthermore, this alkaline urine would aid in maintaining uric acid in solution. In the cloaca, sodium and chloride can be reabsorbed in excess of water, further diluting the urine and conserving these ions. The excretion of ammonium and bicarbonate as the principal ions while sodium and chloride are conserved is equivalent to the production of a more dilute urine, since it permits the excretion of water in excess of sodium.

II. Extrarenal Excretory System

A. Introduction

Since the reptilian nephron is unable to elaborate a hyperosmotic urine, the animal must take in an appropriate amount of osmotically free water each time osmotically active materials are ingested. When fresh water is freely available, the animals experience no difficulty maintaining a constant *milieu interieur*. However, since the kidneys of reptiles cannot conserve excess osmotically free water, species living in arid or marine environments would rapidly enter a state of negative water balance if they had to rely solely on the renal excretory pathway. In these species, the presence of an extrarenal organ capable of excreting excess electrolytes and, at the same time, retaining osmotically free water is important for their survival.

The evolution of extrarenal excretory organs may have occurred separately in each of the three major groups of extant reptiles. The chelonian reptiles (the turtles), which represent a very early offshoot of the stem reptiles, have paired extrarenal excretory organs which are situated posteriorly in the orbits of the eyes. These glands probably represent modified lachrymal glands, but frequently they are referred to more specifically as Harderian glands. A short, stout duct connects each gland to the exterior via an opening in the posterior corner of the eye. Cannulation and direct collection of the fluid from this duct in several species of marine turtles has shown that the fluid contains high concentrations of sodium and potassium (Schmidt-Nielsen and Fänge, 1958; Holmes *et al.*, 1963).

The squamate reptiles (the lizards and snakes) represent much later derivatives from the early reptilian stock. In the lizards, there seems to be little doubt that the extrarenal excretory organ has been developed from the external nasal gland (=dorsal, lateral, or superior nasal gland). This gland lies dorsally or laterally to the nasal cavity and the duct enters the vestibulum, typically posterodorsally, near the anterior end of the nasal cavity (Parsons, 1970). Templeton (1964b), by cannnulation of the nasal gland duct in the false iguana (*Ctenosaura pectinata*) and the chuckwalla (*Sauromalus obesus*), has demonstrated that the gland is indeed the source of a hypertonic extrarenal excretory fluid. Encrustations of salt crystals and samples of fluid collected from the nostrils of several species of terrestrial iguanas and the marine iguana (*Amblyrhynchos cristatus*) have also been shown to contain high concentrations of sodium and potassium (Schmidt-Nielsen and Fänge, 1958; Schmidt-Nielsen *et al.*, 1963; Norris and Dawson, 1964; Dunson, 1969a).

Only two species of snakes, both marine, have been shown to possess extrarenal excretory mechanisms. Uncontaminated extrarenal fluid has not been collected directly from either species, and the precise location of the organ is presently in doubt. The banded sea snake (*Laticauda semifasciata*) has a well developed nasal gland, but the connection with the oral cavity is very indirect and no fluid could be collected from the nasal vestibule of the salt-loaded snake (Taub and Dunson, 1967). The yellow-bellied sea snake (*Pelamis platurus*) possesses no grossly distinguishable nasal gland (Taub and Dunson, 1967). However, both *Laticauda* and *Pelamis* have well developed Harderian glands with ducts opening into the anterior oral cavity (Dunson and Taub, 1967; Taub and Dunson, 1967; Dunson, 1968). Also, salt-loaded sea snakes were observed to deposit drops of fluid onto the substrate whenever the tongue was extended. Closer examination revealed that the fluid was pushed out between the rostral and mental scales by the tongue (Dunson and Taub, 1967; Dunson, 1968). The ion concentrations in the fluid collected from *Laticauda* ranged from 380 to 625 m*M* sodium and from 16.0 to 25.3 m*M* potassium (Dunson and Taub, 1967) and similar ranges of values (346–776 m*M* sodium and 12–37 m*M* potassium) were observed in fluid from *Pelamis* (Dunson, 1968). In both species the $Na^+:K^+$ ratio of this fluid was significantly less than that of the seawater (42.7).

Thus, the evidence, albeit scant and circumstantial, suggests that the extrarenal excretory organ in snakes has been developed from a lachrymal gland. This is in marked contrast to the lizards, where the nasal gland has developed this function. Furthermore, among the squamate reptiles two distinct patterns of extrarenal excretion appear to have evolved. In the lizard, where the nasal glands seem to have been adapted to this function, the extrarenal excretory fluid contains relatively high concentrations of potassium and the $Na^+:K^+$ ratio of the fluid has not been observed to exceed 7.0 (Table V). On the other hand, the extrarenal organs in the sea snakes seem to be associated with glands

TABLE V
CONCENTRATIONS OF ELECTROLYTES IN REPTILIAN SALT GLAND FLUID

Animal	Habitat	Collection conditions	Ion concentrations (m*M*)			$Na^+:K^+$ ratio	Reference
			Na^+	K^+	Cl^-		
Chelonia							
Ridley's turtle (*Lepidochelys olivacea*)	Marine	100 miles from land; no salt load	713	28.8	782	25.0	Dunson, 1969b
Green turtle (*Chelonia mydas*)	Marine	Seawater load	685	20.7	—	37.7	Holmes and McBean, 1964
Loggerhead turtle (*Caretta caretta*)	Marine	Seawater load	909	18.0	—	50.5	Holmes *et al.*, 1963
		NaCl load plus methacholine	732–878	18.0–31.0	810–992	28.3–40.7	Schmidt-Nielsen and Fange, 1958
Diamondback terrapin (*Malaceolemys terrapin*)	Marine	NaCl load	616–784	—	—	—	Schmidt-Nielsen and Fange, 1958
		Seawater acclimated; NaCl load or methacholine injected	682	32.4	—	24.0	Dunson, 1970
	Fresh-water	Freshwater acclimated; methacholine injected	288	—	—	—	Dunson, 1970
Squamata							
Marine iguana (*Amblyrhynchos cristatus*)	Marine	No salt load	1434	235	1256	6.7	Dunson, 1969a
Land iguana (*Conolophus subcristatus*)	Terrestrial	No salt load	692	214	486	3.4	Dunson, 1969a

Green iguana (*Iguana iguana*)	Terrestrial	0–4 hours after NaCl load	507	497	—	1.02	Schmidt-Nielsen *et al.*, 1963
		2 days after NaCl load	728	290	—	2.5	Schmidt-Nielsen *et al.*, 1963
Desert iguana (*Dipsosaurus dorsalis*)	Terrestrial	0–2 hours after NaCl load	494	1387	—	0.36	Schmidt-Nielsen *et al.*, 1963
		2 days after NaCl load	1032	640	—	1.6	Schmidt-Nielsen *et al.*, 1963
Spiney-tailed agamid (*Uromastyx aegyptius*)	Terrestrial	NaCl load	639	1398	—	0.46	Schmidt-Nielsen *et al.*, 1963
Chuckwalla (*Sauromalus obesus*)	Terrestrial	NaCl load	121	379	—	0.32	Templeton, 1964b
		No salt load	—	—	—	0.50	Norris and Dawson, 1964
False iguana (*Ctenosaura pectinata*)	Terrestrial	NaCl load	75	302	—	—	Templeton, 1964b
		NaCl load; <3 month captivity	78	527	487	0.15	Templeton, 1967
		NaCl load; >3 month captivity	439	253	477	1.7	Templeton, 1967
Desert iguana (*Dipsosaurus dorsalis*)	Terrestrial	NaCl load	—	—	—	0.67[a]	Templeton, 1966
		KCl load	—	—	—	0.11[a]	Templeton, 1966
Blue spiney lizard (*Sceloporus cyanogenys*)	Terrestrial	NaCl load	—	—	—	0.11[a]	Templeton, 1966
		KCl load	—	—	—	0.10[a]	Templeton, 1966
Yellow-bellied sea snake (*Pelamis platurus*)	Marine	NaCl load	607	28.0	627	23.6	Dunson, 1968
Banded sea snake (*Laticauda semifasciata*)	Marine	NaCl load	380–625	16.0–25.3	—	25.0–34.8	Dunson and Taub, 1967

TABLE V (*Continued*)

Animal	Habitat	Collection conditions	Ion concentrations (m*M*) Na$^+$	K$^+$	Cl$^+$	Na$^+$:K$^-$ ratio	Reference
Crocodilia							
American crocodile (*Crocodylus acutus*)	Brackish	1.6 weeks 50% seawater	274	10.0	—	27.4	Dunson, 1970
		5.6 weeks 50% seawater	279	7.0	—	40.0	Dunson, 1970
		14 weeks 100% seawater	232	7.0	—	31.0	Dunson, 1970
		20.5 weeks 100% seawater	251	8.0	—	31.0	Dunson, 1970
		20.5 weeks 100% seawater; methacholine injected	266	14.0	—	19.0	Dunson, 1970
		1 week fresh water; methacholine injected	207	9.0	—	20.0	Dunson, 1970
American crocodile (*Crocodylus acutus*) and Asian saltwater crocodile (*Crocodylus porosus*)—mixed population	Brackish	Freshwater acclimated[b]; methacholine injected (11) or NaCl load (1)	245	11.5	—	22.4	Dunson, 1970
		Seawater acclimated[b]; methacholine injected (2), NaCl load (7), or no load (1)	343	22.7	—	18.4	Dunson, 1970

[a] Mean value for collections over periods of 21 days.

[b] Numerals in parenthesis indicate the number of individuals that received treatment.

in the orbit and they secrete a fluid which contains relatively low concentrations of potassium and has a $Na^+:K^+$ ratio which is always greater than 20.0 (Table V). This latter pattern of ion excretion is similar to that of the extrarenal organs in chelonian reptiles, another group of reptiles where the extrarenal excretory organ has probably developed from a lachrymal gland.

The third major evolutionary line of reptiles, the crocodilia, also developed much later than the chelonian reptiles. They have many features in common with the birds, and the modern representatives, the alligators and crocodiles, are considered to be secondarily aquatic. The Asian crocodile (*Crocodylus porosus*) and the American crocodile (*C. acutus*) are found in estuaries, and these species, as well as *C. siamensis* and *Osteolaemus tetraspis,* have been reported in offshore waters (Neill, 1958). None of the alligators has been definitely identified as marine. After salt loading or methacholine treatment, extrarenal secretions have been collected from *C. porosus* and *C. acutus* (Dunson, 1970). Compared with other reptilian groups, the concentrations of sodium and potassium in the extrarenal fluid are low. The $Na^+:K^+$ ratios of the fluids, however, are similar to those in the orbital gland secretions from marine turtles and snakes and are in marked contrast to the $Na^+:K^+$ ratios found in the external nasal gland secretions of lizards (Table V). Dunson (1970) has concluded that the evidence for specialized salt secretion in crocodiles is poor and that the fluid collected in his experiments probably represented the secretions of unspecialized lachrymal glands. There are no data on the composition of unaltered lachrymal secretion in reptiles but the $Na^+:K^+$ ratio of the fluid collected by Dunson was several times greater than that in tears from mammalian lachrymal glands (4.5–10.0).

There is little information on the homologies of cephalic glands in reptiles, birds, and mammals, and therefore a discussion of the evolutionary development of extrarenal excretory organs in these classes of vertebrates is difficult. Furthermore, it is clear that even among the reptiles, meaningful names cannot be assigned at this time to the extrarenal excretory organs that have been identified. Throughout the remainder of the text, therefore, we shall refer to all extrarenal excretory organs in reptiles as "salt glands."

Among the reptiles, the salt glands function in species where the supply of osmotically free water is restricted and the electrolyte intake is relatively great. These conditions exist for animals living in marine or arid terrestrial environment. In all cases, however, the extrarenal excretory system is complementary to the renal excretory pathway, and the two systems function in concert to maintain homeostasis. We shall therefore consider the extrarenal excretory processes in the representatives of each group of reptiles living in these two types of environment.

B. The Marine Environment

Except for the brief period of egg laying, the following five species of turtles are strictly marine: the green turtle (*Chelonia mydas*), the loggerhead turtle (*Caretta caretta*), Ridley's turtle (*Lepidochelys olivacea*), the hawksbill turtle (*Eretmochelys imbricata*), and the leatherback turtle (*Dermochelys coriacea*). In each of these species, the salt gland is located in the orbit. The gland consists of branching tubules radiating from central ducts that open to the exterior by a short duct in the posterior corner of the eye. Although the Chelonian salt glands seem to be relatively smaller than the supraorbital glands of marine birds, they are nevertheless substantial organs. For example, in a loggerhead turtle weighing 130 kg, the combined salt gland weights were 60 gm (Schmidt-Nielsen and Fänge, 1958), and in a smaller specimen weighing 42.3 kg, the combined gland weights were 28.7 gm, i.e., 67.9 mg/100 gm body weight (Holmes *et al.*, 1963). It has also been reported (Holmes *et al.*, 1963) that in two specimens of Ridley's turtle, having a mean body weight of 9.1 kg, the salt glands weighed 7.03 gm (67.0 mg/100 gm body weight) and that the mean salt gland weight of four green turtles weighing 40.2 ± 1.7 kg was 18.2 ± 1.3 gm (45.3 ± 0.9 mg/100 gm body weight).

There is some evidence that, at least for periods up to 2 months, the juvenile green turtle can tolerate a freshwater environment. During this time only small changes occur in the sodium and potassium concentrations in plasma, and the size of the salt gland varies significantly with the environment. After 2 months in fresh water, the salt glands were 38% less in absolute weight and 44% less in relative weights than those of animals kept in seawater (Table VI).

In an attempt to quantitatively study the role of the salt gland in the control of electrolyte balance, a series of juvenile green turtles were given saline loads. During the 6-hour period following the administration of sodium chloride, all of the injected sodium was excreted (Table VII). This excretory rate represented the sum of the renal and extrarenal excretory rates. However, when measurements are made for only short periods, occlusion of the cloaca satisfactorily separates the renal component from the total excretion of electrolytes. Thus, when the sodium and potassium excretory rates of intact turtles and turtles with occluded cloacae were compared following the administration of a saline load, no significant differences were observed (Table VII). It may be de-

TABLE VI

PLASMA AND URINE ELECTROLYTE COMPOSITION AND THE SALT GLAND WEIGHTS OF JUVENILE GREEN TURTLES (*Chelonia mydas mydas*)[a]

Group	Number of turtles	Body weight (gm)	Absolute salt gland weight (mg)	Relative salt gland weight (mg/100 gm body weight)	Plasma electrolyte concentration (mM/liter)	
					Na^+	K^+
Sea water (0–6 months)	9	52.9 ± 4.9	158.4 ± 14.4	303.2 ± 11.6	157.8 ± 1.4	1.48 ± 0.03
Fresh water (4–6 months)	8	59.1 ± 7.2	98.0 ± 16.1[b]	170.4 ± 21.2[c]	130.2 ± 1.0[c]	2.58 ± 0.01[c]

[a] Maintained in sea water from birth to 6 months of age and in fresh water from 4 months after hatching to 6 months of age. (Holmes and McBean, 1964.

[b] $P < 0.02$.

[c] $P < 0.001$, significance of values with respect to the corresponding value for seawater turtles.

TABLE VII

SODIUM AND POTASSIUM EXCRETION BY THE SALINE-LOADED JUVENILE GREEN TURTLE (*Chelonia mydas mydas*)[a]

Treatment	Number of turtles	Body weight (gm)	Initial rate of excretion (μM/hour)		Terminal rate of excretion (μM/hour)		Total excretion (μM)		
			Na^+	K^+	Na^+	K^+	Na^+	K^+	$Na^+:K^+$
Intact	10	103.3 ± 1.7	117.0 ± 16.7	7.83 ± 1.32	63.4 ± 13.1	2.42 ± 0.39	503.0 ± 32.2	22.49 ± 1.35	23.3 ± 1.2
Controls with occluded cloacae	8	114.4 ± 7.2	118.9 ± 16.1	7.29 ± 0.75	98.8 ± 20.7	2.98 ± 0.40	601.9 ± 78.0	22.95 ± 0.93	25.5 ± 2.9
Amphenone	8	102.5 ± 6.3	51.6 ± 9.1[c]	2.75 ± 1.04[c]	37.5 ± 13.1	1.88 ± 0.46	234.6 ± 59.0[c]	14.35 ± 2.06[c]	14.5 ± 2.2[c]
Amphenone plus corticosterone	10	117.8 ± 5.7	298.6 ± 26.9[d]	6.18 ± 1.51	12.2 ± 13.3[b]	0.93 ± 0.42[b]	409.0 ± 44.8	13.55 ± 1.27[c]	30.1 ± 2.1

[a] Each animal received a single intramuscular injection of 500 μM sodium as sodium chloride solution. Amphenone B (5 mg) and corticosterone (2.5 mg) were administered at the time of saline loading. Excretion was measured over a 6 hour period. (Holmes and McBean, 1964.)

[b] $P < 0.02$.

[c] $P < 0.01$.

[d] $P < 0.001$, significance with respect to corresponding value for intact control turtles.

duced, therefore, that under these experimental conditions, the extrarenal organs were the major pathway of sodium and potassium excretion. When amphenone B was administered to the turtles 1 hour prior to saline loading, the amounts of sodium and potassium excreted during the next 6 hours were reduced significantly (Table VII). But, when corticosterone was injected along with amphenone B, the pattern of sodium excretion was similar to that of the untreated intact control turtles; the excretion of potassium in these turtles, however, was not restored to normal (Table VII). In reptiles as well as mammals, amphenone B inhibits adrenal steroidogenesis (Hertz *et al.*, 1955; Rosenfeld and Bascom, 1956; Phillips *et al.*, 1962), and although some steroids may have been synthesized, a relative adrenal insufficiency probably occurred in the amphenone B-treated turtles. These studies suggest that extrarenal excretion of sodium in the salt-loaded turtle is influenced by adrenocortical hormones.

Under natural conditions, the salt glands of marine turtles would function intermittently, presumably under conditions of surplus electrolyte intake. This would occur most commonly in response to the ingestion of food and perhaps seawater. Indeed, there are marked differences between the patterns of sodium and potassium excretion from fed and unfed juvenile turtles (Table VIII). Although only 125 μM sodium was ingested by the fed turtles, they excreted 672 μM sodium during the 5 hour collection period. At the same time, the food contained 212 μM potassium, and yet only 24.6 μM potassium were excreted. In other words, at the end of the 5 hour experimental period, only 11.6% of the ingested potassium had been excreted whereas 528 μM sodium in excess of that ingested had been excreted. This quantity of sodium would be equivalent to drinking 0.5 ml seawater per gram of food eaten. The amount of sodium excreted and the $Na^{+}:K^{+}$ ratio of the ions excreted suggest that the extrarenal pathway was largely responsible for the output of ingested sodium (Table VIII).

Whichever pathway is involved, however, the significance of the high rate of sodium excretion and the concomitant low rate of potassium excretion is obscure. Using larger specimens of *C. mydas* substantial quantities of salt gland fluid can be collected. This fluid contains 684.8 ± 55.5 mM sodium per liter and 20.7 ± 3.6 mM potassium per liter, and repeated collections from the same individual have not revealed significant changes with time in these concentrations (Holmes and McBean, 1964). If we consider a turtle drinking 1 liter of seawater (470 mM sodium and 10 mM potassium), then at the above salt gland fluid concentrations all of the sodium contained in this seawater could be excreted in 686 ml of salt gland fluid. The excretion of all the potassium in

TABLE VIII

SODIUM AND POTASSIUM EXCRETION BY FED AND UNFED GREEN TURTLES (*Chelonia mydas mydas*) OVER A 5 HOUR PERIOD[a]

Treatment	Number of turtles	Body weight (gm)	Initial rate of excretion (μM/hour/100 gm body weight)		Terminal rate of excretion (μM/hour/100 gm body weight)		Total excretion (μM/100 gm body weight		
			Na^+	K^+	Na^+	K^+	Na^+	K^+	$Na^+:K^+$
Unfed	11	93.9 ± 2.3	47.4 ± 4.8	6.45 ± 1.04	4.47 ± 2.34	1.93 ± 0.38	71.1 ± 5.3	13.18 ± 0.81	5.7 ± 0.6
Fed	9	68.1 ± 3.5	130.6 ± 17.6[c]	7.65 ± 1.50	144.8 ± 40.2[c]	4.62 ± 0.91[b]	672.1 ± 88.6[c]	24.62 ± 1.94[c]	27.1 ± 2.2[c]
Fed plus amphenone	9	81.4 ± 6.3	74.9 ± 13.0	6.93 ± 1.75	9.5 ± 11.5	1.50 ± 0.71	127.8 ± 27.9	14.19 ± 1.73	8.5 ± 0.8[b]

[a] The fed animals ate 2.3 ± 0.25 gm shrimp per 100 gm body weight, of the following composition: 788.9 ± 5.8 gm water per kg wet weight, 54.4 ± 1.3 meq Na^+ per kg wet weight and 92.1 ± 3.9 meq K^+ per kg wet weight. Amphenone B was administered as a single intramuscular dose (5 mg) immediately after feeding. (Holmes and McBean, 1964.)

[b] $P < 0.02$.

[c] $P < 0.001$, significance of values with respect to the corresponding value for unfed turtles.

the seawater, however, would require only 483 ml of fluid. The turtle would therefore produce 203 ml of salt gland fluid for which no potassium would be available from the ingested seawater. An examination of sodium and potassium contents of organisms likely to be eaten by the green turtle reveals that they contain high concentrations of potassium and, in contrast to the salt gland fluids, have very low $Na^+:K^+$ ratios (Table IX). It is possible, therefore, that the turtle drinks seawater not only to gain osmotically free water, but also to gain sodium as a "vehicle" for the extrarenal excretion of the potassium contained in the food.

In the group of unfed turtles, much smaller quantities of sodium and potassium were excreted and the $Na^+:K^+$ ratio of the ions excreted was low (Table VIII). Samples of urine drained from the bladders of seawater-maintained juvenile turtles contained 46.6 ± 2.8 m*M* sodium per liter and 7.18 m*M* potassium per liter ($Na^+:K^+$ ratio 6.6 ± 0.5). It is quite possible, therefore, that the unfed turtles were under the conditions of relatively small electrolyte loads and that the kidneys were quite capable of maintaining the electrolyte balance.

Furthermore, it would seem that the extrarenal excretory mechanism comes into play not only under conditions of electrolyte load, but only when the circulating levels of adrenal steroids are adequate. This was indicated by the reduction in the rates of sodium and potassium excretion and the lower $Na^+:K^+$ ratio of the excreted ions when fed turtles were injected with amphenone B (Table VIII).

Some recent studies on two species of marine snakes have established the presence of important extrarenal excretory organs. In *Laticauda*, salt gland secretions were elicited when saline loads of 5 mmoles or more per 100 gm body weight were injected subcutaneously (Fig. 4). An osmotically equivalent load of sucrose also stimulated the extrarenal secretion of sodium, potassium, and chloride, and these ions were secreted in ratios similar to those found in the salt gland secretions of snakes given saline loads (Dunson and Taub, 1967). Thus, a nonspecific osmoreceptor or a "volume" receptor may initiate the onset of salt gland secretion. The limited data available are quite variable, but after saline loading, the salt gland appears to be the dominant pathway for the excretion of ingested sodium, whereas the excretion of potassium seems to be divided equally between the renal and the extrarenal pathways. The relative dominance of the extrarenal and renal pathways with respect to the excretion of sodium and potassium in this species are therefore qualitatively similar to the relative rolès of these systems in chelonian reptiles.

In another species of sea snake (*Pelamis platurus*), the extrarenal

TABLE IX

WATER, SODIUM, AND POTASSIUM COMPOSITION OF VARIOUS FOODS EATEN BY MARINE REPTILES AND BIRDS[a]

Group	Body weight or sample weight (gm)	Water (%)	Electrolytes (meq/kg tissue water)		
			Na^+	K^+	$Na^+:K^+$
Teleosts (6 species)	3.79 ± 0.47	76.77 ± 0.33	75.6 ± 3.0	112.8 ± 4.6	0.68 ± 0.03
Crustaceans (4 species)	1.77 ± 0.23	72.29 ± 0.50	195.9 ± 6.9	116.2 ± 2.6	1.88 ± 0.11
Mollusks (1 species)	7.47 ± 0.46	81.65 ± 0.48	167.4 ± 6.1	104.3 ± 3.4	1.60 ± 0.07
Algae (5 species)	2.44 ± 0.20	86.43 ± 0.49	194.1 ± 8.6	220.9 ± 4.4	0.88 ± 0.17

[a] Adapted from Holmes and McBean, 1964.

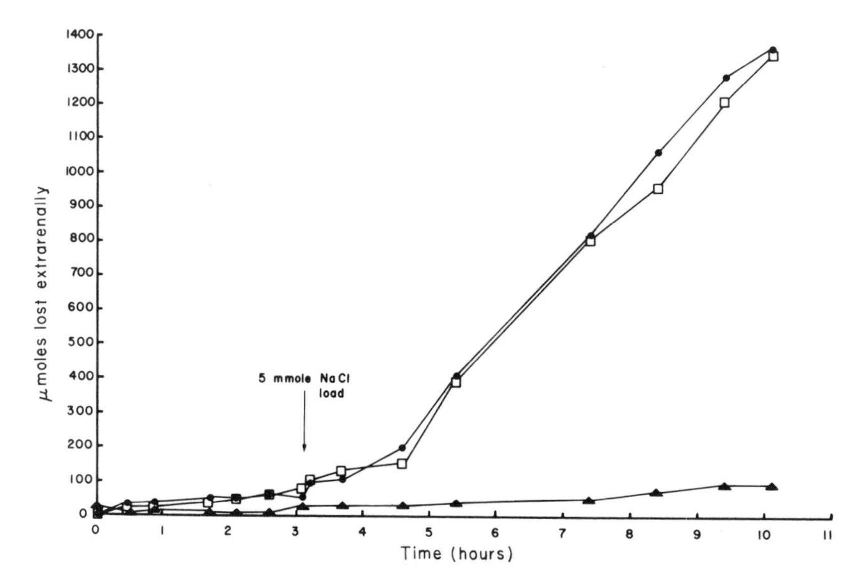

FIG. 4. The effect of subcutaneous salt-loading (5 mmoles sodium chloride) on the extrarenal excretion of electrolytes by a specimen of the banded sea snake, *Laticauda semifasciata;* ● = sodium, ▲ = potassium, □ = chloride (Dunson and Taub, 1967).

excretory mechanism seems to be developed more highly than in *Laticauda.* In contrast to *Laticauda,* whose habit is littoral and may occasionally be terrestrial (Smith, 1926), *Pelamis* is the species most widely distributed and best adapted to life in the sea. The maximum extrarenal sodium and chloride excretory rate observed in *Pelamis* were 200 μmoles/100 gm body weight/hour, and the corresponding rate of potassium excretion ranged from 1 to 9 μmoles/100 gm body weight/hour (Dunson, 1968). These rates of extrarenal excretion were approximately three times the values recorded in *Laticauda* (Dunson and Taub, 1967). The maximum rates of extrarenal sodium and chloride excretion observed in *Pelamis* were directly proportional in each case to the load of sodium chloride administered (Fig. 5). Furthermore, all of the chloride injected into *Pelamis* was excreted extrarenally, and the duration of extrarenal response was proportional to the amount of saline injected (Fig. 6). In *Pelamis* also, the extrarenal response could be elicited by loading the animal with sucrose, but compared with a solution of sodium chloride, twice the molar concentration was required to cause the same rate of extrarenal electrolyte secretion (Dunson, 1968).

When given saline loads or injected with methacholine chloride two species of crocodiles have been shown to secrete a salt gland fluid with a $Na^+:K^+$ ratio comparable to that secreted by turtles and sea snakes

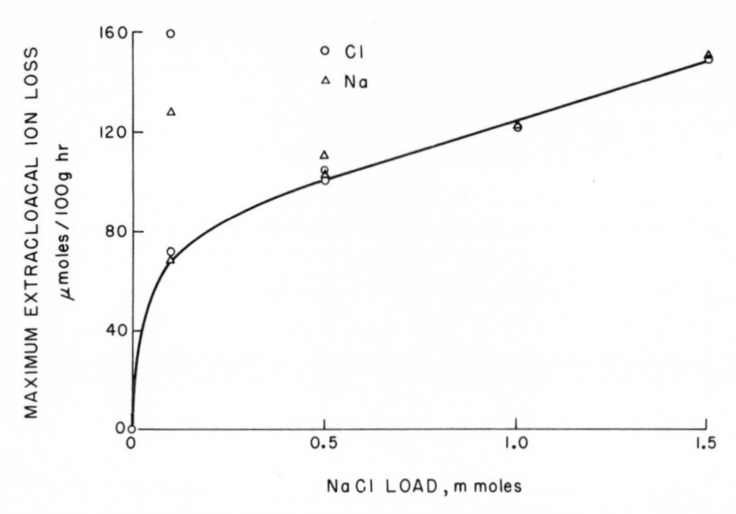

FIG. 5. The relationship between the size of the salt load and the maximum rate of extrarenal sodium and chloride excretion in yellow-bellied sea snakes, *Pelamis platurus* (Dunson, 1968).

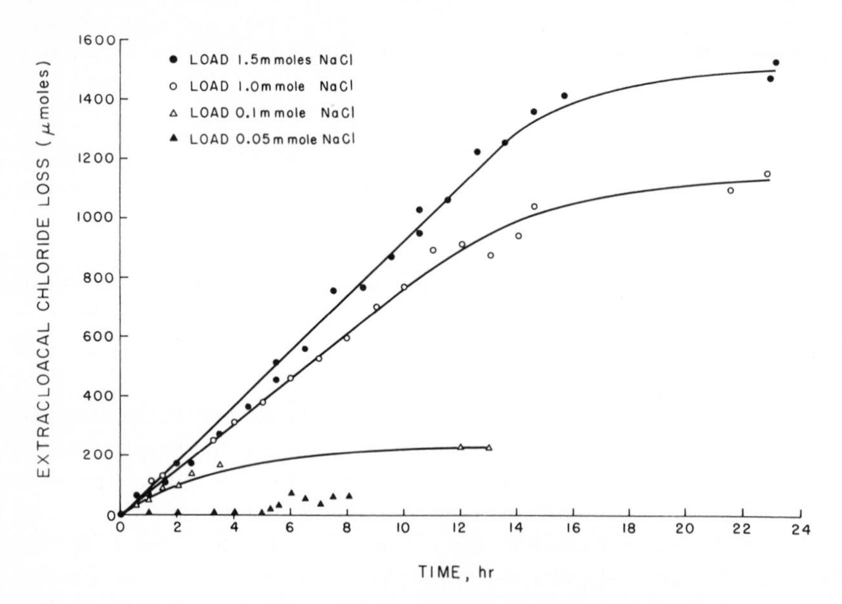

FIG. 6. The effect of different loads of sodium chloride on the duration of the extrarenal excretory response and the amount of chloride excreted by the yellow-bellied sea snake, *Pelamis platurus* (Dunson, 1968).

(Table V). No data are available on the rates of salt gland secretion by either *C. acutus* or *C. porosus*. The juvenile American crocodile (*C. acutus*), judged by changes in body weight, does not appear to be able to withstand immersion in 100% seawater. The five individuals studied by Dunson (1970) showed mean losses in body weight equal to 7% of their initial weights when immersed for 4 days in 100% seawater. The mean initial body weight was restored rapidly when the animals were returned to fresh water, and when maintained subsequently in 25% or 50% seawater for 6 days, no changes in mean body weight occurred. A larger (3.4 kg) American crocodile, from which salt gland secretions were collected, did survive immersion for 5 months in 100% seawater. From these preliminary studies Dunson (1970) has tentatively concluded that although the crocodile may tolerate exposure to a hypertonic seawater medium, the salt glands do not constitute a major pathway of electrolyte excretion.

Among the lizards, only the Galapagos iguana (*Amblyrhynchus cristatus*) is known to inhabit the marine environment. The salt gland fluid from this species contains very high concentrations of electrolytes (Table V), and immediately after capture, the rate of extrarenal excretion has been observed to be quite variable (Dunson, 1969a). This variability between individuals is probably related to the elapsed time between feeding and capture (Table X). The highest rates of sodium, potassium, and chloride secretion in animals presumed to have received a natural salt load during feeding were comparable to the rates recorded after a single subcutaneous injection of 3.56 m*M* sodium chloride (Table X). Indeed, these extrarenal excretory rates were the highest so far recorded in reptiles (Table XI). A typical pattern of extrarenal excretion in response to the presumed intake of food is shown in Fig. 7.

Although the duration of the extrarenal response and the total amount of sodium chloride excreted was proportional to the size of the sodium chloride load administered, no unequivocal relationship was observed between the load of sodium chloride and the initial rate of extrarenal sodium excretion (Table X).

In sharp contrast to other marine reptiles, the extrarenal excretory fluid of the marine iguana is relatively rich in potassium (Table V). The low Na^+:K^+ ratio of the fluid may be characteristic of secretions from the external nasal glands, but it also may have profound adaptive significance when the ion content of the food of the marine iguana is examined. The exact diet of the marine iguana has not been determined, but reports of feeding habits and stomach contents suggest that the diet may be primarily composed of littoral and benthic algae, with the occasional ingestion of crustaceans. Both algae and crustaceans contain relatively high potassium concentrations and have low Na^+:K^+ ratios (Table IX). The secretion of an extrarenal fluid with similarly low Na^+:K^+ ratios, therefore, would favor the rapid excretion of the excess electrolytes in food.

TABLE X

INITIAL RATES OF EXTRARENAL EXCRETION AND TOTAL EXCRETION OF SODIUM CHLORIDE BY THE NASAL GLANDS OF THE MARINE IGUANA (*Amblyrhynchus cristatus*)[a]

Iguana number	Body weight (gm)	Load (m*M* NaCl)	Duration of secretion (hours)	Total NaCl excreted (mmoles)	Initial rates of excretion (μmoles/100 gm/hour)			
					Na^+	K^+	Cl^+	$Na^+:K^+$
18a	119	0[b]	23	3.70	255	50.7	237	5.0
18b	119	0.5	12	0.45	159	7.6	167	20.9
18c	119	1.0	17	1.50	177	25.1	161	7.0
19a	80	0	8	0.18	23	3.1	24	7.4
19b[c]	80	1.0	14	0.80	209	23.0	164	9.1
20	335	0	16	9.50	220	25.7	202	8.6
21a	210	0	8.5	0.46	10	0.7	8	14.3
21b	210	1.0	—	—	82	7.5	62	10.9
21c	210	2.0	12	2.0	146	23.0	149	6.3
23a	356	0	25.5	2.8	30	4.8	27	6.3
23b	356	1.78	22	3.0	100	10.7	78	9.3
23c	356	2.67	22.5	1.8	84	8.8	76	9.5
23d	356	3.56	20	5.6	269	30.0	225	9.0

[a] Immediately after capture and after subcutaneous loading with sodium chloride. (Dunson, 1969a)

[b] No subcutaneous load of sodium chloride administered. Secretions presumed to be due to electrolyte load ingested at last feeding.

[c] Cloacal and extrarenal excretion combined.

TABLE XI

MAXIMUM RATES OF SALT GLAND SECRETION IN REPTILES

Animal	Body weight (gm)	Extrarenal secretory rate (μmoles/100 gm/hour)			Na⁺:K⁺ ratio	Reference
		Na^+	K^+	Cl^-		
False iguana (*Ctenosaura*)	1000	0.96	9.4	—	0.102	Templeton, 1964b
Chuckwalla (*Sauromalus*)	170	3.25	31.0	—	0.105	Templeton, 1964b
Land iguana (*Conolophus*)	428	15.9; 25.5[a]	5.0; 1.6[a]	8.4; 18.4[a]	3.2; 15.9[a]	Dunson, 1969a
Diamondback Terrapin (*Malaclemys terrapin*)	171	26.6	1.0	19.2	26.6	Dunson, 1970
Banded sea snake (*Laticauda*)	300	73	3.3	74	22.1	Dunson and Taub, 1967
Green sea turtle (*Chelonia*)	68	134	4.9	—	27.3	Holmes and McBean, 1964
Yellow-bellied sea snake (*Pelamis*)	100	218	9.2	169	23.7	Dunson, 1968
Marine iguana (*Amblyrhynchus*)	119	255	50.7	237	5.0	Dunson, 1969a

[a] Second figure is result after injection of a NaCl load.

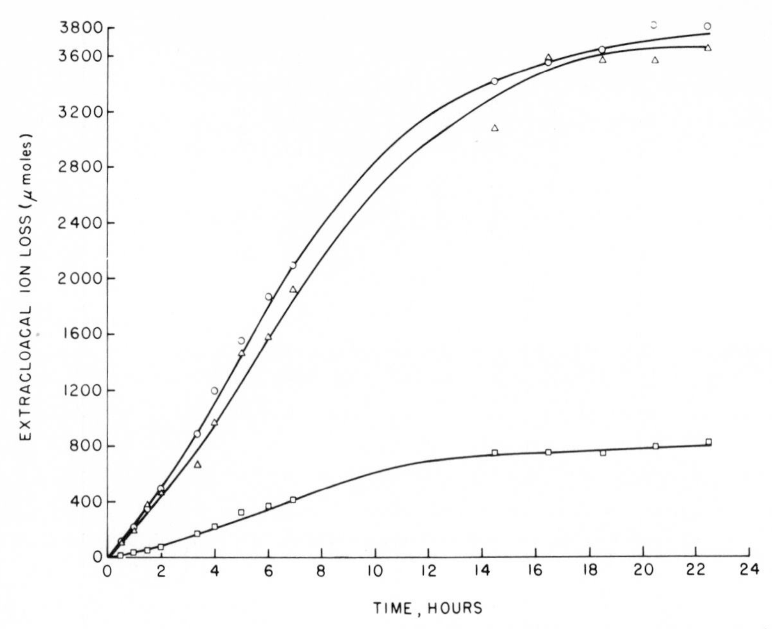

FIG. 7. The extrarenal excretion of electrolytes by the marine iguana, *Amblyrhynchus cristatus,* immediately after capture. The electrolytes excreted during this observation period were presumed to have been derived from ingested food. ○ = sodium, □ = potassium, △ = chloride. (Dunson, 1969b.)

C. THE ARID TERRESTRIAL ENVIRONMENT

The only terrestrial reptiles that have been shown to possess extrarenal excretory organs are lizards. In common with the marine iguana, the salt glands of these reptiles secrete a fluid that is relatively rich in potassium. The administration of methacholine chloride has been shown to initiate the secretory response in the chuckwalla (*Sauromalus obesus*) and the false iguana (*Ctenosaura pectinata*), and the parenteral administration of either sodium chloride or potassium chloride stimulates the extrarenal response in the chuckwalla, the false iguana, the desert iguana (*Dipsosaurus dorsalis*), the land iguana (*Conolophus subcristatus*), the blue spiny lizard (*Sceloporus cyanogenys*), and the Old World desert lizard or the spiny tailed agamid (*Uromastyx aegyptius*) (Schmidt-Nielsen *et al.,* 1963; Norris and Dawson, 1964; Templeton, 1964b, 1966, 1967; Dunson, 1969a).

Studies by Templeton (1964b, 1966) have indicated that the lizard nasal gland has the ability to adjust the relative amounts of sodium and potassium in the excretory fluid dependent upon the relative concentrations of these ions in the food or saline load. Thus, when specimens of

Ctenosaurus were maintained in the laboratory on a diet containing high concentrations of sodium, the concentration of sodium in the nasal gland fluid increased severalfold. This phenomenon was studied more precisely in the desert iguana and the blue spiny lizard by administering intraperitoneal loads (10 μM/gm body weight) of sodium chloride or potassium chloride. In this way, the animals became hypernatremic or hyperkalemic, and the rates of extrarenal excretion of sodium and potassium were subsequently measured.

In the desert iguana (*Dipsosaurus dorsalis*), the rate of potassium excretion in the salt gland fluid increased three- to fourfold during the 3 day period following the injection of either the sodium chloride or the potassium chloride load. The rates of excretion declined during the next 3 days but increased once more when a second load of either sodium chloride or potassium chloride was administered. This time, however, the rate of potassium excretion in the potassium chloride-loaded animals increased to a value approximately five times that observed before the administration of the first potassium chloride load. But the rate of potassium excretion evoked by the second sodium chloride load was similar to that evoked by the first load (Fig. 8). In contrast, the rates of sodium excretion remained unchanged when successive loads of potassium chloride were administered but the successive loads of sodium chloride evoked progressive increases in the rates of sodium excretion (Fig. 8). Thus, the hypernatremic and hyperkalemic conditions evoked corresponding changes in the extrarenal excretory patterns and the rates of sodium or potassium excretion were increased respectively. In the hypernatremic condition, this was achieved by increasing the $Na^+:K^+$ ratio of the salt gland fluid, whereas in the hyperkalemic state, the $Na^+:K^+$ ratio of the salt gland fluid was reduced (Table V).

When similar experiments were conducted on the blue spiny lizards (*Sceloporus cyanogenys*), the excretion of potassium again increased in response to either the sodium chloride or the potassium chloride load (Fig. 9). However, the evoked responses represented much lower excretory rates than those observed in the desert iguana. The rates of sodium excretion increased following the administration of the first load of either sodium chloride or potassium chloride, but they decreased to control levels following subsequent injections of sodium chloride or potassium chloride loads (Fig. 9). No adjustment in the $Na^+:K^+$ ratio of the salt gland fluid was apparent (Table V).

Clearly, the spiny lizard, being a less xerophilic reptile than the desert iguana, does not possess an extrarenal mechanism capable of responding to high loads of sodium. Further, the relative efficiencies of the extrarenal excretory mechanisms in these two species have probably developed

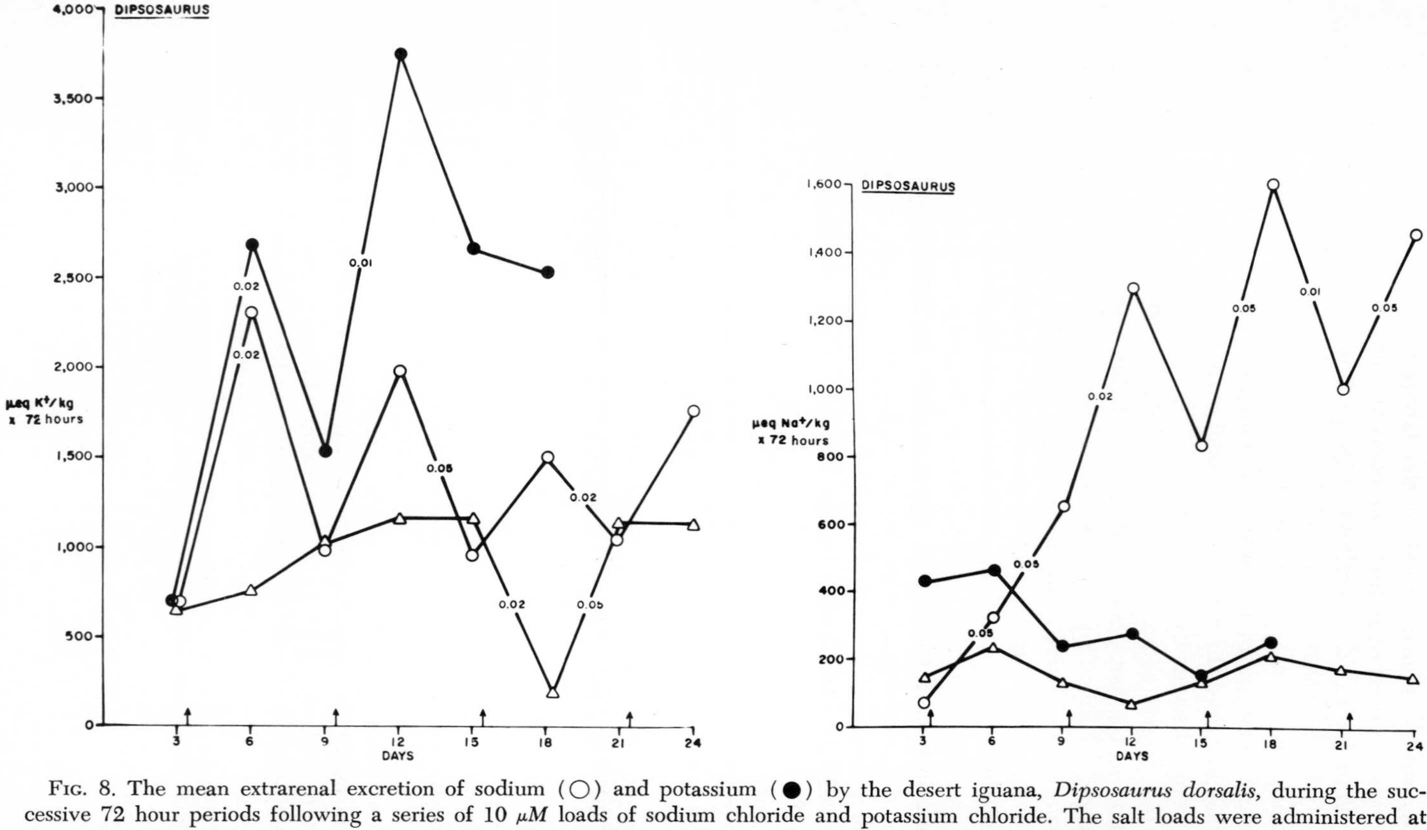

FIG. 8. The mean extrarenal excretion of sodium (○) and potassium (●) by the desert iguana, *Dipsosaurus dorsalis*, during the successive 72 hour periods following a series of 10 μM loads of sodium chloride and potassium chloride. The salt loads were administered at the times indicated by arrows on the abscissa. The significant differences between mean values are indicated by the "p" values which interrupt the lines connecting the respective points on the graph; △ = control. (Templeton, 1966.)

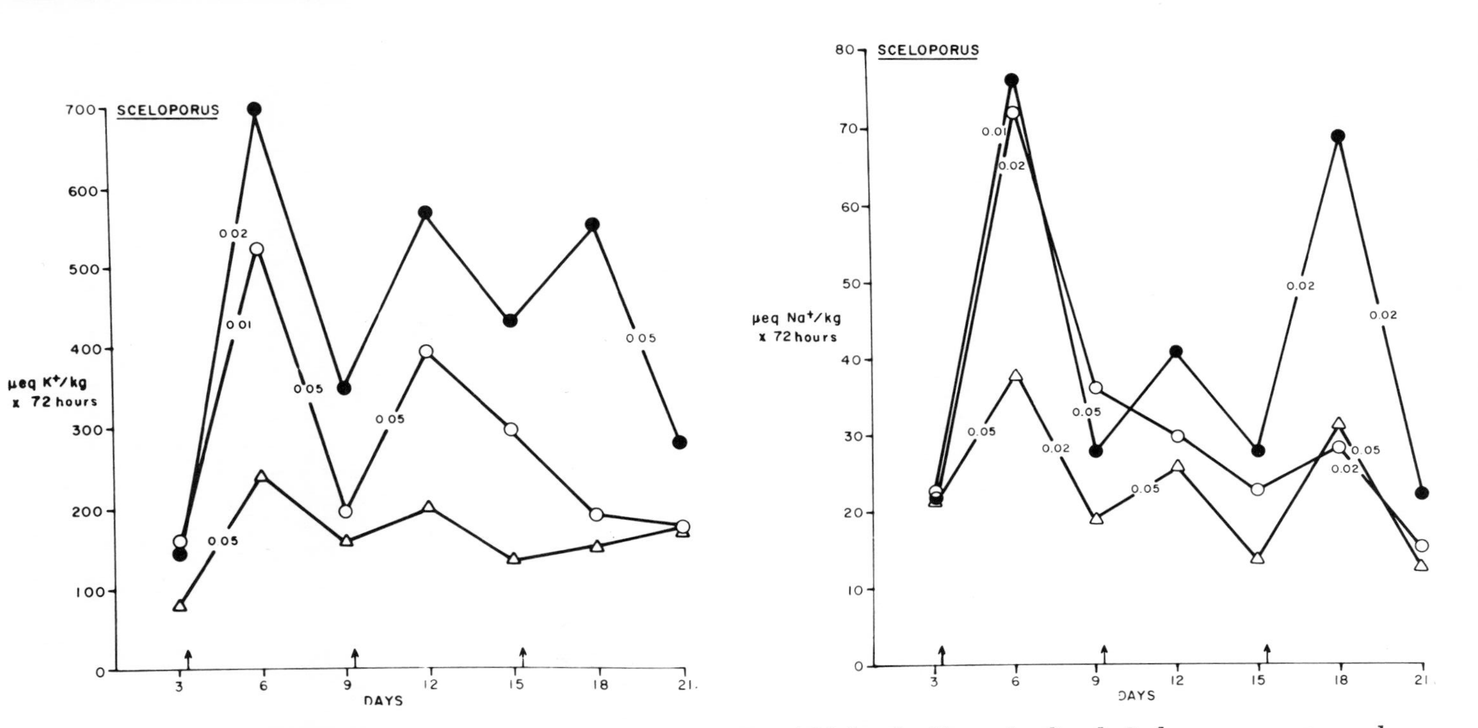

FIG. 9. The mean extrarenal excretion of sodium (○) and potassium (●) by the blue spiny lizard, *Sceloporus cyanogenus*, during the successive 72 hour periods following a series of 10 μM loads of sodium chloride and potassium chloride. The remainder of the legend is the same as for Fig. 8.

in relation to the availability of osmotically free water in their respective environments and the ionic composition of their food. A complete elucidation of this intriguing adaptive phenomenon must await a precise study of the feeding habits of the two species.

ACKNOWLEDGMENTS

The research reported here and the preparation of this review was supported in part by grants from the National Science Foundation to William H. Dantzler (Grants No. GB-11788 and No. GB-28692X) and to W. N. Holmes (Grant No. GB-20806).

REFERENCES

Balagura, S., and Pitts, R. F. (1962). *Amer. J. Physiol.* **203**, 11.
Baze, W. B., and Horne, F. R. (1970). *Comp. Biochem. Physiol.* **34**, 91.
Bentley, P. J. (1959). *J. Physiol. (London)* **145**, 37.
Bentley, P. J., and Schmidt-Nielsen, K. (1965). *J. Cell. Comp. Physiol.* **66**, 303.
Bentley, P. J., and Schmidt-Nielsen, K. (1966). *Science* **151**, 1547.
Berndt, W. O., and Beechwood, E. C. (1965). *Amer. J. Physiol.* **208**, 642.
Braysher, M., and Green, B. (1970). *Comp. Biochem. Physiol.* **35**, 607.
Brodsky, W. A., and Schilb, T. P. (1960). *J. Clin. Invest.* **39**, 974.
Brodsky, W. A., and Schilb, T. P. (1965). *Amer. J. Physiol.* **208**, 46.
Brodsky, W. A., and Schilb, T. P. (1966). *Amer. J. Physiol.* **210**, 987.
Brodsky, W. A., and Schilb, T. P. (1967). *Fed. Proc., Fed. Amer. Soc. Exp. Biol.* **26**, 1314.
Burgess, W. W., Harvey, A. M., and Marshall, E. K. (1933). *J. Pharmacol. Exp. Ther.* **49**, 237.
Clark, N. B. (1965). *Gen. Comp. Endocrinol.* **5**, 297.
Clark, N. B., and Dantzler, W. H. (1972). *Amer. J. Physiol.* **223**, 1455.
Coulson, R. A., and Hernandez, T. (1964). "Biochemistry of the Alligator—A Study of Metabolism in Slow Motion." Louisiana State Univ. Press. Baton Rouge.
Dantzler, W. H. (1964). Ph.D. Thesis, Duke University, Durham, North Carolina.
Dantzler, W. H. (1967a). *Amer. J. Physiol.* **212**, 83.
Dantzler, W. H. (1967b). *Comp. Biochem. Physiol.* **22**, 131.
Dantzler, W. H. (1968). *Amer. J. Physiol.* **215**, 747.
Dantzler, W. H. (1969). *Amer. J. Physiol.* **217**, 1810.
Dantzler, W. H. (1970a). *Comp. Biochem. Physiol.* **34**, 609.
Dantzler, W. H. (1970b). *In* "Hormones and the Environment" (G. K. Benson and J. G. Phillips, eds.), pp. 157–216. Cambridge Univ. Press, London and New York.
Dantzler, W. H. (1971a). *Abstr., Int. Congr. Physiol. Sci., 25th,* Vol. IX, p. 131.
Dantzler, W. H. (1971b). *Physiologist* **14**, 128.
Dantzler, W. H. (1971c). *Comp. Biochem. Physiol.* A **40**, 467.
Dantzler, W. H., and Schmidt-Nielsen, B. (1966). *Amer. J. Physiol.* **210**, 198.
Davis, L. E., and Schmidt-Nielsen, B. (1967). *J. Morphol.* **121**, 255.
Deetjen, P., and Sonnenberg, H. (1965). *Pfluegers Arch. Gesamte Physiol. Menschen Tiere* **285**, 35.
Dessauer, H. C. (1952). *Proc. Soc. Exp. Biol. Med.* **80**, 742.
Diamond, J. M. (1964). *J. Gen. Physiol.* **48**, 15.
Diamond, J. M. (1971). *Fed. Proc., Fed. Amer. Soc. Exp. Biol.* **30**, 6.
Diamond, J. M., and Bossert, W. H. (1967). *J. Gen. Physiol.* **50**, 2061.

Dunson, W. A. (1968). *Amer. J. Physiol.* **215**, 1512.
Dunson, W. A. (1969a). *Amer. J. Physiol.* **216**, 995.
Dunson, W. A. (1969b). *In* "Exocrine Glands," (S. Y. Botelho, F. P. Brooks, and W. B. Shelley, eds.) pp. 83–101. Univ. of Pennsylvania Press, Philadelphia.
Dunson, W. A. (1970). *Comp. Biochem. Physiol.* **32**, 161.
Dunson, W. A., and Taub, A. M. (1967). *Amer. J. Physiol.* **213**, 975.
Elizondo, R. S., and LeBrie, S. J. (1969). *Amer. J. Physiol.* **217**, 419.
Forster, R. P. (1942). *J. Cell. Comp. Physiol.* **20**, 55.
Giebisch, G. (1969). *Nephron* **6**, 260.
Gonzalez, C. F., Shamoo, Y. E., and Brodsky, W. A. (1967a). *Amer. J. Physiol.* **212**, 641.
Gonzalez, C. F., Shamoo, Y. E., Wyssbrod, H. R., Solinger, R. E., and Brodsky, W. A. (1967b). *Amer. J. Physiol.* **213**, 333.
Gonzalez, C. F., Shamoo, Y. E., and Brodsky, W. A. (1969). *Biochim. Biophys. Acta* **193**, 403.
Green H. H., Steinmetz, P. R., and Frazier, H. S. (1970). *Amer. J. Physiol.* **218**, 845.
Hertz, R., Tullner, W. W., Schricker, J. A., Dhyse, F. G., and Hallman, L. F. (1955). *Recent. Progr. Horm. Res.* **11**, 119.
Hiatt, H. H., and Thompson, D. C. (1957). *J. Clin. Invest.* **36**, 557.
Hill, L., and Dawbin, W. H. (1969). *Comp. Biochem. Physiol.* **31**, 453.
Holmes, W. N., and McBean, R. L. (1964). *J. Exp. Biol.* **14**, 81.
Holmes, W. N., Phillips, J. E., and Chester Jones, I. (1963). *Recent Progr. Horm. Res.* **19**, 619.
Junqueira, L. C. U., Malnic, G., and Monge, C. (1966). *Physiol. Zool.* **39**, 151.
Khalil, F. (1948a). *J. Biol. Chem.* **172**, 101.
Khalil, F. (1948b). *J. Biol. Chem.* **172**, 105.
Khalil, F. (1951). *J. Biol. Chem.* **189**, 443.
Khalil, F., and Haggag, G. (1955). *J. Exp. Zool.* **130**, 423.
Khalil, F., and Haggag, G. (1958). *J. Exp. Biol.* **35**, 552.
Klahr, S., and Bricker, N. S. (1964). *Amer. J. Physiol.* **206**, 1333.
Klahr, S., and Bricker, N. S. (1965). *J. Gen. Physiol.* **48**, 571.
Komadina, S., and Solomon, S. (1970). *Comp. Biochem. Physiol.* **32**, 333.
LeBrie, S. J., and Elizondo, R. S. (1969). *Amer. J. Physiol.* **217**, 426.
LeBrie, S. J., and Sutherland, I. D. W. (1962). *Amer. J. Physiol.* **203**, 995.
Lewy, J. E., and Windhager, E. E. (1968). *Amer. J. Physiol.* **214**, 943.
McClanahan, L. (1964). *Comp. Biochem. Physiol.* **12**, 501.
McNabb, F. M. A., and Poulson, T. L. (1970). *Comp. Biochem. Physiol.* **33**, 933.
McWhinnie, D. J., and Cortelyou, J. R. (1968). *Gen. Comp. Endocrinol.* **11**, 78.
Malnic, G., Klose, R. M., and Giebisch, G. (1964). *Amer. J. Physiol.* **206**, 674.
Malnic, G., Klose, R. M., and Giebisch, G. (1966a). *Amer. J. Physiol.* **211**, 529.
Malnic, G., Klose, R. M., and Giebisch, G. (1966b). *Amer. J. Physiol.* **211**, 548.
Maren, T. H. (1967). *Physiol. Rev.* **47**, 595.
Minnich, J. E. (1969). *Amer. Zool.* **9**, 1094.
Minnich, J. E. (1970). *Comp. Biochem. Physiol.* **35**, 921.
Moyle, V. (1949). *Biochem. J.* **44**, 581.
Munsick, R. A. (1966). *Endocrinology* **78**, 591.
Murrish, D. E., and Schmidt-Nielsen, K. (1970). *Science* **170**, 324.
Neill, W. T. (1958). *Bull. Mar. Sci. Gulf Carib.* **8**, 1.
Norris, K. W., and Dawson, W. R. (1964). *Copeia* **4**, 638.

Parsons, T. S. (1970). *In* "Biology of the Reptilia" (C. Gans, ed.), Vol. 2, p. 99. Academic Press, New York.
Perschmann, C. (1956). *Zool. Beitr., Berlin* **2**, 447.
Phillips, J. G., Chester Jones, I., and Bellamy, D. (1962). *J. Endocrinol.* **25**, 233.
Pitts, R. F. (1968). "Physiology of the Kidney and Body Fluids," 2nd ed. Yearbook Publ., Chicago, Illinois.
Porter, P. (1963a). *Res. Vet. Sci.* **4**, 580.
Porter, P. (1963b). *Res. Vet. Sci.* **4**, 592.
Prange, H. D., and Schmidt-Nielsen, K. (1969). *Comp. Biochem. Physiol.* **28**, 973.
Ranges, H. A., Chasis, H., Goldring, W., and Smith, H. W. (1939). *Amer. J. Physiol.* **126**, 103.
Roberts, J. S., and Schmidt-Nielsen, B. (1966). *Amer. J. Physiol.* **211**, 476.
Robin, E. D., Vester, J. W., Murdaugh, H. V., Jr., and Miller, J. E. (1964). *J. Cell. Comp. Physiol.* **63**, 287.
Rogers, L. J. (1966). *Comp. Biochem. Physiol.* **18**, 249.
Rosenfeld, G., and Bascom, W. D. (1956). *J. Biol. Chem.* **222**, 565.
Sawyer, W. H. (1951). *Amer. J., Physiol.* **164**, 457.
Sawyer, W. H., and Sawyer, M. K. (1952). *Physiol. Zool.* **25**, 84.
Sawyer, W. H., Munsick, R. A., and van Dyke, H. B. (1959). *Nature (London)* **184**, 1464.
Schilb, T. P., and Brodsky, W. A. (1966). *Amer. J. Physiol.* **210**, 997.
Schmidt-Nielsen, B. (1971). *Fed. Proc., Fed. Amer. Soc. Exp. Biol.* **30**, 3.
Schmidt-Nielsen, B., and Davis, L. E. (1968). *Science* **159**, 1105.
Schmidt-Nielsen, B., and Forster, R. P. (1954). *J. Cell. Comp. Physiol.* **44**, 233.
Schmidt-Nielsen, B., and Skadhauge, E. (1967). *Amer. J. Physiol.* **212**, 973.
Schmidt-Nielsen, K., and Bentley, P. J. (1966). *Science* **154**, 911.
Schmidt-Nielsen, K., and Fänge, R. (1958). *Nature (London)* **182**, 783.
Schmidt-Nielsen, K., Borut, A., Lee, P., and Crawford, E. (1963). *Science* **142**, 1300.
Scholander, P. F., Hargens, A. R., and Miller, S. L. (1968). *Science* **161**, 321.
Scott, W. N., Shamoo, Y. E., and Brodsky, W. A. (1970). *Biochim. Biophys. Acta* **219**, 248.
Shoemaker, V. H., Licht, P., and Dawson, W. R. (1966). *Physiol. Zool.* **39**, 244.
Smith, M. (1926). "Monograph of the Sea-snakes (Hydrophiidae)." Wheldon and Wesley, London.
Steinmetz, P. R. (1967). *J. Clin. Invest.* **46**, 1531.
Tanner, G. A., and Isenberg, M. T. (1970). *Amer. J. Physiol.* **219**, 889.
Taub, A. M., and Dunson, W. A. (1967). *Nature (London)* **215**, 995.
Templeton, J. R. (1964a). *Physiol. Zool.* **37**, 300.
Templeton, J. R. (1964b). *Comp. Biochem. Physiol.* **11**, 223.
Templeton, J. R. (1966). *Comp. Biochem. Physiol.* **18**, 563.
Templeton, J. R. (1967). *Copeia* **1**, 136.
Thesleff, S., and Schmidt-Nielsen, K. (1962). *J. Cell. Comp. Physiol.* **59**, 31.
Walker, A. M., Hudson, C. L., Findley, T., Jr., and Richards, A. N. (1937). *Amer. J. Physiol.* **118**, 121.
Wirz, H. (1956). *Helv. Physiol. Pharmacol. Acta* **14**, 353.
Woolley, P. (1959). *J. Exp. Biol.* **36**, 453.

CHAPTER 12

Bile Salts in Reptilia

A. R. Tammar

I. Introduction 337
II. Anapsida 337
III. Lepidosauria 339
A. Sauria 339
B. Serpentes 340
IV. Crocodilia 342
V. Significance of Reptilian Bile Salt Distribution 343
VI. Reptilian Bile Salts to June 1971 346
References 350

I. Introduction

Bile salts are secretory products of vertebrate livers and are involved in the digestion and absorption of fat. Cholesterol is assumed to be the universal precursor of bile salts, which are thus steroidal in nature. The final products of cholesterol catabolism in reptiles are bile acids, and the bile salts are conjugates of the bile acids with taurine. (Conjugation with glycine is encountered only in eutherian mammals.)

Primary bile salts are synthesized by the liver directly from cholesterol, and secondary bile salts are those that have been modified by intestinal microorganisms and returned to the liver by the enterohepatic circulation. They are either resecreted by the liver along with the primary bile salts or further modified by liver hydroxylation systems prior to resecretion.

The subject of species differences in bile salts has been extensively reviewed by Haslewood, most recently in 1967 (Haslewood, 1967a,b). In this context C_{24}–5β acids are regarded as advanced and all other bile salts as more primitive. Table I at the end of the chapter has a complete list of references to work on individual species.

II. Anapsida

The biles from eleven species of this sub-class have been examined, two in Japan and nine in this laboratory. Yamasaki and Yuuki (1936) acidified the alkaline hydrolyzate of alligator turtle (*Amyda japonica* = *Trionyx sinensis*) bile and isolated two lactones, which they called trihydroxysterocholanic and trihydroxyisosterocholanic lactones. Kim (1939) isolated trihydroxysterocholanic lactone from the European pond tortoise *Emys orbicularis.* Oxidation of the partially acetylated methyl ester of the parent tetrahydroxysterocholanic acid afforded dinorcholic acid (Kanemitu, 1942a). This suggested that tetrahydroxysterocholanic acid was probably $3\alpha,7\alpha,12\alpha,22\xi$-tetrahydroxy-$5\beta$-cholestan-26-oic acid, although a C_{28} formula was thought possible at the time. Trihydroxysterocholanic lactone from an unspecified turtle (probably *Amyda japonica*) has been analyzed by Amimoto *et al.* (1965) and shown almost conclusively to be the lactone of $3\alpha,7\alpha,12\alpha,22\xi$-tetrahydroxy-$5\beta$-cholestan-26-oic acid (Fig. 1). Trihydroxyisosterocholanic lactone and a further acid, heterocholic acid, isolated from *Amyda japonica* by Kanemitu (1942b) have not been investigated further. Seven of the nine chelonian biles examined in this laboratory afforded *inter alia* an acid which exhibited properties consistent with its being $3\alpha,7\alpha,12\alpha,22\xi$-tetrahydroxy-$5\beta$-cholestan-26-oic acid, but the melting points of the derivatives prepared did not agree with those in the earlier Japanese work. All nine biles contained in addition a mixture of two acids of very similar chromatographic mobility, both considerably less polar than the tetrahydroxy acid referred to above. It is thought, although there is no firm evidence for this, that these two acids might bear the same relationship to the tetrahydroxy acid as chenodeoxycholic or deoxycholic acids do to cholic acid.

Amimoto *et al.* (1965) reported that only a small proportion of the trihydroxysterocholanic and trihydroxyisosterocholanic lactones were present as taurine-conjugated acids in the turtle bile they examined. However, chromatographic examination in this laboratory of seven of the nine chelonian biles referred to above showed that they contained only substances with the polarity of taurine conjugates and no free acids.

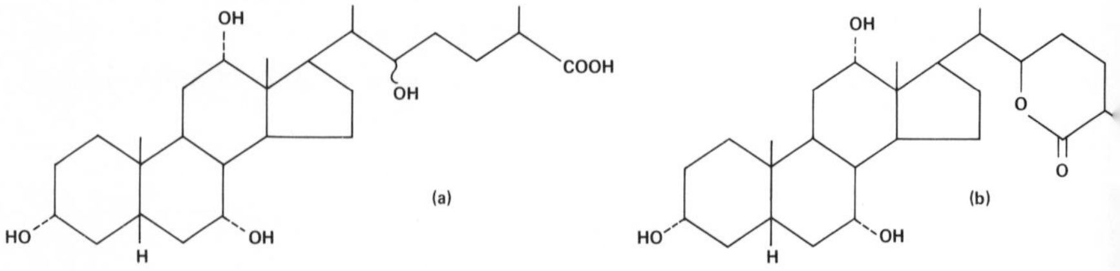

FIG. 1. $3\alpha,7\alpha,12\alpha,22\xi$-tetrahydroxy-$5\beta$-cholestan-26-oic acid; (a) open-chain form, (b) lactone form.

It is possible that the bile examined by Amimoto *et al.* (1965) had been subjected to postmortem microbial attack.

To sum up, chelonian biles yield, on alkaline hydrolysis, substances that do not appear to occur in biles of other species. There is strong evidence that one of these substances is $3\alpha,7\alpha,12\alpha,22\xi$-tetrahydroxy-$5\beta$-cholestan-26-oic acid or its lactone.

III. Lepidosauria

A. Sauria

1. Varandiae and Helodermatidae

Alkaline hydrolysis of *Varanus niloticus* bile (Haslewood and Wootton, 1950) afforded what was at that time an unknown bile acid—varanic acid. Its structure has not been completely determined but Collings and Haslewood (1966) reported that it was almost certainly $3\alpha,7\alpha,12\alpha,24\xi$-tetrahydroxy-$5\beta$-cholestan-26-oic acid (Fig. 2). Taurovaranate is a principal bile salt in the four other varanid species that have been examined, and *Heloderma horridum* has the taurine conjugate of an acid very like varanic acid (Haslewood, 1967b). This latter acid may be a C-24 or C-25 epimer of varanic acid. Thus the bile salts of these two families appear to be taurine conjugates of C_{27}-5β acids.

2. Other Saurian Families

Biles from species of six of the other eighteen saurian families have been examined. In contrast to the Varanidae and Helodermatidae, they are characterized by the presence of C_{24} bile salts, and there is very little evidence of C_{27} acids. Usually either taurocholate (Fig. 3) or tauroallocholate (Fig. 4) predominates to the virtual exclusion of the other; of the six families, three—Iguanidae, Agamidae, and Chamaeleon-

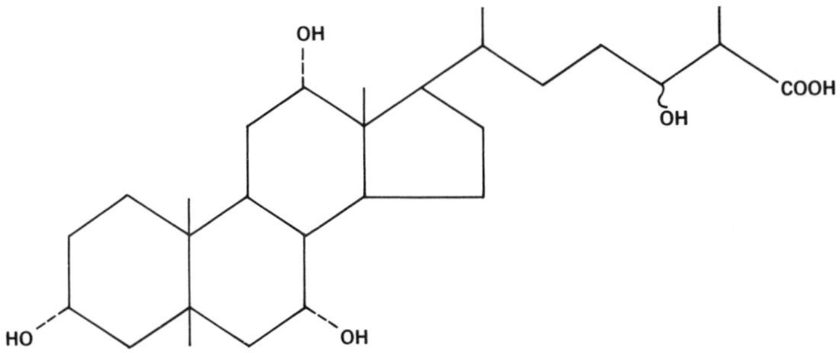

FIG. 2. $3\alpha,7\alpha,12\alpha,24\xi$-tetrahydroxy-$5\beta$-cholestan-26-oic acid.

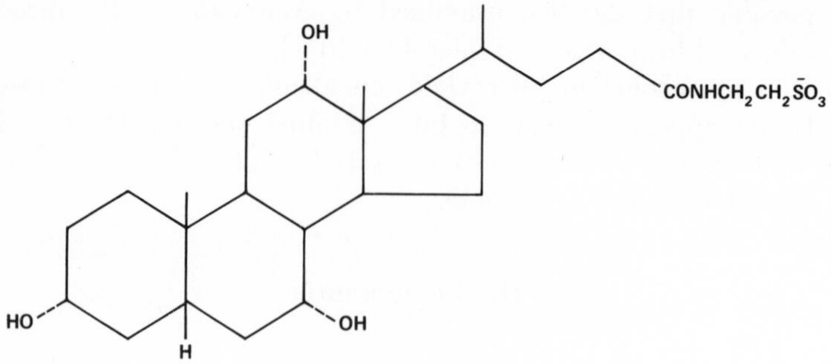

FIG. 3. Taurocholate.

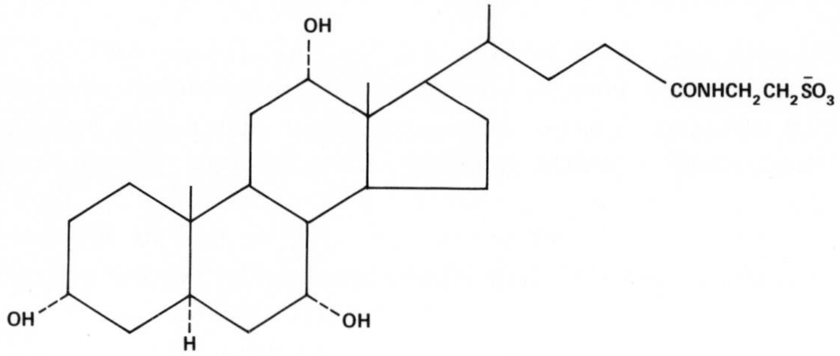

FIG. 4. Tauroallocholate.

tidae—have tauroallocholate, whereas Pygopodidae, Teiidae, and Amphisbaenidae probably have taurocholate.

Thus, of the eight saurian families whose biles have become available for analysis, two have exclusively C_{27}-5β bile salts, three have predominantly C_{24}-5α bile salts and three are "modern" in that they have predominantly C_{24}-5β bile salts.

B. SERPENTES

1. *Boidae*

Haslewood and Wootton (1950) conducted an examination of, *inter alia,* ten snake biles. Three of the biles, all from the family Boidae, afforded a previously unknown acid now known as pythocholic acid. Pythocholic acid, isolated as a lactone, was subsequently obtained from a further six boids (Haslewood and Wootton, 1951; Haslewood, 1951). Pythocholic acid is 3α,12α,16α-trihydroxy-5β-cholan-24-oic acid (Fig. 5) and has now been found in biles of twelve of the thirteen boids examined and probably in one species of Aniliidae (Haslewood, 1967b).

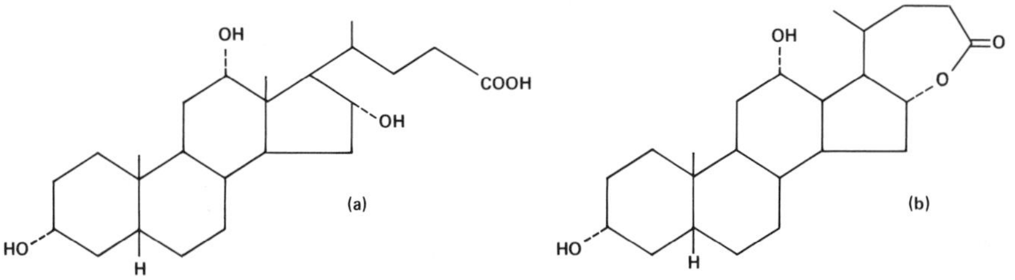

FIG. 5. Pythocholic acid; (a) open-chain form, (b) lactone form.

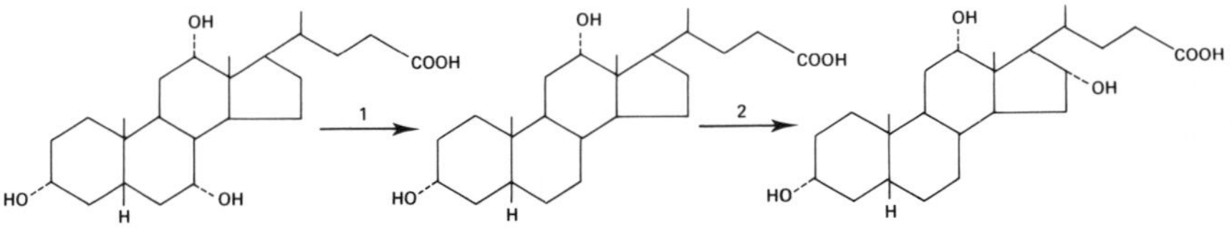

FIG. 6. Pythocholic acid formation from cholic acid; (1) intestinal microorganisms, (2) liver hydroxylation systems.

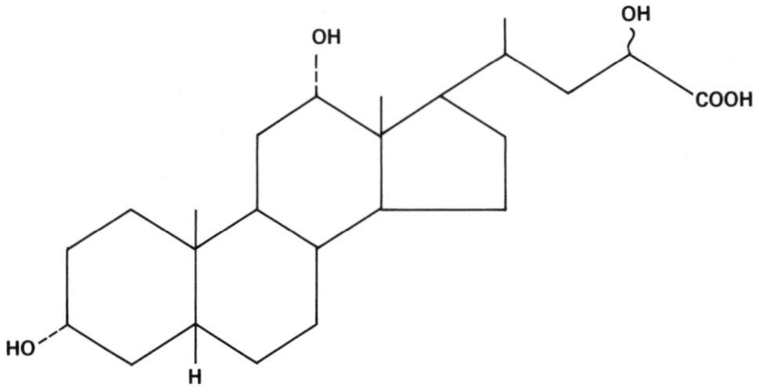

FIG. 7. Bitocholic acid.

Bergström *et al.* (1960) showed that pythocholic acid was a secondary bile acid, formed in python liver by 16α-hydroxylation of deoxycholic acid (3α,12α-dihydroxy-5β-cholan-24-oic acid). The deoxycholic acid was itself formed by the action of intestinal microorganisms on cholic acid (Fig. 6). The ability to hydroxylate deoxycholic acid at position 16α appears to be characteristic of many if not most boid snakes and perhaps of other primitive snake families.

2. *Colubridae and Viperidae*

Haslewood (1961) showed that the principal bile salts of two snakes of the genus *Bitis* were the taurine conjugates of 3α,12α,23ξ-trihydroxy-5β-cholan-24-oic acid, bitocholic acid (Fig. 7) and 3α,7α,12α,23ξ-tetrahydroxy-5β-cholan-24-oic acid. Bile acids hydroxylated at C-23 have since been found in a further eight out of ten species of the subfamily Viperinae, which have been examined (Haslewood, 1967b). In addition, eight of the thirty-eight species of Colubridae that have been examined appear to have only 3α,7α,12α,23ξ-tetrahydroxy-5β-cholan-24-oic acid (Haslewood, 1967b). It would not be surprizing if the formation of bitocholic acid were, like that of pythocholic acid, a response peculiar to certain groups of snakes to deoxycholic acid arriving in the liver via the enterohepatic circulation. The tetrahydroxy acid would arise by incidental 23-hydroxylation of cholic acid. Neither of these ideas has been successfully tested experimentally. Taurocholate is almost a universal constituent of the biles in this group, and indeed it is the only identified component of those biles which do not have 23-hydroxylated bile salts. Tauroallocholate has also been found in minor amounts in species of Viperinae which have taurobitocholate (see Table I).

3. *Other Serpent Families*

Haslewood (1967b) published a list of all ninety-four snake biles examined to that date. The only positively identified component of the thirty biles that do not come into the categories already mentioned is taurocholate. Thus, snake bile salts consist almost exclusively of taurine-conjugated C_{24}-5β bile acids, although in a very few species there are small quantities of 5α bile acids in addition. The most common bile salt is taurocholate, but in isolated groups of snakes, other trihydroxvlated bile acids are found, one of which, pythocholic acid, is a secondary bile acid formed from deoxycholic acid.

IV. Crocodilia

The biles of eight species of the order Crocodilia have been examined, and all have as their principal bile salt the taurine conjugate of 3α,7α,12α-trihydroxy-5β-cholestan-26-oic acid (Fig. 8). Other acids are usually present, and in one case, *Alligator mississipiensis,* 3α,7α,-dihydroxy-5β-cholestan-26-oic acid has been isolated (Dean and Whitehouse, 1966). There is a certain divergence of opinion in the literature as to the stereochemistry at C-25 in these compounds. Haslewood (1952, 1967b) has isolated only the 25*R* isomer of the trihydroxy acid from *A. mississipiensis, Caiman crocodilus, Crocodylus niloticus,* and *Crocodylus acutus.* Mendelsohn and Mendelsohn (1969), however, isolated both 25*R* and 25*S* isomers from *C. niloticus* bile, and Shah *et al.* (1968)

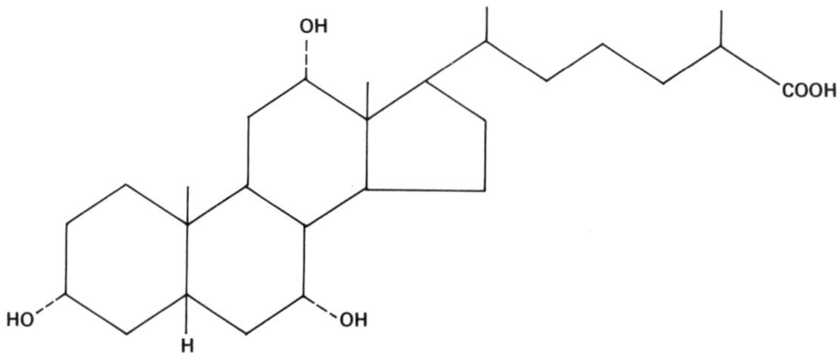

FIG. 8. $3\alpha,7\alpha,12\alpha$,-trihydroxy-5β-cholestan-26-oic acid.

obtained only the 25 S isomer from a further three species. It is reasonable to suppose that oxidation of the C_8 side chain is stereospecific in Crocodilidae as it is in rats (Berséus, 1965), and a possible explanation for finding both isomers is that some racemization takes place during alkaline hydrolysis. Both isomers, however, are effective precursors of cholic acid in rats with bile fistulas (Bridgwater and Lindstedt, 1957). The bile salts of the Crocodilidae are thus taurine-conjugates of C_{27}-5β bile acids with conventional nuclear hydroxylation patterns.

V. Significance of Reptilian Bile Salt Distribution

The idea has been expressed in Chapter 15 of Volume VIII and Chapter 4 of this volume with reference to piscine and amphibian bile salts, that primitive bile salts should be considered in the light of the main mammalian biosynthetic pathway from cholesterol to cholic acid. This pathway, a principal feature of which is that changes to the steroid nucleus precede changes to the side chain, is well treated in the review by Danielsson and Tchen (1968). It should be borne in mind, however, that there is considerable evidence for the existence of a pathway that involves side chain oxidation prior to nuclear modification (Mitropoulos and Myant, 1967; Ayaki and Yamasaki, 1970). Primitive bile acids may be regarded either as intermediates in the cholesterol to cholic acid pathway that cannot be further modified by the species under consideration or alternatively as intermediates specifically modified in a particular species so as to be removed from the main pathway.

Modifications to the steroid nucleus in cholic acid biosynthesis are illustrated in Fig. 9 and start with the 7α-hydroxylation of cholesterol (Fig. 9a) to give 5-cholestene-$3\beta,7\alpha$-diol (Fig. 9b). Oxidation of this gives 7α-hydroxy-4-cholesten-3-one (Fig. 9c), the substrate for the 12α-hy-

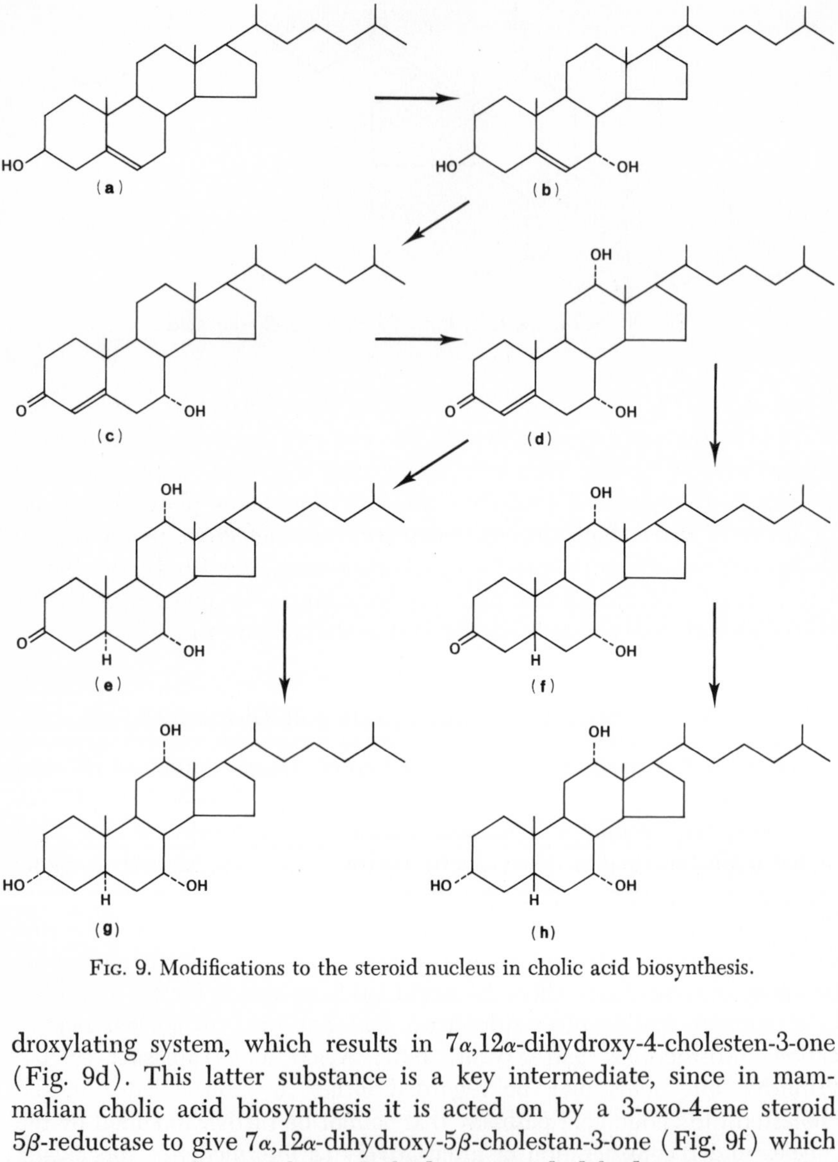

FIG. 9. Modifications to the steroid nucleus in cholic acid biosynthesis.

droxylating system, which results in 7α,12α-dihydroxy-4-cholesten-3-one (Fig. 9d). This latter substance is a key intermediate, since in mammalian cholic acid biosynthesis it is acted on by a 3-oxo-4-ene steroid 5β-reductase to give 7α,12α-dihydroxy-5β-cholestan-3-one (Fig. 9f) which is in turn acted upon by a 3α-hydroxysteroid dehydrogenase to give 5β-cholestane-3α,7α,12α-triol (Fig. 9h). Iguana liver, however, converts 7α,12α-dihydroxy-4-cholesten-3-one to 5α-cholestane-3α,7α,12α-triol (Fig. 9g) via the intermediate formation of 7α,12α-dihydroxy-5α-cholestan-3-one (Hoshita *et al.*, 1968). Thus, only the 5β-reductase appears to be present in chelonians, crocodilians, most snakes and some lizards, and the

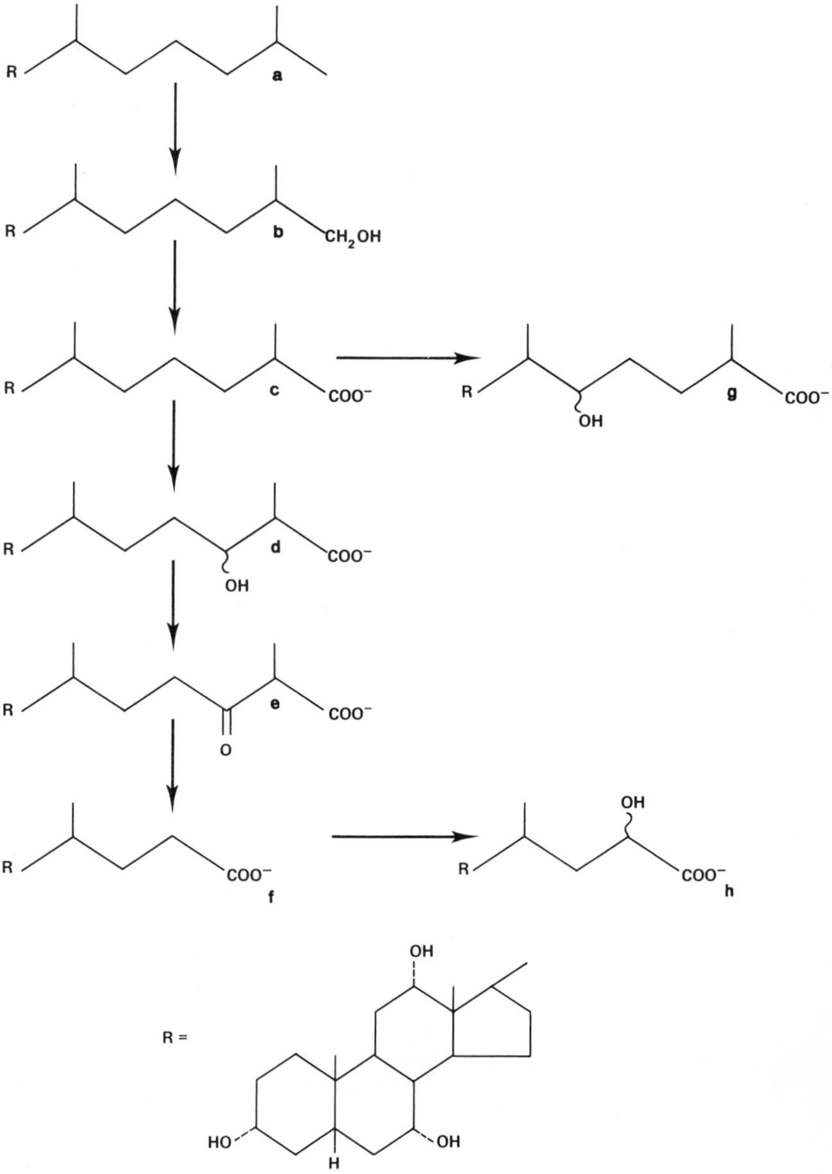

FIG. 10. Side-chain modifications in cholic acid biosynthesis.

5α-reductase system is confined to some lizard families and to some snakes in which it appears to play a very minor role.

Side chain modifications in cholic acid biosynthesis are shown in Fig. 10 where stages a to f represent the main pathway. Stage c represents

$3\alpha,7\alpha,12\alpha$-trihydroxy-5β-cholestan-26-oic acid and is the end of the pathwas in crocodilian species. Chelonians can hydroxylate $3\alpha,7\alpha,12\alpha$-trihydroxy-5β-cholestan-26-oic acid to give $3\alpha,7\alpha,12\alpha,22\xi$-tetrahydroxy-$5\beta$-cholestan-26-oic acid (Fig. 10g), which is removed from the main pathway. Varanic acid-producing lizards, on the other hand, can hydroxylate c at C-24 to give varanic acid (Fig. 10d), which represents an end product of cholesterol catabolism in these species. All other reptilian species can further oxidize varanic acid or its 5α-isomer to give cholic or allocholic acid (Fig. 10f). Pythocholic acid and probably the 23-hydroxylated C_{24} acids (Fig. 10h) arise secondarily by selective hydroxylation of deoxycholic acid in the liver of the relevant species.

Thus, four reptilian orders have been examined. The chelonians may be distinguished by the presence of C_{27}-5β bile acids hydroxylated at C-22. Crocodilians have C_{27}-5β bile acids, which have no side chain hydroxylation. Snakes have C_{24}-5β bile acids with secondary hydroxylation in two groups and sometimes minor amounts of C_{24}-5α acids. Lizards for the most part have C_{24} bile acids, but two families appear to have C_{27}-5β bile acids. Some lizard families have predominantly C_{24}-5α acids, while others have predominantly C_{24}-5β acids. Unfortunately, no tuatara, *Sphenodon punctatum,* bile has yet become available for analysis.

VI. Reptilian Bile Salts to June 1971

Table I, which follows, lists all reptilian bile salts examined up to June 1971. The reader may assume that all acids mentioned in column 2 are taurine conjugated.

TABLE I
REPTILIAN BILE SALTS TO JUNE 1971

Species	Acids found	References
Anapsida		
Chelonia		
Macrochelys temmincki; Clemmys insculpta; Pseudemys ornata	3α,7α,12α,22ξ-tetrahydroxy-5β-cholestan-26-oic acid (probably); other acids	Tammar (1970)
Emys orbicularis	3α,7α,12α,22ξ-tetrahydroxy-5β-cholestan-26-oic acid	Kim (1939)
Testudo vicina	3α,7α,12α,22ξ-tetrahydroxy-5β-cholestan-26-oic acid (probably; trace only); other acids	Tammar (1970)
Testudo graeca; Chelone midas	3α,7α,12α,22ξ-tetrahydroxy-5β-cholestan-26-oic acid (probably); other acids	Tammar (1970)
Amyda japonica (= *Trionyx sinensis*)	3α,7α,12α,22ξ-tetrahydroxy-5β-cholestan-26-oic acid; tetra-hydroxyisosterocholanic acid; heterocholic acid	Yamasaki and Yuuki (1936); Kanemitu (1942a,b); Amimoto *et al.* (1965)
Trionyx triunguis	3α,7α,12α,22ξ-tetrahydroxy-5β-cholestan-26-oic acid (probably); other acids	Tammar (1970)
Trionyx phayrei; Pelusios sinuatus	Unidentified acids	Tammar (1970)
Lepidosauria		
Sauria		
Pygopodidae		
Lialis burtonis	Cholic acid	Haslewood (1967b)
Iguanidae		
Iguana iguana	Allocholic acid; 3α,7α,12α-tri hydroxy-5α-cholestan-26-oic acid	Okuda *et al.* (1968); Hoshita *et al.* (1968)
Dipsosaurus dorsalis	Allocholic acid	Hoshita *et al.* (1968)
Anolis lineatopus	Allocholic acid; cholic acid (trace)	Haslewood (1967b)
Anolis garmani; Anolis grahami; Anolis richardi; Cyclura carinata; Polychrus marmoratus	Allocholic acid; other acids	Haslewood (1967b)
Agamidae		
Amphibolurus barbatus	Allocholic acid; other acids	Haslewood (1967b)
Physignathus lesueri	Allocholic acid; other acids	Haslewood (1967b)
Uromastix thomasi	Allocholic acid; other acids	Haslewood (1967b)
Chameleontidae		
Chameleo muelleri	Allocholic acid; other acids	Haslewood (1967b)
Teiidae		
Ameiva ameiva	Cholic acid; other acids	Haslewood (1967b)
Varanidae		
Varanus niloticus	Varanic acid; other acids	Haslewood and Wootton (1950)
Varanus gouldi; Varanus griseus; Varanus salvator; Varanus varius	Varanic acid; other acids	Haslewood (1967b)
Helodermatidae		
Heloderma horridum	Acid very like varanic acid	Haslewood (1967b)
Amphisbaenidae		
Agamodon anguliceps	Cholic acid; other acids	Haslewood (1967b)

TABLE I (*Continued*)

Species	Acids found	References
Lepidosauria (*Continued*)		
Serpentes		
Typhlopidae		
Typhlops jamaiciensis	Cholic acid	Haslewood (1967b)
Aniliidae		
Cylindrophus rufus	Pythocholic acid (probably)	Haslewood (1967b)
Boidae		
Python reticulatus	Cholic acid; pythocholic acid; $3\alpha,12\alpha$-dihydroxy-7-oxo-5β-cholan-24-oic acid	Haslewood (1951); Kuroda and Arata (1952)
Python molurus; Python sebae	Pythocholic acid	Haslewood and Wootton (1950, 1951)
Boa constrictor constrictor; Boa constrictor imperator	Pythocholic acid	Haslewood and Wootton (1951)
Boa constrictor occidentalis	Pythocholic acid; cholic acid	Haslewood and Wootton (1950, 1951)
Corallus canina	Pythocholic acid; cholic acid	Haslewood and Wootton (1951)
Corallus enhydris	Cholic acid	Haslewood (1967b)
Epicrates cenchria	Pythocholic acid	Haslewood and Wootton (1951)
Eunectes murinus	Pythocholic acid; cholic acid	Haslewood (1951)
Eryx conicus (=Gonglyophis); Eryx jaculus; Eryx johnii	Pythocholic acid	Haslewood (1967b)
Acrochordidae		
Acrochordus javanicus; Chersydrus granulatus	Cholic acid	Haslewood (1967b)
Colubridae		
Anaetulla nasuta (=Dryophis); Boiga blandingi	Cholic acid	Haslewood (1967b)
Boiga dendrophila	Cholic acid	Haslewood and Wootton (1950)
Coluber constrictor mormon; Coluber ravergieri; Coluber viridiflavus; Crotaphopeltis natamboeia; Dipsas variegata trinitas; Dispholidus typus	Cholic acid	Haslewood (1967b)
Drymarchon corais couperi	Cholic acid	Haslewood and Wootton (1950)
Duberria lutrix	Cholic acid	Haslewood (1967b)
Elaphe quadrivirgata	Cholic acid	Iwato and Watanabe (1935)
Elaphe carinata	Cholic acid	Imamura (1940)
Elaphe moellendorfi; Elaphe quatorlineata; Elaphe situla; Elaphe taeniurus; Lampropeltis getulus; Malpolon monspesselana; Natrix natrix; Philodryas olfersi (=Chlorosoma); Philothamnus	Cholic acid	Haslewood (1967b)

TABLE I (*Continued*)

Species	Acids found	References
Lepidosauria (*Continued*)		
hoplogaster (= Chlorophis); Pituophis sayi; Psammophis condanarus; Psammophis sibilans; Ptyas mucosus; Rhamphiophis nostratus; Telescopus fallax (= Tarbophis); Thelotornis kirtlandi; Xenochrophis piscator (= Fowlea)		
Enhydris enhydris (= Hypsirhina); Enhydris pakistanica; Enhydris plumbea; Helicops angulatus; Homalopsis buccata; Pseudoboa cloelia; Pseudoboa neuwiedi; Pseudoboa pelota	$3\alpha,7\alpha,12\alpha,23\xi$-tetrahydroxy-$5\beta$-cholan-24-oic acid	Haslewood (1967b)
Elapidae		
Bungarus multitinctus	Cholic acid	Imamura (1940)
Bungarus fasciatus	Cholic acid	Haslewood (1967b)
Dendroapsis viridis	Cholic acid	Haslewood and Wootton (1950)
Dendroapsis augusticeps; Dendroapsis jamesoni kaimosae; Enhydrina schistosa; Haemachatus haemachatus (= serpedon); Hydrophis cyanocinctus	Cholic acid	Haslewood (1967b)
Naja nivea	Cholic acid	Haslewood and Wootton (1950)
Naja goldi; Naja haje; Naja melanoleuca; Naja naja; Naja nigricollis; Notechis scutatus; Pseudechis porphyriacus	Cholic acid	Haslewood (1967b)
Viperidae		
Viperinae		
Atheris squamiger	$3\alpha,7\alpha,12\alpha,23\xi$-tetrahydroxy-$5\beta$-cholan-24-oic acid; other acids	Haslewood (1967b)
Bitis lachesis (= arietans); Bitis gabonica	$3\alpha,7\alpha,12\alpha,23\xi$-tetrahydroxy-$5\beta$-cholan-24-oic acid; allocholic, bitocholic and (probably) cholic acids	Haslewood (1961)
Bitis nasicornis; Bitis worthingtoni	$3\alpha,7\alpha,12\alpha,23\xi$-tetrahydroxy-$5\beta$-cholan-24-oic acid; other acids	Haslewood (1967b)
Cerastes cerastes; Echis carinatus	Cholic acid	Haslewood (1967b)

TABLE I (*Continued*)

Species	Acids found	References
Lepidosauria (*Continued*)		
Eristocophis macmahoni; Vipera ammodytes; Vipera berus; Vipera palaestinae	$3\alpha,7\alpha,12\alpha,23\xi$-tetrahydroxy-$5\beta$-cholan-24-oic acid; other acids	Haslewood (1967b)
Vipera russelli	$3\alpha,7\alpha,12\alpha,23\xi$-tetrahydroxy-$5\beta$-cholan-24-oic acid; allocholic, bitocholic and (probably) cholic acids	Haslewood (1967b)
Crotalinae		
Agkistrodon piscivorus	Cholic acid	Haslewood (1967b)
Bothrops alternatus; Crotalus terrificus	Cholic acid	Deulofeu (1934)
Crotalus horridus; Crotalus oregonus	Cholic acid	Haslewood and Wootton (1950)
Crotalus adamanteus; Crotalus atrox; Crotalus confluentes; Lachesis muta	Cholic acid	Haslewood (1967b)
Trimeresurus flavoviridis	Cholic acid; deoxycholic acid	Inai and Okada (1964)
Trimeresurus wagleri; Trimeresurus rhombeatus	Cholic acid	Haslewood (1967b)
Archosauria		
Crocodylia		
Alligator mississipiensis	$25R$-$3\alpha,7\alpha,12\alpha$-trihydroxy-5β-cholestan-26-oic acid; $3\alpha,7\alpha$-dihydroxy-5β-cholestan-26-oic acid	Haslewood (1952); Dean and Whitehouse (1966)
Caiman crocodilus	$25R$-$3\alpha,7\alpha,12\alpha$-trihydroxy-5β-cholestan-26-oic acid; other acids	Haslewood (1952)
Caiman latirostris; Caiman sclerops	$25S$-$3\alpha,7\alpha,12\alpha$-trihydroxy-5β-cholestan-26-oic acid	Shah *et al.* (1968)
Crocodylus niloticus	$25R$- and $25S$-$3\alpha,7\alpha,12\alpha$-trihydroxy-5β-cholestan-26-oic acids; other acids	Haslewood (1952); Mendelsohn and Mendelsohn (1969)
Crocodylus johnsonii	$3\alpha,7\alpha,12\alpha$-trihydroxy-5β-cholestan-26-oic acid (probably)	Haslewood and Sjövall (1954)
Crocodylus acutus	$25R$-$3\alpha,7\alpha,12\alpha$-trihydroxy-5β-cholestan-26-oic acid; other acids	Haslewood (1967b)
Gavialis gangeticus	$25S$-$3\alpha,7\alpha,12\alpha$-trihydroxy-5β-cholestan-26-oic acid	Shah *et al.* (1968)

REFERENCES

Amimoto, K., Hoshita, T., and Kazuno, T. (1965). *J. Biochem.* (*Tokyo*) **57**, 565.

Ayaki, Y., and Yamasaki, K. (1970). *J. Biochem.* (*Tokyo*) **68**, 341.

Bergström, S., Danielsson, H., and Kazuno, T. (1960). *J. Biol. Chem.* **235**, 983.

Berséus, O. (1965). *Acta Chem. Scand.* **19**, 325.

Bridgwater, R. J., and Lindstedt, S. (1957). *Acta Chem. Scand.* **11**, 409.

Collings, B. G., and Haslewood, G. A. D. (1966). *Biochem. J.* **99**, 50P.

Danielsson, H., and Tchen, T. T. (1968). *In* "Metabolic Pathways" (D. M. Greenberg, ed.), 3rd ed. Vol. 2, pp. 117–168. Academic Press, New York.

Dean, P. D. G., and Whitehouse, M. W. (1966). *Biochem. J.* **99**, 9P.
Deulofeu, V. (1934). *Hoppe-Seyler's Z. Physiol. Chem.* **229**, 157.
Haslewood, G. A. D. (1951). *Biochem. J.* **49**, 718.
Haslewood, G. A. D. (1952). *Biochem. J.* **52**, 583.
Haslewood, G. A. D. (1961). *Biochem. J.* **78**, 352.
Haslewood, G. A. D. (1967a). *J. Lipid Res.* **8**, 535.
Haslewood, G. A. D. (1967b). "Bile Salts." Methuen, London.
Haslewood, G. A. D., and Sjövall, J. (1954). *Biochem. J.* **57**, 126.
Haslewood, G. A. D., and Wootton, V. (1950). *Biochem. J.* **47**, 584.
Haslewood, G. A. D., and Wootton, V. (1951). *Biochem. J.* **49**, 67.
Hoshita, T., Shefer, S., and Mosbach, E. H. (1968). *J. Lipid Res.* **9**, 237.
Imamura, H. (1940). *J. Biochem. (Tokyo)* **31**, 21.
Inai, Y., and Okada, S. (1964). *Hiroshima J. Med. Sci.* **13**, 329; *Chem. Abstr.* **64**, 5511f.
Iwato, M., and Watanabe, K. (1935). *J. Biochem. (Tokyo)* **21**, 211.
Kanemitu, T. (1942a). *J. Biochem. (Tokyo)* **35**, 173.
Kanemitu, T. (1942b). *J. Biochem. (Tokyo)* **35**, 409.
Kim, C. H. (1939). *J. Biochem. (Tokyo)* **30**, 247.
Kuroda, M., and Arata, H. (1952). *J. Biochem. (Tokyo)* **39**, 225.
Mendelsohn, D., and Mendelsohn, L. (1969). *Biochem. J.* **114**, 1.
Mitropoulos, K. A., and Myant, N. B. (1967). *Biochem. J.* **103**, 472.
Okuda, K., Horning, M. G., and Horning, E. C. (1968). *Proc. Int. Congr. Biochem. 7th, 1967* Abstracts, Vol. IV, p. 721.
Shah, P. P., Staple, E., and Rabinowitz, J. (1968). *Arch. Biochem. Biophys.* **123**, 427.
Tammar, A. R. (1970). Ph.D. Thesis, University of London.
Yamasaki, K., and Yuuki, M. (1936). *Hoppe-Seyler's Z. Physiol. Chem.* **244**, 73.

CHAPTER 13

Seasonal Variations in Reptiles

M. Gilles-Baillien

I. Introduction 353
II. Seasonal Variations in Cold-Climate Reptiles 356
A. Blood 357
B. Urine 362
C. Various Tissues 362
D. Various Membranes 365
III. Seasonal Variations in Reptiles from Dry-Summer Regions 367
IV. Origin of Seasonal Variations 368
A. Temperature 368
B. Photoperiod 370
C. Drought 371
D. Summary 372
V. Concluding Remarks 373
References 374

I. Introduction

Reptiles are classified among poikilothermous animals, their body temperature varying with that of the surroundings. Indeed, temperature can be a matter of life or death when the lethal minimum or maximum is reached (Cowles and Bogert, 1944). Within these limits, temperature ranges can be determined as corresponding to various levels of activity segregating thermal preferences from thermal tolerances (see Guibé, 1970). Within the extreme limits of temperature compatible with activity, however, reptiles are able to control their internal temperatures by various means, mainly behavioral (see Templeton, 1970). Many species can absorb solar radiations (heliotherms). Other species can achieve thermal exchanges by conduction with a substratum at a different temperature from the ambient one (thigmotherms). Also, reptiles can maintain a lower body temperature by hiding from the sun using the shelters provided by their biotope. These forms of thermoregulation are called ecological (since they are closely related to the characteristics of the environment) or behavioral. On the other hand, several mechanisms have been shown to be at play in reptiles to yield what is termed physiological thermoregulation. Some snakes (Benedict, 1932), certain lizards (Cole, 1943), and varans (Bartholomew and Tucker, 1964) are able to elevate their body temperature by several degrees through muscular activity. Some reptiles can change the conductivity of their body by circulatory adjustments. Some others also have the possibility of changing the color of their skin in order to modify the rate of absorption of solar radiation. These last processes are probably mediated through the nervous system (see Bentley, 1971; Schmidt-Nielsen, 1964).

When ecological or physiological thermoregulation is no longer able to maintain internal temperature compatible with activity, the rate of metabolism slows down, and in regions with either cold winters or hot and dry summers, reptiles often undergo either hibernation or estivation. The word estivation has been used to describe the latency affecting poikilotherms as well as mammals and birds under hot but above all dry conditions. Estivation in reptiles, though often mentioned in various species, has not been the subject of detailed or numerous investigations as in amphibians, fishes, or rodents. Yet is seems that estivation is far from being such an elaborate behavior as is the case in hibernation (see Schmidt-Nielsen, 1964).

In contrast, the use of the word hibernation for poikilotherms has raised an objection from workers studying hibernation phenomena in warm-blooded animals. As a matter of fact it has been maintained that reptiles and amphibians are not true hibernators and this on the basis of the following reasons: (1) The wintersleep affecting poikilotherms does not involve hypothermia (Matson, 1946). (2) It is not internally controlled and regulated as in mammals (Cagle, 1950). While the first assumption is obligatorily correct, the second is gratuitous. Nonetheless the debate has incited some authors to propose new words when speaking of the deep winter sleep affecting reptiles. Eisentraut (1933) has recommended "rigidity" (translated from the german word "starre"), while more recently Mayhew (1965b) proposed "brumation" to describe the wintersleep of poikilotherms, specifying that it refers to animals whose sleep has a more complex determinism than a decrease in temperature. The same author made the discrimination among reptiles between "obligatory" and "facultative" hibernators, the latter being able to remain active throughout the year if provided with food and maintained at a high temperature. In fact, it seems that many species of reptiles become sleepy solely according to the temperature, but many others undergo a much deeper sleep the determinism of which is internally controlled, light and temperature being however implicated.

This sounds much like what is described in mammals where "hibernation" is the stock term used to characterize the sleep affecting certain mammals of the Northern Hemisphere. The hibernating mammal is said to be "torpid." These terms are restricted to animals that are unable to move when awakened. In contrast, the bear, for instance, would be called "dormant" and said to undergo "dormancy," because it is able to move and shun or fight the disturbance if awakened (Kayser, 1961). The same gradation in winter latency being found among reptiles, it seems unjustified to forge new words to describe the nature of latency in reptiles on the plea that we are dealing with another animal class. Reptiles whose wintersleep is essentially triggered by a lowering of temperature will be said to be dormant or undergoing dormancy. In these species, a sudden warming up, either artificial or natural causes the animal to awake even in the winter time.

But reptiles living in the Northern Hemisphere in regions affected by hard winters are characterized by a winter sleep that has a more complex determinism: It allows us to speak of hibernation with the same meaning as when considering mammalian hibernators, though some characteristics of this sleep must be different in mammals and in reptiles. This is the case in species whose sleep cannot be interrupted or avoided simply by a high temperature or started at any time of the year. Hibernation here is the consequence of a metabolic preparation which is of intrinsic origin and genetically programmed. However, now it is unquestionable that temperature and photoperiod have to superimpose their part on the metabolic preparation to yield the typical behavior of the hibernating animal (Dessauer, 1953; Mayhew, 1965b; Gilles-Baillien, 1966, 1973c; Florkin and Schoffeniels, 1969). In what is designated by the term "hibernation" we see the consequence of a physiological and hereditary adaptation that allows the animal to survive periods where climatic conditions do not favor active life or growth. This evolutionary adaptation to the environment, allowing the survival of individuals in colder climates, results in an annual metabolic cycle. But to insure the survival of the species, an annual sexual cycle has also appeared in many species but in various forms, all converging, however, to give birth at the most propitious time of the year for the survival of the newly born reptiles. The pattern of testicular activity for instance is generally of the postnuptial type in Chelonia, while in lizards it is of the prenuptial type. In snakes, both patterns can be encountered (Saint-Girons, 1963; Lofts, 1969). The annual metabolic cycle and the sexual one seem to have evolved independently. In *Vipera berus,* for instance, there is a period of hibernation lasting for several months (Vitanen, 1967), while the spermatogenesis of the same species is of the continuous type (Volsoe, 1944) as in species of tropical regions having a fairly constant climate.

As to seasonal variations eventually occurring in species located in regions with hot and dry summers, little is known up till now. It is well established that numerous species undergo estivation and that humidity is critical to the phenomenon. The term dehydration describes the first step of the process leading to estivation. But does estivation involve a genetically controlled metabolic preparation? Too little information is available in this field to attempt any speculation.

To summarize, observed seasonal variations can result from the influence of physical and natural factors such as temperature, light, humidity or drought, availability of food, etc. But they can also be the consequence of the existence of an annual metabolic cycle that has appeared as a hereditary adaptation to the environment. This chapter gives a brief account of some biochemical aspects of these seasonal variations.

II. Seasonal Variations in Cold-Climate Reptiles

In reptiles that must withstand cold winters, it is commonly assumed that the winter season is accompanied by a lack of appetite (Dessauer, 1955a) and often anorexia (Mayhew, 1965b), by a weight loss attributed in major part to dehydration, and by a decreased metabolic rate. The lizard *Anolis carolinensis* gains weight in late summer and fall, then loses it during the winter time (Dessauer, 1953). A weight loss also occurs in the tortoise *Testudo hermanni* in the course of hibernation (Gilles-Baillien, 1973c). In the case of the diamondback terrapin *Malaclemys centrata*, an important weight gain takes place in the animals at the end of the hibernating period when living in seawater. In the same species acclimatized to fresh water, no significant weight variations is recorded in animals undergoing hibernation (Gilles-Baillien, 1973a). In two species of freshwater turtles native to Ontario (*Graptemys geographica* and *Chelydra serpentina*) it is shown that the total body water is significantly lower in summer than in winter (Semple *et al.*, 1969).

As for the metabolic rate in terms of oxygen consumption and carbon dioxide production, in the two lizards *Dipsosaurus dorsalis* (Moberly, 1963) and *Phrynosoma m'calli* (Mayhew, 1965b), a reduced metabolic rate is observed at relatively high temperatures in winter. In contrast, in *Anolis carolinensis* (Dessauer, 1953) as well as in *Caiman latirostris* (Coulson and Hernandez, 1964), no marked effect of the season upon metabolic rate is recorded.

Therefore, it already appears that the pattern of seasonal variations could be different depending on whether the reptile is terrestrial or living in seawater or living in fresh water, and furthermore, within one group differences would exist from one species to the other. This statement can be further illustrated by the analysis of seasonal variations occurring in the blood.

A. Blood

1. Inorganic Ions

In two species of Chelonia, one living in fresh water, *Pseudemys scripta*, the other terrestrial, *Terrapene carolina*, sodium is reported to decrease in the blood of hibernating animals compared to active ones, while chloride and potassium are not influenced significantly (Hutton and Goodnight, 1957). In contrast, in the snake *Vipera aspis*, a species that undergoes a deep winter sleep (Duguy, 1963a), sodium together with potassium remain rather constant in the course of the year (Izard *et al.*, 1961). In the tortoise *Testudo hermanni*, sodium and chloride are enhanced progressively starting in August and reach maximal values at the end of the hibernating period, the arousal resulting in an abrupt decrease of these two ions; the plasma potassium concentration in the same species is not significantly affected (Gilles-Baillien and Schoffeniels, 1965). In the grass snake *Natrix natrix*, the plasma sodium concentration also shows a tendency to increase during winter months (Binyon and Twigg, 1965). Finally, in the lizard *Varanus griseus*, sodium and chloride are reported to decrease during winter while potassium is augmented (Haggag *et al.*, 1965). Similar changes take place in the blood of another lizard, *Uromastyx hardwickii* (Zain-Ul-Abedin *et al.*, 1969). Also, Bentley (1959, 1971) reports that the lizard *Trachysaurus rugosus* has higher sodium and potassium plasma concentrations in summer than in winter. But this animal can undergo dehydration conditions in the summer (Section III) while it does not have to withstand very cold temperature in winter.

The situation is a little clearer in aquatic species. In the softshell turtle *Trionyx spinifer*, the sodium concentration in the blood plasma is decreased by 52% in hibernating animals when compared to the active ones (Dunson and Weymouth, 1965). A decrease in sodium and chloride is observed during the winter months in two other species of Chelonia, *Graptemys geographica* and *Chelydra serpentina* (Semple *et al.*, 1969). Lower sodium, chloride, and potassium concentrations are also recorded in the diamondback terrapin *Malaclemys centrata* when hibernating in fresh water. In contrast, when the same species is allowed to hibernate in seawater (its natural habitat being brackish water), the lowest values of chloride are observed in July and of sodium in July and February; for both sodium and chloride the highest values are obtained at the end of the hibernating period and in spring. As for plasma potassium, its concentration follows the same pattern as in the freshwater-acclimatized turtles—the lowest value during hibernation and the highest in summer (Gilles-Baillien, 1973a).

The evolution of magnesium and calcium concentrations in the blood at the different seasons of the year has also been studied—magnesium because of its well known narcotizing power in various animals, calcium because of its influence on the excitability of nerves and muscles. But here, too, contradictory information is available. While magnesium is shown to increase and phosphorus to decrease during the hibernation of *Pseudemys scripta* and *Terrapene carolina* (Hutton and Goodnight, 1957), magnesium, calcium, and phosphorus are observed to decrease in *Crotalus horridus* (Carmichael and Petcher, 1945). In the case of *Varanus griseus,* calcium and phosphorus, but mainly magnesium are importantly enhanced during the hibernation (Haggag *et al.,* 1965).

2. *Glycemia*

Prado (1946) has recorded no significant seasonal variations in the glycemia of *Bothrops jararaca* and *Bothrops phylodrias* when becoming "inactive." In *Uromastyx hardwickii,* the blood glucose level also remains rather constant throughout the year, this species being reported as dormant during cold months and "not seeming to ingest food" (Zain-Ul-Abedin and Katorski, 1967). The same situation is observed in the case of *Natrix natrix* (Binyon and Twigg, 1965). On the contrary, blood sugar undergoes significant seasonal fluctuation in *Pseudemys scripta* and *Terrapene carolina* and increment being measured during hibernation (Hutton and Goodnight, 1957). And in contrast, *Alligator mississippiensis* (Hopping, 1923; Coulson *et al.,* 1950), *Uromastyx aegyptia* (Khalil and Yanni, 1959), *Vipera aspis* (Agid *et al.,* 1961), and *Varanus griseus* (Haggag *et al.,* 1966a) show the highest value of their glycemia in summer and the lowest in winter. When blood glucose is measured month by month in *Alligator mississippiensis,* the evolution of the values is as follows: a high value in the summer months followed by a drop in October, which is then progressively resumed in the winter and spring months (Coulson and Hernandez, 1964). The same pattern occurs in *Anolis carolinensis* (Dessauer, 1953).

3. *Nitrogenous Compounds*

The nonprotein nitrogen is observed to increase tremendously in the blood of *Varanus griseus* during hibernation. As the excretion of the animal is impaired during hibernation, this increase is attributed to the accumulation of excretory products (Haggag *et al.,* 1966a), but the products involved have not been determined. According to Hutton and Goodnight (1957), uric acid is increased in the blood of *Pseudemys scripta* and *Terrapene carolina* while hibernating. In the same chelonians no significant change affects urea and ammonia levels. In contrast, Carmichael and Petcher (1945) report a decrease in both urea and uric acid when *Crotalus horridus* hibernates. When studying the evolution of the blood osmotic pressure in *Testudo hermanni* during the year, a progressive increase starting in the late summer is observed, which lasts until the end of the hibernating period. The major component responsible for this increase is urea, the concentration of which goes from 4 m*M* in summer to more than 100 m*M* at the end of hibernation (Gilles-Baillien and Schoffeniels, 1965). In *Malaclemys centrata* undergoing hibernation in fresh water, there is an increment in urea concentration in late summer and fall (from 9 m*M* to 40 m*M*) which is partially resumed during hibernation. In the same species kept in seawater, the urea concentration, minimal in July (15 m*M*), starts to increase in September and the following months until it reaches a maximum value (100 m*M*) just before arousal (Gilles-Baillien, 1973a).

Other nitrogenous compounds of the blood are shown to undergo seasonal fluctuations. Campbell and Turner (1937) have observed a tendency to decrease protein nitrogen during the winter months in the turtle *Malaclemys geographica.* The blood serum proteins of *Chrysemys picta* show two maxima, one at the end of the nonfeeding period, the other still higher at the end of the feeding period (Masat and Musacchia, 1965). According to Haggag *et al.* (1966a), the serum proteins of *Varanus griseus* undergo only a slight decrease during hibernation.

Finally free amino acids, though in low concentration in the blood plasma are also shown to vary according to the season in *Testudo hermanni,* the higher values being recorded in summer. A progressive decrease starting in September leads to extremely low values at the end of the hibernating period (Gilles-Baillien, 1969a).

4. Blood Cells

On the basis of indications given essentially by red blood cell count (RBC), hemoglobin content, and hematocrit, some authors have reported hemodilution or hemoconcentration occurring at various periods of the year. For instance, Musacchia and Grundhauser (1958, 1962) have reported that in *Chrysemys picta* a hemodilution occurs during spring and late summer, a hemoconcentration taking place during winter and other parts of the year. These results are corroborated by those obtained in *Pseudemys scripta* (Kaplan, 1960; Kaplan and Rueff, 1960).

In the same *Ps. scripta* as well as in *Terrapene carolina,* Hutton and Goodnight (1957) also report an increased RBC in hibernating animals. This is also the case for *Crotalus horridus* (Carmichael and Petcher, 1945). In contrast, the RBC decreases by 35% during the hibernation of *Varanus griseus,* the hemoglobin content (gm/100 ml blood) also decreasing by 35% (Haggag *et al.,* 1966a). However when the hemoglobin content is expressed per unit volume of erythrocytes, lower values are recorded in winter for *Graptemys geographica* and *Chelydra serpentina* (Semple *et al.,* 1969).

The RBC determined month by month in *Vipera aspis* indicates an annual cycle apparent mostly in males, this cycle being partly masked in females by the reproductive cycle (Duguy, 1963a). In the male, one can summarize the situation by saying that the RBC progressively increases from October until March, very low values being recorded during the two mating periods occurring in spring and autumn; the highest values of the RBC are obtained in summer. When counting the erythroblasts at different months of the year in the same species, an annual cycle in the erythropoiesis has also to be taken in account. Indeed it is maximal in the period preceding hibernation. Then it decreases slowly during hibernation, arousal, and the period of sexual activity in spring. After that it remains practically nonexistent during summer and the beginning of autumn (Duguy, 1963a, 1970a,b). For *Cordylus vittifer* and *Cordylus giganteus,* the percentage of erythroblasts increases just before and just after the hibernating period (Villiers Pienaar, 1962).

The RBC and the hematocrit have also been measured at various periods of the year in *Malaclemys centrata* kept either in fresh water or in seawater (Gilles-Baillien, 1973b). The hematocrit is always lower in the freshwater-acclimatized turtles than in the seawater ones. However, in the freshwater ones, the highest values of the hematocrit are obtained in spring and summer. In seawater turtles, the peak values are recorded at the end of the hibernating period and the lowest in July. The RBC is maximal for both groups of turtles at the end of hibernation. In the seawater group, high values of the RBC are obtained also in September. Establishing the ratio between hematocrit and RBC at the different periods of the year, the mean corpuscular volume of the red blood cell has been determined in the two groups of *Malaclemys centrata.* In freshwater-acclimatized turtles the highest values of the red blood cell volume are recorded during activity and minimal values during hibernation. The reverse situation takes place in seawater animals; a lower volume during the active period which increases during hibernation. These seasonal variations of the volume of the red blood cell together with changes in the inorganic ion content appear to be the result of changes in the permeability characteristics of the red blood cell membrane but also reflect the aptitude of the red blood cell to achieve isosmoticity (Gilles-Baillien, 1973b).

The inorganic ion content of erythrocytes has also been observed to vary according to the seasons in *Graptemys geographica* and in *Chelydra serpentina,* but these variations are far less important than those observed in *Malaclemys centrata* (Semple *et al.*, 1969).

As far as leukocytes are concerned, few authors have studied their annual cycle in reptiles. Nevertheless, in 1923, Michels had already observed that there are very high amounts of eosinophiles in *Clemmys leprosa* when undergoing hibernation. In 1965, Binyon and Twigg reports an important leukopenia to occur during winter in *Natrix natrix.* But it is only through detailed studies of Duguy (1963a,b, 1967, 1970a,b) that a definite annual cycle in the leukocytes of reptiles was ascertained. Essentially, the seasonal variations are particularly marked for lymphocytes and eosinophiles. An investigation carried out on five different species of reptiles native to France suggests that cold-climate reptiles are characterized by a "winter leucogram" and by a "summer leucogram" with no abrupt transition between both. This leads the author to conclude that the modifications of the blood formula are chronologically independent of the onset of hibernation.

Other scattered indications of seasonal variations in the blood of reptiles are worth noting. For instance, in 1923, Hopping pointed out the fact that the oxygen capacity of the blood of *Alligator mississippiensis* varies according to the season. The oxygen unsaturation of the venous blood is slight in winter (20%) and very high in summer (80%), and the carbon dioxide content is 20% higher in early spring than in summer or winter. A seasonal shift in oxygen affinity is also observed in *Dipsosaurus dorsalis* (Pough, 1969). Of particular interest too are the observations of Semple *et al.* (1969) concerning the mean plasma volume in map turtles (*Graptemys geographica*). A smaller plasma volume in winter clearly indicates a sequestration of plasma, since the total body water is higher in winter. But in another turtle hibernating under the same conditions at the same place (*Chelydra serpentina*), the mean plasma volume does not vary in winter.

A most interesting study of the blood coagulation has been initiated by Brambel (1941) and has instigated further investigation. In three species of turtles (*Pseudemys concinna, Pseudemys elegans, Chelydra serpentina*), a marked disturbance of the coagulation mechanism during hibernation is reported, this mechanism being restored at the arousal. An increased blood clotting time is also reported in the hibernating *Uromastyx hardwickii* (Zain-Ul-Abedin and Katorski, 1966). Later, seasonal variations of a metachromatic anticoagulant in plasma of the turtle *Pseudemys scripta* are noted (Jacques and Musacchia, 1961; Jacques, 1963). Finally, it has been shown that the three acid mucopolysaccharide factions of the blood (hyaluronic acid, chondroitin sulfate, and heparin) vary independently as a seasonal function. These variations are different in the two species, *Pseudemys scripta* and *Chrysemys picta,* but it is concluded that increased plasma heparin probably constitutes at least part of an endogenous mechanism insuring the maintenance of hemofluidity during dormancy (Kupchella and Jacques, 1970).

B. Urine

The urine of reptiles undergoing hibernation has not been the subject of numerous investigations. It is striking, however, to note in *Testudo hermanni* how the urine volume increases progressively in the course of hibernation and finally occupies the major part of the abdominal cavity at the end of the hibernating period (Gilles-Baillien, personal observations). The animal stops urinating at the onset of hibernation, and the bladder consequently acts as a reservoir where the animal obtains water to avoid dehydration and stores its excretion products as long as hibernation lasts. The osmotic pressure of the urine of torpid tortoises (*Testudo hermanni*) is always higher than that of active tortoises, and in both conditions the urine is isosmotic to the blood. This higher osmotic pressure observed in the urine of torpid tortoises results mainly from an increase in urea concentration and also to a smaller extent in potassium concentration. In contrast, sodium and chloride are significantly decreased (Gilles-Baillien, 1969a). Therefore, in urine as in blood, urea is the major osmoeffector at work during hibernation. The increase in urea does not necessarily imply an enhanced production; anuria could simply explain its accumulation in the urine and also the parallel increase observed in the blood. However, this does not rule out the possibility that the renal function or that protein catabolism has been modified.

C. Various Tissues

Seasonal variations occurring in tissues have raised much less interest than those taking place in the blood. In several cases we have reported modifications of the blood osmotic pressure in the course of the year. On account of the general and classic principle of isosmoticity between cells and the internal medium, any change in the blood osmotic pressure must be adequately balanced within the intracellular fluid. But modifications occurring in the blood can also result from an altered activity of given tissues or glands.

Chronologically, the first indications of seasonal variations in tissues concern the liver. In *Anolis carolinensis* the liver weight is observed to increase in late summer and fall because of the storage of glycogen and lipids. As the animal consumes its body stores during hibernation, glycogen and lipids that have been accumulated are decreased. The parallel modifications in the blood glucose level can be explained in terms of glycogen storage or release (Dessauer, 1953). Besides the change in glycogen and lipids, there is also a variation in proteins and water. In male *A. carolinensis* the liver is shown to contain twenty times more lipids and thirty times more proteins in fall than in spring. In the liver of females a high level of proteins is recorded as well in the egg-laying season (Dessauer, 1955b). A decrease in the liver glycogen during hibernation is also taking place in *Varanus griseus,* a sharp increase being measured on arousal (Haggag *et al.,* 1966b). Also in *A. carolinensis* an important storage of lipids is taking place in the fat bodies, which are enlarged before hibernation (Dessauer, 1955b).

This view, however, is not supported by the work of Duguy (1962) on *Vipera aspis.* Rather than serving, as long thought, as energetic storage to be consumed during the winter, it is proposed and supported (Volsoe, 1944) that the fat bodies essentially act as storage to be almost totally consumed during the period of sexual activity, their weight being little or not affected during winter (see Guibé, 1970). Recently, however, an important role is again attributed to the fat bodies in the hibernation processus. In *Lacerta vivipara,* abdominal and caudal fat bodies represent one-third of the dry body weight in August and September. At the arousal, they only account for one-sixth of the dry body weight (Avery, 1970). A more refined study performed in *Uromastyx hardwickii* indicates that esterified fatty acids and cholesterol are reduced in the adipose tissue during arousal and activity. Moreover, the lipid pattern changes according to the seasonal physiological state of the lizard (Afroz *et al.,* 1971). As far as muscle is concerned, it is demonstrated that the glycogen content is maintained and even augmented during hibernation of *V. griseus.* In the skeletal muscle of the same species, there is also an enhancement in high-energy phosphate compounds (creatine phosphate and adenosine polyphosphate) at the expense of inorganic phosphorus, which is explained by a depression in the utilization during hibernation (Haggag *et al.,* 1966b). At the subcellular level, an interesting study in *Chrysemys picta* indicates seasonal fluctuation in the oxidative phosphorylation by cardiac muscle mitochondria (Privitera and Mersmann, 1966). As shown above, the seasonal variations so far reported in tissues have only been considered from the point of view of the energetic resources available to the animal when undergoing hibernation. Less interest has been shown in the modifications occurring in tissues in response to changes taking place in the osmolarity of the blood. However we have some information in hand.

In *Ch. picta,* there are indications of seasonal changes in the chemical constituents of various tissues. But according to Musacchia and Grundhauser (1958, 1962) the most significant variable is the water content. In liver, muscle, kidney, and skin, the lowest value of the water content is obtained in midwinter, the highest in late summer. For the liver, a second annual reduction happens in early summer. But when the inorganic ion content (sodium, potassium, chloride) of various tissues is compared in active and in torpid tortoises (*Testudo hermanni*), the highly significant modifications observed are far from being explained by the slight decrease in the water content of the same tissues in the same species (Gilles-Baillien and Schoffeniels, 1965); furthermore, sodium, potassium, and chloride are differentially affected within the same tissue and also in the four different tissues analyzed by us. For instance, in muscle, sodium is increased by 18% and potassium by 28%, while chloride is not modified. In the bladder mucosa, only an increase in the potassium content is observed (16%). In the jejunum mucosa, sodium is increased by 16% and potassium by 51%, while chloride is decreased by 12%. Finally, in the colon mucosa, only an increment of the sodium content (23%) is recorded (Gilles-Baillien, 1969b). These changes, of course, cannot result from a concentration due to water loss. As a matter of fact, it seems that besides a modification in the permeability characteristics of cellular membranes (which will be dealt with further on), another phenomenon is also implicated in these variations. Indeed, the proportion of free and bound sodium and potassium within three epithelia has been investigated in the same species. In the bladder mucosa, the amount of intracellular sodium which is sequestered or tightly bound reaches 31% in winter, while it is only 14% in summer. As for potassium, it represents 44% of total intracellular potassium in summer and 71% in winter (Gilles-Baillien and Bouquegneaux-Tarte, 1972). As far as colon and jejunum mucosa are concerned, the percentage of nonexchangeable sodium and potassium is also significantly augmented (Gilles-Baillien, 1972). These results imply that other inorganic or organic ions or molecules are involved in the increment of osmotic pressure that has to be achieved intracellularly in order to balance the increase of osmotic pressure measured in the blood of torpid tortoises. At the level of jejunum and colon, taurine and urea concentrations are shown to be higher in hibernating tortoises than in the active ones (Gilles-Baillien, 1966).

It is now classically admitted that endocrine glands play an important part in the osmoregulation of blood and intracellular fluid, but little is known up till now about the annual cycle of endocrine glands in reptiles. In *V. aspis,* Duguy (1962, 1963a) reports seasonal modifications occurring in the activity of the hypophysis. It is decreased in autumn, but gonadotropic activity starts again 1 month before arousal. Interrenal and thyroid glands both present a phasis of winter rest. The study of the annual cycle of endocrines has however been more extended in relation to the sexual cycle (see Saint-Girons, 1970).

In the thyroid of *Natrix natrix,* a seasonal cycle is reported in relation to the colloid content and the height of the epithelium (Binyon and Twigg, 1965). But it is rather surprising to note that no detailed investigation of seasonal variations in the production of various hormones by endocrine glands has yet been developed, at least to our knowledge. Indeed, for more than 10 years, intense work has been devoted to the study of the effects of various hormones on different parameters implicated in salt and water economy in poikilotherms (for a recent review, see Bentley, 1971). And among the different seasonal variations in the blood and in tissues reported above, many of them result from a perturbation of the mechanisms involved in salt and water economy. A wide field of investigation is opened in this direction.

D. Various Membranes

It is obvious that when undergoing hibernation the reptile reduces its exchange with the environment, since anorexia and anuria are of general occurrence in the process of hibernation. Furthermore, oxygen consumption and carbon dioxide rejection are also lowered. As far as the tegument is concerned, it has long been thought that the reptilian skin had a very low permeability; it is now evident that this is far from being the case. (Tercafs and Schoffeniels, 1965; Bentley and Schmidt-Nielsen, 1966; Schmidt-Nielsen and Bentley, 1966). But the water permeability of the reptilian skin has not been studied as a seasonal function. However, in a terrestrial reptile facing hibernation, a problem of water economy occurs. It has been proposed that the animal could use its lipids to produce water to compensate for the water loss (Khalil and Abdel-Messeih, 1962), but a much more economical adjustive mechanism exists at the level of the bladder. In *Testudo hermanni,* it is suggested that the bladder can act as a water reservoir during the hibernating period. The water net flux measured *in vitro* across the bladder mucosa is much lower in torpid tortoises than in active ones. But in torpid tortoises, it can be enhanced by antidiuretic hormones such as vasopressin and vasotocin, while this is not the case in active tortoises (Gilles-Baillien, 1967, 1969c). In a closely related species (*Testudo graeca*), Bentley (1962) finds that antidiuretic hormones remain without any effect either on sodium transport or on the water permeability of the bladder, but this only concerns active tortoises. The acquisition of a sensibility to pituitary hormones together with the reduction of the water permeability appears an abrupt phenomena related to the onset of hibernation. Furthermore, since the water flux is measured in identical conditions *in vitro*, one has to assume that the modifications occurring in the bladder mucosa are inherent to the physiological state of the membrane (Gilles-Baillien, 1969c). Another epithelium—the intestinal epithelium—has been shown to be the seat of intense modifications in the course of hibernation. Working for more than 10 years on the intestinal epithelium of *Testudo hermanni*, we had to face annually the problem of seasonal variations. The most striking changes in the permeability characteristics of this epithelium during hibernation are the followings (Gilles-Baillien, 1970):

1. While in active tortoises sodium and chloride appear to be actively transported in both directions at the level of the jejunum, in torpid tortoises only chloride is still actively transported, the active transport mechanism for sodium being inhibited.
2. At the level of the colon, in aroused tortoises an active transport for sodium is postulated at the serosal border toward the blood compartment, while chloride is actively transported in both directions. In torpid tortoises, all these active transport mechanisms are strongly inhibited.
3. When compared to jejunum and colon mucosa of active tortoises, jejunum and colon mucosa of torpid animals are less passively permeable to sodium, potassium, and chloride at the serosal border. In addition, in torpid tortoises the permeability to potassium of the mucosal border seems to be largely increased in the jejunum mucosa and especially in the colon mucosa.
4. Finally, the transfer of a neutral amino acid, L-alanine, toward the blood, maximal during the months of June, July, and the beginning of September, decreases rather abruptly at the end of September, though the tortoises become torpid only 1 month later. The value of this transfer remains low and of the same magnitude as the transfer in the opposite direction during all the period of hibernation. Intestinal transport of amino acids is also reported to vary in the lizard *Uromastyx hardwickii* during hibernation, the rate of transport being highly depressed (Quadri *et al.*, 1970). In the same species the intestinal transfer of glucose and fructose is greatly reduced in hibernating animals, these sugars being no longer transported preferentially (Latif *et al.*, 1967).

In connection with the inhibition of the sodium active transport at the level of jejunum and colon mucosa, it is interesting to note that such an inhibition has been suggested at other levels. Dunson and Weymouth (1965) propose that the very low sodium concentration measured in the plasma of hibernating *Trionyx spinifer* results from the inhibition of the sodium active transport at the level of the pharyngeal villi. However, it must be pointed out that several genera of freshwater turtles lack these pharyngeal villi. A possible inhibition of the sodium active transport is also suggested at the level of the red blood cell membrane in *Malaclemys centrata* undergoing hibernation (Gilles-Baillien, 1973b). Investigations should therefore be undertaken to see if an inhibition of the sodium active transport is of general occurrence in the cells of reptiles undergoing hibernation. On the other hand, the importance of the hormonal control on the sodium active transport during hibernation should receive some attention.

Another interesting field of investigation is suggested by the experiments of Coulson and Hernandez (1964). In *Alligator mississippiensis*, a defect in the utilization of the blood glucose is noted in winter. A possible explanation is that it could reflect a decreased ability of glucose to enter the cell due either to a decreased insulin production or to an increased production of hormones antagonistic to insulin. But further speculation requires additional data. However, the observations so far collected suggest encouraging possibilities of explanation of some seasonal variations at the level of the permeability characteristics of pluricellular or plasma membranes.

III. Seasonal Variations in Reptiles from Dry-Summer Regions

Very little is known about the seasonal variations occurring in reptiles from dry-summer regions. It is commonly taken for granted that various levels of dehydration affect the animal during the summer, the ultimate step leading to estivation. But proper information taken in animals surviving a dry summer and naturally undergoing seasonal variations is scarce, while investigations carried out with animals submitted to dehydration under laboratory conditions are more numerous and often interpreted as conditions encountered in the natural habitat. The latter will, however, be reported in the next section of this chapter for because of the little information we have in hand, it would be premature to assert that an ecological factor is the sole determinant in changes occurring in the animal undergoing a dry summer, especially when an estivation behavior is involved.

Indeed, reptiles have developed more efficient adaptations to heat and dryness than they have to cold, allowing greater activity in the former (Schmidt-Neilsen and Dawson, 1964). That is probably why ecological hot and dry conditions in which the animal has to suspend its activity are rarely reached. However, a true summer sleep is reported to occur in reptiles from central Asia (Kachkarov and Korovine, 1942) and also in reptiles from tropical and desert regions having a dry summer season; e.g., *Heloderma* or *Gopherus* (Woodbury and Hardy, 1948). Estivation has also occasionally been observed in *Lacerta agilis* and *Lacerta viridis* (see Guibé, 1970). The estivation process, in fact, partly represents a defense process against a too intensive water loss. But to our knowledge, the only two observations indicative of a seasonal biochemical variation occurring in reptiles from a dry summer regions both report a higher blood sodium concentration during mid-summer. In *Trachysaurus rugosus,* a remarkable tolerance to a high level of sodium is observed. Cessation of the urine flow would conserve body water, but at the expense of a disturbance of the body fluid composition when the animal undergoes a dry summer season (Bentley, 1959).

In *Amphibolurus ornatus,* sodium retention operates to protect fluid volumes when water is scarce. During mid-summer, the lizard lacks sufficient free water to continue the excretion of electrolytes ingested in their diet (sodium-rich ants). Sodium is instead retained at elevated concentrations in the extracellular fluid, the volume of which is expanded by intracellular water so as to maintain isosmoticity (Bradshaw and Shoemaker, 1967).

Another interesting observation by Larson (1961) that can be reported here states that lizards (*Sceloporus occidentalis*) captured in the summer have a higher critical thermal maximum (CTM) than have spring animals of the same habitat.

IV. Origin of Seasonal Variations

Various factors have been proposed as prime causes responsible for seasonal variations. Ecological factors were of course first honored in this way.

A. Temperature

For years it has been commonly assumed that seasonal variations were the result of the influence of temperature on metabolic processes. Of course, many biological processes obey van't Hoff's law, and their activity is decreased when temperature is lowered. The incidence of

Q_{10} in the case of reptiles is particularly important, as their body temperatures vary according to the environmental temperature. When considering oxygen consumption, for example, it is of course well established that this is decreased by lowering the environmental temperature (see Templeton, 1970). And the role of temperature in the decreased metabolic rate observed in reptiles in the wintertime is evident. But a hibernating species such as *Phrynosoma m'calli* becomes torpid even if maintained at 35°C throughout winter and shows a metabolic rate significantly lower than that of summer animals but comparable to that of animals hibernating in the field at the same period (Mayhew, 1965b).

Now the question is raised as to the effects of temperature on the chemical composition of blood and tissues. In this area, we must pay more attention to the observations made in cold-acclimatized animals in order to see if variations in the composition of blood and tissues are similar or not to those obtained as a seasonal function. For instance, an increase of magnesium in the blood of reptiles undergoing hibernation is more or less successfully observed in reptiles (see Section II). And in 1950, Platner showed that the magnesium content of the blood is increased in cold-acclimatized *Chrysemys elegans*, a loss being concomitantly recorded in muscle and skin. A similar increase in plasma magnesium is also reported in cold-exposed *Testudo ibera* (Munday and Mahy, 1962). But in this same species, the hematocrit plasma sodium and protein levels are shown to remain very constant within the range of 0–37°C (Munday and Mahy, 1965), while we have seen earlier that the same parameters are subjected to important seasonal variations in closely related species of tortoises, in turtles, or other reptiles. A more direct comparison between seasonal variations and the effects of induced cold torpor is provided by the results of Musacchia and his colleagues obtained in the blood of *Chrysemys picta*. First, it is shown that cold torpor essentially results in a hemodilution, a significant decrease being measured in the hematocrit and in the specific gravity of the whole blood. Furthermore, no alteration of the plasma proteins is recorded (Musacchia and Sievers, 1956). In the liver, during cold torpor the lipids are increased while the glycogenolysis is accelerated, but no change is recorded in the blood glucose level (Rapatz and Musacchia, 1957). This is quite at variance with the results obtained in the same turtle during winter. As a matter of fact, a hemoconcentration is taking place in the wintertime (Musacchia and Grundhauser, 1958, 1962). Moreover, the blood proteins are strongly affected by the seasons (Masat and Musacchia, 1965). As for the variations observed in the lipid and glycogen content of the liver and in the blood glucose, those recorded in cold-acclimated *Ch. picta* do not bear comparison at all with those indicated in other reptiles in winter or in reptiles undergoing hibernation (see Section II).

In *Pseudemys scripta* and *Ch. picta*, induced cold torpor and the period of the year have been shown to affect differently the plasma acid mucopolysaccharides, and as far as heparin is concerned, its increase in the wintertime results from an endogenous mechanism at work during dormancy (Kupchella and Jacques, 1970). In the hibernating species *Testudo hermanni*, it seems that a decrease in temperature incites the animal to bury itself. As a matter of fact, when tortoises are maintained at 20°C throughout winter, they do not bury themselves and they appear as only partially torpid; they do not feed, urinate, defecate, but they keep their eyes open and retain some muscular response when disturbed. In these conditions, their torpor could still result from the natural decreased photoperiod to which they are submitted. But lethality within this group is high. Tortoises maintained at 4°C and in the dark bury themselves and start to hibernate if the experiment is performed in autumn. But the same experiment carried out in early spring or in summer causes the animal to die within a few days (Gilles-Baillien, 1966).

In conclusion, temperature can be responsible for some behavioral and biochemical changes that occur in the wintertime but is far from being the only determining factor in the case of hibernating species. The incidence of temperature on seasonal variations has been mostly studied in cold-climate reptiles. In the reptiles from dry-summer regions, the incidence of high temperature has essentially been investigated in relation to the limits of survival.

B. Photoperiod

The part played by photoperiod in the reproductive cycle and in the growth of reptiles need no longer be emphasized (Fox and Dessauer, 1957, 1958; Bartholomew, 1959; Mayhew, 1961, 1964, 1965a; Licht, 1966, 1967, 1968). But how photoperiod is implicated in seasonal variations is harder to trace. However, some interesting data are available that show the importance of photoperiod on the behavior of reptiles. In the lizard *Anolis carolinensis*, food consumption is maximum in summer and minimum in winter, and the variation in appetite seems to be correlated with hours of daylight (Dessauer, 1955a). In *Phrynosoma m'calli* two groups of spring animals, one exposed to a photoperiod of 6 hours per day the other to a photoperiod of 15 hours per day, did not become dormant before the end of autumn when the experiment is terminated. In both groups, however, a lower metabolic rate was obtained comparable to the metabolic rate of the animal normally hibernating at that time of the year (Mayhew, 1965b). In *Testudo hermanni*, a similar experiment is reported with slightly different photoperiods (8 hours per day and 16 hours per day) at a constant temperature of 22°C; but the experiment started in early autumn. The animals exposed to a photoperiod of 8 hours per day became quickly torpid, while those exposed to a photoperiod of 16 hours per day maintain their activity (Gilles-Baillien, 1966). Another group of tortoises exposed to a photoperiod of 16 hours per day since early summer remained active until the end of the hibernating period, consuming food regularly. Within this last group, analyses of the blood composition and of the red blood cell content were performed at different times in winter. The results obtained were comparable to those recorded in control tortoises allowed to hibernate at a low temperature, as far as inorganic ions and urea are concerned. But the tortoises held in a photoperiod of 16 hours per day for 8 months showed pathological signs visible at the level of the liver, pancreas, and gall bladder (Gilles-Baillien, 1973c). In *Alligator mississippiensis,* experiments were designed to study the possible influence of the photoperiod on the blood glucose level, but no effect was recorded (Coulson and Hernandez, 1964).

We are therefore tempted to conclude that in cold-climate reptiles undergoing hibernation, temperature and photoperiod essentially determine the behavior of the animal but are not responsible for all the biochemical variations recorded in the course of a year. Many attempts have been made to define the respective roles of temperature and photoperiod in the onset of hibernation among reptiles (Gilles-Baillien, 1966; Thinès, 1968; Mayhew, 1965b; Aleksiuk, 1970). The disagreement between the different conclusions drawn in these attempts probably results from the fact that different parameters are used to reflect behavioral activity and metabolic depression. At any rate, it means that both factors have a part to play, but at various levels the effects can be intricate, and more experimentation is required to unravel and dissociate the effects of temperature from those of photoperiod.

C. Drought

In reptiles from dry-summer regions, an acute water economy problem occurs when the animal is under natural conditions of dry air and dry food. That is the reason why animals from these regions have often been compared to animals under experimental dehydration. In the field, environmental drought can occur at temperatures compatible with the activity of reptiles. The animal remains active and looks for food, but the lack of water progressively induces dehydration. One could speculate that when an alarming level of dehydration is reached and when temperature becomes too high the animal will stop eating and will try to shelter in order to reduce its water loss. The animal would then undergo a summer latency or estivation. In fact, very little is known about the determinism of estivation in these species.

However it is of interest to report here the known effects of dehydration. In *Chelodina longicollis*, for instance, a 20-day dehydration is reported to affect the type of nitrogenous waste voided, the percentage of uric acid being increased, that of urea being decreased (Rogers, 1966). But this study does not take into account the various factors implicated in the phenomenon—notably the fact that dehydration can involve a longer stay of urine in the bladder. In this case, urea could pass back into the blood at the level of the bladder and equilibrate in all body fluids, including the intracellular fluid, while uric acid precipitates in the bladder. Therefore, the above-mentioned results do not necessarily mean that the nitrogenous excretory pattern has been modified during dehydration. The effects of dehydration have also been studied in *Gopherus agassizii,* a species reported to undergo estivation (Woodbury and Hardy, 1948). This tortoise is highly tolerant to dehydration. Its plasma osmolarity rises markedly under dehydration conditions, and this increase is mainly due to an increment in urea concentration. On the other hand, the renal function of the same tortoise is not significantly affected by dehydration. In contrast the bladder function is important, since the urine enters the bladder only under dehydration conditions while in well hydrated animals the urine runs directly out of the cloaca without entering the bladder (Dantzler and Schmidt-Nielsen, 1966). These results, of course, give only an idea of the seasonal variations that can occur in reptiles under natural dry conditions.

D. Summary

As far as cold-climate reptiles are concerned, we have seen that environmental changes related to the seasons can be responsible for seasonal variations in behavior as well as in the physiological status of the animal. Possibly only ecological factors are at work in species that undergo dormancy (the definition of this term given in Section I). In this case, phenotypic variations involved in acclimatization are responsible for the seasonal variations. But in reptiles that undergo hibernation, these ecological factors can only be superimposed on an annual cycle of the metabolic status that is internally controlled. These reptiles must therefore possess a biological clock, reset every year, which induces cyclic variations of the metabolism (see Florkin and Schoffeniels, 1969). A similar conclusion was formulated to characterize the hibernation of mammals (Pengally and Fischer, 1957). This is why we propose maintaining the term hibernation to characterize the internally controlled winter sleep that affects many reptiles of the cold-climate regions.

The clue to the determination of hibernation, however, is far from being elucidated. Saying that hibernation involves genetically determined adaptations to the environment does not solve the problem. In our opinion, and it is also that of others, the origin of hibernation should be investigated at the level of the endocrine glands and in their production of hormones. But among reptiles, we still lack of valuable information.

V. Concluding Remarks

Seasonal variations reported in the chemistry of cold-climate reptiles are numerous but often contradictory from one species to another. Therefore, in the actual state of our knowledge, it is not possible to outline a general pattern of the seasonal biochemical variations. Nevertheless, it seems that one can expect to have different patterns among the different groups of reptiles. Furthermore, within Chelonia, the group which has been the most extensively studied, the seasonal variations can be entirely different whether the studied animal lives in fresh water or in seawater or whether it is terrestrial. In the future, the study of seasonal variations in a definite species should not be approached without full knowledge of the behavior of the animal during winter. In the case of species showing intermittent dormancy, seasonal variations result from acclimatization to the winter environmental conditions. In the case of animals undergoing hibernation, the seasonal variations reflect an annual metabolic cycle that is genetically determined and that expresses a physiological adaptation to a colder climate. Information concerning the seasonal variations occurring in reptiles from dry-summer regions is scarce. Among the ecological factors that could play a part in estivation, drought is certainly the most important, since it sometimes induces high levels of dehydration. Temperature, lack of availability of food, and photoperiod are probably also implicated in estivation. But seasonal variations and estivation in reptiles from dry-summer regions have been the subject of very few investigations.

In the present state of our knowledge, a more systematic investigation of the seasonal variations occurring in reptiles is necessary for a better understanding of the underlying mechanisms involved. The study of the physiological adaptations implicated in hibernation as well as a better insight into the estivation process offers a wide open investigation field and would throw some light on the broad distribution of reptiles and their colonization of typically different environments.

REFERENCES

Afroz, H., Ishaq, M., and Ali, S. S. (1971). *Proc. Soc. Exp. Biol. Med.* **136**, 894.

Agid, R., Duguy, R., and Saint-Girons, H. (1961). *J. Physiol.* (*London*) **53**, 807.

Aleksiuk, M. (1970). *Can. J. Zool.* **48**, 1155.

Avery, R. A. (1970). *Comp. Biochem. Physiol.* **37**, 119.

Bartholomew, G. A. (1959). *In* "Photoperiodism and Related Phenomena in Plants and Animals," Publ. No. 55, pp. 669–676. *Amer. Ass. Advan. Sci.*, Washington, D.C.

Bartholomew, G. A., and Tucker, V. A. (1964). *Physiol. Zool.* **37**, 341.

Benedict, F. G. (1932). "The Physiology of Large Reptiles with Special Reference to the Heat Production of Snakes, Tortoises, Lizards and Alligators." Carnegie Institution, Washington, D.C.

Bentley, P. J. (1959). *J. Physiol.* (*London*) **145**, 37.

Bentley, P. J. (1962). *Gen. Comp. Endocrinol.* **2**, 323.

Bentley, P. J. (1971). "Zoophysiology and Ecology," Vol. I. Springer-Verlag, Berlin and New York.

Bentley, P. J., and Schmidt-Nielsen, K. (1966). *Science* **151**, 1547.

Binyon, E. J., and Twigg, G. I. (1965). *Nature* (*London*) **207**, 779.

Bradshaw, S. D., and Shoemaker, V. H. (1967). *Comp. Biochem. Physiol.* **20**, 855.

Brambel, C. E. (1941). *J. Cell. Comp. Physiol.* **18**, 221.

Cagle, F. R. (1950). *Ecol. Monogr.* **20**, 31.

Campbell, M. L., and Turner, A. H. (1937). *Biol. Bull.* **73**, 504.

Carmichael, E. B., and Petcher, P. W. (1945). *J. Biol. Chem.* **161**, 693.

Cole, L. C. (1943). *Ecology* (*Brooklyn*) **24**, 94.

Coulson, R. A., and Hernandez, T. (1964). "Biochemistry of the Alligator." Louisiana State Univ. Press, Baton Rouge.

Coulson, R. A., Hernandez, T., and Brazda, F. G. (1950). *Proc. Soc. Exp. Biol. Med.* **73**, 203.

Cowles, R. B., and Bogert, C. H. (1944). *Bull. Amer. Mus. Natur. Hist.* **83**, No. 5, 267.

Dantzler, W. H., and Schmidt-Nielsen, B. (1966). *Amer. J. Physiol.* **210**, 198.

Dessauer, H. C. (1953). *Proc. Soc. Exp. Biol. Med.* **82**, 351.

Dessauer, H. C. (1955a). *Proc. Soc. Exp. Biol. Med.* **90**, 524.

Dessauer, H. C. (1955b). *J. Exp. Zool.* **128**, 1.

Duguy, R. (1962). Ph.D. Thesis, University of Paris.

Duguy, R. (1963a). *Vie Milieu* **14**, 311.

Duguy, R. (1963b). *Bull. Soc. Zool. Fr.* **88**, 99.

Duguy, R. (1967). *Bull. Soc. Zool. Fr.* **92**, 23.

Duguy, R. (1970a). *In* "Biology of the Reptilia" (Gans, ed.), Vol. 3, Chapter 3, pp. 93–109. Academic Press, New York.

Duguy, R. (1970b). *In* "Traité de Zoologie" P.-P. Grassé, ed.), Vol. 14, Part 2, pp. 474–498. Masson, Paris.

Dunson, W. A., and Weymouth, R. D. (1965). *Science* **149**, 67.

Eisentraut, M. (1933). *Mitt. Zool. Mus. Berlin* **19**, 48.

Florkin, M., and Schoffeniels, E. (1969). "Molecular Approaches to Ecology." Academic Press, New York.

Fox, W., and Dessauer, H. C. (1957). *J. Exp. Zool.* **134**, 557.

Fox, W., and Dessauer, H. C. (1958). *Biol. Bull.* **115**, 421.

Gilles-Baillien, M. (1966). *Arch. Int. Physiol. Biochim.* **74**, 328.

Gilles-Baillien, M. (1967). *Ann. Endocrinol.* **28**, 716.
Gilles-Baillien, M. (1969a). *Arch. Int. Physiol. Biochim.* **77**, 427.
Gilles-Baillien, M. (1969b). *Life Sci., Part II* **8**, 763.
Gilles-Baillien, M. (1969c). *Biochim. Biophys. Acta* **193**, 129.
Gilles-Baillien, M. (1970). *Arch. Int. Physiol. Biochim.* **78**, 327.
Gilles-Baillien, M. (1972). *Arch. Int. Physiol. Biochim.* **80**, 789.
Gilles-Baillien, M. (1973a). *J. Exp. Biol.* **59**, 45.
Gilles-Baillien, M. (1973b). *Comp. Biochem. Physiol.* **46A**, 505.
Gilles-Baillien, M. (1973c). *Arch Int. Physiol. Biochim.* **81**, 723.
Gilles-Baillien, M., and Schoffeniels, E. (1965). *Ann. Soc. Roy. Zool. Belg.* **95**, 75.
Gilles-Baillien, M., and Bouquegneaux-Tarte, C. (1972). *Arch. Int. Physiol. Biochim.* **80**, 563.
Guibé, J. (1970). *In* "Traité de Zoologie" (P.-P. Grassé, ed.), Vol. 14, Part 3, pp. 987–1036. Masson, Paris.
Haggag, G., Raheem, K. A., and Khalil, F. (1965). *Comp. Biochem. Physiol.* **16**, 457.
Haggag, G., Raheem, K. A., and Khalil, F. (1966a). *Comp. Biochem. Physiol.* **17**, 335.
Haggag, G., Raheem, K. A., and Khalil, F. (1966b). *Comp. Biochem. Physiol.* **17**, 341.
Hopping, A. (1923). *Amer. J. Physiol.* **66**, 145.
Hutton, K. E., and Goodnight, C. J. (1957). *Physiol. Zool.* **30**, 198.
Izard, Y., Detrait, J., and Boquet, P. (1961). *Ann. Inst. Pasteur, Paris* **100**, 539.
Jacques, F. A. (1963). *Comp. Biochem. Physiol.* **9**, 241.
Jacques, F. A., and Musacchia, X. J. (1961). *Copeia* **2**, 222.
Kachkarov, D. N., and Korovine, E. P. (1942). "La vie dans les déserts." Payot, Paris.
Kaplan, H. M. (1960). *Anat. Rec.* **137**, 369.
Kaplan, H. M., and Rueff, W. (1960). *Proc. Anim. Care Panel* **10**, 63.
Kayser, C. (1961). *Mod. Trends Physiol. Sci.* **8**, 195.
Khalil, F., and Abdel-Messeih, G. (1962). *Comp. Biochem. Physiol.* **6**, 171.
Khalil, F., and Yanni, M. (1959). *Z. Vergl. Physiol.* **42**, 192.
Kupchella, C. E., and Jacques, F. A. (1970). *Comp. Biochem. Physiol.* **36**, 657.
Larson, M. W. (1961). *Herpetologica* **17**, 113.
Latif, S. A., Zain, B. K., and Zain-Ul-Abedin, M. (1967). *Comp. Biochem. Physiol.* **23**, 121.
Licht, P. (1966). *Science* **154**, 1668.
Licht, P. (1967). *J. Exp. Zool.* **165**, 505.
Licht, P. (1968). *J. Exp. Zool.* **166**, 243.
Lofts, B. (1969). *In* "Perspectives in Endocrinology" (E. J. W. Barrington and C. B. Jørgensen, eds.), pp. 239–299. Academic Press, New York.
Masat, R. J., and Musacchia, X. J. (1965). *Comp. Biochem. Physiol.* **16**, 215.
Matson, J. R. (1946). *J. Mammal.* **27**, 203.
Mayhew, W. W. (1961). *Science* **134**, 2104.
Mayhew, W. W. (1964). *Herpetologica* **20**, 95.
Mayhew, W. W. (1965a). *Comp. Biochem. Physiol.* **14**, 209.
Mayhew, W. W. (1965b). *Comp. Biochem. Physiol.* **16**, 103.
Michels, N. A. (1923). *Cellule* **23**, 339.
Moberly, W. R. (1963). *Physiol. Zool.* **36**, 152.
Munday, K. A., and Mahy, B. W. J. (1962). *Int. Congr. Physiol. Sci.* [*Proc.*], *22nd, 1962.* Excerpta Med. Found. Int. Congr. Ser. No. 48, p. 612, 126, 592.

Munday, K. A., and Mahy, B. W. J. (1965). *Life Sci.* **4**, 7.
Musacchia, X. L., and Grundhauser, W. (1958). *Fed. Proc. Fed. Amer. Soc. Exp. Biol.* **17**, Abst. 455.
Musacchia, X. L., and Grundhauser, W. (1962). *Copeia* **3**, 570.
Musacchia, X. L., and Sievers, M. L. (1956). *Amer. J. Physiol.* **187**, 99.
Pengally, E. T., and Fischer, R. C. (1957). *Nature (London)* **180**, 1371.
Platner, W. S. (1950). *Amer. J. Physiol.* **161**, 399.
Pough, F. H. (1969). *Comp. Biochem. Physiol.* **31**, 885.
Prado, J. L. (1946). *Mem. Inst. Butantan Sao Paulo* **19**, 59.
Privitera, C. A., and Mersmann, H. J. (1966). *Comp. Biochem. Physiol.* **17**, 1045.
Quadri, M., Zain, B. K., and Zain-Ul-Abedin, M. (1970). *Comp. Biochem. Physiol.* **36**, 569.
Rapatz, G. L., and Musacchia, X. J. (1957). *Amer. J. Physiol.* **188**, 456.
Rogers, L. J. (1966). *Comp. Biochem. Physiol.* **18**, 249.
Saint-Girons, H. (1963). *Arch. Anat. Microsc. Morphol. Exp.* **52**, 1.
Saint-Girons, H. (1970). *In* "Traité de Zoologie" (P.-P. Grassé, ed.), Vol. 14, Part 3, pp. 681–726. Masson, Paris.
Schmidt-Nielsen, K. (1964). "Desert Animals. Physiological Problems of Heat and Water." Oxford Univ. Press (Clarendon), London and New York.
Schmidt-Nielsen, K., and Bentley, P. J. (1966). *Science* **154**, 911.
Schmidt-Nielsen, K., and Dawson, W. R. (1964). *In* "Handbook of Physiology" (Amer. Physiol. Soc., J. Field, ed.), Sect. 4, pp. 467–480. Williams & Wilkins, Baltimore, Maryland.
Semple, R. E., Sigsworth, D., and Stitt, J. T. (1969). *J. Physiol. (London)* **204**, 39P.
Templeton, J. R. (1970). *In* "Comparative Physiology of Thermoregulation" (G. C. Whittow, ed.), Vol. 1, pp. 167–221. Academic Press, New York.
Tercafs, R. R., and Schoffeniels, E. (1965). *Ann. Soc. Roy. Zool. Belg.* **96**, 9.
Thinès, G. (1968). *Psychol. Belg.* **8**, 131.
Villiers Pienaar, U. (1962). "Hematology of Some South African Reptiles." Witwatersrand Univ. Press, Johannesburg.
Vitanen, P. (1967). *Ann. Zool. Fenn.* **4**, 472.
Volsoe, H. (1944). *Spolia Zool. Mus. Hauniensis* **5**, 1.
Woodbury, A. M., and Hardy, R. (1948). *Ecol. Monogr.* **18**, 145.
Zain-Ul-Abedin, M., and Katorski, B. (1966). *Can. J. Physiol. Pharmacol.* **44**, 505.
Zain-Ul-Abedin, M., and Katorski, B. (1967). *Can. J. Physiol. Pharmacol.* **45**, 115.
Zain-Ul-Abedin, M., Behleem, Z., and Rahman, M. A. (1969). *Pak. J. Biochem.* **2**, 47.

CHAPTER 14

Reptilian Hemoglobins

Bolling Sullivan

I. Introduction 377
II. Hemoglobin Components 378
A. Adult Components 378
B. Fetal Components 384
III. Hemoglobin Function 385
A. Chelonia 385
B. Squamata 388
C. Crocodilia 391
IV. Hemoglobin Structure and Synthesis 393
V. Concluding Comments 394
References 396

I. Introduction

Living reptiles are usually classified into four distinct phylogenetic groups, the Chelonia or turtles, the Squamata or snakes and lizards, the Rhynchocephalia of which only *Sphenodon* remains, and the Crocodilia or gavials, crocodiles, and alligators. Of the five thousand or so living species, only 241 species are not squamates, and most of these are turtles (Porter, 1972). Most snake lineages are distributed throughout the temperate and tropical regions of the world, whereas families of lizards and turtles are often restricted to New or Old World temperate and tropical regions. Living reptiles are the remnants of a group that flourished from the Triassic through the Cretaceous Periods. The surviving groups have settled in a wide array of habitats and show numerous interesting adaptations.

The blood chemistry of reptiles, including their hemoglobins, has been reviewed quite competently by Dessauer (1970). Thus, in this review I will incorporate more recent findings and collate the data from a somewhat different viewpoint. Introductory material and general comments on the structural and functional properties of hemoglobins are contained in the similar chapter on amphibian hemoglobins in this volume and will not be repeated here.

Reptiles have nucleated erythroctyes which are very slowly replaced. The life span of erythrocytes in the box turtle, *Terrepene carolina,* is

600–800 days (Brace and Altland, 1955; Altland and Brace, 1962). Red cell life span is inversely related to temperature (i.e., metabolic activity) and can increase fourfold or more at lower temperatures (Cline and Waldmann, 1962).

Studies of reptilian red cells have been reviewed recently (Saint Girons, 1970); they seem to be rather average in size ($5–15 \times 15–23$ μ). Standard hematological parameters such as blood volume, hemoglobin content, hematocrit, and so forth vary widely in reptiles. These are known to be a function of altitude (Hadley and Burns, 1968; Vinegar and Hillyard, 1972), season (Semple *et al.*, 1970), and diet (Musacchia and Sievers, 1962). Other variations are summarized by Duguy (1970).

II. Hemoglobin Components

Most reptiles have multiple hemoglobins. The early paper electrophoretic studies of Dessauer *et al.* (1957) revealed from one to five components in turtles and lizards. Smaller numbers of components were present in snakes, alligators, and crocodiles. Although electrophoretic techniques have improved considerably since the surveys of reptilian hemolyzates made by Dessauer *et al.* (1957) and Dessauer and Fox (1964), these newer techniques have not been used to repeat earlier work. Methemoglobin and polymerization of hemoglobin tetramers by disulfide bond formation are common occurrences in reptilian hemolyzates (Riggs *et al.*, 1964; Sullivan and Riggs, 1964, 1967a) and must be controlled if meaningful results are to be obtained.

A. Adult Components

It is assumed in this review that the basic structural unit of reptilian hemoglobins is the tetramer ($\alpha_2\beta_2$). Although this structure has not been proved for any reptilian hemoglobin, it is the structure of hemoglobins from almost all vertebrates. At present there is no evidence from studies on reptilian hemoglobins to refute this assumption.

1. *Chelonia*

Dessauer *et al.* (1957) examined hemoglobins from ten species of turtles by paper electrophoresis. They found multiple hemoglobins in all families but the Chelydridae (snapping turtles). They noted trailing areas between the components in emydine samples. Isolation of the "fast" and "slow" hemoglobin components from *Pseudemys scripta* hemolyzates by continuous-flow electrophoresis provided evidence that the two components were interacting. Later studies of these and other species by starch gel electrophoresis have provided better resolution (Dessauer and Fox, 1964; Dozy *et al.*, 1964).

Sullivan and Riggs (1967b) examined hemolyzates from more than fifty chelonian species by starch gel electrophoresis. Prior to electrophoresis, all samples were reduced and alkylated in order to minimize artifacts caused by disulfide polymerization. A diagrammatic comparison of some of these electropherograms is shown in Fig. 1. Hemolyzates from the families Chelydridae and Kinosternidae are not shown but showed multiple components with rather similar electrophoretic properties. The electrophoretic patterns often reflect phylogenetic relationships, and this is seen also in the electrophoretic patterns of the globin chains after removal of the heme (Sullivan and Riggs, 1967b). Newcomer and Crenshaw (1967) studied the serum proteins and hemoglobins of *Geochelone* (*Testudo*) *denticularia* and *G. carbonaria* in an effort to reinforce morphological studies that had elevated the two forms to species status. They found clear differences between hemolyzates from the two species (as had Sullivan and Riggs, 1967b), and noted that hemolyzates from *G. carbonaria* produced a "diffuse, bandless pattern." Of the eight species and subspecies of the Testudinidae examined by Sullivan and Riggs (1967a), hemoglobins from *G. carbonaria* were the only ones that polymerized. No doubt polymerization could account for the diffuse electrophoretic pattern. Lykakis (1971) examined hemolyzates from five European turtles by starch gel electrophoresis and obtained results similar to those of Sullivan and Riggs (1967b). All species had two major hemoglobin bands and four to five minor hemoglobin components.

Hemolyzates from seven species of turtles were examined by starch gel electrophoresis and DEAE-cellulose chromatography (Horton *et al.*, 1972). Their results are in accord with those of previous workers. One unexplained peculiarity of turtle hemoglobins during electrophoresis is shown in their Fig. 2. When hemolyzates from many testudine or emydine species are electrophoresed, they clearly show one slow band and two fast bands in almost equal proportions. If the electrophoresis is continued—Horton *et al.* (1972) stopped many of their experiments at this point—the slower of the two fast bands disappears. In its place is the smear between the two major components first noticed by Dessauer *et al.* (1957). The smear appears to be made up of several minor bands and is present in alkylated and fresh, unalkylated material. I have tried several times, but with little success, to isolate this smear, thinking that the slower of the two fast components was simply dissociating into individual polypeptide chains. Thus, the exact nature of this smearing material is still uncertain and deserves further study. Horton *et al.* (1972) have shown that the slow major component accounts for 60–70% of the hemolyzate, while the fast component accounts for about 30% in emydines. More recent studies of turtle hemolyzates by disk gel electro-

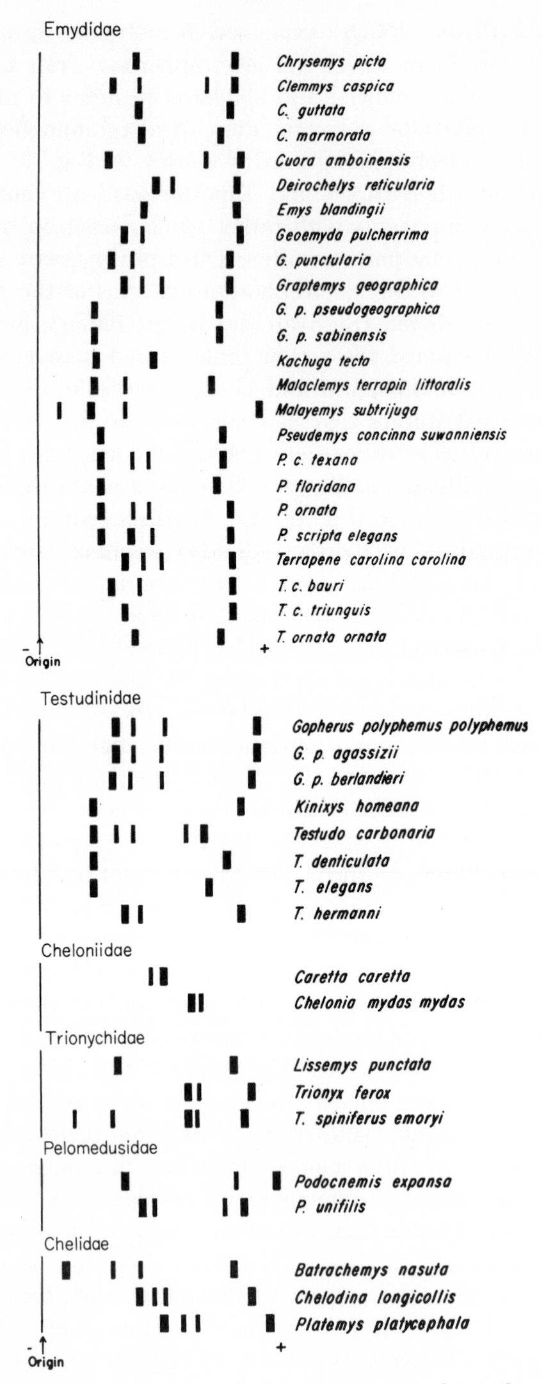

FIG. 1. Diagrammatic comparison of the approximate electrophoretic properties of reduced and alkylated turtle hemoglobins at pH 8.4 (Sullivan and Riggs, 1967b).

phoresis (B. Sullivan, unpublished) indicate that there are two major hemoglobin bands (one fast, one slow) with up to seven minor bands in emydines.

Hemoglobin polymorphism has been reported in *Chrysemys picta* (Manwell and Schlesinger, 1966). These workers noted variation in the slow major component and found different allelic frequencies in populations from Minnesota and Wisconsin versus populations from Arkansas and Louisiana. The morphs were shown to have different oxygen affinities, and this was correlated with adaptation to anoxic conditions. The incidence of hemoglobin polymorphism in turtles appears to be rather low, but it is difficult to know how many samples have been studied. I have examined (B. Sullivan, unpublished results) over seventy individuals of the diamond back terrapin (*Malaclemmys terrapin centrata*) and found no polymorphism. Manwell and Schlesinger (1966) report similar results for *Pseudemys scripta.* It would be of interest to examine large populations of other species to determine whether or not turtles are as polymorphic at their hemoglobin loci as other groups of reptiles and amphibians seem to be.

2. *Squamata*

a. Lacertilia. Dessauer *et al.* (1957) and Dessauer and Fox (1964) studied a number of hemolyzates from lizards. Many had multiple hemoglobins. Additional studies by Guttman (1970a,b, 1971) using starch gel electrophoresis have revealed that lizard hemolyzates are much simpler than those of turtles. These studies are summarized in Fig. 2. Several points should be noted. Most species appear to have one or two bands (Guttman has not always differentiated between major and minor components). Several species are polymorphic for hemoglobins; *Sceloporus undulatus* and *Gerrhonotus multicarinatus* are particularly polymorphic. In spite of Guttman's statement to the contrary, no allelic morphs are being inherited in a clear, Mendelian fashion. They may all be inherited in a straightforward manner, but the evidence doesn't support this yet, perhaps because sufficiently large samples have not been obtained. Guttman (1971) concluded from his survey that the electrophoretic patterns of lizard hemolyzates were phylogenetically inconsistent and therefore of little taxonomic value. It should be pointed out that the charge on hemoglobin molecules is not conservative and may be of no more taxonomic value than the shape of a scale or bone or length of an appendage. On the other hand, the amino acid sequences of the constituent polypeptide chains are probably of considerable taxonomic value, and the number of hemoglobin loci can be of value. This latter character appears

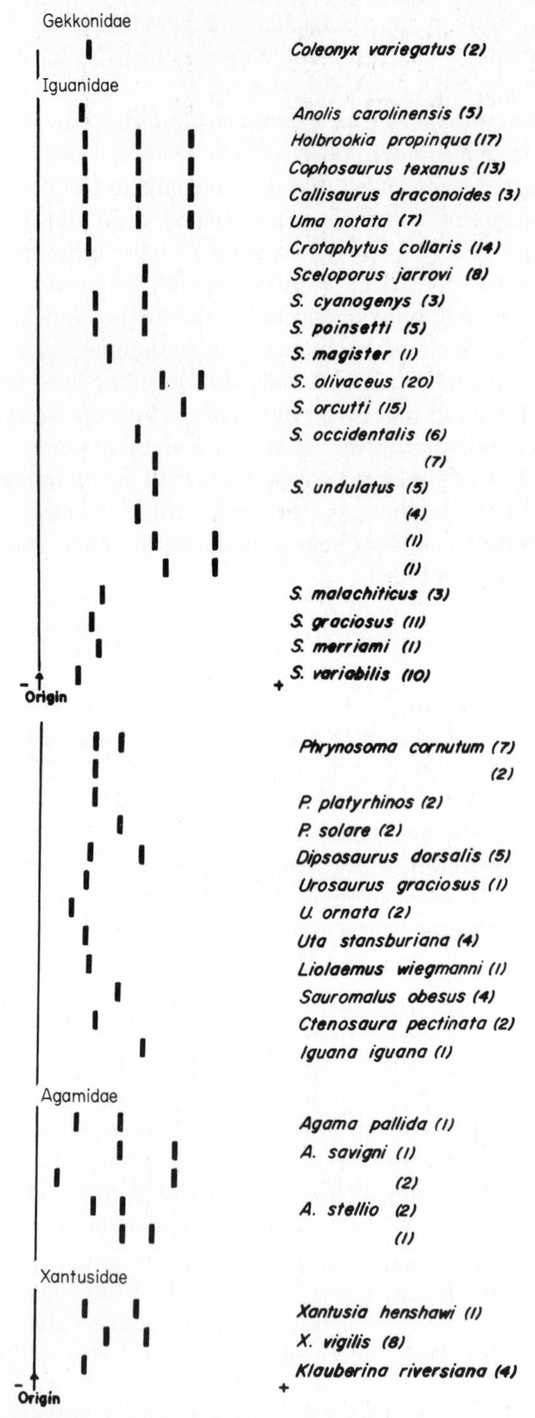

FIG. 2. Diagrammatic comparison of the approximate electrophoretic properties of lizard hemoglobins at pH 8.5 (Guttman 1970a,b, 1971).

to be quite variable in lizards, unlike the situation in turtles, at least at the present level of resolution. Gorman has studied the electrophoretic patterns of hemoglobins of species of *Anolis* in the Caribbean as a supplement to more conventional ecological and systematic studies (Gorman and Dessauer, 1965, 1966; Gorman *et al.*, 1971). Most species show a single hemoglobin on starch gel electrophoresis, but hybrids between species showed both parental bands. Webster *et al.* (1972) resolved two hemoglobin bands in their studies of hemolyzates from four *Anolis* species. Hemoglobin polymorphism was present in two of these species.

b. Ophidia. Dessauer *et al.* (1957) examined hemolyzates from some forty species of snakes. They found some variation from species to species, but because of the very high isoelectric points peculiar to snake hemoglobins, their resolution was severely limited. Snakes clearly have multiple hemoglobins (Sullivan, 1967).

Schwantes (1972) studied the electrophoretic properties of hemolyzates from 24 species and subspecies of the Boidae, Viperidae, and

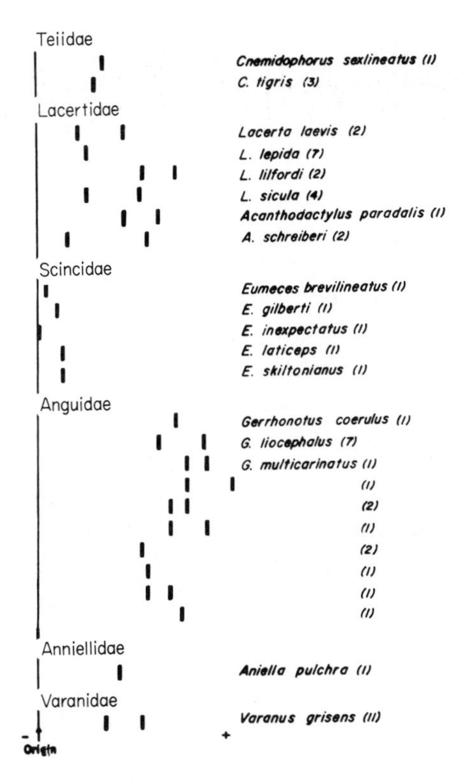

Fig. 2, continued.

Colubridae. As described by other investigators many species had cathodal components indicative of unusually high isoelectric points. Most species had multiple hemoglobin components, but none seemed to polymerize to octameric or larger aggregates. Intraspecific variation in hemoglobin was found for *Boa constrictor amarali* and *Crotalus durissus terrificus* hemolyzates, but some expected phenotypes were clearly missing. This is reminiscent of earlier work on polymorphic hemoglobins in lizards, which also deviate strongly from a Mendelian pattern of inheritance.

3. *Rhynchocephalia*

Hemolyzates from *Sphenodon* have not been examined.

4. *Crocodilia*

Dessauer *et al.* (1957) examined hemolyzates from three crocodilians—*Alligator mississippiensis, Caiman latirostris,* and *Crocodylus acutus.* Each species showed a single band which was species specific in its electrophoretic mobility. Later studies (B. Sullivan, unpublished) have shown that *Alligator* hemolyzates contain one major and two or more minor hemoglobin bands.

B. Fetal Components

The distribution of fetal hemoglobins in vertebrate species is erratic. Even among primates there are species that seem to lack a specific fetal hemoglobin (Sullivan, 1971). On the other hand, all five vertebrate classes contain species alleged to have fetal hemoglobins. The structural and evolutionary relationships among the various fetal hemoglobins would make an interesting study.

The most concrete evidence for the existence of fetal hemoglobins in reptiles comes from the work of Pough (1969). He reports that juvenile *Dipsosaurus* have a single hemoglobin, but adults contain a second, slower-migrating band. He also found (Pough, 1971) that the electrophoretic properties of *Thamnophis sirtalis* hemolyzates change continuously with increasing size of the animal. McCutcheon (1947) reported an increase in oxygen affinity of blood from turtle embryos (*Malaclemmys terrapin centrata*). Similar results were obtained by Manwell (1955) in his study of *Thamnophis.* Ramsey (1941) noted that the amount of alkali-resistant hemoglobin in alligator hemolyzates drops after 2 years of age. Detailed genetic and structural studies are needed to confirm the presence of distinct fetal components in reptiles, but there is every reason to believe that such components are present in most if not all groups of reptiles.

III. Hemoglobin Function

The functional properties of several reptilian hemolyzates and purified hemoglobin components have been studied. As mentioned in the chapter on amphibian hemoglobins, there are a number of pitfalls in measuring the oxygen affinity of hemoglobin solutions. These are discussed in more detail in Antonini and Brunori (1971) and Riggs (1970). It should be pointed out that Prado (1946), Pough (1969), and Sullivan and Riggs (1964) found lizards, snakes, and turtles in nature with very high amounts of methemoglobin (ferrihemoglobin). When large (greater than 5%) amounts of methemoglobin are present, they should, when possible, be removed or converted to ferrohemoglobin prior to measuring the affinity for oxygen. Furthermore, it has been demonstrated that many reptilian hemolyzates have multiple hemoglobins. In order to correctly evaluate oxygen-binding data, one should determine the equivalence or nonequivalence of the various components. Third, it is apparent that numerous reptiles use phosphates to regulate the oxygen affinity of their hemoglobins. Normal preparation of hemolyzates partially removes these molecules and experiments should be interpreted with this in mind.

A. Chelonia

The oxygen-binding properties of hemolyzates from a number of turtle species were studied by Wilson (1939) and Sullivan and Riggs (1967c). Eleven species were studied in some detail by these latter workers who showed that turtle hemolyzates generally have lower oxygen affinities than do mammalian hemolyzates. The oxygen affinity at a fixed pH varied from species to species, as did the pH dependence of oxygen affinity (Bohr effect). Heme–heme interactions were pH dependent and often maximal in the physiological range. The variation of n with pH was species specific. For instance in *Gopherus polyphemus*, n varies from 1.0 near pH 5 to 3.0 at pH 7. In *Clemmys guttata*, n was near 1.0 at all pH values (5–8). Bohr effects varied from −0.27 for hemolyzates from *Malayemys subtrijuga* to −0.77 for hemolyzates from *Trionyx ferox*. Similar but more extensive data on hemolyzates from *Dermatemys mawii* are shown in Fig. 3 (B. Sullivan, unpublished). This species has a very high Bohr effect (−0.94). Heme–heme interactions are pH dependent, rising abruptly from 2 at pH values below 6.5 to near 3 at pH values of 7 and above. The pH dependence of n could be attributable to the fact that multiple components are present. If these components have different oxygen affinities and Bohr effects, then discontinuous Hill plots would result and cooperativity would vary with the magnitude of the differences in oxygen offinity of the

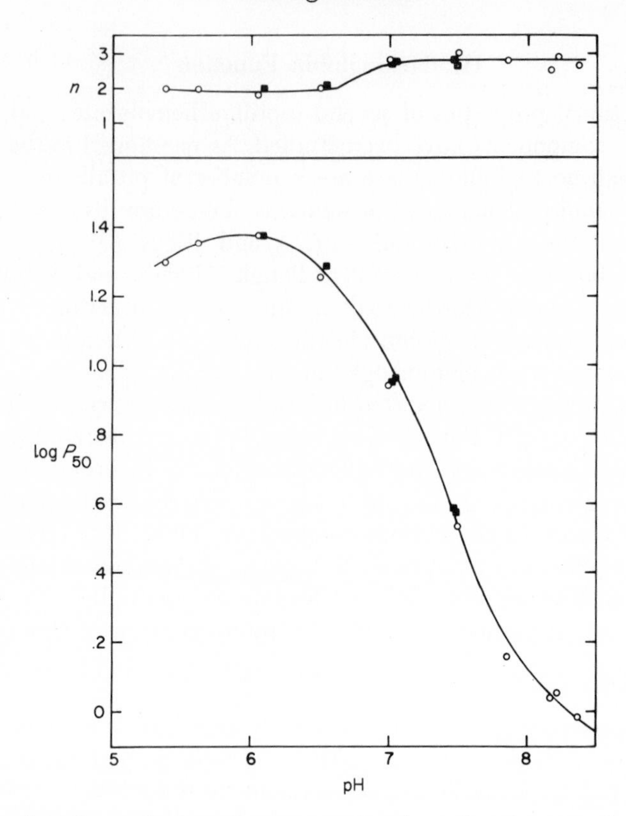

Fig. 3. The effects of pH and mercaptoethanol on the oxygen-binding properties and heme–heme interaction value, n, of dialyzed hemolyzates from the turtle *Dermatemys mawii*. Temperature 20°C; ○ = 0.1 *M* phosphate; ■ = 0.1 *M* phosphate, 0.1 *M* MSH.

components. If one makes a mixture of two hemoglobins with identical interaction values but different affinities, the resulting mixture will always have a lowered n value (relative to the isolated hemoglobins) unless the hemoglobins are interacting. There is ample evidence that the components do differ in their oxygen affinities, but Hill plots are surprisingly linear above 20% saturation. Continuity may result from interaction between the components, but this has not been proved.

Lenfant *et al.* (1970) studied the oxygen-binding properties of blood from two South American turtles. They found a low affinity (26 mm Hg at pH 7.6, 25°C) for blood from *Chelys fimbriata* and a higher affinity (14–15 mm Hg at pH 7.6, 25°C) for blood from the more terrestrial species *Testudo tabulata*. Several workers have tried to correlate the oxygen affinity of turtle blood with activity patterns. Present data do not support these correlations, but the effects of phosphates were not considered, and this omission may have obscured any relationships.

It has been shown by Riggs (1966) that polymerization does not affect the functional properties of frog hemoglobin. Figure 3 shows the effect of adding mercaptoethanol to *Dermatemys* hemolyzates. Hemoglobins in this hemolyzate do polymerize. However, when polymer is dissociated by mercaptoethanol, it leaves the binding properties unchanged. Reduction of disulfide bridges and alkylation with iodoacetic acid or ethyleneimine also do not modify the binding properties (B. Sullivan, unpublished). Thus, unlike human hemoglobin which is affected by alkylation, modification of sulfhydryl groups is without effect on the oxygen-binding properties of frog and turtle hemoglobins.

In the previous experiment, turtle hemolyzates were not stripped of adhering phosphate molecules. Because turtle hemolyzates have such low oxygen affinities, one might not expect them to respond sharply to the presence of adenosine triphosphate (ATP) or inositol hexaphosphate (IHP). The latter compound was thought originally to be limited to bird and turtle erythrocytes (Rapoport and Guest, 1941) but has since been found in other vertebrates. Although it is often noted that bird and turtle erythrocytes contain IHP, in fact only one turtle species has been studied (Table I). The effect of stripping and addition of

TABLE I
ORGANIC PHOSPHATES IN REPTILIAN ERYTHROCYTES[a]

Species	Phosphate concentration (mg/100 ml blood)		
	ATP	IHP	Organic acid-soluble phosphate
Turtles			
Chelydra serpentina	19.4	27.6	58.3
Lizards			
Heloderma suspectum	90.0	0	122.8
Snakes			
Agkistrodon piscivorus	121.3	0	165.3
Natrix taxispilota	99.4	0	127.6
Lampropeltis getulus	96.3	0	118.2
Alligators			
Alligator mississippiensis			
15 cm	8.7	0	21.7
60 cm	9.8	0	13.8

[a] Rapoport and Guest, 1941).

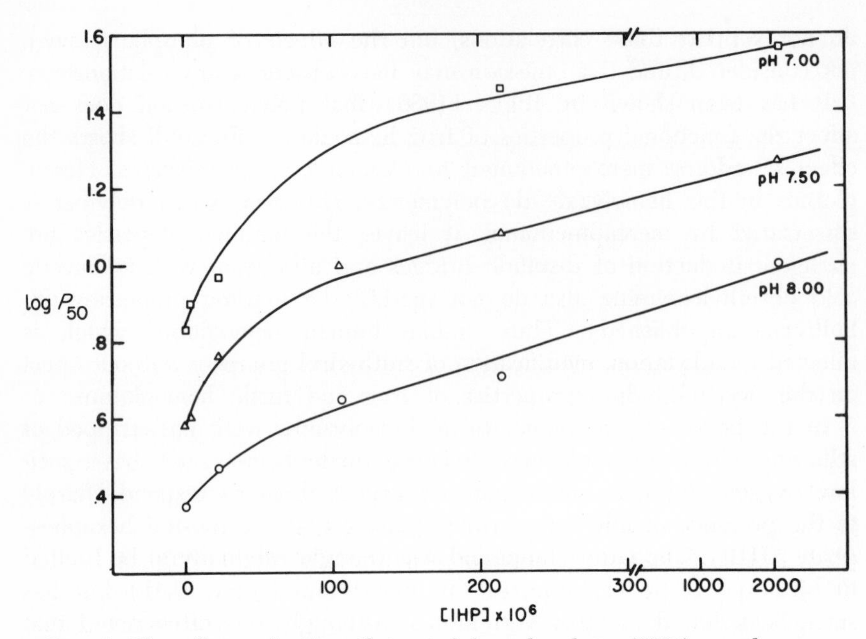

FIG. 4. The effects of pH and inositol hexaphosphate (IHP) on the oxygen affinity of stripped hemolyzates from the turtle, *Malaclemmys terrapin centrata.* Temperature 20°C, 0.05 *M* tris, 0.1 *M* sodium chloride buffers.

IHP to hemolyzates from the diamond backed terrapin, *Malaclemmys terrapin centrata,* are shown in Fig. 4 (B. Sullivan, unpublished). As with human hemolyzates, the magnitude of the decrease in oxygen affinity by phosphate is pH dependent. This probably reflects not only the ionization of a positive group(s) on the protein but may also reflect the ionization of one of the phosphate groups. IHP is a better inhibitor than ATP, but both are effective.

B. SQUAMATA

1. Lacertilia

The only recent studies of the oxygen-binding properties of lizard bloods are those of Pough (1969) and Wood and Moberly (1970). Older studies are by Dill *et al.* (1935) and Edwards and Dill (1935). Pough reported that in lizards the oxygen affinity is correlated with body size and adapted to the temperature at which the species is normally active. A seasonal shift in oxygen affinity was observed for blood from *Dipsosaurus dorsalis.* The oxygen affinity of bloods from *Sceloporus occi-*

dentalis at 0 and 6000 feet elevation were identical. Similarly, Dawson and Poulson (1962) found no correlation between oxygen capacity of lizard bloods and altitude (but see Vinegar and Hillyard, 1972; Hadley and Burns, 1968). Blood from the five species studied by Pough responded quite differently to changes in temperature, and this was interpreted in terms of adaptation to preferred temperatures. However, these samples contained 2–50% methemoglobin (average near 20%) as they were drawn from the animals. This raises the question of the relationship between measurements made in the presence of considerable methemoglobin and measurements made in its absence. Methemoglobin is known to cause an increase in the oxygen affinity and decrease in heme–heme interactions. The problem of methemoglobin formation is greatest at elevated temperatures. It is probably significant that of the species studied by Pough, blood from the species with the least amount of methemoglobin showed the greatest temperature dependence, and blood from the species with the greatest amount of methemoglobin showed the smallest temperature effect. This behavior is predicted if methemoglobin is affecting oxygen affinity. The temperature dependence of P_{50} for iguana (*Iguana iguana*) blood is linear and resembles more closely the dependence observed for other vertebrate bloods and hemoglobin solutions.

Iguana blood has a P_{50} of 51 mm Hg at 35°C and pH 7.4, which is considerably lower than the P_{50} values obtained for other species of lizard blood by Pough (1969). The shape of the binding curve was temperature dependent at 90% saturation levels but not at 50% or 10% levels of oxygen saturation. The shape of the binding curve was also temperature dependent for bloods from *Uma notata inornata* and *Sceloporus occidentalis* but not *Dipsosaurus dorsalis* nor *Gerrhonotus multicarinatus* (Pough, 1969). Higher n values were obtained at elevated temperatures and measured 3.7–4.3. These values are probably the highest known for vertebrate hemoglobins and should be confirmed (especially since many of Pough's binding curves are drawn from three experimental points). The Bohr effect of *Iguana* blood was −0.52 (Wood and Moberly, 1970), which is similar to that of human blood. Pough (1969) noted varying degrees of sensitivity to changes in P_{CO_2} in his study. Apparently, blood from *Gerrhonotus multicarinatus* has little or no Bohr effect.

The only studies of oxygen binding by lizard hemolyzates are those of Dawson (1960). He found that 2% solutions of hemoglobin from *Eumeces obsoletus* had P_{50} values of 19 mm Hg at 28°C and 24 mm Hg at 33°C. Heme–heme interaction values were 2.2 at both temperatures. Experiments were done in $\frac{1}{15}$ M phosphate buffers at pH 7.2. Lizard hemolyzates do not appear to have the high oxygen affinities so characteristic of snake hemolyzates. Lizard erythrocytes appear to contain significant amounts of ATP but only a single species has been studied (Table I).

2. *Ophidia*

There have been few recent studies of the oxygen-binding properties of snake hemoglobins. Sullivan (1967) studied hemolyzates of *Natrix taxispilota*, a water snake. He found that heme–heme interactions were pH independent, dropping slightly below 3.0 only at pH values below 5.5 and above 8.0. Similar results are shown in Fig. 5 (B. Sullivan, unpublished), where the oxygen affinity of hemolyzates from the cottonmouth (*Agkistrodon*) and the rattlesnake (*Crotalus*) are plotted as a function of pH. These hemolyzates also show a slight drop in n at pH values above 8 and below 5.5 and the high oxygen affinity characteristic of *Natrix* hemolyzates.

Oxidation of *Natrix* hemoglobin causes it to polymerize, although only a quarter of the hemolyzate formed polymer (Sullivan, 1967). Alkylation of the sulfhydryl groups with iodoacetamide prevented polymerization

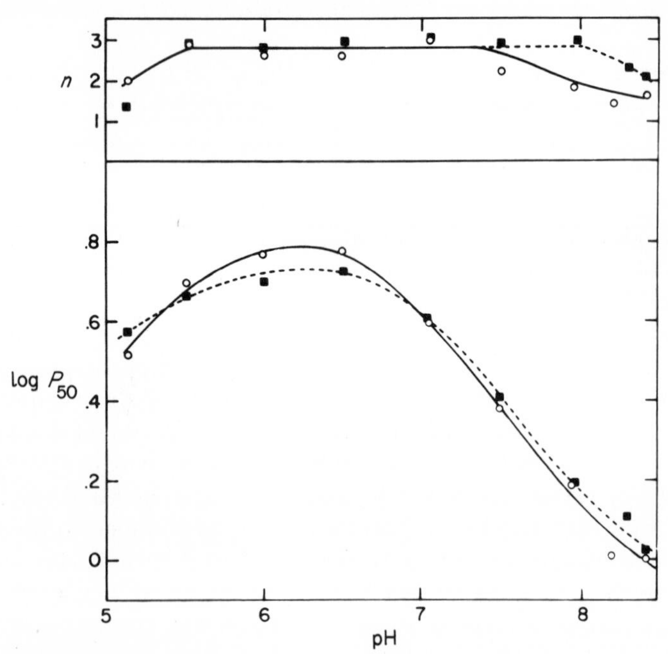

FIG. 5. The effects of pH on the oxygen-binding properties and heme–heme interaction value, n, of dialyzed hemolyzates from two snake species, *Agkistrodon piscivorus* (○) and *Crotalus horridus* (■).

and slightly increased the oxygen affinity, but it effectively eliminated heme–heme interactions (Sullivan, 1967). Dilution of snake hemoglobin causes a slight lowering of n and an increase in oxygen affinity. The effect is somewhat greater than for human hemoglobin, but less than for cat hemoglobin, which is known to dissociate.

Greenwald (1971) studied the effect of temperature on the oxygen-binding properties of blood from the Gopher snake (*Pituophis catenifer*). Blood from this snake species was more strongly temperature dependent than lizard bloods (Pough, 1969). Sullivan (1967) reported that the temperature dependence of oxygen binding by *Natrix* hemolyzates was slightly greater than that of human hemolyzates. However, the P_{50} for Gopher snake blood at pH 7.4 and 20°C was near 30 mm Hg whereas the P_{50} values of dialyzed (versus dilute salt solutions) snake hemolyzates were nearer 3 mm Hg under these conditions (Sullivan, 1967, and unpublished). The effects of organic phosphates on snake hemoglobins have not been measured, but must be large indeed.

Pough (1971) reported that the electrophoretic properties of garter snake (*Thamnophis sirtalis sirtalis*) hemolyzates change with increasing length of individuals. As previously reported by Manwell (1955), the oxygen affinity decreases with increasing size of the individual. However, diluted hemolyzates from individuals of various lengths have identical oxygen affinities. This certainly implies that there is a change in the organic phosphate content during maturation of the species or that new components are being synthesized and they respond differently to phosphates.

Snake erythrocytes are known to contain very large amounts of ATP and no IHP (Rapoport and Guest, 1941). In this regard they are similar to lizard hemolyzates but quite different from those of turtles and alligators (Table I).

C. Crocodilia

The oxygen-binding properties of alligator and crocodile blood were studied by Dill and Edwards (1931, 1935). They found moderately low oxygen affinities (21.5 mm Hg at 20°C for alligator and 40 mm Hg at 29°C for crocodile) and moderately high Bohr effects (−0.80 and −0.76, respectively). Wilson (1939) studied the properties of alligator blood and hemoglobin solutions. In phosphate buffers, alligator hemoglobin had a very high oxygen affinity and very low Bohr effect. Similar data (B. Sullivan, unpublished) are shown in Fig. 6. However, for blood the oxygen affinity was low and the Bohr effect was very large. These findings point to a very large effect on oxygen affinity caused

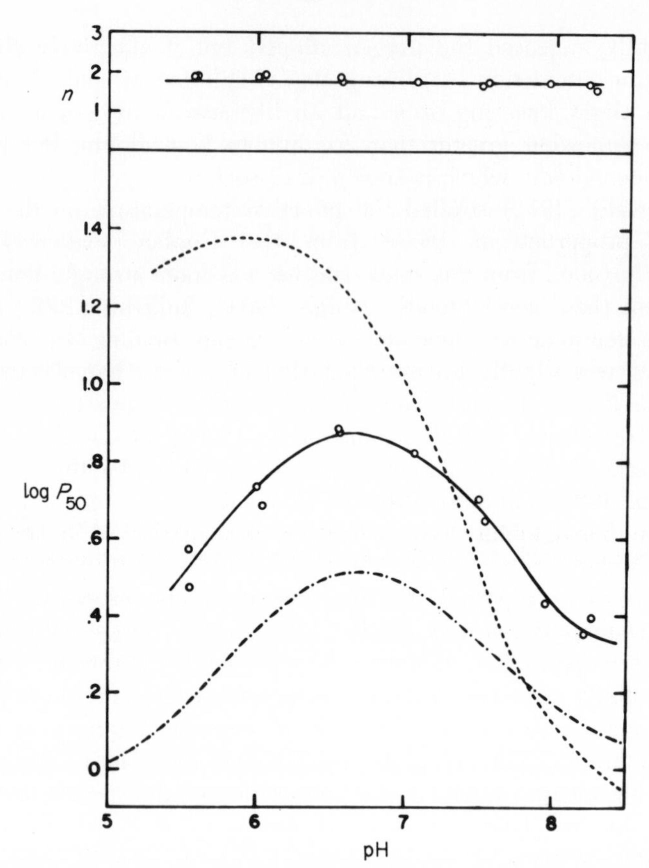

FIG. 6. The effects of pH on the oxygen-binding properties and heme–heme interaction value, *n*, of dialyzed hemolyzates from the alligator *Alligator mississippiensis* (———). Temperature 20°C., 0.1 *M* phosphate buffers. The oxygen-binding properties of dialyzed hemolyzates from the turtle *Dermatemys mawii* (----) and the snake *Natrix taxisipilota* (–·–·–) (Sullivan, 1967) are shown for comparison.

by some intracellular metabolite. Since the experiments of Wilson and those in Fig. 6 were done in phosphate buffers, one questions the role of phosphate as the active species. Furthermore, the concentration of organic phosphates (in the form of ATP) is surprisingly low in alligator erythrocytes (Table I). The ratio of ATP to hemoglobin tetramer is only about 1:10 (assuming 9 gm of hemoglobin and 9 mg of ATP per 100 ml of blood). Thus, the substance regulating oxygen affinity in alligators is very efficient, but unknown.

Heme–heme interaction values are independent of pH and rather low at 1.8 (Fig. 6). In most respects, alligator hemolyzates resemble those of snakes rather than those of lizards or turtles.

IV. Hemoglobin Structure and Synthesis

There are surprisingly little data on the structural properties of reptilian hemoglobins. Svedberg and Hedenius (1934) first reported the presence of polymeric hemoglobins in reptiles (*Chrysemys picta, Lacerta vivipara,* and *Coluber constrictor*). Riggs *et al.* (1964) showed that polymer resulted from the formation of disulfide bridges between hemoglobin tetramers. Polymerization was promoted by oxidizing conditions and could be reversed by incubation with a disulfide reducing reagent such as mercaptoethanol. Although polymerization normally occurs after hemolysis, it does occur in turtles *in vivo* during periods of environmental stress (Sullivan and Riggs, 1964). Polymeric hemoglobins in turtles seem to have subunit-dissociation and oxygen-affinity properties similar to unpolymerized hemoglobins (Sullivan and Riggs, 1967d; B. Sullivan, unpublished).

The distribution of polymerizing hemoglobins in turtle hemolyzates was surveyed by Sullivan and Riggs (1967a) and Horton *et al.* (1972). Hemolyzates from species in the families Chelydridae, Kinosternidae, Platysternidae, Emydidae, Trionchidae, and Chelidae contain hemoglobins that polymerize to varying extents. Hemolyzates from most species in the families Testudinidae, Cheloniidae, and Pelomedusidae do not contain hemoglobins that polymerize. Hemoglobin polymers (mostly dimers formed by two hemoglobin tetramers, but also higher order aggregates) isolated by gel filtration seem to contain all the major electrophoretic components (Sullivan and Riggs, 1967a; B. Sullivan, unpublished; but see Ramirez and Dessauer, 1957). This implies that the globin chain common to several major components (Dozy *et al.,* 1964; Sullivan and Riggs, 1967b) may contain the active sulfhydryl group.

At least one hemoglobin component in hemolyzates from *Lacerta vivipara, Coluber constrictor,* and *Natrix taxispilota* will polymerize (Svedberg and Hedenius, 1934; Sullivan, 1967). Nothing more is known about the distribution of polymerizing hemoglobins in other reptilian hemolyzates. The functional significance of polymer formation is unknown (if in fact it has any biological significance). Dessauer (1970) has suggested that disulfide bridged hemoglobin could act as a receptor for hydrogen ions. It seems likely, though, that hemoglobin polymerization *in vivo* has no functional significance but only reflects the inability of reptilian erythrocytes to maintain a reducing environment. Hemoglobin polymerization *in vitro* occurs in hemolyzates from selected species of most other vertebrate groups and reflects the availability of cysteinyl residues.

Duget *et al.* (1971, 1973) have begun a structural analysis of the single hemoglobin found in *Viper aspis*. The amino terminal 34 residues of the α-chain are shown in Fig. 7. When compared to chicken, fish, and mammalian α-chains, *Viper* α-chain is remarkably distinct. Except for threonine at position 28, there are no homologies that link avian and reptilian α-chains. Furthermore, *Viper* α-chain is as different from avian (*Gallus*) α-chain as it is from mammalian (human) α-chain. It is as different from human α-chain as is the α-chain of carp. The rest of this sequence should be very interesting. Manwell and Schlesinger (1966) fingerprinted tryptic digests of the slow and fast electrophoretic components of hemolysates from *Pseudemys scripta*. Both components showed approximately thirty peptide spots and differed by ten peptides. This is consistent with the presence of two polypeptides in each hemoglobin component, one of which is shared. Electrophoresis of globin chains from fast and slow components gives similar results (Dozy *et al.*, 1964; Sullivan and Riggs, 1967b). Dessauer and Sutton (1964) have fingerprinted several reptilian hemolyzates, but the details of their studies have not been published. N-terminal amino acids of hemolyzates from *Clemmys caspica* (Christomanos and Pavlopulu, 1968) and a snake hemoglobin (Ozawa and Satake, 1955) have been reported, but are of little value, since they were done on whole hemolyzates. Similar data are available for the amino acid compositions of whole hemolyzates from several species (summarized in Dessauer, 1970). It is hoped that additional structural studies will soon be done on purified reptilian hemoglobins. The wide variation in functional properties and large differences in oxygen affinity between blood and hemoglobins solutions hint that structure–function relationships will be most interesting.

There have been very few studies of hemoglobin synthesis in reptiles. Both bleeding and phenylhydrazine injections are efficient stimuli for erythropoiesis (Altland and Thompson, 1958; Hirschfeld and Gordon, 1961, 1964). Efrati *et al.* (1970) found evidence of hemopoiesis in bone marrow of *Agama stellio*, but not in the spleen, kidney, liver, or testes. Englebert and Young (1970) suggest that significant numbers of new cells arise as perturbations from the nuclei of mature erythrocytes. This rather unorthodox method of erythrocyte production deserves further study. Unlike the situation reported for frogs, there are no indications of seasonal changes in the production of erythrocytes throughout the year (Efrati *et al.*, 1970; Kaplan and Rueff, 1960).

V. Concluding Comments

Reptilian hemoglobins are not as well characterized as those of amphibians. The oxygen-binding properties of reptilian hemolyzates and

	1 5 10 15
Man	NH_2–Val–Leu–Ser–Pro–Ala–Asp–Lys–Thr–Asn–Val–Lys–Ala–Ala–Trp–Gly–Lys–Val–Gly–Ala–
Chicken	— — — Asn — — — Asn — — — Gly–Ile –Phe–Thr — Ile –Ala–Gly
Viper	— — — Glu–Asp — — Asn–Arg — Arg–Thr –Ser–Val — — Asn–Pro–Glu–
Carp	Ac –Ser — — Asp–Lys — — ALa–ALa — — Ile — — Ala — Ile –Ser–Pro–

20 25 30 34	
His–Ala–Gly–Glu–Tyr–Gly–Ala–Glu–Ala–Leu–Glu–Arg–Met–Phe–Leu	Man
— — Glu — — — — — Thr — — — — — Ile	Chicken
Leu–Pro — — — — Ser — Thr — Thr — — — Ala	Viper
Lys — Asp–Asp–Ile — — — — — Gly — — Leu–Thr	Carp

FIG. 7. A comparison of the α-chain of *Viper* hemoglobin with α-chains from carp, chicken, and man. Sequence differences are shown; identities to human α-chain are not.

bloods may well include the extremes in affinity among terrestrial vertebrates. The remarkable adaptations of turtles to anoxic conditions, the adaptations of lizards and snakes to a wide range of environmental temperatures, and the interesting variations in circulation (White, 1970) and pH (Howell *et al.*, 1970) found in reptiles all point to the possibility of unusual functional and structural properties of their hemoglobins. Perhaps the most interesting aspect of studies completed to date is the unusually large difference between the oxygen-binding properties of reptilian hemoglobins in solution and the oxygen-binding properties of these same hemoglobins in erythrocytes in blood. An understanding of these differences will undoubtedly do much to bridge the gap between biochemical and physiological studies of oxygen transport. We know almost nothing about hemoglobin structure or its synthesis in reptiles. Clearly, much remains to be learned about hemoglobins in this interesting group of vertebrates.

REFERENCES

Altland, P. D., and Brace, D. C. (1962). *Amer. J. Physiol.* **203,** 1188.

Altland, P. D., and Thompson, E. C. (1958). *Proc. Soc. Exp. Biol. Med.* **99,** 456.

Antonini, E., and Brunori, M. (1971). "Hemoglobin and Myoglobin in their Reaction with Ligands." Amer. Elsevier, New York.

Brace, K. C., and Altland, P. D. (1955). *Amer. J. Physiol.* **183,** 91.

Christomanos, A. A., and Pavlopulu, C. (1968). *Enzymologia* **34,** 51.

Cline, M. J., and Waldmann, T. A. (1962). *Amer. J. Physiol.* **203,** 401.

Dawson, W. R. (1960). *Physiol. Zool.* **33,** 87.

Dawson, W. R., and Poulson, T. L. (1962). *Amer. Midl. Natur.* **68,** 154.

Dessauer, H. C. (1970). *In* "Biology of the Reptilia" (C. Gans, ed.), Vol. 3, pp. 1–72. Academic Press, New York.

Dessauer, H. C., and Fox, W. (1964). *In* "Taxonomic Biochemistry and Serology" (C. A. Leone, ed.), pp. 625–647. Ronald Press, New York.

Dessauer, H. C., and Sutton, D. E. (1964). *Fed. Proc., Fed. Amer. Soc. Exp. Biol.* **23,** 474.

Dessauer, H. C., Fox, W., and Ramirez, J. R. (1957). *Arch. Biochem. Biophys.* **71,** 11.

Dill, D. B., and Edwards, H. T. (1931). *J. Biol. Chem.* **90,** 515.

Dill, D. B., and Edwards, H. T. (1935). *J. Cell. Comp. Physiol.* **6,** 243.

Dill, D. B., Edwards, H. T., Bock, A. V., and Talbott, J. H. (1935). *J. Cell. Comp. Physiol.* **6,** 37.

Dozy, A. M., Reynolds, C. A., Still, J. M., and Huisman, T. H. J. (1964). *J. Exp. Zool.* **155,** 343.

Duguet, M., Chauvet, J.-P., and Acher, R. (1971). *FEBS Lett.* **18,** 185.

Duguet, M., Chauvet, J.-P., and Acher, R. (1973). *FEBS Lett.* **29,** 10.

Duguy, R. (1970). *In* "Biology of the Reptilia" (C. Gans, ed.), Vol. 3, pp. 93–109. Academic Press, New York.

Edwards, H. T., and Dill, D. B. (1935). *J. Cell. Comp. Physiol.* **6,** 21.

Efrati, P., Nir, E., and Yaari, A. (1970). *Isr. J. Med. Sci.* **6,** 23.

Engelbert, V. E., and Young, A. D. (1970). *Can. J. Zool.* **48**, 209.
Gorman, G. C., and Dessauer, H. C. (1965). *Science* **150**, 1454.
Gorman, G. C., and Dessauer, H. C. (1966). *Comp. Biochem. Physiol.* **19**, 845.
Gorman, G. C., Licht, P., Dessauer, H. C., and Boos, J. O. (1971). *Syst. Zool.* **20**, 1.
Greenwald, O. E. (1971). *Comp. Biochem. Physiol. A.* **40**, 865.
Guttman, S. I. (1970a). *Comp. Biochem. Physiol.* **39**, 563.
Guttman, S. I. (1970b). *Comp. Biochem. Physiol.* **34**, 569.
Guttman, S. I. (1971). *J. Herpetol.* **5**, 11.
Hadley, N. F., and Burns, T. A. (1968). *Copeia* **4**, 737.
Hirschfeld, W. J., and Gordon, A. S. (1961). *Anat. Rec.* **139**, 306.
Hirschfeld, W. J., and Gordon, A. S. (1964). *Amer. Zool.* **4**, 305.
Horton, B., Fraser, R., Dupourque, D., Bailey, D., and Chernoff, A. (1972). *J. Exp. Zool.* **180**, 373.
Howell, B. J., Baumgardner, F. W., Bondi, K., and Rahn, H. (1970). *Amer. J. Physiol.* **218**, 600.
Kaplan, H. M., and Rueff, W. (1960). *Proc. Anim. Care Panel* **10**, 63.
Lenfant, C., Johansen, K., Petersen, J. A., and Schmidt-Nielsen, K. (1970). *Resp. Physiol.* **8**, 261.
Lykakis, J. J. (1971). *Comp. Biochem. Physiol. B* **39**, 83.
McCutcheon, F. H. (1947). *J. Cell. Comp. Physiol.* **29**, 333.
Manwell, C. P. (1955). Master's Thesis, University of Washington, Seattle.
Manwell, C. P., and Schlesinger, C. V. (1966). *Comp. Biochem. Physiol.* **18**, 627.
Musacchia, X. J., and Sievers, M. L. (1962). *Trans. Amer. Microsc. Soc.* **81**, 198.
Newcomer, R. J., and Crenshaw, J. W. (1967). *Copeia* **2**, 481.
Ozawa, H., and Satake, K. (1955). *J. Biochem. (Tokyo)* **42**, 641.
Porter, K. R. (1972). "Herpetology." Saunders, Philadelphia, Pennsylvania.
Pough, F. H. (1969). *Comp. Biochem. Physiol.* **31**, 885.
Pough, F. H. (1971). *Amer. Zool.* **11**, 207.
Prado, J. L. (1946). *Science* **103**, 406.
Ramirez, J. R., and Dessauer, H. C. (1957). *Proc. Soc. Exp. Biol. Med.* **96**, 690.
Ramsey, H. J. (1941). *J. Cell. Comp. Physiol.* **18**, 369.
Rapoport, S., and Guest, G. M. (1941). *J. Biol. Chem.* **138**, 269.
Riggs, A. (1966). *In* "International Symposium of Comparative Hemoglobin Structure" (A. Christomanos and D. J. Polychronakos, eds.) pp. 126–128, Triantafylo, Thessaloniki, Greece.
Riggs, A. (1970). *In* "Fish Physiology" (W. S. Hoar and D. J. Randall, eds.), Vol. 4, pp. 209–252. Academic Press, New York.
Riggs, A., Sullivan, B., and Agee, J. R. (1964). *Proc. Nat. Acad. Sci. U.S.* **51**, 1127.
Saint Girons, M.-C. (1970). *In* "Biology of the Reptilia" (C. Gans, ed.), Vol. 3, pp. 73–91. Academic Press, New York.
Schwantes, A. R. (1972). "Hemoglobinas e haptoglobinas em seypentes (Squamata, Reptilia)." Ph.D. thesis, Universidade Federal do Rio Grande do Sol.
Semple, R. E., Sigsworth, D., and Stitt, J. T. (1970). *Can. J. Physiol. Pharmacol.* **48**, 282.
Sullivan, B. (1967). *Science* **157**, 1308.
Sullivan, B. (1971). *In* "Comparative Genetics in Monkeys, Apes, and Man" (A. B. Chiarelli, ed.), pp. 213–256. Academic Press, New York.
Sullivan, B., and Riggs, A. (1964). *Nature (London)* **204**, 1098.
Sullivan, B., and Riggs, A. (1967a). *Comp. Biochem. Physiol.* **23**, 437.

Sullivan, B., and Riggs, A. (1967b). *Comp. Biochem. Physiol.* **23**, 449.
Sullivan, B., and Riggs, A. (1967c). *Comp. Biochem. Physiol.* **23**, 459.
Sullivan, B., and Riggs, A. (1967d). *Biochim. Biophys. Acta* **140**, 274.
Svedberg, T., and Hedenius, A. (1934). *Biol. Bull.* **66**, 191.
Vinegar, A., and Hillyard, S. D. (1972). *Comp. Biochem. Physiol.* **A43**, 317.
Webster, T. P., Selander, R. K., and Yang, S. Y. (1972). *Evolution* **26**, 523.
White, F. N. (1970). *Fed. Proc., Fed. Amer. Soc. Exp. Biol.* **29**, 1149.
Wilson, J. W. (1939). *J. Cell. Comp. Physiol.* **13**, 315.
Wood, S. C., and Moberly, W. R. (1970). *Resp. Physiol.* **10**, 20.

CHAPTER 15

Endocrinology of Reptilia—The Pituitary System

Paul Licht

I. Introduction 399
A. Pituitary System 399
B. Gross Morphology and Cytology of the Reptilian Pituitary Gland .. 400
C. Components of the Adenohypophysial System 401
D. Biochemistry of Pituitary Hormones 402
II. Growth Hormone (GH) and Prolactin (PL) 403
A. Definitions 403
B. Physiological Actions of Mammalian GH and PL 404
C. Effects of Hypophysectomy on Growth 408
D. Assay and Biochemical Characterization of Reptilian GH and PL ... 409
III. Gonadotropins and Reproduction 413
A. Evidence for a Pituitary–Gonadal Axis 413
B. Characterization of Reptilian Gonadotropins 420
C. Endocrine Gonad—Sex Steroids 423
IV. Adrenocorticotropin (ACTH) and Corticoid Hormones 428
A. Interrenal Glands 428
B. Evidence for a Pituitary–Interrenal Axis 433
C. Biochemistry of ACTH 436
V. Thyrotropin (TSH) 437
A. Biosynthesis of Thyroid hormone (Thyroxine) 437
B. Evidence for a Pituitary–Thyroidal Axis 440
C. Biochemistry of TSH 441
VI. Concluding Remarks 442
References 444

I. Introduction

A. Pituitary System

Reptiles possess a complex endocrine system including diverse hormone-producing glands. There has been a tendency to assume that reptilian hormones generally resemble those of the well studied mammals in structure and function. While gross similarities are certainly evident among vertebrate hormones, a detailed examination of what little is known about the reptilian endocrines reveals that they show many distinctive features. Certainly, specific details about the endocrinology of reptiles is required to fully understand reptilian physiology. Because of their interesting phylogenetic position as the "primitive" amniotes, information on reptilian hormones is also crucial for the understanding of hormonal evolution.

Although there are many discrete endocrine tissues scattered throughout the vertebrate body, they are by no means independent of one another. In fact, a large part of the vertebrate endocrine system is controlled by a single "master" endocrine gland—the anterior pituitary gland, or adenohypophysis. In view of the general importance of the pituitary gland and its satellite endocrine tissues, and because there have been numerous recent advances into the chemistry and physiology of the hormones related to this part of the endocrine system, the pituitary system has been chosen as the focal point for this chapter.

B. Gross Morphology and Cytology of the Reptilian Pituitary Gland

The pituitary gland of reptiles, like that of all other vertebrates, is located below the brain, directly under the hypothalamus and close to the median eminence; the pituitary being situated in a depression of the basisphenoid bone called the sella turcica. The relatively small size of the gland tends to belie its importance. Experience in my laboratory has shown that in most small reptiles, the gland rarely exceeds a few milligrams in weight; in very large species, like the snapping turtle *Chelydra serpentina,* the weight of the anterior pituitary is only about 15 mg, and it is about 100 mg in a 10 foot alligator. The cytology and morphology of the reptilian pituitary have been extensively reviewed by Saint Girons (1967, 1970) and Wingstrand (1966), and only a few details pertinent to the present discussion need be considered here.

The important anatomical feature to recognize is that the pituitary gland of reptiles is composed of three distinct segments—a posterior, intermediate, and anterior region. The posterior region, the neurohypophysis, develops as an evagination of the hypothalamus and is distinct in structure and function from the other two portions of the gland (see review by Follett, 1970; Skadhauge, 1969). The intermediate and anterior segments of the pituitary comprise the adenohypophysis. This is derived from a buccal pouch (Rathke's pouch) from the epithelium of the mouth and is composed of secretory cells, in contrast to the neuronal tissue and glial elements of the neurohypophysis. The size of the intermediate segment (pars intermedia) of the reptilian adenohypophysis correlates with the color-changing habits of the species; it is thought to produce only a single hormone (melanocyte-stimulating hormone or MSH) which is involved in color change. The pars distalis, the anterior-most portion of the adenohypophysis, is the site of at least five distinct hormones with a diversity of actions.

C. Components of the Adenohypophysial System

Figure 1 summarizes the five major components of the endocrine system revolving around the hormones of the pars distalis based on the classic mammalian pattern. A sixth possible factor, lipotropin, identified in the mammalian pituitary, is not included since there is no information on this hormone in nonmammalian species. The terminology for the pituitary hormones—derived largely from knowledge of their actions in mammals—and the standard abbreviations shown in Fig. 1 have been adopted for the present discussion.

Preliminary evidence for the existence of the five major types of hormones in the reptilian pars distalis is derived from cytological studies involving over a hundred species, representing all four orders of Reptilia. Such studies have demonstrated that there are at least five distinct tinctorial cell types in the reptilian pars distalis (reviewed by Saint Girons, 1970). Numerous attempts have been made to relate each of these types

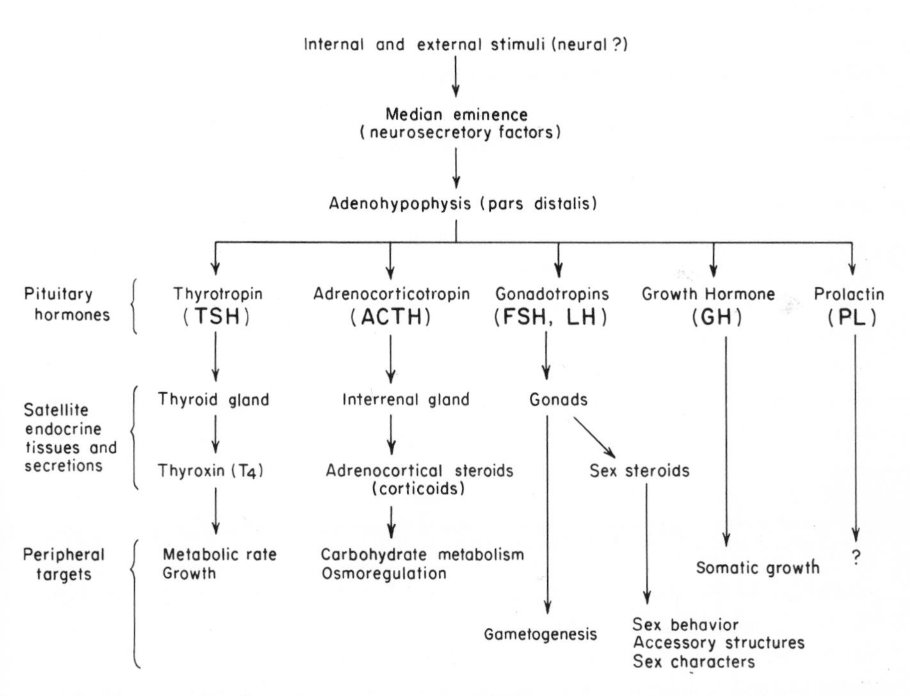

Fig. 1. Generalized scheme for the relationships of the anterior pituitary gland (adenohypophysis) and various physiological functions in vertebrates based on information derived largely from mammalian studies. Peripheral effects for PL are omitted, since their actions in mammals cannot be readily extrapolated to reptiles.

of cell to the production of a particular hormone, largely on the basis of apparent correlations between activities of the various cell types and various target tissues (e.g., Saint Girons, 1970; Eyeson, 1970). An important feature of the reptilian pars distalis is that each cell type tends to be restricted to either the rostral or caudal half of the gland. This zonation has allowed further testing of the tentative identification of functional cell types by comparing distributions of cells with the distribution of hormones as determined by bioassay. While such physiological studies are consistent with the hypothesis that each type of cell is probably associated with production of a particular hormone, these data suggest that several of the tentative identifications of cell function based on histological studies may be in error (see Licht and Pearson, 1969a; Licht and Nicoll, 1969; Licht and Bradshaw, 1969; Licht and Rosenberg, 1969; Licht and Papkoff, 1972; Licht, 1974). Details of these investigations are discussed under the separate sections on each hormone.

An important aspect of the pituitary system that is not detailed in Fig. 1 is the component involved in the regulation of hormonal secretion by the pituitary. It is well established in mammals that the pars distalis is ultimately under the control of factors that are produced in certain regions of the hypothalamus, especially in the region of the median eminence. These factors are neurosecretory products that are transmitted to the pituitary gland via the portal circulation that enters the rostral tip of the pars distalis. While these neurosecretory factors have not been isolated in reptiles, there is reasonably good evidence for such hypothalamic control for several pituitary hormones in reptiles (Lisk, 1967; Callard and Willard, 1969; Nicoll *et al.*, 1970; Callard and Chester Jones, 1971). There is also evidence that negative feedback inhibition between the products of the satellite glands and the trophic hormones of the pituitary is involved in the regulation of several pituitary hormones, and that this feedback may operate through the hypothalamus. Detailed consideration of this aspect of pituitary regulation is considered to be beyond the scope of the present review.

D. Biochemistry of Pituitary Hormones

Although the present volume is intended to emphasize chemical aspects of reptilian physiology, available information on the chemistry of reptilian endocrines is far too limited to permit a meaningful review to be based solely on this aspect of the system. Few direct chemical studies have been made on reptilian hormones, and most knowledge of their structure is derived from inference from mammalian hormonal structure. Thus, it is assumed that the hormonal secretions of the adenohypophysis are either proteins or polypeptides. In contrast, the hormonal secretions of some of the satellite glands controlled by the pituitary (e.g., adrenals and gonads) are steroids.

One valuable approach toward elucidating the structure of hormones in lieu of direct chemical study is to examine their biological activities in relation to the better studied hormones of other vertebrates, especially mammals. This indirect approach to the biochemistry of hormones is related to the general concept of zoological specificity. It is based upon the assumption that some insight into the extent of overlap in the structure of a particular type of hormone can be gained by studying the activity of the hormones of one species in another species. A high degree of species specificity in hormonal activity implies that either the structure of the hormone or the receptor sites have become highly differentiated in different species. Accordingly, attention will be focused on the ability of reptilian hormones to stimulate nonreptilian species and the ability of nonreptilian hormones to stimulate reptiles.

II. Growth Hormone (GH) and Prolactin (PL)

A. Definitions

Growth hormone (GH), as its name implies, is defined primarily by its effect in stimulating lean body growth. Intensive comparative study of the actions of prolactin (PL) (see reviews by Mazzi, 1969; Bern and Nicoll, 1968) has revealed that this hormone exhibits a wide range of activities among vertebrates; for example, it is involved in osmoregulation (freshwater survival) in many teleosts, water drive in newts, crop-sac hypertrophy in pigeons (a standard bioassay), and mammary gland growth and lactation in mammals. Most of these distinctive activities are clearly not relevant to the reptilian system. The activity that appears to be important for reptiles is one that is closely related to that classically ascribed to GH, namely, its effects on metabolism and growth.

In some mammals, notably the primates, the degree of overlap in the structure and function of PL and GH is so great that there was controversy regarding whether they were, in fact, two separate molecular entities. The high degree of structural similarity in primates has led to the speculation that the two hormones may have originated from a common ancestral molecule (Bewley and Li, 1970; Fellows *et al.*, 1970). Thus, the extent of structural and functional overlap between PL and GH in lower vertebrates is of special interest for the understanding of hormonal evolution.

The lack of a simple direct relationship with a satellite endocrine gland, as in the case of other pituitary hormones, and the complex nature of the physiological processes in which PL and GH appear to be involved greatly complicate the study of these two hormones. Most of the information on their potential actions in reptiles is based on responses of reptiles to injections of exogenous (mammalian) hormones. Much of the evidence for the existence of these hormones in the reptilian pituitary is derived from bioassays of reptilian pituitary extracts in nonreptilian species.

B. Physiological Actions of Mammalian GH and PL

The physiological effects of highly purified mammalian PL and GH provide considerable insight into the potential role of these two hormones in reptiles. However, in reviewing these data, several problems should be borne in mind. Few studies employed hypophysectomized animals; thus, the possibility of interaction with endogenous hormones cannot be ruled out. Only a few reptilian species have been examined, mostly lizards. Most studies were based on ovine PL and bovine GH, and recent comparative studies on the physiology of vertebrate PL and GH (see Section D below) have shown that results obtained with the hormone of a heterologous species may give a misleading impression about the physiological actions of the species' own hormone. This was particularly striking in the case of the somatotropic properties of turtle PL.

1. Body Growth

DiMaggio (1961) reported one of the first major studies of mammalian GH on body growth in a reptile, using the lizard *Anolis carolinensis*. DiMaggio claimed that they were functionally hypophysectomized (i.e., short photoperiods were used to suppress endogenous GH production), but the absence of endogenous GH in such intact animals has not been proven. His results indicated seasonal changes in the responsiveness to GH, but in general, this hormone was shown to promote lean growth. GH caused increased nitrogen metabolism (excretion) only in fed animals. Increased lean growth was attributed to improved efficiency of protein utilization, since no increase in appetite was observed; however, the latter observation is not consistent with all other studies of GH in reptiles (Licht and Jones, 1967; Licht, 1967).

In subsequent studies with the same lizard, it was found that ovine PL (in doses lower than those used by DiMaggio in his GH studies) promoted lean body growth in the young of both sexes (P. Licht, unpublished) and in adult males (Licht and Jones, 1967; Licht, 1967). In these cases, PL had multiple growth-promoting actions. In some seasons (fall–winter) there was a marked hyperphagia that undoubtedly contributed to increased growth rates (Licht and Jones, 1967), but growth effects were still evident even in seasons when such hyperphagia was absent or when a controlled diet was used (Licht and Jones, 1967; Licht, 1967). In the fall, PL caused a decline in the rate of oxygen consumption, and this may have contributed to increased growth rates independent of food intake. In all of these experiments, PL consistently caused a decrease in hepatic lipid stores, presumably channeling this energy into lean growth.

Comparison of the effects of bovine GH and ovine PL in juveniles of another lizard, *Lacerta sicula,* (Licht and Hoyer, 1968), generally confirmed the results observed in *Anolis.* In *Lacerta,* both hormones caused hyperphagia with an attendant acceleration of lean growth; bovine GH was slightly more effective than ovine PL at the same dose level. Also, the two hormones were observed to have a differential action on lipid stores—PL reduced hepatic lipids as observed in *Anolis,* whereas GH had its main action on abdominal fat pads. In baby snapping turtles, PL increased lean growth with splanchnomegaly of heart, kidney, and spleen to about the same extent that GH did (Nichols, 1973).

While the above data indicate that PL may be a potentially important growth hormone in reptiles, several problems exist in evaluating its action. Despite its marked somatotropic effect in juvenile and adult male *Anolis,* even relatively high doses of ovine PL have virtually no lean growth effect in adult females of this species (P. Licht, unpublished). Unfortunately, it is not known if these females are responsive to GH. More important, Meier (1969) reported that the actions of PL in this lizard (and in numerous other vertebrates) varied diurnally; i.e., the nature of the response to the hormone varied with the time of day that injections were given. At certain injection times (especially at midday), PL appeared to stimulate fattening rather than lean growth. These results implied that the action of the hormone may vary depending on the time of day that it is released from the pituitary. Unfortunately, Meier did not quantify appetite and other aspects of the metabolism of the lizards in these experiments. No variations in the nature of the growth response were observed in *Anolis* injected at three different times of day when standardized food intake was used to eliminate the effects of hyperphagia (P. Licht, unpublished). The possibility of diurnal variations in the effects of GH have not been explored in reptiles.

In addition to the major aspects of growth discussed above, PL and GH have been observed to exert a variety of other metabolic actions that may be generally related to the phenomenon of growth. For example, ovine PL led to the regeneration of significantly heavier tails in intact *A. carolinensis*, although it did not affect rate of elongation during regeneration (Licht, 1967). GH alone has not been tested in this connection. PL accelerated molting in several lizard species, apparently due to an acceleration of the proliferative phase of the epidermal cycle, as opposed to a reduction in the resting phase (reviewed by Maderson *et al.*, 1970). In this respect, the effects of PL resemble those of TSH or thyroxine. Although PL may exert an effect on the thyroid (Licht, 1967), it probably has a more direct action on the skin, since PL continued to enhance molting in thyroidectomized lizards (Maderson *et al.*, 1970). The effect of GH on molting has not been examined in lizards, but it may inhibit sloughing in snakes (Maderson *et al.*, 1970). More information is also required on the actions of PL and GH in reptiles other than lizards, since, for example, the endocrine relationships of sloughing in snakes differ markedly from those observed in lizards (Maderson *et al.*, 1970).

2. *Blood Glucose*

In addition to body growth, another distinctive action of GH in mammals and several other vertebrates is its role in the regulation of blood glucose. The basic action of GH is to elevate blood glucose (hyperglycemia), a condition normally associated with insulin deficiency (diabetes). There is evidence for this "anti-insulin" action of GH in a wide array of reptiles, including chelonians, squamates, and crocodilians. Interspecific differences are apparent in the magnitude of response to various treatments but are difficult to evaluate because of the lack of standardization of doses and experimental conditions (e.g., temperature, feeding conditions, length of treatment). Much of the data on this subject has been reviewed by Coulson and Hernandez (1964) and Penhos *et al.* (1967), and only the general aspects of the problem will be considered here.

In intact reptiles, administration of mammalian GH generally has only a minor or no hyperglycemic action (see Vladescu *et al.*, 1970), but it invariably aggravates the diabetic hyperglycemia following pancreatectomy. The results of experiments involving hypophysectomy are consistent with these data, removal of the pars distalis generally causing a moderate fall in blood glucose. In one species of turtle, hypophysectomy also reduced the rate of hyperglycemia after pancreatectomy, and hypophysectomized turtles showed an increased sensitivity to insulin and less increase in blood glucose levels after oral administration of glucose.

The mechanism of action of GH on blood sugar has not been studied in detail. In several turtles, degenerative histological changes in β-cells of the pancreas suggested that insulin production was reduced by GH treatment (Cardeza, 1957; Marques, 1955). It is also likely that the action of GH is normally associated with secretions from the adrenal cortex (=interrenal gland), since the diabetogenic actions of GH in reptiles can be greatly exaggerated by simultaneous administration of glucocorticoids such as cortisone and hydrocortisone. Surprisingly, there have been no studies of the interaction of GH with corticosterone, the naturally occurring corticoid in reptiles (see Section IV). Coulson and Hernandez (1964) were unable to demonstrate a reduction in glucose utilization in alligators rendered hyperglycemic with GH and cortisone.

These studies leave little doubt that GH serves in the regulation of blood glucose levels in reptiles. Unfortunately, the possible actions of PL in connection with blood glucose regulation in reptiles have not been examined. In view of the large degree of overlap between GH and PL in other aspects of growth and metabolism, one might expect PL to be involved. Further information on the respective actions of the two trophic hormones might help distinguish their roles in reptiles.

3. *Reproduction*

Both GH and PL have been implicated in reproduction in lizards, and limited evidence suggests that they may have distinctive actions in this regard. PL was considered to be an antigonadal hormone in both sexes of several vertebrates, but studies on males of two lizards, *Anolis carolinensis* (Licht and Jones, 1967; Licht, 1967) and *Lacerta sicula* (Licht and Hoyer, 1968), failed to show any action of PL on testicular activity. More recently, Callard and Ziegler (1970) reported that administration of mammalian PL inhibited the action of exogenous gonadotropin on ovarian growth in female lizards (*Dipsosaurus dorsalis*). A similar action was observed in another lizard, *Sceloporus cyanogenys,* although the effect of PL was less pronounced in this species (Callard *et al.,* 1972). In these cases, PL appeared to be acting at the gonad level, blocking the response to gonadotropin. In contrast, GH was found to be required for an exogenous gonadotropin or estrogen to be effective in promoting ovarian growth in *D. dorsalis* (Callard and Ziegler, 1970). Since hypophysectomy also reduced the response to gonadotropins and estrogen, the authors argued that endogenous GH shows this interaction in ovarian growth.

In female *A. carolinensis,* hypophysectomy had no influence on the response to exogenous gonadotropin and, in intact animals, PL (even in relatively large doses) had little or no effect on endogenous or exogenous gonadotropin activity (P. Licht and R. E. Jones, unpublished). Thus, there may be interspecific differences in the role of PL and GH in reproduction. However, it should be noted that the studies reporting an effect used pregnant mares serum gonadotropin (a placental hormone), while those showing no effect were based on pituitary follicle-stimulating hormone. Thus, the possibility must be considered that the type of gonadotropin used may be at the basis of the apparent interspecific differences.

4. *Osmoregulation*

The action for which PL is best known in fish (teleosts) is the regulation of plasma electrolytes, such as sodium; the hormone is frequently indispensable for survival in fresh water. A single study in a lizard, *Dipsosaurus dorsalis*, suggested that PL may function in the regulation of water and electrolyte balance in reptiles (Cahn *et al.*, 1970). PL alone had little effect, but when given with an adrenal steroid—corticosterone—it was able to restore blood water and sodium levels to normal in hypophysectomized lizards.

C. Effects of Hypophysectomy on Growth

Because most of the physiological responses to PL and GH involve complex processes depending on several different hormones, hypophysectomy does not provide precise data on the presence or actions of these two hormones. The effects of hypophysectomy in connection with several of the potential actions of PL and GH have already been discussed. Unfortunately, few hypophysectomy studies were specifically designed to examine the dependence of growth on the pituitary, and in many cases, temperatures were too low to allow meaningful observation of growth (or lack thereof).

Extensive experience with *Anolis carolinensis* in my laboratory has demonstrated that a marked decline in appetite (hypophagia) followed hypophysectomy at warm temperatures. Animals had to be hand fed to maintain positive energy balance and lean body growth was rarely observed. Hypophysectomized *Anolis* usually showed increased fattening if fed sufficiently, but if starved, fat metabolism did not differ from intact starved animals. Changes in nitrogen metabolism, as determined by uric acid excretion, were not observed in either fed or starved hypophysectomized *Anolis* (D. H. Gist and P. Licht, unpublished). Thus, hypophysectomy did not cause a clear change in protein metabolism in unfed animals.

Cieslak (1945) indicated that garter snakes, *Thamnophis sirtalis*, fared well after hypophysectomy, but he did not present growth data. My own experience with several species of *Thamnophis* showed that survival was good in adults after hypophysectomy even at warm temperatures, but appetite was poor, and even with hand feeding, there was little or no growth. Wright and Chester Jones (1957) reported that lizards, *Agama agama,* maintained a constant body weight for 30 days after hypophysectomy but growth values were not given for controls. In the same study, hand fed snakes, *Natrix natrix,* maintained weights after hypophysectomy along with controls in winter, but the operated animals lost significant amounts of weight relative to controls in summer.

Turtles often appear to live for long periods after hypophysectomy, but few data are available on growth rates. Wagner (1955) reported that hypophysectomized *Chrysemys d'Orbignyi* seemed to eat more than controls, but food intake was not quantified, and data for body weight indicated a weight loss in operated animals compared with a weight gain in controls. Measurements of linear (lean) growth were not reported in this study. I have found that young adult turtles, *Pseudemys scripta,* continued to feed well after hypophysectomy and maintained body weight as well as intact animals when fed *ad libitum.* There are no data on the effects of hypophysectomy on growth processes in crocodilians.

These few experiments dealing with the relationship between growth and the pituitary are clearly inconclusive. They suggest that growth is at least partly dependent on the pituitary in some reptiles, but they certainly do not provide proof of either GH or PL in the reptilian pituitary. The lack of experiments on the effects of hypophysectomy in young reptiles is most unfortunate, as such data could provide valuable information on pituitary growth factors.

D. Assay and Biochemical Characterization of Reptilian GH and PL

The best evidence for the presence of GH and PL in the reptilian pituitary comes from assay of reptilian pituitaries in nonreptilian species. These data also provide insights into certain biochemical properties of the reptilian hormones.

The presence of a "prolactin-like" principle in the adenohypophysis of several species of lizard, snake, turtle, and crocodilian has been demonstrated by the pigeon crop-sac bioassay (Grignon and Herlant, 1959; Nicoll and Bern, 1965; Licht and Nicoll, 1969; Eyeson, 1970; Nicoll and Nichols, 1971); and pituitaries from a few species of reptile have also been shown to exhibit prolactin-like activity in the newt water-drive test, mammotropic bioassay (Nicoll and Bern, 1969), and cutaneous yellowing reaction of the fish *Gillichthyes* (Sage and Bern, 1970). Thus, reptilian PL appears to show essentially the full spectrum of activities exhibited by other vertebrate PL. This contrasts markedly with the PL of fish which is essentially inactive in birds and mammals (see review by Bern and Nicoll, 1968).

Pituitary extracts from a turtle, lizard, and crocodilian stimulated growth of tibial epiphysial cartilage in the hypophysectomized rat (Hayashida, 1970), the assay classically used to define growth hormone. The potencies of these diverse reptilian pituitaries were all virtually identical. Using a growth test in juvenile post-metamorphic toads, distinct proteins were identified as GH in the pituitaries of several representatives of the same three orders of reptiles (Nicoll and Licht, 1971). Enemar and von Mecklenberg (1962) reported that reptilian pituitaries caused body growth in frog tadpoles, but these results can also be interpreted as indicating PL activity (Bern and Nicoll, 1968).

In an elegant comparative immunochemical study of vertebrate GH, Hayashida (1970) demonstrated that extracts from reptilian pituitaries exhibited distinct immunochemical cross-reactivity with a monkey antiserum prepared to rat GH. GH from the three orders of reptiles and from birds all showed approximately the same degree of immunochemical relatedness to rat GH. As a group, the immunochemical relatedness of reptilian and avian hormones to rat GH was less than that shown by other mammalian GHs, but greater than that shown by amphibian and teleost GH (Fig. 2). These studies provide some insight into the extent of evolution that has occurred among reptilian and other vertebrate GHs.

None of the above studies prove that PL and GH exist as separate molecular entities in the reptile. The best evidence for their separateness was obtained by electrophoretic analyses (Nicoll and Nichols, 1971; Nicoll and Licht, 1971). When examined on a polyacrylamide gel electrophoretic system, pituitary extracts of snake, lizard, turtle, and crocodilian yielded prominent and characteristic slow-migrating and fast-migrating protein bands (Fig. 3). The fast-migrating band was found to exhibit PL activity by the pigeon crop-sac assay (Nicoll and Nichols, 1971). The slow-migrating protein band was identified as GH by using the toad growth test. Similar analyses on representative amphibians, birds, and mammals showed a comparable situation, indicating that the electrophoretic mobilities of PL and GH are generally similar in all tetrapods. The PL in the two turtles studied was somewhat unusual in that it also exhibited a pronounced somatotropic activity in the toad test, whereas this was not found for the PL of the other reptilian species examined (Fig. 3). Thus, at least in the case of the chelonian hormones, there appeared to be considerable overlap between the activities of PL

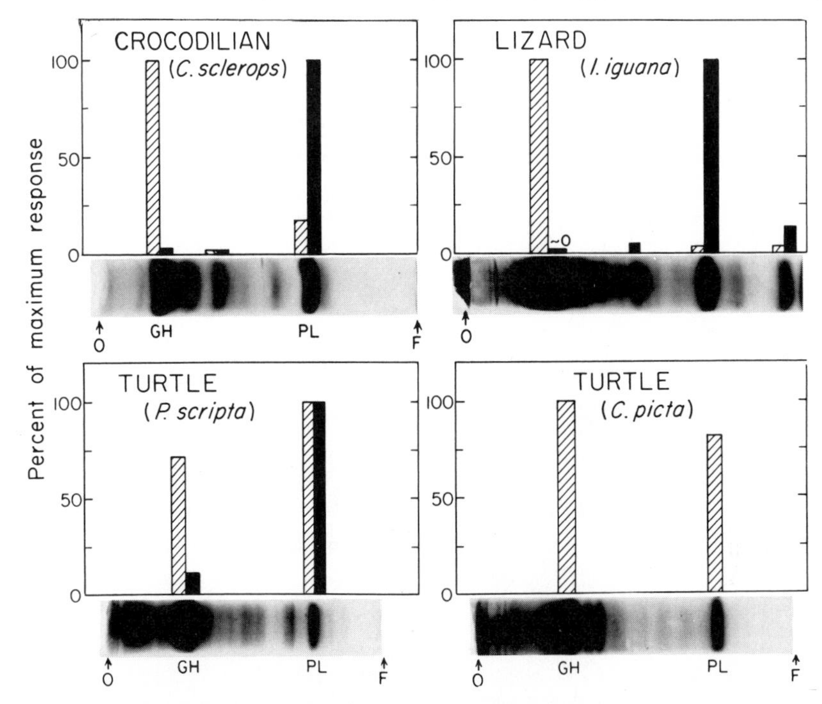

FIG. 2. Electrophoretic and biological characteristics of reptilian pituitary growth hormone (GH) and prolactin (PL). Polyacrylamide disc electrophoretic columns of homogenates of pars distalis stained in aniline blue-black are shown for four species; O-origin; F-ion front. Eluates from regions corresponding to the prominent proteins characteristic of the pars distalis were tested for biological activity in toad growth test (hatched bars) or pigeon crop-sac assay (solid bars) to identify GH and PL, respectively. Based on data from Nicoll and Licht (1971) for growth hormones and Nicoll and Nichols (1971) for prolactin.

and GH; however, the GH did not have PL activity in the crop-sac assay.

The existence of separate GH and PL in reptiles is also evident from their location in the pars distalis. Pigeon crop-sac bioassay and electrophoretic analysis demonstrate that PL is confined almost entirely to the rostral zone, whereas GH occurs predominantly in the caudal zone (Licht and Nicoll, 1969). Sage and Bern (1970) also found that the rostral zone of a lizard and turtle pars distalis contained the majority of the *Gillichthyes* skin-yellowing activity. Eyeson (1970) reported that in an *Agama* lizard PL (identified by pigeon crop-sac activity) was concentrated in the caudal half of the pars distalis, but his bioassay data consisted of weak responses in only a few test birds. Except for

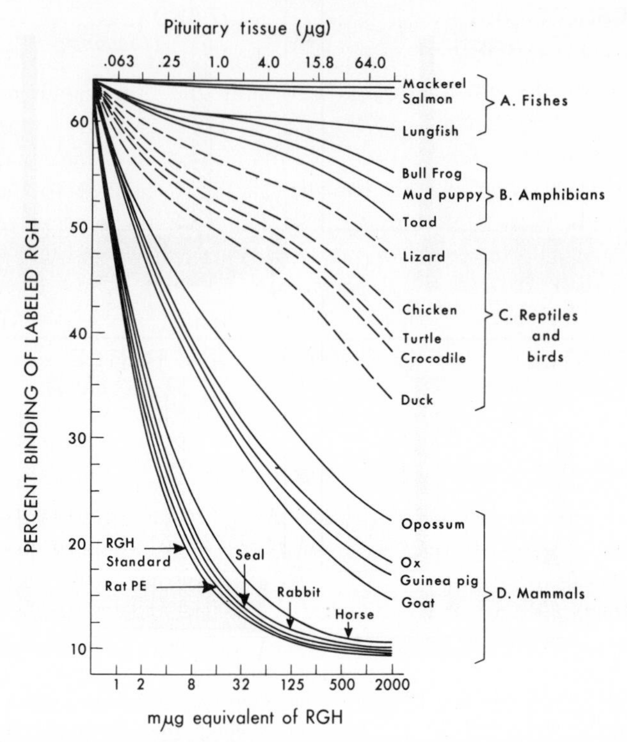

FIG. 3. Immunochemical relatedness of growth hormone in pituitary extracts from representatives of various vertebrate classes based on radioimmunoassay with monkey antiserum to rat growth hormone (RGH). Reproduced from Hayashida (1970).

the last report, the distribution of the two hormones coincides with two acidophilic cell types (erythrosinophils in the rostral zone and orangeophils in the caudal zone), and it seems likely that these are the cells that produce the two hormones (Licht and Nicoll, 1969).

The most important information on the chemical structure of reptilian GH comes from the recent purification studies on the snapping turtle *Chelydra serpentina.* Papkoff and Hayashida (1972) demonstrated that this reptilian GH could be isolated in highly purified form by chemical fractionation procedures, e.g., ion exchange chromatography on Amberlite IRC-50 and gel filtration on Sephadex G-100, that had been proven effective in isolating mammalian GH. The turtle GH behaved similarly to mammalian hormones during fractionation. Furthermore, characterization of the turtle hormone indicated marked similarities to mammalian GHs in regard to amino terminal acids and total amino acid composition. The determination of amino acid sequence to confirm the degree of structural similarity remains to be done. In the same fractionation studies, comparison of an avian (duck) GH also showed marked similarities to the turtle GH. These and the immunochemical data (Hayashida, 1970; Papkoff and Hayashida, 1972) indicate a general conservatism in the structures among reptilian, avian, and mammalian GHs.

Unfortunately, reptilian PL has proven more difficult to isolate than GH in pituitary fractionation studies. A gross inactivation of the PL is typically encountered during the fractionation procedures that have yielded GH (H. Papkoff, C. S. Nicoll and P. Licht, unpublished). Thus, the important comparisons between the chemistry of the GH and PL of these nonmammalian forms still remain to be done.

III. Gonadotropins and Reproduction

A. Evidence for a Pituitary–Gonadal Axis

Two functions of the vertebrate gonad must be recognized to understand the hormonal control of reproduction. In addition to the germinal epithelium in which the production of gametes (sperm and eggs) occurs, the gonads of both sexes contain endocrine tissues that produce a series of steroids (sex steroids) that control many extragonadal aspects of the reproductive system (accessory sexual structures, secondary sexual characteristics, sexual behavior, etc.). Both activities of the gonads appear to be under the control of the adenohypophysis. According to the classic scheme proposed for mammals, two separate pituitary hormones are involved in these functions: gametogenesis is primarily dependent on follicle stimulating hormone (FSH) and gonadal steroidogenesis on luteinizing hormone (LH). LH is also considered to be responsible for gamete release from the gonads (spermiation and ovulation). Although this view is oversimplified even for mammals (see van Tienhoven, 1968), it has had a major impact on reptilian endocrinology. The question of whether separate FSH and LH molecules exist in reptiles is an important one for understanding the physiology of reptilian reproduction and the evolution of vertebrate gonadotropins.

1. *Presence and Actions of Endogenous Gonadotropic Factors*

There is ample evidence from hypophysectomy experiments in male turtles (Wagner, 1955; Combescot, 1955, 1958; Licht, 1972a), lizards (Wright and Chester Jones, 1957; Pandha and Thapliyal, 1966; Licht and Pearson, 1969a; Reddy and Prasad, 1970a,b, 1971; Eyeson, 1970, 1971; Dufaure, 1970), and snakes (Takewaki and Hatta, 1941; Cieslak, 1945; Vivien and Stenger, 1955; Licht, 1972b) that spermatogenesis and testicular steroidogenesis are dependent on hormones produced by the pituitary. The details of the rate of testicular involution following hypophysectomy appear to vary among species. However, rather than actual interspecific variations, many of these differences are probably due to insufficient data on the pattern of regression and the use of different experimental conditions (temperature, stage of spermatogenic cycle, etc.). For example, in hypophysectomized *Anolis carolinensis*, testicular regression is completed within 10–14 days at 32°C, whereas there is little change after 3 weeks at 20°C (Licht and Pearson, 1969a).

Data for the effects of hypophysectomy on reproduction in female reptiles is more limited than for males, but a pituitary–gonadal axis is evident. Ovarian development is arrested; atresia of yolky follicles and involution of accessory sexual structures, such as the oviducts, typically occur after removal of the pars distalis (Dodd, 1960; Licht, 1970; Eyeson, 1970; Callard and Ziegler, 1970; Callard *et al.*, 1972).

The presence of a gonadotropic principle in the pituitary of reptiles was further confirmed by showing that homoplastic implantation of pituitary tissue prevented gonadal atrophy following hypophysectomy in several snakes and lizards (Dodd, 1960; Licht and Rosenberg, 1969). The presence of gonadotropic activity in turtle and snake pituitaries has also been ascertained by assay in hypophysectomized lizards (Fig. 4).

The above studies do not demonstrate whether the different activities of the gonads are controlled by one or two gonadotropins. Several modifications of the basic techniques of hypophysectomy and assay of pituitary extracts have been employed to examine this question. Cytological studies on numerous reptilian species, representing the three major orders, have led to the conclusion that there are two separate gonadotropic producing cells (an FSH and LH cell) in the pars distalis (Saint Girons, 1970; Eyeson, 1970). Unfortunately, these conclusions are based largely on an indirect approach in which correlations were sought between different tinctorial cell types and various reproductive changes (e.g., gametogenesis versus spermiation or ovulation), starting with the assumption that the various reproductive processes must be dependent on the activity of two distinct gonadotropins. In any case, according to these cytological studies, in all reptiles, the presumed LH cell is confined almost entirely to the rostral zone of the pars distalis; the distribution of the presumed FSH cell is more irregular (it tends to be located medially). If these conclusions are correct (and if two hormones with separate functions exist), then it should be possible to demonstrate both quantitatively and qualitatively different gonadotropic activities in different regions of the pars distalis.

A technique of partial adenohypophysectomy whereby selected regions of the pars distalis are removed showed a quantitative relationship between testicular activity and the amount of pars distalis present in *A. carolinensis,* but there was no evidence of differential activity in different regions of the gland (Licht and Pearson, 1969a; Licht and Rosenberg, 1969). Similar results were obtained when hypophysectomized *Anolis* were treated with extracts prepared from the rostral and caudal halves of the pars distalis of the same species. Both halves of the gland were equally effective in maintaining gonadal function (Licht and Rosenberg, 1969). Extracts from rostral and caudal halves of the pituitary of a turtle, *Pseudemys scripta,* were tested in hypophysectomized *Anolis* with the same results (Licht and Papkoff, 1972). These data further confirm the presence of a gonadotropic factor(s) in the pars distalis, but they suggest that gonadotropic activity is uniformly distributed in the two halves of the gland and give no evidence of two different kinds of gonadotropins. There has been some recent success in the chemical isolation of gonadotropic factors from the reptilian pituitary (discussed below).

2. *Actions of Nonreptilian Gonadotropins*

Studies involving the responses of reptiles to heterologous (nonreptilian) gonadotropins constitute important evidence for the potential roles of FSH and LH in the control of reproduction. Since the degree of potency and specific actions exhibited by such hormones presumably reflect the extent to which their structures (at least with regard to biologically active sites) resemble those of the reptilian molecules, inferences regarding the chemical nature of the reptilian gonadotropin(s) may also be derived from these data.

Most early studies with mammalian gonadotropins used rather crude preparations in large doses (reviewed by Dodd, 1960). Some of these dealt with two of the mammalian "pregnancy" hormones that are produced by the placenta—human chorionic gonadotropin and pregnant mare's serum gonadotropin (for details, see Panigel, 1956; Combescot, 1955, 1958; Licht, 1967; Jones, 1969; Reddy and Prasad, 1970b; Callard and Ziegler, 1970; Callard *et al.*, 1972; Eyeson, 1971; Dodd, 1960). These placental gonadotropins were active in reptiles, but since they contain both FSH and LH activities (by mammalian bioassay) and have no obvious homology in reptiles, the data provide little insight into the reptilian FSH and LH activities.

One might expect to find qualitative differences in the response of reptiles to purified preparations of FSH and LH from mammals (or any other vertebrate) if these two molecules existed separately in reptiles. Some recent investigations have sought to demonstrate such a differential effect of the two hormones. In interpreting these results,

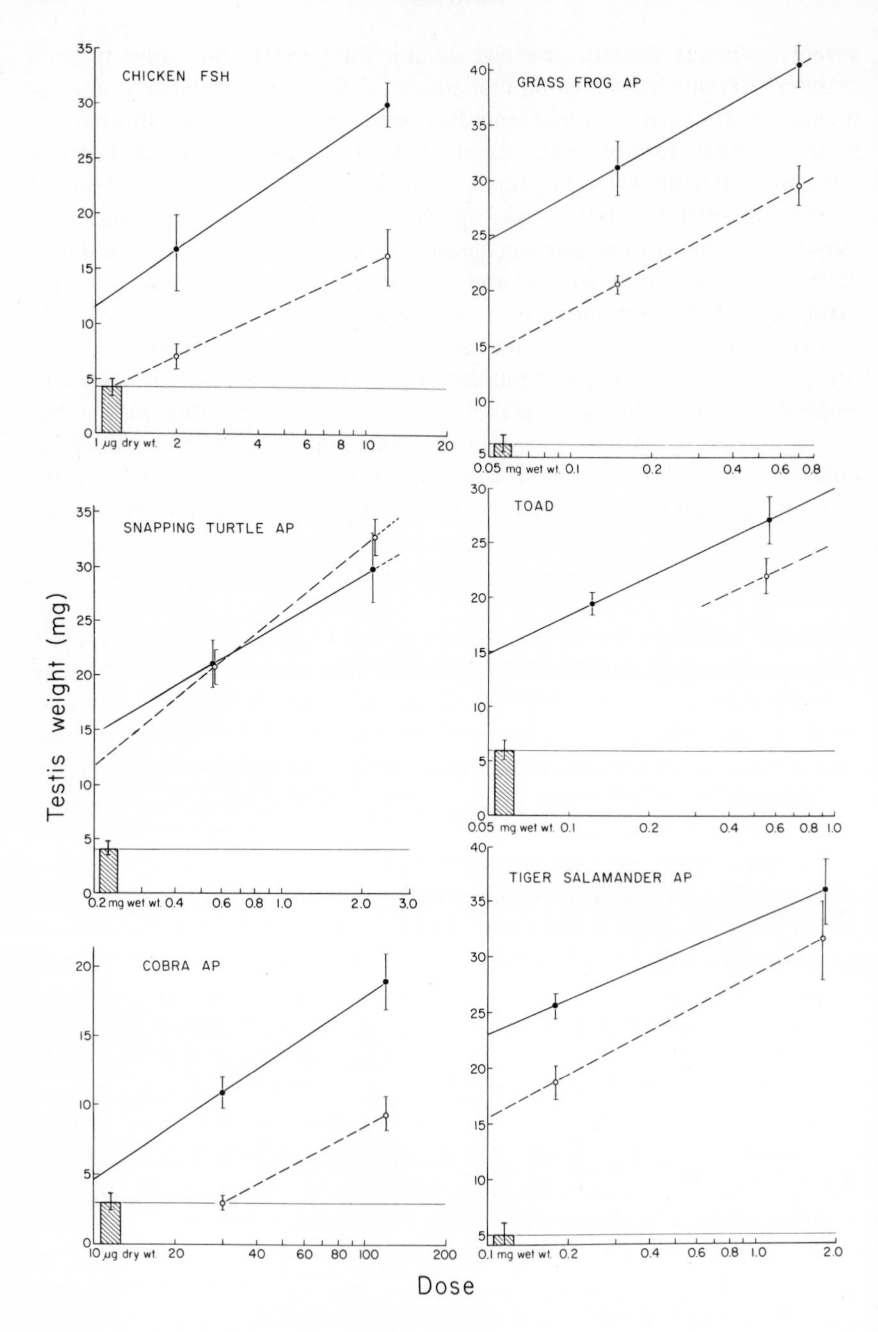
CHICKEN FSH
GRASS FROG AP
SNAPPING TURTLE AP
TOAD
COBRA AP
TIGER SALAMANDER AP
Testis weight (mg)
Dose
1 µg dry wt.
0.05 mg wet wt.
0.2 mg wet wt.
10 µg dry wt.
0.1 mg wet wt.

several important factors, such as dosage and experimental temperature, must be taken into account. For example, in two lizards, *Anolis carolinensis* and *Xantusia vigilis*, spermatogenic and steroidogenic responses to mammalian gonadotropins showed differential temperature sensitivities. At warm body temperatures (e.g., 30°C), hormones caused an extremely rapid rate of sex steroid secretion and spermatogenesis (sperm may be derived from spermatogonia in 3 weeks). In contrast, stimulation of steroid secretion could still be obtained within a few weeks of treatment at 20°C, but there was essentially no sperm production; months were required for the production of only spermatocytes (Licht and Pearson, 1969b; Licht, 1972a). Similar temperature effects were observed in females; steroidogenesis, as evidenced by oviducal enlargement, could be stimulated by hormones within a few weeks, but ovarian follicle enlargement was effectively blocked for over a month at 20°C, in contrast to the production of eggs within a few weeks at 30°C (Licht, 1972a). The gametogenic and steroidogenic tissues also showed different dose thresholds (Licht and Pearson, 1969b). In *Anolis*, low doses of FSH caused normal follicle enlargement and ovulation, but high doses of FSH induced abnormal ovarian growth and blocked ovulation, a condition that could not be corrected by the addition of LH (Licht, 1970).

Della Corte *et al.* (1966) concluded that equine LH and porcine FSH had differential effects on testes of the lizard *Lacerta sicula;* during the period of winter stasis, LH appeared to stimulate the steroid dehydrogenase activity of the interstitium (indicating a steroidogenic action), while FSH was only mildly stimulatory to spermatogenesis. These results, however, are difficult to interpret, since small sample sizes were used, animals were not hypophysectomized, doses were massive, specific activity and purity of hormones were not defined, and the studies were done at very low temperatures. In summer at warm temperatures, porcine FSH promoted the growth of testes, spermatogenesis, and interstitial cell activity (Della Corte and Cosenza, 1965).

FIG. 4. Tests for neuraminidase sensitivity of gonadotropic activity in homogenates or extracts from the pars distalis of various vertebrates based on the maintenance of testicular weight in hypophysectomized lizards (*Anolis carolinensis*). Solid curves show the activity of control preparations and dashed curves are for neuraminidase-treated portions of the same preparation; except in the toad, each curve is based on a two-point assay, with the mean and standard error shown for each dose (five to seven lizards per dose). Mean values for saline-injected hypophysectomized lizards in each test are shown by the cross-hatched vertical bars at the lower left. Neuraminidase treatment caused a highly significant reduction in activity (by 50–90%), except in the case of the snapping turtle pituitary homogenate. Based on Licht and Papkoff (1972).

Both ovine FSH and LH exhibited essentially the same broad spectrum of actions in *A. carolinensis*; they promoted testicular growth, spermatogenesis, spermiation, and androgenesis (Licht and Pearson, 1969b). However, the specific activity of the most highly purified FSH molecule was considerably more potent (about 2000-fold) than that of the best available LH in all these activities. Similar discrepancies between the actions of the two ovine gonadotropins were also observed in a gecko, *Hemidactylus flaviviridis* (Reddy and Prasad, 1970a,b), and a xantusid, *X. vigilis* (Licht, 1973). LH appeared completely ineffective in the gecko, but only a single, relatively low dose was examined. Studies in the lizard *Agama agama* also demonstrated that FSH was considerably more effective than LH in maintaining testis growth and spermatogenesis after hypophysectomy (Eyeson, 1970, 1971). However, LH was considered to be the more effective in stimulating steroidogenesis based on the epithelial height of the epididymis (Eyeson, 1971). Several difficulties exist in interpreting these data. Hormones were tested at only one dose with only three animals per group. Further, height of the epithelium may be a misleading index of androgens, because the epithelium often becomes stretched due to sperm in the lumen; more such stretching would be expected with FSH treatment due to the greater stimulation of spermatogenesis. Finally, histological examination of interstitial cells indicated that FSH produced as much or greater stimulation than did LH (Eyeson, 1971).

Tests of ovine hormones in hypophysectomized snakes, *Thamnophis sauritus* and *T. sirtalis*, and hypophysectomized turtles, *Pseudemys scripta*, yielded results similar to those reported in lizards. FSH was highly effective in promoting testicular growth and steroid secretion, whereas LH was completely ineffective at the dose tested in the turtle and had a much lower specific activity in the snakes (Licht, 1972b,c).

Work in female reptiles is much more limited than for males and is confined entirely to a few species of lizard. In general, the results parallel those detailed for the male reptiles. Ovine FSH was far more effective than ovine LH in promoting ovarian growth (vitellogenesis) and ovarian sex steroid production (Ferguson, 1966; Jones, 1969; Licht, 1970; Licht and Hartree, 1971; Eyeson, 1970). FSH alone induced ovulation in lizards (Licht, 1970; Turner *et al.*, 1973). The effects of human FSH and LH (Licht and Hartree, 1971) and porcine FSH and LH (Licht and Papkoff, 1972) in *A. carolinensis* were the same as those obtained with ovine hormones.

There was no evidence of a synergism between FSH and LH in any of the above studies. The potency of highly purified FSH preparations makes it unlikely that LH contamination contributed to the action of the FSH (Licht and Pearson, 1969b). Studies with highly purified LH and removal of FSH contaminants by antibody treatment confirm that mammalian LH also has intrinsic activity in the lizard that is independent of FSH (Licht and Papkoff, 1973a).

Mammalian FSH and LH have recently been shown to consist of two chemically distinct subunits. When tested in mammalian bioassays, these subunits exhibit little or no gonadotropic activity but can be recombined to yield normal activity. In *A. carolinensis,* the subunits of LH were also without appreciable gonadotropic activity, being less than about 10% as active as the native LH molecule. However, studies with the subunits of FSH yielded very different results. One of the subunits, α, was essentially inactive, but the other, β, was as potent as the native molecule; chemical combination with α did not increase the activity of β (Licht and Papkoff, 1971). The active β subunit is considered to be the hormone specific molecule, since the α subunit shares many structural characteristics with the α subunit of LH (and also TSH). These results indicate that even though the lizard is highly sensitive to mammalian FSH, it is not necessarily responding in the same way that the mammal does.

Preliminary studies with FSH and LH purified from chicken pituitaries also indicated that both molecules have similar broad activities on gametogenic and steroidogenic processes in both sexes of *A. carolinensis* but that FSH is the more potent in all regards (Licht and Hartree, 1971). The relative potency of the avian FSH was about tenfold greater in the lizard than in mammals, indicating some zoological specificity among the vertebrate gonadotropins; i.e., avian and reptilian gonadotropins are more similar than are the avian and mammalian hormones. The specificity is also evident in studies with teleost material. All evidence available on piscine gonadotropin indicates that these vertebrates possess only a single pituitary gonodotropin. Purified salmon gonadotropin was found to be effective in promoting spermatogenesis and androgenesis in *Anolis,* and in fact, the lizard may be the highest vertebrate in which piscine material stimulates spermatogenesis (Licht and Donaldson, 1969). Work in my laboratory has also demonstrated that spermatogenesis and androgenesis in the lizard can be stimulated with pituitary extracts from diverse anuran and urodele amphibians (e.g., Fig. 4), and complete testicular maintenance can also be accomplished with anuran (bullfrog) FSH alone (Licht and Papkoff, 1974). Thus, the reptile appears to be a "universal" recipient with regard to vertebrate gonadotropins and it is probably primarily an FSH-responsive system.

B. Characterization of Reptilian Gonadotropins

1. Chemical Studies

We have seen that considerable biological similarity (i.e., cross-reactivity between species) exists among all of the vertebrate gonadotropins when the reptile is used as the test animal. These physiological studies suggest that reptilian gonadotropin resembles the mammalian and avian FSH molecule more than the LH molecule, and that FSH is capable of stimulating all of the gonadal functions in both sexes (at least in lizards). These findings must now be viewed in relation to recent data regarding the chemical properties of reptilian gonadotropins and their relation to other vertebrate gonadotropins.

a. Neuraminidase Sensitivity. Mammalian FSH is rich in sialic acid, and this portion of the molecule is considered essential for its biological activity (see Geschwind, 1969). Removal of sialic acid, readily accomplished with the enzyme neuraminidase, destroys most of the biological activity of FSH. In contrast, mammalian LH either lacks sialic acid or this portion of the molecule is not involved in its biological activity. Tests using *Anolis carolinensis* to assay gonadotropins (Licht and Papkoff, 1972) confirmed that neuraminidase treatment reduced the activity of mammalian FSH (e.g., rat, porcine, ovine); however, this sensitive assay demonstrated that inactivation was not complete (only about 80%). Likewise, chicken FSH, salmon gonadotropin, and crude pituitary homogenates from several frogs, toads, and salamanders (but not newts) were inactivated by this enzyme, when tested in the lizard. In this connection, an especially interesting situation exists among reptilian hormones.

Crude pituitary homogenates from *Anolis* and a snake (cobra) showed an inactivation of about 80% after neuraminidase treatment (Fig. 4). This same inactivation was subsequently observed with other squamates—rat snakes (*Ptyas*) and rattlesnakes (*Crotalus*) (P. Licht, unpublished). In contrast, early tests (Licht and Papkoff, 1972) with three genera of turtle (*Kinosternum, Pseudemys,* and *Chelydra*) revealed that the gonadotropins in these reptiles were unaffected by neuraminidase (e.g., Fig. 4). These neuraminidase studies suggested a marked divergence in the biochemistry of gonadotropins between the chelonian and squamate reptiles; the turtles appeared to be divergent from almost all other vertebrates with regard to their independence of sialic acid for potency. However, we have subsequently discovered that the three families of Cryptodiran turtles included above are not representative of Chelonia. Members of the more primitive suborder Pleurodira (side-neck turtles) and many of the other Cryptodira such as sea turtles (*Chelonia*) and soft-shells (*Trionyx*) showed marked sensitivity to neuraminidase. Thus, apparently only one major assemblage of the Cryptodira, coincidentally represented by the three families included in the earlier studies, has lost the sialic acid from the gonadotropin; i.e., a marked divergence in hormone structure is evident within the Chelonia and this correlates with the concepts of phylogeny in this group.

b. Chemical Fractionation of Gonadotropins. The relative inactivity of reptilian gonadotropins in standard mammalian bioassays (see below) and difficulties in working with small quantities of pituitary glands have previously hindered isolation of reptilian gonadotropic hormones. Recent studies in my laboratory made in conjunction with Dr. Harold Papkoff of the Hormone Research Laboratory, University of California in San Francisco, have overcome some of these problems and allowed purification of reptilian gonadotropins. In a preliminary study, using an ammonium sulfate fractionation that was effective in separating an FSH and LH from mammalian pituitaries, we were only able to isolate a single gonadotropin from the turtle *Pseudemys scripta* (Papkoff and Licht, 1972). The bulk of the turtle hormone was precipitated by 0.6 saturated ammonium sulfate with virtually no activity in the more concentrated 0.6–0.8 saturated fraction; in mammals, this latter fraction contains most of the FSH. These same results were obtained with ammonium sulfate in subsequent studies on pituitaries from snapping turtle (Licht and Papkoff, 1973b), alligator, and rattlesnake (P. Licht, H. Papkoff, and S. W. Farmer, unpublished). Ion exchange chromatography on Sephadex G-100 also failed to separate two gonadotropins from the snapping turtle glands. This turtle hormone was soluble in a solution of 10% ammonium acetate–40% ethanol, which suggests that it is glycoprotein in nature like the mammalian gonadotropins.

These early studies suggested the existence of a single gonadotropin in the turtle. However, when the turtle material was subjected to a combination of ion exchange chromatography on Amberlite IRC-50 and DEAE-cellulose and the products tested with a variety of bioassays (testicular maintenance in lizards, spermiation in frogs, radiophosphorus uptake in chick testes and ovulation in excised frog ovaries), clear evidence was obtained for the existence of two distinct gonadotropins. Furthermore, their behavior suggested homologies with mammalian FSH and LH. For example, the unadsorbed fraction on Amberlite IRC-50 was highly potent in the *Anolis* lizard assay (for FSH) but inactive in anuran ovulation (an LH assay); whereas the reverse was found for one of the adsorbed fractions (pH 6 eluate). Additional separation was accomplished with DEAE-cellulose; the LH-like activity was largely unadsorbed (eluted with 0.03 *M* buffer); whereas the FSH-like activity was adsorbed (Licht and Papkoff, 1973b). The two mammalian counterparts show similar chromatographic behaviors in these systems.

Amino acid analysis of the turtle LH shows marked similarities with the composition of mammalian LH. Several preliminary studies, including N-terminal analysis, sedimentation coefficient at different pH's and counter-current distribution patterns, have also suggested that the turtle LH may consist of subunits that are homologous to those of mammalian LH (H. Papkoff and P. Licht, unpublished). Further evidence for chemical homologies among turtle, avian, and mammalian LH's was obtained by immunological analyses (P. Licht, B. K. Follett, and B. Goldman, unpublished). For example, antisera raised against chicken LH and the β-subunit of bovine LH will block the biological activity of turtle LH; the turtle LH will also compete with the homologous LH in radioimmunoassay with these antisera and with an antiserum against the ovine LH molecule.

These findings cast a new light on the question of whether one or two gonadotropins exist in reptiles. Despite the indirect physiological evidence suggesting a single hormone, at least some chelonian reptiles clearly possess an FSH- and LH-like hormone. The specific physiological roles of these two in the turtle still remains unresolved; the turtle FSH acts as a complete gonadotropin when tested in another reptile, the lizard. While these chemical studies seem to negate conclusions of earlier physiological tests, these results for snapping turtles should not be generalized too quickly. Our preliminary attempts to fractionate two gonadotropins from squamate reptiles (snakes) have thus far been unsuccessful; in particular, it has been difficult to demonstrate the existence of LH activity (see next section).

2. *Biological Characterization*

Information on the biological activity of reptilian gonadotropins in nonreptilian species should provide additional insights into the comparative biochemistry of these hormones. Of particular interest is the question of whether FSH and LH activities can be demonstrated. Unfortunately, the reptilian hormones tend to be relatively inactive in standard mammalian bioassays for these hormones. Several attempts to show an activity of turtle pituitaries in the Steelman–Pohley rat assay for FSH have been unsuccessful (Faiman *et al.*, 1967; Licht and Papkoff, 1973b), and tests for LH by ovarian ascorbic acid depletion in rats (Parlow assay) have been equivocal. Faiman *et al.* (1967) reported a positive response for the pituitaries of snapping turtles, but they gave no details or statistical values. This result contrasts sharply with repeated tests in our laboratory in which *Chelydra serpentina* and *Pseudemys scripta* glands were inactive at very high doses (Papkoff and Licht, 1972; Licht and Papkoff, 1973b). Bratcher and Kent (1971) reported that pituitaries from adult female lizards *Anolis carolinensis* gave a positive response in the Parlow LH assay, but their results were equivocal, since responses were marginal and failed to show a dose dependency. We have occasionally observed such marginal responses with turtle and snake pituitaries, but they are rarely repeatable, and we tend to regard them as nonspecific reactions in the assay. Turtle gonadotropic fractions were relatively potent in an *in vitro* assay based on progestin secretion by cultured primate granulosa cells (Channing *et al.*, 1974), and they were also active in lactic acid production in excised rat ovaries (S. W. Farmer, unpublished). Thus, there is not a complete species specificity for reptilian hormones in the mammalian system, but these *in vitro* systems are not specific for FSH or LH.

Turtle and crocodilian pituitaries, including a wide array of species, are relatively potent in ovulation tests with anuran ovaries *in vitro* (Licht and Papkoff, 1973b; Licht, 1974). Tests with mammalian and avian hormones indicate that this assay is highly specific for LH and hence may be taken as evidence for reptile LH; in fact, this assay was important in isolating the LH from the snapping turtle (Licht and Papkoff, 1973b). These findings are consistent with Witschi's (1955) report that turtle pituitaries are rich in LH based on the weaver finch feather color test. However, in marked contrast to the above findings, I have been unable to demonstrate any ovulating activity in anuran ovaries with the pituitaries from snakes or lizards. Tests with very high concentrations of glands from several lizard and snake species have been consistently negative (Licht, 1974). Thus, either the squamate reptiles lack an LH or there is a sharp difference in the species specificity of anuran ovaries among reptiles. In either case, these results point to a major divergence in the LH-like hormones among the different reptilian orders. The lack of an LH in squamates would be consistent with findings that the lizard and snake seem to be unresponsive to LH; they respond to various types of FSHs (mammal, bird, turtle, frog) as a complete hormone.

C. Endocrine Gonad—Sex Steroids

Under the influence of pituitary gonadotropins, certain tissues within the gonads secrete a series of hormones that have a wide variety of physiological actions on the reproductive system. Unlike the pituitary gonadotropins, which are proteins, these hormones, like those of the interrenal glands, are steroids. The gonadal hormones are usually referred to as the sex steroids because of their actions connected with reproduction and "sexuality."

1. Biosynthesis of Gonadal Steroids

The chemistry of the gonadal steroid hormones and their biosynthetic pathways are relatively well known for mammalian species. The compounds of particular interest here are the androgens, estrogens, and progestogens. These are derived from cholesterol with pregnenolone and progesterone, which are important intermediate products in the biogenesis of all three final products. The principal steroidal products of the testes are androgens, and those from the ovaries are estrogens and progestogens, but there is evidence for all three types being produced in the gonads of both sexes. Available data indicate that the major steroid products and biosynthetic pathways in the reptiles are similar in many respects to those of mammals.

Almost all the reptilian studies have relied on *in vitro* incubation of gonadal tissue. Tam and Phillips (1969) demonstrated large quantitative seasonal changes in the biosynthetic abilities of cobra testes. (This finding was not unexpected in light of the pronounced seasonal cycles of gonads of almost all temperate-zone reptiles.) Thus, the quantities and ratios of substances produced by *in vitro* incubations must be interpreted cautiously.

Androgens, estrogens, and progestogens have been identified in incubates of snake and lizard testes. The major androgens are probably testosterone and androstenedione, both produced from progesterone (Tam and Phillips, 1969; Callard, 1967; Lupo di Prisco *et al.*, 1967a,b). Gottfried *et al.* (1967) failed to detect testosterone production from endogenous precursors in cobra testes, but their experiments were performed in April when testosterone production may be relatively low (Tam and Phillips, 1969). Estrone has also been demonstrated in a snake (Gottfried *et al.*, 1967) and lizard (Lupo di Prisco *et al.*, 1967a).

The ovary of the lizard, *Lacerta,* showed all the major steps in the pathways for estrogen production, with estrone and estradiol-17β being the major products (Chieffi, 1967; Lupo di Prisco, *et al.*, 1967a). Callard and Leathem (1965) focused attention on the synthesis of progesterone from pregnenolone in ovaries of snakes to determine their potential importance in pregnancy. They demonstrated that in an egg-laying species, the ovaries synthesized much less progesterone than in viviparous (livebearing) forms. In the latter, the ovaries of pregnant animals, especially in late pregnancy, synthesized more progesterone than those of nonpregnant individuals of the same species. Callard *et al.* (1972) cited preliminary findings in which circulating progesterone levels in the snake,

Natrix, rose from 1 ng/ml plasma during the early follicular growth phase to 5 ng/ml just prior to ovulation and in early pregnancy. Bragdon *et al.* (1954) have reported similar increases in plasma progesterone in pregnant snakes based on bioassay of plasma. Corpora lutea from snapping turtles (*Chelydra*) have the capacity to convert cholesterol to progesterone *in vitro* (Klicka and Mahmoud, 1973).

For a comprehensive picture of reptilian steroid synthesis, we still need more comparative work, studies on changes related to seasonal and physiological activity, data on the nature of circulating steroids, and tests of the effects of altering temperature.

2. *Physiological Actions of Sex Steroids*

Androgens (such as testosterone) in males and estrogens (such as estradiol) in females are frequently referred to as the male and female sex hormones, because they tend to differentially stimulate changes in reproductive structures and behavior that are typically sex specific. Progestogens appear to play a secondary role in this regard. Successful experiments have been conducted in many species of reptiles in which males were feminized or females masculinized by the alternate sex steroid (see Dodd, 1960; Kehl and Combescot, 1955). However, it should be borne in mind that both types of sex steroid are found naturally in both sexes and may also be produced by the adrenal gland (Gottfried *et al.*, 1967). This review of Dodd (1960) serves as an excellent guide to the general literature for reptiles and only a few special problems that have received recent attention will be discussed here.

a. Androgens. Generally, androgens are the primary regulators of the development of the accessory sexual structures. In squamate reptiles, one important site of action is the sexual segment of the kidney, a portion of the renal collecting duct system. This segment is lacking in chelonians and crocodilians and the ducts are generally more responsive in lizards than in snakes (see recent review by Prasad and Sanyal, 1969). The female sex hormones do not stimulate the renal sex segment, although they may stimulate the epididymis (Prasad and Sanyal, 1969). The male sex steroids may also influence secondary sexual characteristics in males, such as femoral pores (see especially Cole, 1966; Chiu *et al.*, 1970) and body coloration, and various aspects of sexual behavior, such as territoriality and courtship. The extent of dependence of these characters on gonadal secretions varies with species. In some cases, the early development of the sexual character may be steroid dependent, but thereafter steroids are not required for maintenance.

Androgens may also influence nonreproductive processes such as skin sloughing. Testosterone increased the frequency of sloughing in the gecko (Chiu *et al.*, 1970), although gonadotropins (which stimulate testosterone) had no affect on sloughing in *Anolis carolinensis* (Maderson and Licht, 1967). In snakes, estrogens and testosterone have been reported to inhibit sloughing indefinitely (Maderson *et al.*, 1970).

Considerable controversy still exists regarding whether androgens have a direct action on gametogenesis. It has been suggested that androgens stimulate spermatogenesis and thus represent a mechanism by which the effects of gonadotropins are mediated. However, the results of most of the early reports on this subject are open to question. For example, the most frequently cited study is that of Forbes (1941) in which testosterone was reported to stimulate spermatogenesis in intact lizards (*Sceloporus floridanus*) on the basis of an apparent increase in the number of sperm in the seminiferous tubules. However, the testes were producing sperm actively at the start of treatment, and Forbes' own data showed that testosterone caused a reduction in testicular weight. These results can be interpreted as showing a decline in gonadotropin production (due to negative feedback of steroids) with a concomitant reduction in the rate of sperm release. Ramaswami and Jacob (1963) reported that testosterone had no effect in the lizard *Uromastix* but stimulated the testes of juvenile *Varanus;* no data were presented in support of the latter statement. Ramaswami and Jacob (1965) also reported that testosterone stimulated some increase in testicular size and promoted sperm production in juvenile (intact) crocodilians; but the authors' photomicrographs showed only slight spermatocyte production. Since the animals were not hypophysectomized in any of the above studies, the role of endogenous pituitary gonadotropins cannot be ruled out. In intact *Lacerta sicula,* testosterone suppressed the normal seasonal initiation of spermatogenic recrudescence (Licht *et al.*, 1969).

Testosterone treatment has generally had no or only slight effects on the maintenance of spermatogenesis in hypophysectomized lizards. It was ineffective in the gecko, *Hemidactylus flaviviridis* (Reddy and Prasad, 1970a,b), and the agamid, *Agama agama* (Eyeson, 1971). In another gecko, *Gekko gecko* (Lofts and Chiu, 1968), and an iguanid, *Anolis carolinensis* (Licht and Pearson, 1969a), testosterone slowed the rate of testicular involution but did not maintain normal spermatogenesis.

One possible basis for the discrepancies among studies is that insufficient doses were used or that studies were not detailed enough to observe slight effects (e.g., in the rate of regression). Since testosterone is normally synthesized within the testis, and probably within the seminiferous tubules in conjunction with Sertoli cells (see Lofts, 1969), very high systematic doses may be required to elevate intratesticular concentrations to effective levels. Implants of testosterone into the gonads might prove informative; preliminary experiments of this kind in *Anolis* (P. Licht, unpublished) proved negative.

b. Estrogens and Progestogens. The physiological actions of estrogens and progestogens, especially with regard to ovarian activities, are still poorly understood. There is ample evidence from female lizards that estrogens are an important factor in the control of vitellogenesis. Estradiol caused an enlargement of the liver of several lizards, presumably due to the production of vitellogenic proteins (Hahn, 1967; Callard *et al.*, 1972). In the lizard *Dipsosaurus dorsalis,* the pituitary gland and probably growth hormone appear to be required for this estrogenic effect (Callard and Ziegler, 1970). Hepatic materials are then transported via the blood to the ovary for yolking of the developing eggs prior to ovulation. Urist and Schjeide (1961) presented data on changes in blood composition associated with this vitellogenic action of estradiol in male turtles. Estradiol caused mobilization of fat in several lizards (Hahn, 1967; La Pointe, 1969). The development of the oviduct in preparation for ovulation seemed to be largely under the control of estrogens (Callard *et al.*, 1972; La Pointe, 1969; Wilkinson, 1966), although exogenous androgens also caused stimulation of oviducts (Prasad and Sanyal, 1969).

The role of progesterone is much less clear. In the viviparous lizard *Sceloporus cyanogenys,* progesterone inhibited ovarian growth and ovulation, possibly through an inhibition of hepatic vitellogenesis (Callard *et al.*, 1972); but it had no effect on ovulation in *Anolis carolinensis* (R. E. Jones, unpublished data). Progesterone had little or no effect on oviducal weight in several lizards (La Pointe, 1969; Callard *et al.*, 1972; Wilkinson, 1966), but careful histological examination revealed that in the viviparous lizard *Xantusia vigilis,* estrogen and progesterone may act on different regions in the oviduct (Yaron, 1970). Progesterone has also been reported to induce histological changes in another viviparous lizard, *Zootoca* (Panigel, 1956), and at least one egg-laying lizard, *Uromastix* (Kehl, 1944). Thus, the lack of large increases in oviducal weight with progesterone may give a misleading impression of its role in oviducal development. Progesterone was more effective than estrogen for inducing sexual body coloration in female collared lizards *Crotaphytus* (Cooper and Ferguson, 1972). Much more information is needed on the dose response characteristics of both steroids and on their possible interaction (synergism) in reptiles.

Space permits only mention of another important problem, yet unresolved, in the physiology of the two female sex steroids relating to their role in pregnancy (for reviews and recent data, see Dodd, 1960; van Tienhoven, 1968; Knobil and Sandler, 1963; Callard *et al.*, 1972). The development and persistence of a distinct corpus luteum that appears to secrete steroids during pregnancy (or at least they are produced in ovarian tissue) lead to the supposition that such steroids might be important in maintaining pregnancy. The differences in progesterone synthesis by ovaries of pregnant and nonpregnant snakes and increases in circulating progesterone during pregnancy have already been mentioned. These data indirectly suggest an involvement of ovarian progesterone secretion in pregnancy. However, many of the attempts to demonstrate a clear dependence of pregnancy on the ovary or corpus luteum have been unsuccessful.

IV. Adrenocorticotropin (ACTH) and Corticoid Hormones

A. Interrenal Glands

1. Definition of ACTH

The major action of adrenocorticotropin (ACTH) is the stimulation of the adrenal cortex, and any discussion of the properties of this pituitary hormone requires consideration of the physiology of the adrenal gland. In mammals, the adrenal gland consists of two distinct regions, an outer cortical zone and an inner medullary region. The medulla is comprised of chromaffin tissue and elaborates adrenaline-like substances (catecholamines); this tissue is essentially independent of the pituitary gland. In contrast, the outer cortical tissue is the site of production of a variety of steroid hormones (called corticoids or corticosteroids), and it is on this tissue that the pituitary ACTH has its primary actions—a general long-term maintenance and growth of the cortex as well as the shorter term action of regulating steroid release. In fact, since the adrenal cortex stores very little hormone, the increase of steroid secretion by ACTH also involves a relatively rapid (short-term) increase in the rate of steroidogenesis in the tissue. The increase in size of the cortex under the trophic influence of ACTH presumably is related to an increased steroidogenic capability of the gland as a whole.

The distinctive nature of the cortical and medullary tissue in the adrenal is clearly seen in lower vertebrates (e.g., Elasmobranchs), where the two types of tissue exist as separate glands. In these forms, the cortical tissues comprise what is known as the interrenal gland. In reptiles, the cortical and chromaffin (medullary) types of tissue occur together in a single discrete gland, but the arrangement of the two tissues is very different from that found in mammals. There is considerable intermingling between the two types of tissues, their pattern of mixing and proportions showing wide interspecific variation [see Gabe (1970) for a detailed review of adrenal morphology and histochemistry in reptiles]. Although the two tissues exist together, it is generally agreed that the term "interrenal gland" is preferable to "adrenal" when referring specifically to the steroidogenic cortical tissues, and this terminology has been adopted for the present discussion.

2. Chemistry of Steroidogenesis in the Interrenal Gland

The steroids produced by the interrenal gland and their biosynthetic pathways in reptiles have received considerable attention in recent years, and the chemistry of these hormones is now relatively well known. Although the number of reptilian species that have been examined is small, they represent all three major orders and no major interspecific differences are evident. Since these data have been reviewed extensively in recent years (Phillips and Bellamy, 1963; Nandi, 1967; Mialhe and Koch, 1969; Sandor, 1969; Huang *et al.*, 1969), they will be considered only briefly here.

The steroid hormones secreted by the adrenal gland of mammals may be grouped into three major classes on the basis of their physiological activities—sex steroids, glucocorticoids, and mineralocorticoids. The sex steroids are identical to those characteristically secreted by the gonads and their chemistry has already been discussed in Section IIIC. Although small amounts of sex hormones—androgens (testosterone) and estrogens (estrone)—have been identified in incubates of interrenal tissue from the cobra (Gottfried *et al.*, 1967), their physiological importance has not been evaluated. The two types of corticoids are the major products of the interrenal glands. Glucocorticoids are involved in carbohydrate metabolism and include compounds such as cortisol, cortisone, and corticosterone. Mineralocorticoids have an effect on mineral (electrolyte) balance and include steroids such as aldosterone and deoxycorticosterone. Representatives of both types of steroids have been identified in the interrenal glands of reptiles.

All of these steroid hormones, including the wide variety of sex hormones produced by the gonads, share many common features in their chemical composition and biosynthesis. They are all basically derived from conversion of cholesterol via pregnenolone. Metabolic analyses of *in vitro* incubates of reptilian interrenal tissues (with endogenous and exogenous precursors) indicate that corticosterone and, to a slightly lesser extent, 18-hydroxycorticosterone are the principle types of glucocorticoids produced by tissues of snakes, lizards, turtles, and crocodilians. Aldosterone appears to be the main mineralocorticoid in reptiles. The major synthetic pathway for the corticoids appears to be pregnenolone → progesterone → 11-oxycorticosterone → corticosterone, with 18-hydroxycorticosterone and aldosterone being derived from conversions of the corticosterone. This pattern of biosynthesis resembles that demonstrated for amphibians and birds but is different from that of some mammalian species in which cortisol is the primary glucocorticoid (see Sandor, 1969).

Unfortunately, there is little information on the nature of circulating steroids in reptiles; thus, the physiological significance of the compounds extracted from or secreted by the interrenal tissue *in vitro* is not entirely clear. Phillips and Chester Jones (1957) measured relatively large amounts of corticosterone in the effluent venous blood from the interrenal gland in the snake *Natrix natrix*. They also reported small amounts (about 5% of corticosterone levels) of cortisol. Cortisol was also tentatively identified in the blood of a turtle (Chester Jones *et al.*, 1959). More recently, Bradshaw and Fontaine-Bertrand (1970) demonstrated that corticosterone could be readily measured in saurian blood by a sensitive competitive-binding radioassay. The technique depends upon corticosterone being the major circulatory corticoid and has been employed to measure plasma corticosterone levels under different physiological conditions in the lizard *Anolis carolinensis* (Licht and Bradshaw, 1969). Thus, it seems likely that corticosterone is indeed a major steroid secreted by the reptilian interrenal gland, but the significance of cortisol, which is barely or not detectable in interrenal incubates, remains obscure. Circulating aldosterone has not been positively identified in reptiles.

3. *Physiological Actions of Corticoids*

Only a very few studies have directly examined the mineral and glucocorticoid actions of the interrenal steroids in reptiles. In general, the limited available data suggest that the corticosteroids are involved in both types of activities, but it is by no means clear that two distinct classes of steroid compounds can be distinguished on the basis of their actions on electrolyte and carbohydrate regulation.

a. Water and Electrolyte Regulation (Mineralocorticoids). General knowledge of osmoregulation in reptiles suggests that if the adrenal steroids are active in this phenomenon, they might exert their effects at several distinct sites of salt exchange: kidneys, nasal salt glands, bladder and oral mucosa (not all species possess all sites).

Mineralocorticoid activity of corticosteroids related to renal function was recently demonstrated in a snake (*Natrix cyclopion*) and a turtle (*Chrysemys picta*) by surgical adrenalectomy. Elizondo and Le Brie (1969) effected functional adrenalectomy in the snake by occluding the circulation to the gland. In water-loaded (diuretic) snakes, such adrenal insufficiency resulted in a fall in plasma sodium, potassium, and chloride. The change in sodium could be explained by increased renal sodium excretion (specifically a decline in sodium resorption in the proximal tubules of the nephron). The fall of potassium was the reverse of what is usually observed in mammals, and this could not be accounted for by altered renal handling of the ion.

Administration of aldosterone and corticosterone had little effect in the intact snake, possibly due to the already high endogenous levels of these hormones (Elizondo and Le Brie, 1969). However, when snakes were sodium loaded—a condition that should suppress endogenous aldosterone—exogenous aldosterone caused a significant decrease in urine flow and ion excretion via increased tubular resorption (Le Brie and Elizondo, 1969). These results are consistent with the observations on adrenalectomized snakes and suggest that aldosterone acts to conserve sodium in the snake. Unfortunately, corticosterone was not examined in this system.

Surgical adrenalectomy in the turtle (Butler and Knox, 1970) caused a fall in plasma sodium similar to that observed in the snake, but the turtle showed a rise rather than a fall in plasma potassium. The change in ions was attributed to altered renal clearance since no change in muscle potassium was observed, but renal function was not directly examined in the turtle. Unfortunately, the ability of adrenal steroids to correct the effects of adrenalectomy was not examined.

Aldosterone was shown to increase the active transport of salts across the bladder of a tortoise *in vitro* (Bentley, 1962). This action could be important in the chelonian study mentioned above, but no attempts have been made to assess this action *in vivo,* and the possibility of a similar action in other reptiles has not been explored.

An involvement of corticoids in extrarenal salt regulation, specifically in connection with the nasal salt glands, has been demonstrated in two reptilian species. Interestingly, the proposed action of the adrenal steroids is essentially opposite in the two cases.

In a marine turtle, *Chelonia mydas,* chemical blockage of adrenal steroidogenesis by amphenone caused marked reduction in nasal gland secretion of both sodium and potassium in turtles loaded with sodium chloride (Holmes and McBean, 1964). Administration of corticosterone to animals with adrenal insufficiency returned nasal salt secretion to normal levels; unfortunately, aldosterone was not studied. Thus, the adrenal steroids appear to promote salt excretion in the nasal gland of this turtle, an action that is the opposite from that proposed for the renal system in which the adrenal steroids favor sodium retention.

In an herbivorous desert lizard, *Dipsosaurus dorsalis,* Templeton *et al.* (1968) concluded, on the basis of surgical adrenalectomy and administration of aldosterone in intact and adrenalectomized animals, that this mineralocorticoid promoted sodium retention while allowing potassium excretion in the nasal salt gland. V. H. Shoemaker and S. D. Bradshaw (unpublished data) also found that aldosterone virtually abolished sodium excretion by the nasal gland in intact *Dipsosaurus,* even when the lizards were given a sodium chloride load. There was little or no effect on glandular potassium excretion. ACTH and corticosterone had the same type of sodium conserving effect as aldosterone on nasal gland secretion (V. H. Shoemaker and S. D. Bradshaw, unpublished); thus, the response to corticosterone was opposite to that observed in the marine turtle.

Plasma electrolyte balance was not examined in the above studies on nasal gland excretions. However, Chan *et al.* (1970) reported that hypophysectomy increased plasma sodium and decreased potassium in *D. dorsalis.* Aldosterone caused a further elevation in plasma sodium in the hypophysectomized lizard suggesting a sodium-conserving action. Corticosterone alone had little effect, but when given with prolactin, blood composition (electrolyte and water content) was returned to normal levels; thus, the action of corticosterone is opposite to that of aldosterone. The sites of action (e.g., kidney or nasal salt glands) were not studied. These findings on plasma sodium are inconsistent with those obtained in connection with the nasal salt gland response to corticoids in the same species.

In contrast to the rise in plasma sodium in *Dipsosaurus* (Chan *et al.,* 1970), hypophysectomy had essentially no effect on plasma electrolytes in an *Agama* lizard and a snake (Wright and Chester Jones, 1957). Plasma sodium declined in one study on a hypophysectomized freshwater turtle (Butler and Knox, 1970) but showed no change in another (Trobec and Stanley, 1971). Thus, the effects of hypophysectomy on mineral balance in reptiles are somewhat confusing and contradictory.

Available data on water and electrolyte balance in reptiles allow the conclusion that the corticoids play some role in osmoregulation, but it would be premature to attempt any generalizations about their patterns of action. It is even difficult to determine whether interspecific differences exist in the actions of corticoids, since the lack of standardization in experimental protocols precludes meaningful interspecific comparison. For example, diets (salt and water intakes) were not standardized among studies, even for a single species, yet the effectiveness of mineralocorticoids in snakes was shown to vary with the basic salt balance of the animal (see Elizondo and Le Brie, 1969; Le Brie and Elizondo,

1969). Interspecific differences in the physiological actions of corticoids and their relation to diet are of special interest for reptiles, since species may face very different types of salt stresses (e.g., marine turtles normally encounter a sodium load, while the herbivorous desert lizard faces a potassium load). It is important to have more comprehensive studies on the actions of corticoids on different salt-regulating sites (e.g., kidney and salt glands) in each species.

b. Carbohydrate Regulation (*Glucocorticoids*). There is evidence that steroids from the interrenal may be involved in carbohydrate regulation in turtles, snakes, lizards, and crocodilians in a manner similar to that found in mammals. Unfortunately, all such studies have utilized cortisone or hydrocortisone; and corticosterone, the naturally occurring steroid in reptiles, has not been tested. Yet in their studies on nasal salt glands in *Dipsosaurus*, V. H. Shoemaker and S. D. Bradshaw (unpublished) found that corticosterone had physiological effects but cortisol did not. Most of these data are reviewed in connection with blood sugar studies involving growth hormones (see Section II).

Butler and Knox (1970) demonstrated a drop in hepatic glycogen levels following hypophysectomy and adrenalectomy, the latter having the greater effect. Consistent with these findings was a rise in hepatic glycogen following β-methasone treatment (which would be expected to decrease endogenous corticoids). Unfortunately, these experiments were carried out using relatively low body temperatures, and the slow metabolic changes may have obscured or slowed the normal corticoid effects. Hypophysectomy also caused a significant decrease in hepatic glycogen (3.5 to 0.6 mg %) within 1 week in fasted *Anolis carolinensis* kept at warm temperatures (P. Licht and D. H. Gist, unpublished data). Single daily injections of ACTH were without effect on blood glucose in the turtle *Emys orbicularis*, but continuous infusions of ACTH caused marked hyperglycemia (Vladescu, 1965).

There have been no simultaneous studies of blood sugar and hepatic glycogen in reptiles. Furthermore, alterations in hepatic glycogen and blood glucose levels in hypophysectomized animals cannot be taken as evidence for an ACTH-mediated action of corticosteroids, since at least two other pituitary hormones, prolactin and growth hormone, may have important effects on these systems (see Section II).

B. Evidence for a Pituitary–Interrenal Axis

1. In Vivo Studies

Attempts to demonstrate a pituitary–interrenal axis in reptiles have relied heavily upon indirect, chemical methods for altering ACTH production. These techniques are based upon the premise that there is a negative feedback regulation between the level of circulating corticoids and ACTH secretion. Accordingly, injection of exogenous steroids should lower endogenous ACTH production, and alternatively, chemical blockage of endogenous steroid release by the interrenals should lead to compensatory increases in pituitary ACTH secretion. This assumption of a negative feedback between steroids and ACTH is based largely on evidence derived from mammalian studies, but cells in the reptilian pars distalis appear to respond to manipulation of plasma corticoids in the expected fashion. Del Conte (1969) observed increased activity of certain cells in the rostral pars distalis in the lizard *Cnemidophorus lemniscatus* treated with metapyrone, a chemical inhibitor of corticosteroid secretion; and similar changes were observed in *Lacerta sicula* following surgical adrenalectomy (P. Licht and H. E. Hoyer, unpublished data). Since the location of these cells corresponded to the distribution of ACTH in the reptilian pars distalis as determined by bioassay (Licht and Bradshaw, 1969), these histological findings support the hypothesis of a compensatory increase in ACTH production following reduction in circulating corticoids.

Evaluation of ACTH levels has usually been based upon histological or gravimetric criteria for assessing responses in interrenal activity [reviews by Knobil and Sandler (1963), Gabe *et al.* (1964), Ramaswami (1967), Miahle and Koch (1969), and Chan *et al.* (1970) summarize these findings in detail]. There is evidence in representatives of all three major reptilian orders that the interrenal tissues respond to manipulation of the pituitary gland. One of the earliest studies on hypophysectomy in snakes (Schaeffer, 1933) also showed that replacement therapy with homoplastic pituitaries prevented the degenerative changes in the interrenals. Miller (1952), Wright and Chester Jones (1957), and Chan *et al.* (1970) demonstrated that mammalian ACTH alone appeared capable of preventing the regression of the interrenal tissue in surgically hypophysectomized lizards. These *in vivo* studies support the view that some pituitary trophic factor is required for the maintenance of the interrenal tissues. They do not rule out the possibility that several different pituitary hormones may affect interrenal function.

An important limitation of studies that rely upon histological and gravimetric data for evaluating interrenal activity is that they do not demonstrate the extent to which the release of specific steroids is controlled by ACTH. This has been an important question, because studies of steroidogenesis *in vitro* discussed below frequently failed to find an effect of ACTH on steroidogenesis. Final proof of an ACTH control of interrenal function requires direct measurements of steroid secretion.

Phillips and Chester Jones (1957) reported high levels of corticosterone in the effluent blood from the interrenal gland of an intact snake following injection of a large dose of mammalian ACTH. However, since they did not report corticosterone levels for controls, the effectiveness of the ACTH cannot be evaluated. More extensive studies on circulating corticosterone in *Anolis carolinensis* (Licht and Bradshaw, 1969) confirmed the results of histological studies. There was an abrupt decline in plasma corticosterone following surgical hypophysectomy or injection of dexamethasone (a steroid known to block pituitary ACTH secretion in other vertebrates). Further, a single injection of purified mammalian ACTH caused a dose-dependent rise in circulating corticosterone in these dexamethasone-blocked lizards. Both the magnitude and pattern of response to ACTH were highly temperature dependent.

Gist and de Roos (1966) demonstrated the presence of ACTH activity in the pituitary of a crocodilian (*Caiman*) by bioassay on chicken adrenal glands, but they were unable to detect an effect of the crocodilian pituitary on steroidogenesis in crocodilian interrenal tissue *in vitro.* However, using another reptile (the dexamethasone-blocked *Anolis*) as an assay animal, Licht and Bradshaw (1969) found ACTH activity in acid extracts of pituitaries from several lizards, a turtle, and the caiman; i.e., all caused corticoid release *in vivo.* In all cases, the vast majority of this ACTH activity was confined to the rostral zone of the pars distalis.

2. *In Vitro Studies*

Many of the *in vitro* studies dealing with the biosynthetic pathways of the corticoids have simultaneously attempted to elucidate the action of ACTH on steroidogenic rates. However, despite all the evidence for an ACTH dependence of the reptilian interrenal gland *in vivo,* attempts to demonstrate an action of ACTH *in vitro* have been largely unsuccessful. In fact, the surprising lack of success in most such studies led Miahle and Koch (1969) to conclude that steroidogenesis in the reptilian interrenal may show significantly less dependence on ACTH than in other vertebrates, a conclusion that is clearly inconsistent with the results of *in vivo* studies. Several factors may account for the discrepancy between the effects of ACTH *in vivo* and *in vitro.*

Huang *et al.* (1969) showed that a careful kinetic approach was necessary to study the actions of ACTH on corticosterone production by cobra tissue *in vitro.* Their finding that mammalian ACTH caused both an increased synthesis and metabolism of the steroid, resulting in a net increase in turnover, suggested that the failure of others to demonstrate an ACTH effect may have been due to an inadequate sampling procedure. Incubation temperatures employed in this study, and in the only other one in which ACTH appeared to effect steroidogenesis *in vitro* (Leloup-Hatey, 1968), were relatively high. Temperature has a marked effect on steroidogenesis *in vivo* (Licht and Bradshaw, 1969), and the failure of ACTH to stimulate steroidogenesis *in vitro* may thus have been due to the use of suboptimal temperatures.

It seems reasonable to conclude that the secretion of corticosterone by the reptilian interrenal gland shows much the same dependence on pituitary ACTH as in mammals. The exact mode of action (e.g., site in the biosynthetic pathway) remains to be established. There is still no good information on the regulation of other corticoids, especially aldosterone, by ACTH in the reptile. The fact that the effects of hypophysectomy in reptiles are often less severe than adrenalectomy in connection with activities associated with mineralocorticoids (e.g., Butler and Knox, 1970; Chan *et al.*, 1970), suggests some autonomy in corticoid secretion.

C. Biochemistry of ACTH

Although there have been no chemical studies of reptilian ACTH, several of the physiological studies discussed above indicate some resemblance among ACTH molecules from diverse vertebrates. For example, mammalian ACTH is an effective corticotropic factor in reptiles, and reptilian ACTH is effective in birds (Gist and de Roos, 1966). Since reptilian material has not been tested in mammals, little can be said about potential phylogenetic specificity in ACTH. Further indirect information on the structure of reptilian ACTH may be derived from the known relationship between the structure and function of mammalian ACTH and another pituitary hormone, melanocyte-stimulating hormone (MSH).

Studies in a variety of vertebrates, including *Anolis carolinensis* (Burgers, 1960), have shown that MSH exists in several biologically active forms, and there is considerable similarity in the amino acid sequence of certain of the forms of MSH in animals as unrelated as elasmobranchs and mammals (Lowry and Chadwick, 1970). These phylogenetic comparisons suggest that α-MSH may represent the most primitive molecule. Furthermore there is a striking similarity in amino acid analysis of mammalian α-MSH and ACTH—both molecules are relatively short straight chain polypeptides (see Geschwind, 1967). Associated with this overlap in structure is an overlap in function, since ACTH also exhibits MSH activity (e.g., skin darkening). The MSH activity of ACTH is readily apparent in *A. carolinensis*, which provides a sensitive bioassay for MSH. Reptilian ACTH also possesses considerable MSH activity when tested in reptiles. In attempting to localize ACTH in the pars distalis from diverse reptiles (Licht and Bradshaw, 1969), it was found that preliminary identifications of the fractions containing ACTH could be made on the basis of their darkening effect in *Anolis*. Since ACTH was shown to be restricted to the rostral zone of the pars distalis in all reptiles tested, there should have been no contamination of these fractions with MSH, which is presumably restricted to the pars intermedia; i.e., ACTH and MSH were separated by the caudal half of the pars distalis. The alternate overlap in activity is less clear, and it will be interesting to determine whether reptilian MSH possesses a significant amount of corticotropic activity in any vertebrate.

V. Thyrotropin (TSH)

A. Biosynthesis of Thyroid Hormone (Thyroxine)

Pituitary thyrotropic activity or TSH is defined primarily by the regulation of hormone production by the thyroid gland. The reptilian thyroid gland is a follicular structure situated in the throat near the sternum. TSH may have general long-term trophic effects and short-term actions involving the immediate synthesis and/or release of thyroidal hormones. Lynn (1970) has reviewed in detail the morphology and histology of the reptilian thyroid as well as the major physiological actions known for thyroidal hormones in reptiles. The present discussion will accordingly be limited to the chemistry of thyroidal hormone and its relation to the pituitary gland.

The thyroid of reptiles, like all other vertebrates, is characterized by its avidity for iodide, iodide being removed from the blood and used in the synthesis of the thyroidal hormone, thyroxine. This concentrating of iodide has been demonstrated by numerous radioiodide (^{131}I) studies involving species of all three reptilian orders. Although information on the metabolism of iodine in the thyroid is limited to one species of turtle, *Pseudemys floridana*, (Shellabarger *et al.*, 1956) and two lizards, *Sceloporus occidentalis* and *Anolis carolinensis* (Kobayashi and Gorbman, 1959; Licht and Rosenberg, 1969), the data suggest that the biosynthesis of thyroxine in reptiles follows the general pattern established for other vertebrates. Iodine is typically converted to organically bound iodine in thyroglobulin in the thyroid, and a series of reactions then occurs on the tyrosine and thyronine residues of this protein. Chromatographic analysis of hydrolyzates of the turtle and lizard thyroid gland is consistent with the expected pattern (Fig. 5). The iodine is found to be distributed among a series of compounds, including monoiodotyrosine (MIT), diiodotyrosine (DIT), triiodothyronine (T_3), and

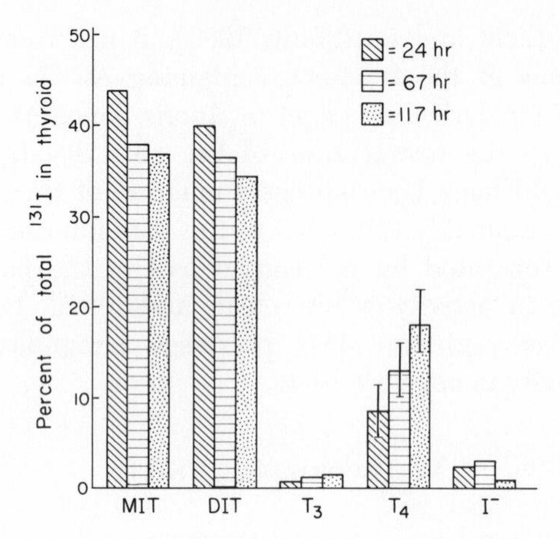

FIG. 5. The distribution of radioiodine in thyroidal homolyzates of adult male lizards (*Anolis carolinensis*). Vertical bars show the mean concentration of various iodoproteins (with 95% confidence limits shown for T_4) at three time intervals after a single intraperitoneal injection of 10 μCr ^{131}I in groups of eight lizards maintained at a body temperature of 30°C in midsummer. Twenty-four hour values are based on Licht and Rosenberg (1969), both studies were undertaken under similar conditions and results should be comparable. Abbreviations: MIT-monoiodotyrosine; DIT-diiodotyrosine; T_3-triiodothyronine; T_4-tetraiodothyronine or thyroxine; I^--inorganic iodide.

tetraiodotyronine (thryoxine or T_4). T_3 normally occurs in very low concentrations relative to T_4 in the thyroid (Fig. 5).

The release of hormone from the gland presumably requires the liberation of the T_4 from thyroglobulin by proteolysis. However, there have been no direct studies on this aspect of T_4 production in reptilian tissues. There is also little information on the nature of the circulating thyroid hormone. T_4 has been identified in the blood (e.g., Licht and Rosenberg, 1969), but the possible presence of circulating T_3 has not been examined. Tanabe *et al.* (1969) discussed the nature of the protein binding of circulating T_4 in a diversity of reptiles and compared the protein carriers with those of other vertebrates.

While the pattern of thyroidal iodine metabolism in the few reptiles studied conformed to the typical mammalian pattern, the two early studies on the turtle (Shellabarger *et al.*, 1956) and lizards (Kobayashi and Gorbman, 1959) led Gorbman (1963, p. 300) to conclude that a " . . . remarkable aspect of thyroid function in adult reptiles is the slowness of thyroxine production . . . ." This conclusion was based largely on the observation that MIT, DIT, and T_4 appeared to form only very slowly after radioiodine uptake and that there appeared to be a very slow loss of radioiodine from the gland, suggesting a low rate of T_4 release. However, subsequent studies using one of the same lizard species, *A. carolinensis,* (Licht and Rosenberg, 1969), suggested that the apparently sluggish radioiodine metabolism in these early studies was probably due to the use of subnormal temperatures. When maintained at normal, preferred, body temperatures (about 30°C), the lizard showed a rapid production of the iodinated tyrosines and T_4 (Fig. 5); in fact, labeled T_4 was readily detected in the circulation within 24 hours after the injection of ^{131}I. Walker (1973) recently showed a similar situation in a *Sceloporus* lizard. Lynn *et al.* (1965) also demonstrated that temperatures between 15° and 35°C had pronounced effects on the rates of thyroidal ^{131}I uptake and release (hence, T_4 turnover) in this lizard. Shellabarger *et al.* (1956) reported similar temperature effects in a turtle, and Turner and Tipton (1972a) demonstrated the same effects in a snake.

It is interesting that the peripheral actions of T_4 (especially on oxidative metabolism) in lizards are also highly temperature dependent (Maher, 1961, 1965; Maher and Levedahl, 1959; Wilhoft, 1966; Turner and Tipton, 1972b). Since few studies on thyroidal iodine uptake and metabolism in reptiles have taken these important multiple effects of temperature into account, meaningful interspecific comparisons among reptiles and other vertebrates (e.g., Sembrat, 1963) are not possible in most cases.

A few comparative studies on the proteins involved in the iodine metabolism of the thyroid including two species of tortoise (*Testudo graeca* and *T. hermanni*) and a marine turtle (*Thalassochelis caretta*) raise some interesting evolutionary questions about thyroglobulins. Data of Salvatore *et al.* (1965) and Roche *et al.* (1968) suggested that nonhomeothermic vertebrates, including these turtles, might have an unusual composition of iodoproteins in the thyroid. The normal thyroglobulin has an ultracentrifugal sedimentation constant of 19 S, and this represented about 85% of the total iodoprotein in all tetrapods examined. However, in nonhomeotherms like the turtles, there were also relatively large amounts of heavier (27 S) and especially lighter (12 S) components, the latter representing about 11% of total iodoprotein. Radioiodine studies indicated that the iodine becomes distributed among the three components in about the same ratios as indicated by protein analysis.

The physiological significance of the relatively high proportion of the smaller (12 S) component is not known, but Roche *et al.* (1968) speculated that it may be characteristic of "sluggish" thyroids, and that it may represent a step in the synthesis of the normal 19 S thyroglobulin. While Roche *et al.* (1968) argued that the 12 S is a stable component, the possibility that it is an artifact arising from degradation of the 19 S component during chemical analysis has not been fully eliminated. Also, there have been insufficient kinetic studies to assess fully the role of these small iodoproteins in thyroglobulin synthesis.

B. Evidence for a Pituitary-Thyroidal Axis

It is well established that the activity of the thyroid gland in reptiles is at least partially dependent on hormonal factors produced by the pituitary. Most early studies (see review by Lynn, 1970) utilized indirect histological criteria to study thyroidal activity. These showed that hypophysectomy caused regression in the thyroidal epithelium, whereas injection of pituitary extracts of reptiles or purified (mammalian) TSH prevented these regressive changes or caused hypertrophy of the gland. That these morphological changes reflected altered hormonal synthesis was subsequently established by radioiodine studies. Hypophysectomy caused a decline in iodine uptake (Nussbaum, 1963; Licht and Rosenberg, 1969), and this was associated with a depression in the rate of T_4 synthesis (Licht and Rosenberg, 1969). Alternatively, injections of exogenous (mammalian) TSH or extracts of reptilian pituitary glands stimulated an increased rate of thyroidal ^{131}I uptake (Shellabarger *et al.*, 1956; Nussbaum, 1963; Licht and Rosenberg, 1969; Buckingham, 1970), and this was related to increased T_4 synthesis in the gland (Licht and Rosenberg, 1969). Physiological studies also indicated that TSH increased levels of circulating T_4 (Maher and Levedahl, 1959).

While the above histological and physiological studies establish the dependence of thyroxine production on TSH, most of these data deal with relatively long-term effects. In contrast, very little is known about potential short-term actions of TSH on the thyroid. The fall in thyroidal activity in *Anolis carolinensis* following hypophysectomy, even at high temperatures, was found to be relatively slow, ^{131}I uptake being almost normal for about a week (Licht and Rosenberg, 1969). In mammals and birds, TSH can be assayed by measuring the release of radioiodine from the thyroid or increase in circulating T_4 within 24 hours or less after TSH administration. There have been no such short-term studies on reptiles. Injection of TSH for several days had little effect on radioiodine release from thyroids of young alligators (Waterman, 1961), and equivocable results were obtained in the turtle (Shellabarger *et al.*, 1956). Preliminary attempts to show short-term responses to mammalian TSH in *A. carolinensis* have been unsuccessful (P. Licht, unpublished). Thus, the possibility exists that the thyroid has a greater autonomy and that the gland responds less rapidly to TSH in the reptile than in the mammal.

Another class of evidence for a pituitary–thyroid axis in reptiles is obtained from studies involving the feedback relationship between the two glands. As in mammals, inhibition of T_4 production (by chemical blockage) led to a compensatory hypertrophy of the thyroid gland or elevated ^{131}I uptake, suggesting elevated TSH release. Alternatively, injection of T_4 led to the reduction in thyroidal epithelial height or a decrease in ^{131}I uptake (Lynn, 1970). This evidence may be interpreted as indicating a negative feedback between the thyroid, via circulating T_4, and pituitary TSH secretion. The evidence cited does not by itself prove that the alteration of thyroid activity is due to a pituitary factor. However, such an explanation seems likely, since associated changes are evident in specific cell types (thyrotropes) in the pituitary gland. In fact, this is probably the best known aspect of reptilian pituitary cytology; e.g., the thyrotropes are one of the few types of cells studied by experimental techniques utilizing ultrastructural observations to verify changes in secretory activity (Forbes, 1971).

C. Biochemistry of TSH

1. *Chemical Studies*

Mammalian TSH has been identified as a glycoprotein consisting of two chemically distinct subunits. There have been no direct studies on the chemical structure of reptilian TSH, but in preliminary attempts to fractionate hormones from a reptilian (snapping turtle) pituitary, TSH activity was confined to glycoprotein fractions along with gonadotropins (Licht and Papkoff, 1973b). TSH activity in a lizard pituitary was shown to be heat labile, a characteristic that is common to many but not all vertebrate TSHs (Dodd *et al.*, 1963).

2. *Biological Studies*

Indirect information on the structure of reptilian TSH also comes from studies of the zoological specificities of vertebrate TSHs. Histological and radioiodine uptake studies have shown that the reptilian thyroid is responsive to mammalian, avian, and amphibian TSH (Gorbman, 1946; Lynn, 1970; Licht and Hartree, 1971). Among reptiles, the lizard thyroid was stimulated by extracts from the pituitaries of turtles (P. Licht, unpublished). It was not clear from these studies whether the reptile was more or less sensitive to some of these TSHs than were other vertebrates, but there was some evidence (Gorbman, 1946) that the lizard was less responsive than the fish and amphibian to the TSHs of avian and amphibian species. Teleost TSH was apparently without effect in the lizard (Licht and Hartree, 1971; Gorbman, 1946) and mammals (Fontaine, 1969), although it did stimulate the amphibian thyroid (Gorbman, 1946).

Studies of the effectiveness of reptilian TSH in nonreptilian species are very limited and the results conflicting. Fontaine (1969) reported that pituitary extracts of a turtle and crocodilian were highly effective in promoting thyroidal activity in a fish (trout), but the same extracts were inactive when tested in a mammal (mouse). On the other hand, Dodd *et al.* (1963) were able to detect slight TSH activity in pituitary extracts of a lizard by using a different mammalian (mouse) bioassay. It has not been determined whether the failure of reptilian pituitary to stimulate the mammalian thyroid was due to zoological specificity (i.e., incompatibility between the reptilian TSH and mammalian receptor) or simply to a low titer of TSH in the reptilian pituitaries used in these assays. In any case, the relative inactivity of reptilian TSH in mammals does not fit into an expected pattern of zoological specificity, since the TSHs of amphibians and fish are relatively effective in the mammal (Dodd and Dodd, 1969; Fontaine, 1969). Pituitary extracts from the snapping turtle stimulate uptake of radiophosphorus by the thyroid of day-old chicks, and dose response curves are parallel with those of mammalian TSH (Licht and Papkoff, 1973b).

More careful comparisons of the relative responses of reptiles to nonreptilian TSH and vice versa are required before any generalizations can be made about the evolution of TSH or thyroidal receptor sites. At present, we can only conclude that reptiles possess a TSH that is probably glycoprotein in nature, but that there are important chemical differences between reptilian and mammalian TSHs.

VII. Concluding Remarks

Sufficient evidence exists to conclude that the reptilian pituitary system is probably based on the same six major categories of hormonal activities that are found in the mammalian adenohypophysis—melanocyte-stimulating hormone, prolactin, growth hormone, thyrotropin, adrenocorticotropin, and gonadotropin. The preliminary nature of the biochemical studies on the reptilian pituitary hormones prohibits extensive generalization, but it seems reasonable to conclude that these hormones are polypeptides of the same types as in mammals. However, it is also clear that the reptilian pituitary hormones show several structural and functional features that distinguish them from the mammalian pattern.

The degree of structural similarity between specific mammalian and reptilian hormones has not been ascertained, but it seems likely that future biochemical study will probably reveal some major dissimilarities in some groups of reptiles. The gonadotropins of reptiles and mammals may show a major divergence. In fact, there is enough information for some hormones, such as gonadotropin and prolactin, to suggest that major dichotomies in structure may even exist within the Reptilia, especially between the Chelonia (turtles) and Squamata (lizards and snakes). Thus, the need for caution in extrapolating from studies on a single reptilian species must be emphasized. Evaluation of the evolutionary import of the divergence between reptiles and mammals or among reptiles must await additional comparative studies on other nonmammalian vertebrates.

Limited tests of the biological cross-reactivity between the hormones of reptiles and other vertebrates also suggest some major evolutionary changes in endocrine physiology. In general, the reptiles seem responsive to all of the pituitary hormones from higher vertebrates and to some (e.g., gonadotropins) from all vertebrates; but they are relatively insensitive to others (e.g., thyrotropin, TSH) from lower vertebrates. There is a notable lack of the reciprocal type of tests, i.e., biological activity of reptilian hormones in nonreptilian species. There is some indication that mammals are relatively unresponsive to reptilian hormones such as gonadotropins and TSH. It is not clear to what extent this zoological specificity of hormonal action is a result of evolution in the structure of the hormone molecules or in the receptor sites.

The preliminary evidence for zoological specificity in pituitary hormone action should be critically examined in evaluating the information on the physiological actions of hormones. It cannot be overstressed that the vast majority of hormonal studies in reptiles have been based on the use of mammalian (primarily ungulate) hormone preparations. Until purified reptilian hormones can be obtained, many basic questions regarding the physiological actions of hormones must remain unanswered. This seems to be a particularly important problem in the physiology of gonadotropin (e.g., do reptiles possess one or two hormones?) and prolactin (e.g., is it a growth hormone?) but less so for the study of the steroid hormones of the satellite glands such as the gonad and interrenal. However, in the latter case, special attention must be given to the selection of hormones, since the major secretory (or circulating) end products of these glands may differ between reptiles and mammals. In fact, too few species have been examined to rule out the possibility of interspecific differences in circulatory steroids among reptiles.

In so far as the results obtained with exogenous (nonreptilian) hormones reflect the physiological actions of the endogenous molecules, it may be concluded that the control of the satellite glands by the pituitary is typical of the pattern observed in mammals. However, it should be stressed that this pattern only emerges if the reptiles are studied under certain environmental conditions. The pronounced effect of temperature on the responses to the pituitary hormones has been stressed, and such thermal effects may presumably alter the response to other hormones. Future studies of hormonal physiology in reptiles will, we hope, recognize the importance of experimental conditions and attempt to control and examine them in greater detail.

ACKNOWLEDGMENTS

I am grateful to Drs. C. S. Nicoll and M. Maher for reviewing parts of this manuscript, and I am especially indebted to my wife, Barbara M. Licht, for her assistance in various phases of writing.

The preparation of this review and many of the unpublished data discussed herein were supported in part by NSF grant GB-22642.

REFERENCES

Bentley, P. J. (1962). *Gen. Comp. Endocrinol.* **2**, 323–328.

Bern, H. A., and Nicoll, C. S. (1968). *Recent Progr. Horm. Res.* **24**, 681–720.

Bewley, T. A., and Li, C. H. (1970). *Science* **168**, 1361–1362.

Bradshaw, S. D., and Fontaine-Bertrand, E. (1970). *Comp. Biochem. Physiol.* **36**, 37–48.

Bragdon, D. E., Lazo-Wasem, E. A., Zarrow, M. X., and Hisaw, F. L. (1954). *Proc. Soc. Exp. Biol. Med.* **86**, 477–480

Bratcher, J. B., and Kent, G. C. (1971). *Proc. Soc. Exp. Biol.* **136**, 839–841.

Buckingham, M. B. (1970). *Gen. Comp. Endocrinol.* **14**, 178–183.

Burgers, A. C. J. (1960). *Acta Endocrinol. (Copenhagen), Suppl.* **51**, 329–330 (abstr.).

Butler, D. G., and Knox, W. H. (1970). *Gen. Comp. Endocrinol.* **14**, 551–566.

Callard, I. P. (1967). *J. Endocrinol.* **37**, 105–106.

Callard, I. P., and Leathem, J. H. (1965). *Arch. Anat. Microsc. Morphol. Exp.* **54**, 35–48.

Callard, I. P., and Willard, E. (1969). *Gen. Comp. Endocrinol.* **13**, 460–467.

Callard, I. P., and Ziegler, H. (1970). *J. Endocrinol.* **47**, 131–132.

Callard, I. P., Bayne, C. G., and McConnell, W. F. (1972). *Gen. Comp. Endocrinol.* **18**, 175–194.

Cardeza, A. F. (1957). *Rev. Soc. Argent. Biol.* **33**, 67–79.

Chan, D. K. O., Callard, I. P., and Chester Jones, I. (1970). *Gen. Comp. Endocrinol.* **15**, 374–387.

Channing, C. P., Licht, P., Papkoff, H., and Donaldson, E. M. (1974). *Gen. Comp. Endocrinol.* (in press).

Chester Jones, I., Phillips, J. G., and Holmes, W. N. (1959). *In* "Comparative Endocrinology" (A. Gorbman, ed.), pp. 582–612. Wiley, New York.

Chieffi, G. (1967). *Proc. Int. Congr. Horm. Steroids, 2nd, 1966* Excerpta Med. Found. Int. Congr. Ser. No. 132, pp. 1047–1057.

Chiu, K. W., Lofts, B., and Tsui, H. W. (1970). *Gen. Comp. Endocrinol.* **15**, 12–19.

Cieslak, E. S. (1945). *Physiol. Zool.* **18**, 299–329.

Cole, C. J. (1966). *Herpetologica* **22**, 199–206.

Combescot, C. (1955). *C. R. Soc. Biol.* **149**, 1969–1971.

Combescot, C. (1958). *C. R. Soc. Biol.* **152**, 1077–1079.
Coulson, R. A., and Hernandez, T. (1964). "Biochemistry of the Alligator." Louisiana State Univ. Press, Baton Rouge.
Creaser, C. W., and Gorbman, A. (1939). *Quart. Rev. Biol.* **14**, 311–331.
Del Conte, E. (1969). *Experientia* **25**, 1330–1332.
Della Corte, F., and Cosenza, L. (1965). *Atti Soc. Peloritana, Sci. Fis. Mat. Natur.* **11**, Suppl., 113–122.
Della Corte, F., Angelini, F., and Cosenza, L. (1966). *Atti Soc. Peloritana, Sci. Fis. Mat. Natur.* **12**, 643–650.
DiMaggio, A. (1961). Ph.D. Dissertation, Louisiana State University, Baton Rouge.
Dodd, J. M. (1960). *In* "Marshall's Physiology of Reproduction" (A. S. Parkes, ed.), 3rd ed., Vol. 1, Part 2, pp. 417–582. Longmans, Green, New York.
Dodd, J. M., and Dodd, M. H. I. (1969). *Colloq. Int. Cent. Nat. Rech. Sci.* **177**, 277–286.
Dodd, J. M., Ferguson, F. M., Dodd, M. H. I., and Hunter, R. B. (1963). *In* "Thyrotropin" (S. C. Werner, ed.), pp. 3–28. Thomas, Springfield, Illinois.
Dufaure, J.-P. (1970). *C. R. Acad. Sci.* **270**, 1145–1148.
Elizondo, R. S., and Le Brie, S. J. (1969). *Amer. J. Physiol.* **217**, 419–425.
Enemar, A., and von Mecklenberg, C. (1962). *Gen. Comp. Endocrinol.* **2**, 273–278.
Eyeson, K. (1970). *Gen. Comp. Endocrinol.* **14**, 357–367.
Eyeson, K. N. (1971). *Gen. Comp. Endocrinol.* **16**, 342–355.
Faiman, L., Ryan, R. J., Greslin, J. G., and Reichert, L. E., Jr. (1967). *Proc. Soc. Exp. Biol. Med.* **125**, 1232–1234.
Fellows, R. E., Hurley, T. W., and Brady, K. L. (1970). *Fed. Proc., Fed. Amer. Soc. Exp. Biol.* **29**, 579 (abstr.).
Ferguson, G. W. (1966). *Copeia*, 495–498.
Follett, B. K. (1970). *Int. Encyl. Pharmacol. Ther.* **1**, 321–350.
Fontaine, Y.-A. (1969). *Acta Endocrinol. (Copenhagen)*, **60**, Suppl. 136, 1–154.
Forbes, M. S. (1971). *Gen. Comp. Endocrinol.* **16**, 452–464.
Forbes, T. R. (1941). *J. Morphol.* **68**, 31–70.
Gabe, M. (1970). *In* "Biology of the Reptilia" (C. Gans, ed.), Vol. 3, pp. 263–318. Academic Press, New York.
Gabe, M., Martoja, M., and Saint Girons, H. (1964). *Annee. Biol.* **3**, 303–376.
Geschwind, I. I. (1967). *Amer. Zool.* **7**, 89–108.
Geschwind, I. I. (1969). *Gen. Comp. Endocrinol., Suppl.* **2**, 180–189.
Gist, D. H., and de Roos, R. (1966). *Gen. Comp. Endocrinol.* **7**, 304–313.
Gorbman, A. (1946). *Univ. Calif. Berkeley, Publ. Zool.* **51**, 229–243.
Gorbman, A. (1963). *In* "Comparative Endocrinology" (U. S. von Euler and H. Heller, eds.), Vol. **1**, pp. 291–324. Academic Press, New York.
Gottfried, H., Huang, D. P., Lofts, B., Phillips, J. G., and Tam, W. H. (1967). *Gen. Comp. Endocrinol.* **8**, 18–31.
Grignon, G., and Herlant, M. (1959). *C. R. Soc. Biol.* **153**, 2032–2034.
Hahn, W. E. (1967). *Comp. Biochem. Physiol.* **23**, 83–93.
Hayashida, T. (1970). *Gen. Comp. Endocrinol.* **15**, 432–452.
Holmes, W. N. and McBean, R. L. (1964). *J. Exp. Biol.* **41**, 81–90.
Huang, D. P., Vinson, G. P., and Phillips, J. G. (1969). *Gen. Comp. Endocrinol.* **12**, 637–643.
Jones, R. E. (1969). *J. Exp. Zool.* **171**, 217–222.
Kehl, R. (1944). *Rev. Can. Biol.* **3**, 131–219.
Kehl, R., and Combescot, C. (1955). *Mem. Soc. Endocrinol.* **4**, 57–74.

Klicka, J., and Mahmoud, I. Y. (1973). *Steroids* **21**, 483–495.
Knobil, E., and Sandler, R. (1963). *In* "Comparative Endocrinology" (U. S. von Euler and H. Heller, eds.), Vol. 1, pp. 477–491. Academic Press, New York.
Kobayashi, H., and Gorbman, A. (1959). *Annot. Zool. Jap.* **32**, 179–184.
La Pointe, J. L. (1969). *J. Endocrinol.* **43**, 197–205.
Le Brie, S. J., and Elizondo, R. S. (1969). *Amer. J. Physiol.* **217**, 426–430.
Leloup-Hatey, J. (1968). *Comp. Biochem. Physiol.* **26**, 997–1013.
Licht, P. (1967). *Gen. Comp. Endocrinol.* **9**, 49–63.
Licht, P. (1970). *Gen. Comp. Endocrinol.* **14**, 98–106.
Licht, P. (1972a). *Gen. Comp. Endocrinol., Suppl.* **3**, 477–488.
Licht, P. (1972b). *Gen. Comp. Endocrinol.* **19**, 273–281.
Licht, P. (1972c). *Gen. Comp. Endocrinol.* **19**, 282–289.
Licht, P. (1973). *Comp. Biochem. Physiol.* A **45**, 7–20.
Licht, P. (1974). *Gen. Comp. Endocrinol.* (in press).
Licht, P., and Bradshaw, S. D. (1969). *Gen. Comp. Endocrinol.* **13**, 226–235.
Licht, P., and Donaldson, E. M. (1969). *Biol. Reprod.* **1**, 307–314.
Licht, P., and Hartree, A. S. (1971). *J. Endocrinol.* **53**, 329–349.
Licht, P., and Hoyer, H. E. (1968). *Gen. Comp. Endocrinol.* **11**, 338–347.
Licht, P., and Jones, R. E. (1967). *Gen. Comp. Endocrinol.* **8**, 228–244.
Licht, P., and Nicoll, C. S. (1969). *Gen. Comp. Endocrinol.* **12**, 526–535.
Licht, P., and Papkoff, H. (1971). *Gen. Comp. Endocrinol.* **16**, 586–593.
Licht, P., and Papkoff, H. (1972). *Gen. Comp. Endocrinol.* **19**, 102–114.
Licht, P., and Papkoff, H. (1973a). *Gen. Comp. Endocrinol.* **20**, 172–176.
Licht, P., and Papkoff, H. (1973b). *Gen. Comp. Enodcrinol.* (in press).
Licht, P., and Pearson, A. K. (1969a). *Biol. Reprod.* **1**, 107–119.
Licht, P., and Papkoff, H. (1974). *Endocrinology* (in press).
Licht, P., and Pearson, A. K. (1969b). *Gen. Comp. Endocrinol.* **13**, 367–381.
Licht, P., and Rosenberg, L. L. (1969). *Gen. Comp. Endocrinol.* **13**, 439–454.
Licht, P., Hoyer, H. E., and van Oordt, P. G. W. J. (1969). *J. Zool.* **157**, 469–501.
Lisk, R. D. (1967). *Gen. Comp. Endocrinol.* **8**, 258–266.
Lofts, B. (1969). *Gen. Comp. Endocrinol., Suppl.* **2**, 147–155.
Lofts, B., and Chiu, K. W. (1968). *Arch. Anat., Histol., Embryol.* **51**, 409–418 (abstr.).
Lowry, P. J., and Chadwick, A. (1970). *Nature (London)* **226**, 219–222.
Lupo di Prisco, C., Chieffi, G., and Delrio, G. (1967a). *Experientia* **23**, 73–75.
Lupo di Prisco, C., Delrio, G., and Cheiffi, G. (1967b). *Gen. Comp. Endocrinol.* **9**, 109 (abstr.).
Lynn, W. G. (1970). *In* "Biology of the Reptilia" (C. Gans, ed.), Vol. 3, pp. 201–234. Academic Press, New York.
Lynn, W. G., McCormick, J. J., and Gregosek, J. C. (1965). *Gen. Comp. Endocrinol.* **5**, 587–595.
Maderson, P. F. A., and Licht, P. (1967). *J. Morphol.* **123**, 157–172.
Maderson, P. F. A., Chiu, K. W., and Phillips, J. G. (1970). *Mem. Soc. Endocrinol.* **18**, 259–284.
Maher, M. J. (1961). *Amer. Zool.* **1**, 461 (abstr.).
Maher, M. J. (1965). *Gen. Comp. Endocrinol.* **5**, 320–325.
Maher, M. J., and Levedahl, B. H. (1959). *J. Exp. Zool.* **190**, 169–190.
Marques, M. (1955). *Rev. Soc. Argent. Biol.* **31**, 177–183.
Mazzi, V. (1969). *Boll. Zool.* **36**, 1–60.
Meier, A. (1969). *Gen. Comp. Endocrinol., Suppl.* **2**, 55–62.

Miahle, C., and Koch, B. (1969). *Colloq. Int. Cent. Nat. Rech. Sci.* **177**, 103–134.

Miller, M. R. (1952). *Anat. Rec.* **113**, 309–324.

Nandi, J. (1967). *Amer. Zool.* **7**, 115–133.

Nichols, C. W., Jr. (1973). *Gen. Comp. Endocrinol.* **21**, 219–224.

Nicoll, C. S., and Bern, H. A. (1965). *Endocrinology* **76**, 156–160.

Nicoll, C. S., and Licht, P. (1971). *Gen. Comp. Endocrinol.* **17**, 490–507.

Nicoll, C. S., and Nichols, C. (1971). *Gen. Comp. Endocrinol.* **17**, 300–310.

Nicoll, C. S., Fiorindo, R. P., McKennee, C. T., and Parsons, J. A. (1970). *In* "Hypophysiotropic Hormones of the Hypothalamus: Assay and Chemistry" (J. Meites, ed.), pp. 115–144. Williams & Wilkins, Baltimore, Maryland.

Nussbaum, N. (1963). *Anat. Rec.* **145**, 340 (abstr.).

Pandha, S. K., and Thapliyal, J. P. (1966). *Naturwissenschaften* **51**, 201–202.

Panigel, M. (1956). *Ann. Sci. Natur., Zool. Biol. Anim.* [12] **18**, 569–668.

Papkoff, H., and Hayashida, T. (1972). *Proc. Soc. Expt. Biol. Med.* **140**, 251–255.

Papkoff, H., and Licht, P. (1972). *Proc. Soc. Exptl. Biol. Med.* **139**, 372–376.

Penhos, J. C., Wu, C. H., Reitman, M., Sodero, E., White, R., and Levine, R. (1967). *Gen. Comp. Endocrinol.* **8**, 32–43.

Phillips, J. G., and Bellamy, D. (1963). *In* "Comparative Endocrinology" U. S. von Euler and H. Heller, eds.), Vol. 1, pp. 208–257. Academic Press, New York.

Phillips, J. G., and Chester Jones, I. (1957). *J. Endocrinol.* **16**, iii (abstr.).

Prasad, M. R. N., and Sanyal, M. K. (1969). *Gen. Comp. Endocrinol.* **12**, 110–118.

Ramaswami, L. S. (1967). *Proc. Int. Congr. Horm. Steroids, 2nd., 1966* Excerpta Med. Found. Int. Congr. Ser. No. 132, pp. 1084–1093.

Ramaswami, L. S., and Jacobs, D. (1963). *Naturwissenschaften* **50**, 453–454.

Ramaswami, L. S., and Jacob, D. (1965). *Experientia* **21**, 206–207.

Reddy, P. R. K., and Prasad, M. R. N. (1970a). *Gen. Comp. Endocrinol.* **14**, 15–24.

Reddy, P. R. K., and Prasad, M. R.N. (1970b). *J. Exp. Zool.* **174**, 205–214.

Reddy, P. R. K., and Prasad, M. R. N. (1971). *Gen. Comp. Endocrinol.* **16**, 288–297.

Roche, J., Salvatore, G., Sena, L., Aloj, S., and Covelli, I. (1968). *Comp. Biochem. Physiol.* **27**, 67–82.

Sage, M., and Bern, H. A. (1970). *Amer. Zool.* **10**, 499 (abstr.).

Saint Girons, H. (1967). *Ann. Sci. Natur. Zool. Biol. Anim.* [12] **9**, 230–308.

Saint Girons, H. (1970). *In* "Biology of the Reptilia" (C. Gans, ed.), Vol. 3, pp. 135–200. Academic Press, New York.

Salvatore, G., Sena, L., Viscidi, E., and Salvatore, M. (1965). *Curr. Top. Thyroid Res., Proc. Int. Thyroid Conf., 5th, 1965* pp. 193–206.

Sandor, T. (1969). *Gen. Comp. Endocrinol., Suppl.* **2**, 284–298.

Schaeffer, W. H. (1933). *Proc. Soc. Exp. Biol. Med.* **30**, 1363–1365.

Sembrat, K. (1963). *Folia Biol.* (*Prague*) **11**, 473–481.

Shellabarger, C. J., Gorbman, A., Schatzlein, F. C., and McGill, D. (1956). *Endocrinology* **59**, 331–339.

Skadhauge, E. (1969). *Colloq. Int. Cent. Nat. Rech. Sci.* **177**, 63–68.

Takewaki, K., and Hatta, K. (1941). *Annot. Zool. Jap.* **20**, 4–8.

Tam, W. H., and Phillips, J. G. (1969). *Gen. Comp. Endocrinol.* **13**, 117–125.

Tanabe, Y., Ishii, T., and Yoshinori, T. (1969). *Gen. Comp. Endocrinol.* **13**, 14–21.

Templeton, J. R., Murrish, D., Randall, E., and Mugaas, J. (1968). *Amer. Soc. Zool.* **8**, 818–819 (abstr.).

Trobec, T. N., and Stanley, J. G. (1971). *Gen. Comp. Endocrinol.* **17**, 479–482.

Turner, F. B., Licht, P., Thrasher, J. D., Medica, P. A., and Lannom, J. L., Sr. (1973). *Proc. Nat. Symp. Radioecol. 3rd,* (in press).

Turner, J. E., and Tipton, S. R. (1972a). *Gen. Comp. Endocrinol.* **18**, 195–197.
Turner, J. E., and Tipton, S. R. (1972b). *Gen. Comp. Endocrinol.* **18**, 98–101.
Urist, M. R., and Schjeide, A. O. (1961). *J. Gen. Physiol.* **44**, 743–756.
van Tienhoven, A. (1968). "Reproductive Physiology of Vertebrates." Saunders, Philadelphia, Pennsylvania.
Vivien, J., and Stenger, C. (1955). *C. R. Soc. Biol.* **149**, 1042–1045.
Vladescu, C. (1965). *Rev. Roum. Biol., Ser. Zool.* **10**, 123–128.
Vladescu, C., Baltac, M., and Trandaburu, T. (1970). *Ann. Endocrinol.* **31**, 863–868.
Wagner, E. M. (1955). *Acta Physiol. Lat. Amer.* **5**, 219–228.
Walker, R. F. (1973). *Gen. Comp. Endocrinol.* **20**, 137–143.
Waterman, A. J. (1961). *Amer. Zool.* **1**, 475 (abstr.).
Wilhoft, D. C. (1966). *Gen. Comp. Endocrinol.* **7**, 445–451.
Wilkinson, R. F., Jr. (1966). *Diss. Abstr.* **26**, 5612.
Wingstrand, K. G. (1966). *In* "The Pituitary Gland" (G. W. Harris and B. T. Donovan, eds.), Vol. 1, pp. 58–126. Univ. of California Press, Berkeley.
Witschi, E. (1955). *Mem. Soc. Endocrinol.* **4**, 149–163.
Wright, A., and Chester Jones, I. (1957). *J. Endocrinol.* **15**, 83–99.
Yaron, Z. (1970). *Amer. Zool.* **10**, 492–493 (abstr.).

CHAPTER 16

Venoms of Reptiles

Findlay E. Russell and Arnold F. Brodie

I. Introduction 449
II. General Characteristics of Venom 450
III. Low Molecular Weight Venom Components 455
IV. Venom Enzymes 465
References 474

I. Introduction

Of the more than 3500 species of snakes, less than 400 are known to be venomous or of danger to man. Most of these species are found in the family Elapidae (the cobras, coral snakes, mambas, and kraits); Viperidae (the puff adder, Gaboon viper, Russell's viper, horned vipers of the Sahara, saw-scaled vipers, and the European vipers); Crotalidae (the rattlesnakes, copperheads, and water moccasins of North America, the bushmaster of South America, and certain primitive pit vipers of Asia); and the Hydrophidae (the sea snakes). A few venomous species are found in the Colubridae, a family of snakes that exhibits a number of transitional characteristics between the nonvenomous and venomous snakes. The most highly developed venom apparatus found in the colubrids is that of the boomslang, a rear-fanged tree snake of Africa. Only two of the more than 3000 species of lizards are considered venomous. They are the Gila monster of the southwestern United States and northern Mexico, and the escorpion of the west part of central Mexico.

Snake venoms must be considered products of adaptation in the evolution of these reptiles. While their development as complex mixtures is not well understood, it appears that along with the evolution of the fang and venom gland, the venom has reached an equally high degree of development. The favorable production of many enzymes having a diversity of biological activities, associated for the most part with digestive functions and lethal and paralyzing properties, has made the venomous snakes far more versatile in their prey catching and feeding habits than their nonvenomous counterparts. What role adaptation to the venom has played in the armament of the venomous snake's prey is difficult to say; but it seems likely that, as in all of the animal kingdom, some animals have developed an active immunity to certain of the venoms (Juratsch and Russell, 1971), and that this phenomenon may reflect changes in the chemistry of reptilian venoms. The venoms of some snakes also have a defensive function, and in the case of the spitting cobras, and perhaps the poisonous lizards, this function may be of considerable or prime importance.

Although we have not yet determined the specific biological functions for all the fractions of any snake venom, nor have we completely characterized these poisons, evidence to date indicates that there is a remarkable design in their organization. One is continually impressed when studying their chemistry and pharmacology by how often he finds an activity related to the design of the venom, that, had he studied the animal more carefully, he might have anticipated. This is not to say that we have clearly established that all venom fractions contribute in an effective way to the role of poison, but it is nonetheless remarkable how many of them do. And it seems best to reserve judgement on those that do not until our techniques for isolating and characterizing these toxins and their biological activities are a bit more advanced.

Finally, it must be conceded that to speculate on what may have evolved chemically in a snake venom over a period of twenty million years could be somewhat trying, since the venom must have represented adaptations at every period when competition and survival necessitated its further modification. Although this subject has not been explored in any depth, the reader will find a good review of the possible evolution of the venom apparatus, and to a certain extent on the venoms of reptiles, in the works of Klauber (1956), Bogert and Del Campo (1956), Minton and Minton (1969), Bellairs (1969), Gans and Elliott (1968), and Kochva and Gans (1970). Table I lists a few of the more clinically important venomous snakes and certain data on their venoms. It can be seen that there is as much as a thousandfold difference in the lethal activity of some of these poisons.

II. General Characteristics of Venom

Most venoms studied to date appear to be mixtures, and the reptile venoms are no exception. Although the more biologically active fractions of some animal venoms are steroids, alkaloids, mucopolysaccharides, furans, or formic acid, snake venoms are perhaps the most complex, containing ten or more enzymes, several active nonenzymatic proteins, and at least two or three peptides, as well as a number of other substances. Not only are the snake venoms complex in their chemical structure, but they are remarkably diversified and complicated in their modes of action. Some snake venoms have an effect, either directly or indirectly, on almost every organ system, and few are tissue specific, although some may have a more marked effect on one or another organ system or tissue. No reptile venom should ever be considered, as is so often done, as a neurotoxin or a cardiotoxin or a hemotoxin and thus permit its other biological properties to be minimized or dismissed. Many errors in the development of our understanding of the properties of venoms have been perpetrated because of this oversimplified attempt at classification. It is best to consider all reptile venoms as complex mixtures capable of producing one or more deleterious changes in several organ systems or tissues, and able to provoke some of these changes concurrently (Russell, 1960).

In considering the biological effects of reptile venoms, one cannot overlook the autopharmacological response. Whether this is initiated by a specific venom component or not is not known, but whatever the stimulating substance(s) is (are), the response of the organism to the toxin can sometimes be so marked that it far overshadows the direct effects of the venom. It takes of the order of 150 simultaneous bee stings to cause death in a nonsensitized human, but the sting of a single bee can be fatal to the sensitized individual. Persons bitten by venomous snakes sometimes become so sensitive to the snake's venom that even a minor envenomation precipitates anaphylactic shock. The release of such autopharmacological substances as histamine, bradykinin, and adenosine, for example, can be so overwhelming that their effects may be far more deleterious and far more sudden than those of the venom.

Investigations on the chemistry of snake venoms are further complicated by the fact that qualitative as well as quantitative differences in the composition of these poisons may exist not only from species to species within the same genus, but also from individual to individual within the same species. A venom may vary in the individual animal at different times of the year or under different environmental conditions, or even with the age and perhaps the sex of the snake.

Finally, it must be admitted that scrupulous attention to species separation in the pooling of venom samples by collectors has not always been given, with the result that a number of discrepancies in the literature on the chemistry of these poisons can be attributed directly to the use of different venoms. One of us has witnessed a sample of a viper venom being added to that of a elapid venom and the same being dried in a desiccator and subsequently labeled as a species-specific elapid venom. We have seen *Bothrops atrox* venom mixed with *Bothrops nummifer* venom and sold as the former. A degree of caution should

TABLE I
VENOMS OF SOME IMPORTANT DANGEROUS SNAKES[a]

Species	Average length of adult (inches)	Approximate yield, dry venom (mg)	Intraperitoneal LD_{50} (mg/kg)	Intravenous LD_{50} (mg/kg)
North America				
Rattlesnakes (*Crotalus*)				
Eastern diamondback (*C. adamanteus*)	32–65	370–700	1.89	1.68
Western diamondback (*C. atrox*)	30–65	175–320	3.71	4.20
Red diamond (*C. ruber ruber*)	32–52	120–350	6.69	3.70
Timber (*C. horridus horridus*)	32–54	75–150	2.91	2.63
Prairie (*C. viridis viridis*)	32–46	35–100	2.25	1.61
Southern Pacific (*C. v. helleri*)	32–48	75–150	1.60	1.29
Great Basin (*C. v. lutosus*)	32–46	75–150	2.20	—
Mojave (*C. scutulatus*)	22–40	50–90	0.23	0.21
Sidewinder (*C. cerastes*)	18–30	18–40	4.00	—
Moccasins (*Agkistrodon*)				
Cottonmouth (*A. piscivorus*)	30–50	90–145	5.11	4.00
Copperhead (*A. contortrix*)	24–36	40–70	10.50	10.92
Coral snakes (*Micrurus*)				
Eastern coral snake (*M. fulvius*)	16–28	2–6	0.97	—
Central and South America				
Rattlesnakes (*Crotalus*)				
Tropical (*C. durissus terrificus*)	20–48	20–40	0.30	—
New World pit vipers (*Bothrops*)				
Barba amarilla (*B. atrox*)	46–80	70–160	3.80	4.27
Bushmaster (*Lachesis mutus*)	70–110	280–450	5.93	—
Europe				
Vipers				
European viper (*Vipera berus*)	18–24	6–18	0.80	0.55

Africa				
Vipers				
Puff adder (*Bitis lachesis*)	30–48	130–200	3.68	—
Saw-scaled viper (*Echis carinatus*)	16–22	20–35	—	2.30
Mambas (*Dendroaspis*)				
Green mamba (*D. angusticeps*)	50–72	60–95	—	0.45
Asia				
Cobras (*Naja*)				
Indian cobra (*N. naja*)	45–65	170–325	0.40	0.40
Kraits (*Bungarus*)				
Common krait (*B. caeruleus*)	36–48	8–20	—	0.09
Vipers (*Vipera*)				
Russell's viper (*V. russelli*)	40–50	130–250	—	0.08
Pit vipers (*Agkistrodon*)				
Malayan pit viper (*A. rhodostoma*)	25–35	40–60	—	6.20
Australia				
Tiger snake (*Notechis scutatus*)	30–56	30–70	0.04	—
Indo-Pacific				
Sea snakes				
Beaked sea snake (*Enhydrina schistosa*)	30–48	7–20	—	0.01

[a] From Russell (1967).

be exercised when obtaining venoms from any source, even a reputable chemical firm, which often does not know the conditions under which the venom has been collected and processed. It is regrettable to note that even chemical firms do not always carefully label their venom products (Russell *et al.*, 1963b).

In spite of these concerns, our knowledge of the chemistry of reptile venoms is encouraging. For many of the snake venoms we now have some definitive data on their composition and on their modes of action. We have been able to isolate and assay many snake venom fractions, and indeed, now employ quite a few in the practice of medicine and as tools in biology. The greatest advances in our knowledge on these poisons have been made during the past six or seven years. This has not necessarily been due to the development and use of far more advanced techniques, but, in part, due to more careful consideration of the chemical evolution of the venom, that is, the design of the venom in the adaptation of the reptile.

There was a time in our thinking when all of the biological effects of reptile venoms were attributed to the enzymatic activities of these poisons. This is not difficult to understand, since our techniques for the study of enzymatic activities had progressed much more rapidly than those for the nonenzymatic proteins. It is now generally recognized that perhaps the lethal and some of the more deleterious effects of snake venoms are attributable to the nonenzymatic components, some of which are peptides. Some biologically active components of snake venoms may not be proteins (Minton, 1968).

These peptides or polypeptides, which are often called neurotoxins because they are said to produce a nondepolarizing type of neuromuscular block, tend to be basic in character with a pH above 8.5, and are usually quite heat stable and dialyzable. Most of them appear to have a molecular weight in the vicinity of 7000, although smaller particle sizes have been reported (Fischer and Kabara, 1967). They also appear to be rich in disulfide bonds and remarkably similar in their amino acid composition. Most of the elapid polypeptides are rich in lysine, aspartic acid, threonine, glutamic acid, and glycine. The peptide of the venom of the rattlesnake *Crotalus viridis helleri* is very rich in lysine and contains large amounts of glycine and proline (J. W. Dubnoff and F. E. Russell, unpublished results). As previously noted, these peptides are responsible for the greater part of the lethal activity of the venom, as well as for some of its other important biological activities. It is not clear, however, wherein lie the specific sites and mechanisms of action, in spite of considerable literature to the contrary. They are said to be poor antigens, but this, too, should be questioned.

III. Low Molecular Weight Venom Components

Although several of the small proteins of snake venoms were being studied three decades ago (Micheel *et al.*, 1937; Ghosh *et al.*, 1941; Slotta and Frankel-Conrat, 1938) and later (Sasaki, 1957), it was not until the mid-sixties that definitive chemical experiments were carried out on these proteins and peptides. The sequence data for some of the polypeptides isolated from snake venom are shown in Table II. The authors are deeply indebted to Professor N. Tamiya for his assistance in preparing this material, and express their appreciation to Dr. D. L. Eaker and Dr. D. P. Botes for their suggestions and advice. Dr. D. L. Eaker notes (personal correspondence) that the principal neurotoxin of the venom of *Naja naja naja* of West Pakistan is yet a fourth variant, differing from the "*N. naja siamensis* 3 toxin" with respect to residue 28 (32 in the homology chart), which is Gly rather than Ala.

Tamiya and his colleagues (1967), at the First International Symposium on Animal Toxins in 1966, presented a paper on the chromatography, crystallization, electrophoresis, ultracentrifugation, and amino acid composition of the venom of the sea snake *Laticauda semifasciata*. Almost all the lethal activity of the poison was recovered as two toxins, erabutoxin a and b, using carboxymethylcellulose chromatography. Thirty percent of the proteins of the crude venom were erabutoxins. The homogeneity of the crystalline toxins was demonstrated by rechromatography, disk electrophoresis, and ultracentrifugation. Sixty-one amino acids were found in each of the toxins, and erabutoxin b was found to be more basic than erabutoxin a. No alanine or methionine were present, and one of the three arginine residues was at the N-terminus. The crystals demonstrated by these investigations were thought to be sulfate salts of the basic toxins. The molecular weights, by ultracentrifugation, were approximately 7430 for each of the toxins. Using the same methods, these workers also demonstrated that laticotoxin a from the sea snake *Laticauda laticauda* had sixty-two amino acids, and that this toxin had a molecular weight, by ultracentrifugation, of 6520. These molecular weight values agree with those calculated from the amino acid compositions of the toxins. The three toxins were ultracentrifugally nondispersed.

In another paper, Tamiya and Arai (1966) reported sixty-two amino acids for erabutoxin a and b, and again demonstrated that the two toxins had the same amino acid composition, except that the former had one more aspartic acid residue and one less histidine residue than the latter. Sato and Tamiya (1971) presented the amino acid sequences of these toxins. Erabutoxin b was reduced, then S-carboxymethylated

TABLE II

SEQUENCE DATA ON SOME POLYPEPTIDES OF SNAKE VENOMS

			1	2	3	4	5	6	7	8	9	10	11	12	13	14	15	16	17	18	19
1.	Erabutoxin a	*L. semifasciata*	Arg	Ile		Phe	Asn	Gln	His	Ser	Ser	Gln	Pro	Gln	Thr	Thr	Lys	Thr		Pro	Ser
2.	Erabutoxin b	*L. semifasciata*	Arg	Ile		Phe	Asn	Gln	His	Ser	Ser	Gln	Pro	Gln	Thr	Thr	Lys	Thr		Pro	Ser
3.	α toxin	*N. nigricollis*	Leu	Glu		His	Asn	Gln	Gln	Ser	Ser	Gln	Pro	Pro	Thr	Thr	Lys	Thr		Pro	—
4.	Cobrotoxin	*N. naja atra*	Leu	Glu		His	Asn	Gln	Gln	Ser	Ser	Gln	Thr	Pro	Thr	Thr	Thr	Gly		Ser	Gly
5.	Toxin α and δ	*N. haja & N. nivea*	Leu	Gln		His	Asn	Gln	Gln	Ser	Ser	Gln	Pro	Pro	Thr	Thr	Lys	Thr		Pro	—
6.	Toxin β	*N. nivea*	Met	Ile		His	Asn	Gln	Gln	Ser	Ser	Gln	Arg	Pro	Thr	Ile	Lys	Thr		Pro	—
7.	Toxin IV	*H. haemachatus*	Leu	Glu		His	Asn	Gln	Gln	Ser	Ser	Gln	Thr	Pro	Thr	Thr	Gln	Thr		Pro	—
8.	Toxin II	*H. haemachatus*	Leu	Glu	Cys	His	Asn	Gln	Gln	Ser	Ser	Gln	Pro	Pro	Thr	Thr	Lys	Ser	Cys	Pro	—
9.	α-Bungarotoxin	*B. multicinctus*	Ile	Val		His	—	Thr	Thr	Ala	Thr	Ile	Pro	Ser	Ser	Ala	Val	Thr		Pro	Pro
10.	Toxin α	*N. nivea*	Ile	Arg		Phe	—	—	—	Ile	Thr	Pro	Asp	Val	Thr	Ser	Glu	Ala		Pro	Asp
11.	Toxin A	*N. naja*	Ile	Arg		Phe	—	—	—	Ile	Thr	Pro	Asp	Ile	Thr	Ser	Lys	Asp		Pro	Asn
12.	Toxin 3	*N. naja naja*	Ile	Arg		Phe	—	—	—	Ile	Thr	Pro	Asp	Ile	Thr	Ser	Lys	Asp		Pro	Asn
13.	Toxin 4	*N. naja naja*	Ile	Arg		Phe	—	—	—	Ile	Thr	Pro	Asp	Ile	Thr	Ser	Lys	Asp		Pro	Asn
14.	Toxin 3	*N. n. siamensis*	Ile	Arg		Phe	—	—	—	Ile	Thr	Pro	Asp	Ile	Thr	Ser	Lys	Asp		Pro	Asn
15.	Cardiotoxin	*N. n. atra*	Leu	Lys		—	Phe	—	—	Lys	Leu	Val	Pro	Leu	Phe	Tyr	Lys	Thr		Pro	Ala
16.	Cytotoxin I	*N. naja*	Leu	Lys		—	Phe	—	—	Lys	Leu	Ile	Pro	Leu	Ala	Tyr	Lys	Thr		Pro	Ala

	20	21	22	23	24	25	26	27	28	29	30	31	32	33	34	35	36	37	38	39	40	41	42	43	44	45	46	47	48
1.		Ser	Glu	Ser			Asn		Gln		Ser		—	Phe	—	—	—			Thr	Ile	Ile	Glu	Arg			—	—	Gly
2.		Ser	Glu	Ser			His		Gln		Ser		—	Phe	—	—	—			Thr	Ile	Ile	Glu	Arg			—	—	Gly
3.		Glu	Thr	Asn			Lys		Val		Arg		—	His	—	—	—			Thr	Ile	Ile	Glu	Arg			—	—	Gly
4.		Glu	Thr	Asn			Lys		Arg		Arg		—	His	—	—	—			Tyr	Arg	Thr	Glu	Arg			—	—	Gly
5.		Glu	Thr	Asn			Lys		Arg		Arg		—	His	—	—	—			Ser	Ile	Thr	Glu	Arg			—	—	Gly
6.		Glu	Thr	Asn			Lys		Arg		Arg		—	His	—	—	—			Thr	Ile	Ile	Glu	Arg			—	—	Gly
7.		Glu	Thr	Asn			Lys		Gln		Ser		—	His	—	—	—			Ser	Arg	Thr	Glu	Arg			—	—	Gly
8.	Gly	Asp	Thr	Asn	Cys	Tyr	Asn	Lys	Arg	Trp	Arg	Asp	—	His	—	—	—	Arg	Gly	Thr	Ile	Ile	Glu	Arg	Gly	Cys	—	—	Gly
9.		Glu	Asn	Leu			Arg		Met		Cys		Ala	Phe	Cys	Ser	Ser			Lys	Lys	Val	Glu	Leu			Ala	Ala	Thr
10.		—	His	Val			Thr		Met		Cys		Asn	Phe	Cys	Gly	Met			Lys	Arg	Val	Asp	Leu			Ala	Ala	Thr
11.		—	His	Val			Thr		Thr		Cys		Gly	Phe	Cys	Ser	Ile			Lys	Arg	Val	Asp	Leu			Ala	Ala	Thr
12.		—	His	Val			Thr		Thr		Cys		Gly	Phe	Cys	Ser	Ile			Lys	Arg	Val	Asp	Leu			Ala	Ala	Thr
13.		—	His	Val			Thr		Thr		Cys		Gly	Phe	Cys	Ser	Ser			Lys	Arg	Val	Asp	Leu			Ala	Ala	Thr
14.		—	His	Val			Thr		Thr		Cys		Ala	Phe	Cys	Ser	Ile			Lys	Arg	Val	Asp	Leu			Ala	Ala	Thr
15.		Lys	Asn	Leu			—		Met	Phe	Met	Val	Ala	—	—	Thr	—	Pro	Lys	Val	Pro	Val	Lys	Arg			Ile	Asp	Val
16.		Lys	Asn	Leu			—		Met	Tyr	Met	Val	Ser	—	—	Asn	—	Lys	Thr	Val	Pro	Val	Lys	Arg			Ile	Asp	Val

	49	50	51	52	53	54	55	56	57	58	59	60	61	62	63	64	65	66	67	68	69	70	71	72	73	74	75
1.			Thr	Val	Lys	Pro	Gly	Ile	Lys	Leu	Ser			Glu	Ser	Glu	Val		Asn	Asn							
2.			Thr	Val	Lys	Pro	Gly	Ile	Lys	Leu	Ser			Glu	Ser	Glu	Val		Asn	Asn							
3.			Thr	Val	Lys	Pro	Gly	Ile	Lys	Leu	Asn			Thr	Thr	Asp	Lys		Asn	Asn							
4.			Ser	Val	Lys	Asn	Gly	Ile	Glu	Ile	Asn			Thr	Thr	Asp	Arg		Asn	Asn							
5.			Ser	Val	Lys	Lys	Gly	Ile	Glu	Ile	Asn			Thr	Thr	Asp	Lys		Asn	Asn							
6.			Ser	Val	Lys	Lys	Gly	Val	Gly	Ile	Tyr			Lys	Thr	Asp	Lys		Asn	Arg							
7.			Thr	Val	Lys	Pro	Gly	Ile	Lys	Leu	Lys			Thr	Thr	Asp	Arg		Asn	Lys							
8.	Cys	Pro	Thr	Val	Lys	Pro	Gly	Ile	Asn	Leu	Lys	Cys	Cys	Thr	Thr	Asp	Arg	Cys	Asn	Asn							
9.			Ser	Lys	Lys	Pro	Tyr	Glu	Glu	Val	Thr			Ser	Thr	Asp	Lys		Asn	His	Pro	Pro	Lys	Arg	Gln	Pro	Gly
10.			Lys	Val	Lys	Pro	Gly	Val	Asp	Ile	Lys			Ser	Arg	Asp	Asp		Asp	Pro	Phe	Pro	Thr	Arg	Lys	Arg	Ser
11.			Thr	Val	Arg	Thr	Gly	Val	Asp	Ile	Gln			Ser	Thr	Asp	Asp		Asp	Thr	Phe	Pro	Thr	Arg	Lys	Arg	Pro
12.			Thr	Val	Arg	Thr	Gly	Val	Asp	Ile	Gln			Ser	Thr	Asp	Asn		Asn	Pro	Phe	Pro	Thr	Arg	Lys	Arg	Pro
13.			Thr	Val	Arg	Thr	Gly	Val	Asp	Ile	Gln			Ser	Thr	Asp	Asn		Asn	Pro	Phe	Pro	Thr	Arg	Lys	Arg	Pro
14.			Thr	Val	Lys	Thr	Gly	Val	Asp	Ile	Gln			Ser	Thr	Asp	Asn		Asn	Pro	Phe	Pro	Thr	Arg	Lys	Arg	Pro
15.			Lys	Ser	Ser	Leu	Val	Leu	Lys	Tyr	Val			Asn	Thr	Asp	Arg		Asn								
16.			Lys	Asn	Ser	Leu	Val	Leu	Lys	Tyr	Glu			Asn	Thr	Asp	Arg		Asn								

Snake	Residues	References
1	62	Sato and Tamiya, 1971; Endo *et al.*, 1971
2	62	Sato and Tamiya, 1971; Endo *et al.*, 1971
3	61	Eaker and Porath, 1967
4	62	Yang *et al.*, 1969
5	61	Botes *et al.*, 1971
6	61	Botes, 1971
7	61	Strydom and Botes, 1971
8	61	Strydom and Botes, 1971
9	74	Mebs *et al.*, 1971
10	71	Botes, 1971
11	71	Nakai *et al.*, 1971
12	71	Karlsson *et al.*, 1971
13	71	Karlsson *et al.*, 1971
14	71	Karlsson *et al.*, 1971
15	60	Narita and Lee, 1970
16	60	Hayashi *et al.*, 1971 (and N. Tamiya, personal communication, 1972)

and hydrolyzed with trypsin. Seven tryptic fragments were isolated using column chromatography and paper electrophoresis. Some of the fragments were further hydrolyzed with α-chymotrypsin, pepsin, Nagarse, Proctase A or Proctase B. The amino acid sequences of the fragment peptides were determined by subtractive Edman degradation. From the tryptic digest of reduced S-carboxymethylated and trifluoroacetylated erabutoxin b, two fragments were isolated. From the amino acid composition of the fragments and terminal sequence studies of the reduced and S-carboxy-methylated erabutoxin b, the sequence of the seven tryptic fragments was determined.

The tryptic digestion of reduced and S-carboxymethylated erabutoxin a gave similar fragments, only one of which was different from the corresponding fragment in erabutoxin b. The amino acid sequence analysis of the fragment peptide showed the difference between the two erabutoxins was that the former had asparagine and the latter had histidine at position 26.

It was found that when erabutoxin a was partially hydrolyzed with enzymes and sulfuric acid and the peptides separated by column chromatography and paper electrophoresis, the amino acid analyses of the sulfur-containing peptides and their oxidized components showed the four disulfide bonds between half-cystine residues at positions 3 and 24, 17 and 41, 43 and 54, and 55 and 60, from the N-terminus (Endo *et al.*, 1971).

Subsequently, Tu *et al.* (1971) described two toxins from the venom of the sea snake *L. semifasciata,* which they isolated on Sephadex and carboxymethylcellulose columns. Purity was established by isoelectric focusing, electrophoresis, sedimentation velocity, and sedimentation equilibrium. The toxins were isolated in a crystalline form, and their lethal effect was found to be increased five- to sixfold when compared to the crude venom. The molecular weight, as determined by amino acid composition, gel filtration, and sedimentation equilibrium, was approximately 6800 for both toxins. They had similar amino acid compositions, with toxin a containing sixty-two amino acid residues and toxin b containing sixty-one. No free sulfhydryl groups were present, and no significant change in lethality was observed when the lysine and arginine residues were modified. Toxin a contained 8.2% of the total protein, while toxin b contained 8.0%.

The venom of the Formosan cobra, *Naja naja atra,* has been studied by a number of workers in Taiwan and Japan. At the symposium in 1966, Su *et al.* (1967) reported on the isolation of cobra "neurotoxin" from this snake. The toxin was separated by repeated fractionation with ammonium sulfate. The final product was polypeptidic in nature, and was approximately seven times more lethal than the crude venom. A product said to be identical to this in both its lethal and certain other pharmacological characteristics was also obtained by starch gel electrophoresis. When the ammonium sulfate fraction was subjected to the same electrophoretic procedure, it concentrated as a single peak, with but little increase in its lethal property. These workers believed that this toxin was "essentially similar" to the "neurotoxin" described by Sasaki (1957) and the "cobrotoxin" described by Yang (1965). They also note that apart from the neuromuscular blocking properties of cobra neurotoxin, the toxin has cardiotoxic, hemolytic, anticoagulant, and local tissue effects (Peng, 1951; Su *et al.*, 1967).

In the rat phrenic nerve–diaphragm preparation, the toxin caused a fivefold greater decrease in the indirectly elicited contractions than the same dose of crude venom. It also differed from the whole venom in that it did not cause contracture of the diaphragm nor inhibition of the muscle response to direct stimulation. It also failed to reduce the release of acetylcholine from the phrenic nerve endings. This study indicated to the authors that the site of action was the motor end plates. The blocking effect, unlike that produced by the whole venom, was reversed to a considerable extent by neostigmine or by the removal of the toxin.

The chick biventer cervicis muscle contracted in the presence of crude cobra venom but not in the presence of the neurotoxin, which, however, caused a decrease in the contractile response of the muscle to both acetylcholine and direct stimulation. In the frog rectus abdominis muscle preparation, several changes were elicited by the toxin. The authors suggested that there was a competitive antagonism between the neurotoxin and acetylcholine. In the frog motor nerve–sartorius muscle preparation, the toxin caused changes consistent with those observed in the rat diaphragm. The authors concluded that cobra neurotoxin and *d*-tubocurarine have properties in common. In another group of experiments, employing intracellular microelectrodes, they noted that the qualitative effects of cobra neurotoxin on the end plate potentials appeared to be identical to those produced by *d*-tubocurarine.

Yang *et al.* (1969) found that "cobrotoxin," a crystalline toxin from the venom of *N. naja atra,* which Yang (1965) had previously described, contained fifteen amino acids and was devoid of alanine, methionine, and phenylalanine. The minimal molecular weight, calculated from amino acid composition of the sixty-two amino acid residues, was 6949. The toxin consisted of a single peptide chain cross-linked intramolecularly by four disulfide bonds. Leucine was found in the amino terminal position and asparagine in the carboxy terminal position.

Lee *et al.* (1968) isolated a toxin from the venom of *N. naja atra,* which they called cobra venom "cardiotoxin." It accounts for the greatest part of the venom (20–40%); it is heat stable and has a pHi above 12. Fifty-two to sixty-one residues of the fifteen to seventeen amino acids are arranged as a nonenzymatic, dialyzable, single-chain polypeptide with a molecular weight of 5840–6912, lacking free sulfhydryl groups, and cross-linked intramolecularly by three or four disulfide bonds. The toxin is rich in basic hydrophobic amino acids and lysine, the former composing 25% of the molecule and the latter 20% (Narita and Lee, 1970). These authors note that the tertiary structure of the cardiotoxins is closely homologous to those of the "neurotoxins." Jiménez-Porras (1970) states that the two toxins can be differentiated, in that the cardiotoxin has larger amounts of branched-chain, hydrophobic amino acids and lysine, as well as by an absence of sensitivity to tryptic hydrolysis.

This cardiotoxin causes contracture of the chick biventer cervicis muscle followed by paralysis and a similar response in the frog sartorius muscle and the rat diaphragm. This effect is said to be due to irreversible depolarization of the cell membrane. It is suggested that the toxin's ability to cause systolic arrest of the isolated frog heart and the rat atrium is due to the same mechanism. The cardiotoxin also produces a contraction of the guinea pig ileum, vasoconstriction in the ear vessels of the rabbit, and local irritation of the rabbit conjunctiva and rat hind paw. In the cat, the toxin causes a decrease in systolic blood pressure and various electrocardiographic changes. Lee *et al.* (1968) conclude that cardiotoxin exerts its effects on various kinds of cells, causing irreversible depolarization of the cell membrane.

A toxin designated as toxin α, from the venom of the Egyptian cobra, *Naja haje haje,* was isolated by gradient chromatography on Amberlite CG-50 and further purified by gel filtration on Sephadex G-50. Homogeneity was verified by free boundary electrophoresis, acrylamide gel electrophoresis, sedimentation velocity, amino acid analysis, and end group analysis. It was found to consist of a single polypeptide chain of sixty-one amino acid residues and was cross-linked by four disulfide bridges. Alanine, methionine, and phenylalanine were absent. It was termed a "neurotoxin." No evidence was presented for its neurotoxic characteristics (Botes and Strydom, 1969).

Subsequently, these two workers isolated two toxins, toxins II and IV, from the venom of the South African Ringhals, *Hemachatus haemachatus,* by gradient chromatography on Amberlite CG-50, with further purification by gel filtration on Sephadex G-50. Studies showed formula weights of 6838 and 6831. Each toxin contained sixty-one amino acid residues and both were devoid of alanine, methionine, and phenylalanine. There were twelve differences in the amino acid sequences of the two toxins (Strydom and Botes, 1971). Toxin α, from the venom of the cobra *Naja nigricollis*, termed a "neurotoxin" on the basis that its chemical structure resembled other "neurotoxins," was isolated on Sephadex G-75, followed by ion exchange chromatography on Amerlite IRC-50 or Bio-Rex 70 (Karlsson *et al.*, 1966). It was found to be analytically and immunoelectrophoretically homogeneous, and in mice had an LD_{100} of 75–80 μg/kg. The molecular weight, 6787, obtained by equilibrium ultracentrifugation, was only 2% higher than that calculated from the amino acid composition. The toxin contained eight residues of half-cystine and no free sulfhydryl groups (Eaker and Porath, 1967).

Boquet *et al.* (1967), before the First International Symposium on Animal Toxins in 1966, demonstrated that "toxin α" of *N. nigricollis* venom was similar to the "neurotoxin" isolated by Sasaki (1957), both having estimated molecular weights of about 6000. Under the same conditions, cross precipitation by heterologous sera (*N. naja* and *N. nigricollis*) showed that the phospholipase and the neurotoxin of both of these venoms were antigenic substances of very similar constitution.

Several biologically active proteins and peptides have been separated from the venom of the banded krait *Bungarus multicinctus.* Three of these, α-, β-, and γ-bungarotoxin, have a particular effect on the peripheral nervous system, and all inhibit neuromuscular transmission. α-bungarotoxin, the most abundant polypeptide, is said to act postsynaptically through an irreversible combination with acetylcholine receptors of the motor end plates. The other two apparently act presynaptically and produce Wedensky inhibition in the rat diaphragm. The α-bungarotoxin is a β structure, while the β fraction has an α structure (Chang and Lee, 1963; Lee and Chang, 1966).

It can be seen that some of these toxins are very similar. They are single-chain polypeptides composed of sixty-one or sixty-two residues of fifteen or sixteen amino acids, and they have molecular weights between 6000 and 7000. In general, they are rich in arginine, lysine, and the amide forms of glutamic and aspartic acids, and they are relatively devoid of the more hydrophobic amino acids. They are strongly basic and, as pointed out by Jiménez-Porras (1970), the primary structure of some shows repeated clusters of several residues of the same amino acid, or similar ones, along the chain.

The four cystine bridges that cross-link the polypeptide chain, located near the ends of the molecule, account for the high sulfur content of these toxins. The integrity of the disulfide bonds appears essential for the more important biological properties of the toxins. The central part of the molecule of most of the elapid and hydrophiid toxins contains all of the tyrosine and most of the basic amino acid residues, but no cystine, proline, nor the few hydrophobic branched-chain residues of leucine, isoleucine and valine (Jiménez-Porras, 1970).

It has been suggested that these structural features, particularly the uncrossed-linked, hydrophilic central loop may be the biologically active site for the neurotoxic components of the polypeptide (Eaker and Porath, 1967). This appears to be the only region where a considerable degree of α-helical structure could exist (Botes and Strydom, 1969). The contention is supported by the optical rotatory dispersion and circular dichroism data (Yang *et al.*, 1967, 1968; Jiménez-Porras, 1970).

Using carboxymethylcellulose columns, Ramsey *et al.* (1972) fractionated the venom of the coral snake *Micrurus fulvius*. One heterogeneous and six homogeneous fractions were obtained, with the heterogeneous and two of the homogeneous fractions accounting for 25% each of the recovery rate. These three fractions were lethal at a dose level similar to that of the crude venom and possessed anticoagulant activity. All fractions exhibited strong phospholipase A activity and hemolyzed red blood cells, but no fractions exhibited proteolytic activity.

In 1966, at the symposium previously noted, Moroz *et al.* (1967) described a toxin, "viperotoxin" from the venom of the Palestinian viper, *Vipera palestinae*. This substance was termed a "neurotoxin" on the basis of its ability to cause paresis, jerky body movements, convulsions, and paresis of the hind limbs, as signs in mice injected with lethal doses of the protein (Gitter *et al.*, 1957). It was separated by ion exchange chromatography and differential salt precipitation and was found to be homogeneous by ultracentrifugal, electrophoretic, and immunochemical criteria. Its molecular weight was approximately 11,600. Viperotoxin contained one hundred eight amino acid residues and had one polypeptide chain cross-linked intramolecularly by three disulfide bridges. Lysine was its amino terminal acid and proline was in the carboxy terminal position.

The first important studies on the nonenzymatic fractions of crotalid venoms were those by Slotta and Frankel-Conrat (1938). These workers separated a crystalline protein, "crotoxin," from the venom of the rattlesnake *Crotalus terrificus terrificus* (*Crotalus durissus terrificus*). It contained the toxic nonenzymatic protein plus hyaluronidase, phospholipase, and possibly several other enzymes. Following removal of phospholipase A, crotoxin was further separated as "crotactin" and a second component, possibly "crotamine." Subsequently, Gonçalves (1956) obtained three fractions which had specific biological activity: (1) crotamine, with a molecular weight of 10,000–15,000; (2) proteolytic enzyme; and (3) neurotoxin, which corresponds to crotoxin. Neelin (1963), using two-dimensional starch gel electrophoresis, demonstrated at least twenty zones in the venom of the rattlesnake *Crotalus adamanteus*. The low content of fast cationic components was said to be consistent with the reported absence of strongly basic, dialyzable crotamine for this species.

Approximately ten fractions were observed on starch gel electrophoresis of *Crotalus atrox* venom by Jiménez-Porras (1967), and Glenn *et al.* (1963) demonstrated sixteen precipitin systems in double diffusion agar columns with the same venom. At the Western Pharmacological Society meeting in 1969, Dubnoff and Russell (1970a) presented evidence of the presence of a lethal nonenzymatic protein having a molecular weight of approximately 30,000 and one or two peptides of approximately 6000 in the venom of the Southern Pacific rattlesnake *Crotalus viridis helleri*. Subsequently, Dubnoff and Russell (1970a,b), using Sephadex-50, ion exchange chromatography on carboxymethylcellulose and on IRC-50 resin, isolated two peptides with molecular weights of about 6000 each and a protein of approximately 30,000 from the venom of the rattlesnake *C. v. helleri*. The peptides moved as cations on cellulose acetate at pH 8.6. There were sixty-three amino acid residues. These three fractions could easily be distinguished from each other and from the crude venom, not only by their chemical characteristics but also by their pharmacological properties, as shown in Table III.

More recently, Bonilla and Fiero (1971) have isolated highly basic proteins from the venom of three species of rattlesnakes—*C. viridis viridis, C. horridus horridus* and *C. horridus atricaudatus*. The proteins were separated by recycling adsorption chromatography using Bio-Gel P-2 and by ion exchange on carboxymethylcellulose. These proteins were found to constitute less than 10% of the total protein of the crude venom. They were of low molecular weight, isoelectric points above pH 10.8, and showed pharmacological properties similar to those of *C. viridis helleri* (Dubnoff and Russell, 1970a). Amino acid analyses of the three venoms gave fifty-six to sixty-one, thirty-eight to forty-two, and fifty-two to fifty-four residues, respectively. The subcutaneous LD_{50} in mice was found to be 41, 43, and 32 mg/kg body weight, respectively.

The relationships between certain of the physiochemical properties of these toxins and specific modes or sites of action are becoming increasingly apparent. However, it still seems wise, particularly in view of the synergistic and antagonistic effects of individual venom fractions within the complex, and the clinical implications often unleashed from what has too frequently been a very limited understanding of the function of the venom, to guard against oversimplified, and often overzealous labeling until adequate assays on appropriate biological preparations

TABLE III

SIGNS AND LD_{50} IN MICE FOLLOWING INTRAVENOUS INJECTION OF *Crotalus viridis helleri* VENOM FRACTIONS

Venom fraction	Immediate signs	Signs after 1–3 minutes	Signs after 3–15 minutes	Time to death	LD_{50} (mg/kg)
Crude venom	Hypoactive; extensor thrust hind limbs; flattening, blanching ears	Extensor thrust and paresis hind limbs; flattening, blanching ears	Extensor thrust and marked paresis hind limbs; cyanosis, exophthalmos; superficial respirations; prostration	10–15 minutes	1.98
Protein	Hypoactive; some weakness hind limbs; flattening, blanching ears	Hypoactive; some weakness hind limbs; flattening, blanching ears; cyanosis	Extensor thrust and paresis hind limbs; cyanosis, exophthalmos; superficial respirations; prostration	1–7 minutes	1.51
Peptide I	Marked extensor thrust hind limbs; curling front limbs and paws under body; prostration	Prostration	—	90 seconds	0.35
Peptide II	Hyperactive and ataxia; eyes closed; flattening, blanching ears	Cyanosis; exophthalmos; prostration	—	2 minutes	0.37

have been carried out. As one of us has pointed out elsewhere, we have yet to find a "neurotoxin" whose physiopharmacological activities are limited to a highly specific anatomical receptor site, as is too often claimed. There is no doubt that there is a target site for these various fractions, but in such a complicated system as the biological one, it is quite difficult to accept the theory that a toxin must be site specific, even though its chief activity may certainly be so. The mellowing of opinion on the biological properties of tetrodotoxin, for one, is an example of the caution that should be exercised when researchers feel obligated to put things right with their chemistry sets and squid axons.

The separation of snake venoms into single or more often compound fractions, as coagulants, anticoagulants, hemorrhagins, hemolysins, neurotoxins, nerve growth factors, etc., has been described by numerous workers. Many of these pharmacological classifications, as already noted, are unfortunate, because the components in question are not pure or single fractions or because the other biological properties have not been adequately determined. Nevertheless, the rather impressive amount of literature on these substances, and the fact that studies have been done with single fractions and do show evidence of some specific pharmacological activity, make it necessary to note this area of twilight. The reader is directed to the review of Jiménez-Porras (1970) for what has been a very courageous attempt to divide pharmacological activities on some sort of a biochemical basis. However, the pharmacologist faced with a title such as "Snake Venom Proteases and the Blood Coagulation, Shocking, Hemorrhagic and Myonecrotic Effects" is likely to find himself in a corresponding state of shock. Jiménez-Porras has done a fine job seeking to establish chemical–pharmacological relationships, but the fact remains that much of our work does not yet lend itself to such clear-cut deductions and conclusions. It seems best, therefore, that in this short review, and in these disciplines, we present a table of selected references from which the reader can obtain some direction in these areas of study (Table IV). We regret that space does not permit us to note all of the important studies on these topics, and that there is an obvious overlapping in pharmacological–chemical considerations.

IV. Venom Enzymes

The enzymes of snake venoms have certainly been more thoroughly studied in the past and their modes of action more carefully defined than the other components of snake venoms. It is not difficult to understand why earlier workers attributed most, if not all, of the deleterious biological effects of the crude venom to one or another of the enzymes. The techniques for studying enzymatic activity were far advanced over those for protein chemistry in general, and thus biochemists and pharmacologists aligned specific toxicological properties, including lethality, with one or several enzymes, and usually without sufficient or acceptable data. Some investigators were quite adamant in insisting that the "neurotoxic" effects of *Naja* venoms, for instance, were due entirely to cholinesterase, that the proteinases were responsible for all of the cytolytic effects of venoms, that hyaluronidase was the sole spreading factor, etc. Even today, some reviews on snake venoms still reflect these misconceptions. It is well to repeat, reptile venoms are complex mixtures capable of producing one or many deleterious biological effects, sometimes concurrently, and that these effects can be complicated by synergistic and antagonistic activities evoked by the different components, as well as by the autopharmacological response of the envenomated organism, be it man or beast.

Reviews on the enzymes and their activities in reptile venoms can be found in the contributions by Zeller (1948, 1951), Braganca (1955), Kaiser and Michl (1958), Condrea and de Vries (1965), Boquet (1966), Russell (1967), Tu *et al.* (1967), Sarkar and Devi (1968), and Jiménez-Porras (1970). In Table IV are listed some of the more important enzyme activities of snake venom.

Proteinases have been found in most snake venoms. This trypsin-like enzyme causes the digestion of tissue proteins and peptides. All Crotalidae venoms so far examined appear to be rich in proteolytic enzyme activity. In general, the Viperidae venoms have lesser amounts of proteinase than the Crotalidae, while the Elapidae and Hydrophiidae venoms either have no proteolytic activity or very little. In the crotalids and viperids there does not appear to be a direct relationship or pattern relating to the amounts of proteinase in the various species (Tu *et al.*, 1966). If one considers the data from experimental studies that demonstrate proteinase activity and tissue digestion and compares these with the clinical picture, or even with the cytolytic changes provoked in experimental animals following envenomation, it can be seen that there is not a consistent relationship between the amount of proteinase in a snake venom and the degree of cytolytic damage; even though those venoms rich in proteinase activity tend to produce the more deleterious local reactions in the clinical cases (Russell, 1967).

Most crotalid and viperid venoms have a high content of heat labile, acidic proteinases and heat-stable nonproteolytic amino acid-ester hydrolases. Venoms rich in protease activity also exhibit high phospholipase A activity. Some venoms contain proteinases similar to trypsin,

TABLE IV
SELECTED REFERENCES ON CERTAIN SNAKE VENOM PROPERTIES

Pharmacology, general
- Kaiser and Michl, 1958
- Meldrum, 1965
- Russell, 1967
- Kornalik, 1967
- Russell and Puffer, 1970

Pharmacology, cardiovascular
- Lee, 1941
- Russell *et al.*, 1962
- Halmagyi *et al.*, 1965
- Vital Brazil *et al.*, 1968
- Vick *et al.*, 1967

Pharmacology, nervous system
- Morrison and Zamecnik, 1950
- Russell and Bohr, 1962
- Parnas and Russell, 1967
- Lee, 1971
- Rosenberg, 1971

Autopharmacology
- Rocha e Silva *et al.*, 1949
- Russell, 1965
- Ferreira, 1965
- Rothschild, 1967
- Diniz, 1968

Uncoupling and reverse acceptor control
- Ghosh and Chatterjee, 1948
- Aravindakshan and Braganca, 1959
- Elliott and Gans, 1967

Coagulation and anticoagulation defects
- Rosenfeld *et al.*, 1959
- Reid *et al.*, 1963
- Devi, 1968
- Gitter and de Vries, 1968
- Denson, 1969

Thrombocytopenia
- Gitter *et al.*, 1960
- Rechnic *et al.*, 1962
- Klibansky *et al.*, 1966
- La Grange and Russell, 1970
- Lyons, 1971

Direct and indirect lytic factors
- Raudonat and Holler, 1958
- Condrea *et al.*, 1964
- Slotta *et al.*, 1967
- Aloof-Hirsch *et al.*, 1968
- Slotta and Vick, 1969

Immunology and antivenins, experimental
- Minton, 1957
- Russell, 1961
- Keegan *et al.*, 1962
- Minton, 1967
- Munjal and Elliott, 1972

Immunology and antivenins, clinical
- Christensen, 1955
- Russell and Lauritzen, 1966
- Stanić, 1969
- Russell and Puffer, 1970
- World Health Organization, 1971

Nerve growth factors
- Levi-Montalcini, 1952
- Cohen and Levi-Montalcini, 1956
- Levi-Montalcini and Cohen, 1956
- Banks *et al.*, 1968
- Angeletti, 1968

while others have proteinases that differ remarkably from trypsin, and it seems apparent that there are several proteases in some venoms.

Although there has been some question of the role proteinases play in the deleterious blood changes provoked by snake venoms (Russell, 1967), the present concept that proteases are completely free from hemorrhagic properties must also be viewed with some reserve. There are still some problems related to the complex hemorrhagic phenomena involved, when a whole venom is studied, that cannot be easily dismissed on the basis of data obtained from studies on single venom fractions or enzymes. Venom proteases may be completely free from hemorrhagic properties, and venom hemorrhagins may be devoid of proteolytic (and other enzyme) activities, if our definition of hemorrhagic properties is carefully limited, but it seems likely that the proteases may play some coagulant, anticoagulant, fibrinolytic, antithromboplastic, and plasminogenic role in the whole venom.

L-Amino acid oxidase has been found in over seventy snake venoms, although it appears to be absent in those of the sea snakes (Russell and Scharffenberg, 1964). It catalyzes the oxidation of L-α-amino acids and of α-hydroxy acids. It is the most active of the known amino acid oxidases, and ophio-L-amino acid oxidase is probably a group of homologous enzymes. The enzyme is very unstable at pH 9.0 in the absence of substrate amino acid (Paik and Kim, 1967); it contributes to the yellow color of most snake venoms.

L-Amino acid oxidase is a glycoprotein, having a molecular weight of approximately 13,000, with two moles of FAD per mole of enzyme, and intrachained disulfide bridges and free sulfhydryl groups. According to De Kok and Rawitch (1969), it is a noncovalent dimer formed by two polypeptide subunits each having molecular weights of 70,000 and being very similar in their amino acid composition. The three possible combinations of the two peptides account for the three isoenzymes, which are characterized by their electrophoretic mobility but are otherwise indistinguishable. This enzyme is one of the few known to be reversibly inactivated by freezing (Curti *et al.*, 1968). This has been attributed to limiting conformational changes at the active flavin site, a feature that indicates that each of the isomers exists in multiple molecular forms (Wellner and Hayes, 1968).

Few enzymes are as capable of attacking so many different substances as L-amino acid oxidase. Zeller (1951) suggests that the enzyme is probably not a toxic component of snake venoms, but that its action is integrated with the digestive function of the venom. If this oxidase is responsible for the activating power of snake venoms, it is a nonhydrolytic digestive enzyme. The enzyme is not present in venoms produced from organs that have not developed from digestive glands, such as those of the bee and scorpion (Russell, 1967).

Ophio-L-amino acid oxidase does not contribute to the profound fall in systemic arterial pressure produced by the crude venom. It has no effect on neuromuscular transmission, and while its intravenous LD_{50} in mice is 9.13 mg per kilogram body weight (about one-eighth as toxic as the crude venom), this lethal effect may be due to other components separated by the method generally employed (Wellner and Meister, 1960). Even if the enzyme were pure, this LD_{50} might not be significant

TABLE V
SOME ENZYMES OF REPTILE VENOMS

Enzyme	Systematic name	References
Proteinases	—	Maeno *et al.*, 1959; Yang, 1960; Pfleiderer and Sumyk, 1961; Tu *et al.*, 1967
L-Arginine-ester hydrolases	—	Suzuki, 1968
Transaminase	—	Tsai, 1961
Hyaluronidase	Hyaluronate lyase	Slotta and Ballester, 1954; Jaques, 1956; Barme and Detrait, 1959
L-Amino acid oxidase	L-amino acid: O_2 oxidoreductase	Singer and Kearney, 1950; Zwisler, 1965; Wellner and Hayes, 1968; De Kok and Rawitch, 1969
Acetylcholinesterase	Acetylcholine acetyl hydrolase	Iyengar *et al.*, 1938; Zeller, 1947; Christensen and Anderson, 1967
Phospholipase A	Phosphatide acyl hydrolase	Suzuki *et al.*, 1958; Wakui and Kawashi, 1959; Condrea and De Vries, 1965
Phospholipase B	Lysolecithin acyl hydrolase	Doery and Pearson, 1964
Phospholipase C	Phosphatidylcholine cholinephosphohydrolase	Vidal Breard and Elias, 1950
Ribonuclease	ATP pyrophosphohydrolase	Gulland and Walsh, 1945; Taborda *et al.*, 1952a; Tsai, 1961
Deoxyribonuclease	Deoxylribonucleate 3′-nucleotidohydrolase	Taborda *et al.*, 1952b; Bowman and Kaletta, 1956; Nikolskaya and Budowski, 1962
Phosphodiesterase	Orthophosphoric diester phosphohydrolase	Sinsheimer and Koerner, 1952; Björk, 1963; Russell *et al.*, 1963a
Phosphomonoesterase	—	Vasilenko, 1963; Richards *et al.*, 1965; Tu and Chua, 1966
5′-Nucleotidase	5-Ribonucleotide phosphohydrolase	Kaye, 1955; Kara and Sormova, 1962; Bragdon and McManus, 1952
Adenosine triphosphatase	ATP pyrophosphohydrolase	Zeller, 1950; Kaye, 1960; Pfleiderer and Sumyk, 1961
Alkaline phosphatase	Orthophosphoric monoester phosphohydrolase	Suzuki *et al.*, 1958; Sulkowski *et al.*, 1963
Acid phosphatase	Orthophosphoric monoester phosphohydrolase	Jiménez-Porras, 1970
Nucleotide pyrophosphatase	Dinucleotide nucleotidohydrolase	Suzuki *et al.*, 1960; Zeller, 1950
Exopeptidases	Dipeptide and tripeptide hydrolases	Michl and Molzer, 1965; Wagner and Prescott, 1966; Tu *et al.*, 1967

if one considers that in the particular venom studied, the enzyme accounted for less than 3% of the total weight of the crude material; thus, the oxidase contributes less than 1% of the total lethal effect of the venom. Antivenin prepared against this venom neutralized the lethal effect of the L-amino acid product studied (Russell *et al.,* 1963b).

Snake venom cholinesterase is a true acetylcholinesterase which is able to hydrolyze a variety of acidic esters, including acetylcholine ester and noncholine ester, propionate esters, and the higher esters of butyric and caproic acids. It has been identified in the venoms of at least fifty snakes (Russell and Scharffenberg, 1964). As a whole, elapid venoms are rich in this enzyme, although it is apparently lacking in *Naja nigricollis, Dendroaspis* sp., *Micrurus nigrocinctus,* and *Pseudoechis collettii* (Christensen and Anderson, 1967; Jiménez-Porras, 1970). Viperid and crotalid venoms either do not contain the enzyme or possess it in only small amounts. A heat-stable acetylcholinesterase-inactivating factor has been demonstrated in some cobra venoms.

As the earlier studies on elapid venoms showed the marked effects of these poisons on neuromuscular activity, and as these venoms were shown to contain large amounts of cholinesterase, it was thought that this enzyme was responsible for the "curare-like effect" of the whole venom, and the death of the experimental animal (Iyengar *et al.,* 1938; Bovet-Nitti and Bovet, 1943). It is now generally thought that cholinesterase is not the principal component responsible for the neuromuscular deficit provoked by the crude venom, either in experimental animals or man. Furthermore, the enzyme does not contribute significantly to the lethal activity of elapid venoms. What part the enzyme plays in the overall effects of a crude venom is not known. As previously noted, the function of acetylcholinesterase is to hydrolyze acetycholine, which is involved in the transmission of nerve impulses. It is known that some venoms free of this enzyme increase the cholinesterase activity of intact squid nerve, possibly by altering permeability and thus allowing an excess of substrate (Rosenberg and Dettbarn, 1964), but the relationships and significances here are not clear.

Phospholipases catalyze the hydrolysis of lipids. Snake venom phospholipase A is a highly stable enzyme which catalyzes the hydrolysis of one of the fatty acyl ester bonds of diacyl phosphatides, releasing both saturated and unsaturated fatty acids and the strongly hemolytic component lysolecithin (Condrea and de Vries, 1965; Russell, 1967).

The enzyme is present in almost all snake venoms, which are one of its richest sources, and while it may occur as a single component, in most snake venoms two or more forms have been found (Iwanga and Kawauchi, 1959; Saito and Hanahan, 1962). *Crotalus adamanteus* and *C. atrox* have two forms of phospholipase A, each having molecular weights of approximately 30,000. The α and β forms of phospholipase A_2 have been shown to be composed of two subunits, each having molecular weights of approximately 15,000. It appears that the dimer is the enzymatically active form of these proteins (Wells and Hanahan, 1969; Wells, 1971).

It is generally thought that the hemolysis produced by snake venoms is due to phospholipase A (Zeller, 1951; Slotta and Borchert, 1954). This may be provoked by direct action in which the phospholipids of the red cell membrane are hydrolyzed, or it may be provoked through indirect action in which lysolecithin is produced from plasma lecithin by the phospholipase. Direct lytic venoms, such as those of most elapids, lyse washed red blood cells and produce intravascular hemolysis, while the indirect lytic venoms, such as those of certain vipers, appear to be able to lyse red blood cells and produce intravascular hemolysis but apparently only in the presence of added serum or lecithin. The reaction, however, may be influenced by the protective effect of the plasma proteins, the possible sensitivity of the red cells to lysolecithin, or through the influence of the spleen (de Vries *et al.*, 1962; Condrea and de Vries, 1965).

Venom phospholipase A is known to release histamine from the rat diaphragm (Habermann, 1957) and a slow-contracting substance, which produces hemolysis, from certain other tissues (Meldrum, 1965). It probably also plays some role in the disruption of the normal sequence of electron transfer by phospholipids. Its role in altering nerve, muscle, or neuromuscular junction conduction and possible central nervous system activity is a consideration that has elicited much controversy. Certainly, its ability to destroy or alter certain phospholipids in nerve tissues essential to electron transfer could make it a "neurotoxin" (Russell, 1967).

It has been claimed by some workers that this enzyme is *the* neurotoxic component of snake venoms (Braganca and Quastel, 1953; Aravindakshan and Braganca, 1959). However, this seems quite unlikely, since most of the definitive studies during the past few years have shown that the fractions more specifically responsible for the deleterious changes in nerve conduction are nonenzymatic proteins of relatively low molecular weight. It is quite possible, of course, that venom phospholipase A facilitates the penetration of neuropharmacologically active components into nerve tissue and in this way contributes to the neurological deficit (Russell, 1967).

Phospholipase B catalyzes the hydrolysis of lysolecithin to glycerophosphocholine and fatty acid. It has been found in a number of snake venoms (Doery and Pearson, 1964). Phospholipase C catalyzes the hydrolysis of phosphatidylcholine to a diglyceride and choline phosphate. It has been found in the venom of *Bothrops alternatus* (Vidal Breard and Elias, 1950). The specific pharmacological activities of these two enzymes in snake venoms are not known.

Hyaluronidase, an enzyme that degrades hyaluronic acid, a polymer of *N*-acetylglucosamine and glucuronic acid, and certain other substrates in connective and other tissues, has been found in every snake venom so far examined with the exception of that of *Naja nigricollis.* It hydrolyzes the hyaluronic acid gel of the spaces between cells and fibers, particularly in connective tissue, and thus reduces the viscosity of these tissues. This breakdown in the hyaluronic acid barrier probably allows other fractions of the venom to penetrate the involved tissues. The enzyme appears to be related to the extent of the edema and swelling caused by the venom, but to what degree it contributes to these signs is not known. It does not, as once supposed, contribute in any effective way to the venom-induced hemorrhage.

Snake venoms contain several phosphatases, a group of enzymes with alkaline pHi values that are involved in the hydrolysis of phosphate bonds in nucleotides. Snake venoms are the richest source of these enzymes in nature. They may be found in multiple molecular forms or isozymes. The more important of this group are phosphomonoesterase, phosphodiesterase, 5′-nucleotidase, ATPase, NADase, and endonuclease.

Nonspecific phosphomonoesterases are a group of enzymes having different pH optima but the common property of being able to hydrolyze phosphate bonds in nucleotides. They are usually present in higher concentrations in elapid venoms than in crotalid and viperid venoms. Their specific role in snake venom has not been clearly established.

Phosphodiesterases have been found in almost all snake venoms. The enzyme has properties of an orthophosphoric diester phosphohydrolase, which also releases 5′-nucleotide from polynucleotides, thus acting as an exonuclease. As with phosphomonoesterase, there may be several phosphodiesterases having similar properties in a single venom. Some preparations (Williams *et al.*, 1960) of the enzyme have been shown to produce an immediate and profound fall in systemic arterial pressure following intravenous injection. In mice, the same preparation has an intravenous LD_{50} or 4.0 mg per kilogram body weight, which is quite toxic. These actions may be due to the enzyme, or they may be due to the presence of some other protein in the separation, since several bands were seen (Russell *et al.*, 1963a) when the material obtained by this method was subjected to disk electrophoresis.

5′-Nucleotidase (AMPase) is a common constituent of all snake venoms, and it is quite dissimilar from that obtained from other sources. Again, it appears to be a group of enzymes exhibiting different molecular properties depending on the specific venom source. In most instances it is the most active phosphatase in the venom. It is a 5′-ribonucleotide phosphohydrolase that catalyzes the hydrolysis of 5′-mononucleotides, yielding the ribonucleoside and orthophosphate. In some snake venoms, two or three 5′-nucleotidases have been found (Maeno and Mitsuhashi, 1961), and although the enzyme has been separated as a toxic component, it seems likely that the electrophoretic mobility of the enzyme was the same as that of an unidentified lethal fraction (Master and Rao, 1963).

It can be seen that snake venoms may contain a variety of enzymes, that the enzymes may occur in different forms, even within the same species or snake, that they vary in number from family to family as well as between different species of the same family, and even within individual snakes, and at different times of the year; and that while there are certain relationships between their presence or absence within snake venoms as a whole, there are notable exceptions of rules when one considers the enzymes from the standpoint of the taxonomy of the snake. As a whole, viperid and crotalid snake venoms either have no or little acetylcholinesterase, while most elapids are rich in this enzyme, although there are taxonomically scattered exceptions. L-Amino acid oxidase is found in all viper venoms so far examined, with the exception of one geographically distinct group of *Vipera aspis* in France, and it is absent in the sea snakes and most elapids and present in only small amounts in certain other elapids. Most crotalid venoms have high proteolytic activity, while viperid snakes usually have less, and elapid and sea snake venoms have little or none. Phospholipase A is a common constituent of almost all snake venoms, but it is richest in the crotalid snakes. Hyaluronidase is distributed throughout almost all snake venoms, the most notable exceptions being *N. nigricollis* and certain of the sea snakes.

The use of snake venom components in medicine, particularly as therapeutic agents, has only been superficially studied. No group of naturally occurring substances, with the possible exception of the marine toxins, has a greater potential value as a source for new drugs than the snake venoms. Aside from the individual fractions themselves, combinations of two or more of these components have been shown to have most interesting and unpredictable biological properties, most of which we are only now beginning to understand. In addition, the value of venoms as tools in biology, particularly molecular biology, has already been demonstrated, and again, this concern is in its infancy. Finally, the use of venom components as aids in taxonomy, as criteria for establishing species or genus relationships, is promising.

REFERENCES

Aloof-Hirsch, S., de Vries, A., and Berger, A. (1968). *Biochim. Biophys. Acta* **154**, 53.

Angeletti, R. H. (1968). *J. Chromatogr.* **37**, 62.

Aravindakshan, I., and Braganca, B. M. (1959). *Biochim. Biophys. Acta* **31**, 463.

Banks, B. E. C., Banthorpe, D. V., Berry, A. R., Davis, H. S., Doonan, S., Lamont, D. M., Shipolini, R., and Vernon, C. A. (1968). *Biochem. J.* **108**, 157.

Barne, M., and Detrait, J. (1959). *C. R. Acad. Sci.* **248**, 312.

Bellairs, A. (1969). "The Life of Reptiles," 2 vols. Weidenfeld & Nicolson, London.

Björk, W. (1963). *J. Biol. Chem.* **238**, 2487.

Bogert, C. M., and Del Campo, R. M. (1956). *Bull. Amer. Mus. Natur. Hist.* **109**, 1.

Bonilla, C. A., and Fiero, M. K. (1971). *J. Chromatogr.* **56**, 253.

Boquet, P. (1966). *Toxicon* **3**, 243.

Boquet, P., Izard, Y., Jouannet, M., and Meaume, J. (1967). *In* "Animal Toxins" (F. E. Russell and P. R. Saunders, eds.), pp. 293–298. Pergamon, Oxford.

Botes, D. P. (1971). *J. Biol. Chem.* **246**, 7383.

Botes, D. P., and Strydom, D. J. (1969). *J. Biol. Chem.* **244**, 4147.

Botes, D. P., Strydom, D. J., Anderson, C. G., and Christensen, P. A. (1971). *J. Biol. Chem.* **246**, 3132.

Bovet-Nitti, F., and Bovet, D. (1943). *Ann. Inst. Pasteur, Paris* **69**, 309.

Bowman, H. G., and Kaletta, U. (1956). *Nature (London)* **178**, 1394.

Braganca, B. M. (1955). *In* "Neurochemistry." (K. A. C. Elliott, I. H. Page, and J. H. Quastel, eds.), pp. 612–630. Thomas, Springfield, Illinois.

Braganca, B. M., and Quastel, J. H. (1953). *Biochem. J.* **53**, 88.

Bragdon, D. E., and McManus, J. F. A. (1952). *Quart. J. Microbiol. Sci.* **93**, 391.

Chang, C. C., and Lee, C.-Y. (1963). *Arch. Int. Pharmacodyn. Ther.* **144**, 241.

Chen, H-C., and Ouyang, C. (1967). *Toxicon* **4**, 235.

Christensen, P. A. (1955). "South African Snake Venoms and Antivenoms." SAfr. Inst. Med. Res., Johannesburg.

Christensen, P. A., and Anderson, C. G. (1967). *In* "Animal Toxins" (F. E. Russell and P. R. Saunders, eds.), pp. 223–234. Pergamon, Oxford.

Cohen, S., and Levi-Montalcini, R. (1956). *Proc. Nat. Acad. Sci. U.S.* **42**, 571.

Condrea, E., and de Vries, A. (1965). *Toxicon* **2**, 261.

Condrea, E., de Vries, A., and Mager, J. (1964). *Biochim. Biophys. Acta* **84**, 60.

Curti, B., Massey, V., and Zmudka, M. (1968). *J. Biol. Chem.* **243**, 2306.

De Kok, A., and Rawitch, A. B. (1969). *Biochemistry* **8**, 1405.

Denson, K. W. E. (1969). *Toxicon* **7**, 5.

Devi, A. (1968). *In* "Venomous Animals and Their Venoms" (W. Bücherl, E. Buckley, and V. Deulofeu, eds.), Vol. 1, pp. 119–165. Academic Press, New York.

de Vries, A., Kirschmann, C., Klitansky, C., Condrea, E., and Gitter, S. (1962). *Toxicon* **1**, 19.

Diniz, C. R. (1968). *In* "Venomous Animals and Their Venoms" (W. Bücherl, E. Buckley, and V. Deulofeu, eds.), Vol. 1, pp. 217–227. Academic Press, New York.

Doery, H. M., and Pearson, J. E. (1964). *Biochem. J.* **92**, 599.
Dubnoff, J. W., and Russell, F. E. (1970a). *Proc. West. Pharmacol. Soc.* **13**, 98.
Dubnoff, J. W., and Russell, F. E. (1970b). *Abstr. Int. Symp. Anim. Plant Toxins, 2nd,* p. 41.
Eaker, D. L., and Porath, J. (1967). *Jap. J. Microbiol.* **11**, 353.
Elliott, W. B., and Gans, C. (1967). *In* "Animal Toxins" (F. E. Russell and P. R. Saunders, eds.), pp. 235–236. Pergamon, Oxford.
Endo, Y., Sato, S., Ishii, S., and Tamiya, N. (1971). *Biochem. J.* **122**, 463.
Ferreira, S. H. (1965). *Brit. J. Pharmacol.* **24**, 163.
Fischer, G. A., and Kabara, J. J. (1967). *In* "Animal Toxins" (F. E. Russell and P. R. Saunders, eds.), pp. 283–292. Pergamon, Oxford.
Gans, C., and Elliott, W. B. (1968). *Advan. Oral Biol.* **3**, 45.
Ghosh, B. N., and Chatterjee, A. (1948). *Indian Chem. Soc.* **25**, 359.
Ghosh, B. N., De, S. S., and Chowdhurry, D. K. (1941). *Indian J. Med. Res.* **29**, 367.
Gitter, S., and de Vries, A. (1968). *In* "Venomous Animals and Their Venoms" (W. Bücherl, E. Buckley, and V. Deulofeu, eds.), Vol. 1, pp. 359–401. Academic, New York.
Gitter, S., Kochwa, S., de Vries, A., and Leffkowitz, M. (1957). *Amer. J. Trop. Med. Hyg.* **6**, 180.
Gitter, S., Levi, G., Kochwa, S., de Vries, A., Rechnic, J., and Casper, J. (1960). *Amer. J. Trop. Med.* **9**, 391.
Glenn, W. G., Mallette, W. G., Fitzgerald, J. B., Crockett, A. T. K., and Glass, T. G., Jr. (1963). *In* "Venomous and Poisonous Animals and Noxious Plants of the Pacific Region" (H. L. Keegan and W. V. Macfarlane, eds.), pp. 415–426. Pergamon, Oxford.
Gonçalves, J. M. (1956). *In* "Venoms," Publ. No. 44, pp. 261–274. Amer. Ass. Advan. Sci., Washington, D.C.
Gulland, J. M., and Walsh, E. O. (1945). *J. Chem. Soc., London* p. 172.
Habermann, E. (1957). *Naunyn-Schmiedabergs Arch. Exp. Pathol. Pharmakol.* **230**, 538.
Halmagyi, D. F. J., Starzecki, B., and Horner, G. J. (1965). *J. Appl. Physiol.* **20**, 709.
Hayashi, K., Takechi, M., and Sasaki, T. (1971). *Biochem. Biophys. Res. Commun.* **45**, 1357.
Iwanga, S., and Kawachi, S. (1959). *J. Pharm. Soc. Jap.* **79**, 582.
Iyengar, N. K., Sehra, K. B., Mukerji, B., and Chopra, R. N. (1938). *Curr. Sci.* **7**, 51.
Jaques, R. (1956). *In* "Venoms," Publ. No. 44, pp. 291–293. Amer. Ass. Advan. Sci., Washington, D.C.
Jiménez-Porras, J. M. (1967). *In* "Animal Toxins" (F. E. Russell and P. R. Saunders, eds.), pp. 307–321. Pergamon, Oxford.
Jiménez-Porras, J. M. (1970). *Clin. Toxicol.* **3**, 389.
Juratsch, C. E., and Russell, F. E. (1971). *Herpeton* **6**, 1.
Kaiser, E., and Michl, H. (1958). "Die Biochemie der tierschen Gifte." Deuticke, Vienna.
Kara, J., and Sormova, Z. (1962). *Collect. Czech. Chem. Commun.* **27**, 506.
Karlsson, E., Eaker, D. L., and Porath, J. (1966). *Biochim. Biophys. Acta* **127**, 505.
Karlsson, E., Arnberg, H., and Eaker, D. (1971). *Eur. J. Biochem.* **21**, 1.
Kaye, M. A. G. (1955). *Biochim. Biophys. Acta* **18**, 456.

Kaye, M. A. G. (1960). *Biochim. Biophys. Acta* **38**, 34.

Keegan, H. L., Whittemore, F. W., and Maxwell, G. R. (1962). *Copeia* **2**, 313.

Klauber, L. M. (1956). "Rattlesnakes. Their Habits, Life Histories, and Influence on Mankind." Univ. of California Press, Berkeley.

Klibansky, C., Ozcan, E., Joshua, H., Djaldetti, M., Bessler, H., and de Vries, A. (1966). *Toxicon* **3**, 213.

Kochva, E., and Gans, C. (1970). *Clin. Toxicol.* **3**, 363.

Kornalik, F. (1967). "Animal Toxins." State Health Publ., Prague.

La Grange, R. G., and Russell, F. E. (1970). *Proc. West. Pharmacol. Soc.* **13**, 99.

Larson, P. R., and Wolff, J. (1968). *J. Biol. Chem.* **243**, 1283.

Lee, C.-Y. (1941). *Jap. J. Med. Sci. 4* **14**, 200.

Lee, C.-Y. (1971). *In* "Neuropoisons: Their Pathophysiological Actions" (L. L. Simpson, ed.), Vol. 1, pp. 21–70. Plenum, New York.

Lee, C.-Y., and Chang, C. C. (1966). *Mem. Inst. Butantan, São Paulo* **33**, 55.

Lee, C.-Y., Chang, C. C., Chiu, T. H., Chiu, P. J. S., Tseng, T. C., and Lee, S. Y. (1968). *Naunyn-Schmiedabergs Arch. Pharmakol. Exp. Pathol.* **259**, 360.

Levi-Montalcini, R. (1952). *Ann. N.Y. Acad. Sci.* **55**, 330.

Levi-Montalcini, R., and Cohen, S. (1956). *Proc. Nat. Acad. Sci. U.S.* **42**, 695.

Lyons, W. J. (1971). *Toxicon* **9**, 237.

Maeno, H., and Mitsuhashi, S. (1961). *J. Biochem.* (*Tokyo*) **50**, 434.

Maeno, H., Mitsuhashi, S., Sawai, Y., and Okonogi, T. (1959). *Jap. J. Microbiol.* **3**, 131.

Master, R. W. P., and Rao, S. S. (1963). *Biochim. Biophys. Acta* **71**, 416.

Mebs, D. (1968). *Hoppe-Seyler's Z. Physiol. Chem.* **349**, 1115.

Mebs, D. (1970). *In* "Bradykinin and Related Kinins" (F. Sicuteri, M. Roche e Silva, and N. Black, eds.), pp. 107–116. Plenum, New York.

Mebs, D., Narita, K., Iwanaga, S., Samejima, Y., and Lee, C. Y. (1971). *Biochem. Biophys. Res. Commun.* **44**, 711.

Meldrum, B. S. (1965). *Pharmacol. Rev.* **17**, 393.

Micheel, F., Dietrich, H., and Bishoff, G. (1937). *Hoppe-Seyler's Z. Physiol. Chem.* **249**, 157.

Michl, H., and Molzer, H. (1965). *Toxicon.* **2**, 281.

Minton, S. A., Jr. (1957). *Amer. J. Trop. Med. Hyg.* **6**, 1097.

Minton, S. A., Jr. (1967). *Toxicon* **5**, 47.

Minton, S. A., Jr. (1968). *Toxicon* **6**, 93.

Minton, S. A., Jr., and Minton, M. R. (1969). "Venomous Reptiles." Scribners, New York.

Moroz, C., de Vries, A., and Sela, M. (1967). *In* "Animal Toxins" (F. E. Russell and P. R. Saunders, eds.), pp. 303–306. Pergamon, Oxford.

Morrison, L. R., and Zamecnik, P. C. (1950). *Arch. Neurol. Psychiat.* **63**, 367.

Munjal, D., and Elliott, W. B. (1972). *Toxicon* **10**, 47.

Nakai, K., Sasaki, T., and Hayashi, K. (1971). *Biochem. Biophys. Res. Commun.* **44**, 893.

Narita, K., and Lee, C. Y. (1970). *Biochem. Biophys. Res. Commun.* **41**, 339.

Neelin, J. M. (1963). *Can. J. Biochem. Physiol.* **41**, 1073.

Nikol'skaya, I. I., and Budowski, E. I. (1962). *Vopr. Med. Khim.* **8**, 73.

Paik, W. K., and Kim, S. (1967). *Biochim. Biophys. Acta* **139**, 49.

Parnas, I., and Russell, F. E. (1967). *In* "Animal Toxins" (F. E. Russell and P. R. Saunders, eds.), pp. 401–415. Pergamon, Oxford.

Peng, M. T. (1951). *Mem. Fac. Med., Nat. Taiwan Univ.* **1**, 200.

Pfleiderer, G., and Sumyk, G. (1961). *Biochim. Biophys. Acta* **51**, 482.

Ramsey, H. W., Snyder, G. K., Kitchen, H., and Taylor, W. J. (1972). *Toxicon* **10**, 67.

Raudonat, H. W., and Holler, B. (1958). *Naunyn-Schmiedebergs Arch. Exp. Pathol. Pharmakol.* **233**, 431.

Rechnic, J., Trachtenberg, P., Casper, J., Moroz, C., and de Vries, A. (1962). *Blood* **20**, 735.

Reid, H. A., Chan, K. E., and Thean, P. C. (1963). *Lancet* **1**, 621.

Richards, G. M., Du Vair, G., and Laskowski, M. (1965). *Biochemistry* **4**, 501.

Rocha e Silva, M., Beraldo, W. T., and Rosenfeld, G. (1949). *Amer. J. Physiol.* **156**, 261.

Rosenberg, P. (1971). *In* "Neuropoisons: Their Pathophysiological Actions. (L. L. Simpson, ed.), Vol. 1, pp. 111–137. Plenum, New York.

Rosenberg, P., and Dettbarn, W. D. (1964). *Biochem. Pharmacol.* **13**, 1157.

Rosenfeld, G., Hampe, O. G., and Kelen, E. M. (1959). *Mem. Inst. Butantan, Sao Paulo* **29**, 143.

Rothschild, A. M. (1967). *Experientia* **23**, 741.

Russell, F. E. (1960). *Amer. J. Med. Sci.* **239**, 1.

Russell, F. E. (1961). *Toxicon* **3**, 438.

Russell, F. E. (1965). *Toxicon* **2**, 277.

Russell, F. E. (1967). *Clin. Pharmacol. Ther.* **8**, 849.

Russell, F. E., and Bohr, V. C. (1962). *Toxicol. Appl. Pharmacol.* **4**, 165.

Russell, F. E., and Lauritzen, L. (1966). *Trans. Roy. Soc. Trop. Med. Hyg.* **60**, 797.

Russell, F. E., and Puffer, H. W. (1970). *Clin. Toxicol.* **3**, 433.

Russell, F. E., and Scharffenberg, R. S. (1964). "Bibliography of Snake Venoms and Venomous Snakes." Bibliographic Assoc., West Covina, California.

Russell, F. E., and Buess, F. W., and Strassberg, J. (1962). *Toxicon* **1**, 5.

Russell, F. E., Buess, F. W., and Woo, M. Y. (1963a). *Toxicon* **1**, 99.

Russell, F. E., Buess, F. W., Woo, M. L., and Eventov, R. (1963b). *Toxicon* **1**, 229.

Russell, F. E., Timmerman, W. F., and Meadows, P. E. (1970). *Toxicon* **8**, 63.

Saito, K., and Hanahan, D. J. (1962). *Biochemistry* **1**, 521.

Sarkar, N. K., and Devi, A. (1968). *In* "Venomous Animals and Their Venoms" (W. Bücherl, E. E. Buckley, and V. Deulofeu, eds.), Vol. 1, pp. 167–216. Academic Press, New York.

Sasaki, T. (1957). *J. Pharmacol. Soc. Jap.* **77**, 848.

Sato, S., and Tamiya, N. (1971). *Biochem. J.* **122**, 453.

Singer, T. P., and Kearney, E. B. (1950). *Arch. Biochem.* **29**, 190.

Sinsheimer, R. L., and Koerner, J. F. (1952). *J. Biol. Chem.* **198**, 293.

Slotta, K., and Ballester, A. (1954). *Mem. Inst. Butantan, Sao Paulo* **26**, 311.

Slotta, K., and Borchert, P. (1954). *Mem. Inst. Butantan, Sao Paulo* **26**, 297.

Slotta, K., and Frankel-Conrat, H. (1938). *Nature (London)* **142**, 213.

Slotta, K., and Vick, J. A. (1969). *Toxicon* **6**, 167.

Slotta, K., Gonzalez, J. D., and Roth, S. C. (1967). *In* "Animal Toxins" (F. E. Russell and P. R. Saunders, eds.), pp. 369–377. Pergamon, Oxford.

Stanić, M. (1969). *Toxicon* **6**, 287.

Strydom, A. J. C., and Botes, D. P. (1971). *J. Biol. Chem.* **246**, 1341.

Su, C., Chang, C. C., and Lee, C. Y. (1967). *In* "Animal Toxins" (F. E. Russell and P. R. Saunders, eds.), pp. 259–267. Pergamon, Oxford.

Sulkowski, E., Björk, W., and Laskowski, M., Sr. (1963). *J. Biol. Chem.* **238**, 2477.

Suzuki, T. (1968). *Mem. Inst. Butantan, Sao Paulo* **33**, 389.

Suzuki, T., Iwanaga, S., and Kawachi, S. (1958). *J. Pharmacol. Soc. Jap.* **78**, 568.

Suzuki, T., Iizuka, K., and Murata, Y. (1960). *J. Pharmacol. Soc. Jap.* **80**, 868.

Taborda, A. R., Taborda, L. C., Williams, J. N., Jr., and Elvehjem, C. A. (1952a). *J. Biol. Chem.* **194**, 227.

Taborda, A. R., Taborda, L. C., Williams, J. N., Jr., and Elvehjem, C. A. (1952b). *J. Biol. Chem.* **195**, 207.

Tamiya, N., and Arai, H. (1966). *Biochem. J.* **99**, 624.

Tamiya, N., Arai, H., and Sato, S. (1967). *In* "Animal Toxins" (F. E. Russell and P. R. Saunders, eds.), pp. 249–258. Pergamon, Oxford.

Tsai, F. F. (1961). *Fukuoka Acta Med.* **52**, 47.

Tu, A. T. (1971). *In* "Neuropoisons: Their Pathophysiological Actions" (L. L. Simpson, ed.), pp. 87–109. Plenum, New York.

Tu, A. T., and Chua, A. (1966). *Comp. Biochem. Physiol.* **17**, 297.

Tu, A. T., Passey, R. B., and Tu, T. (1966). *Toxicon* **4**, 59.

Tu, A. T., Toom, P. M., and Murdock, D. S. (1967). *In* "Animal Toxins" (F. E. Russell and P. R. Saunders, eds.), pp. 351–362. Pergamon, Oxford.

Tu, A. T., Hong, B., and Solie, T. N. (1971). *Biochemistry* **10**, 1295.

Vasilenko, S. K. (1963). *Biokhimiya* **28**, 602.

Vick, J. A., Ciuchta, H. P., and Manthei, J. H. (1967). *In* "Animal Toxins" (F. E. Russell and P. R. Saunders, eds.), pp. 269–282. Pergamon, Oxford.

Vidal Breard, J. J., and Elias, V. E. (1950). *Arch. Farm. Bioquim. Tucuman* **5**, 77.

Vital Brazil, O., Farina, R., Yoshida, L., and de Oliveira, A. V. (1968). *Mem. Inst. Butantan Sao Paulo* **33**, 993.

Wagner, F. W., and Prescott, J. M. (1966). *Comp. Biochem. Physiol.* **17**, 191.

Wakui, K., and Kawachi, S. (1959). *J. Pharmacol. Soc. Jap.* **79**, 1177.

Wellner, D., and Hayes, M. B. (1968). *Ann. N.Y. Acad. Sci.* **151**, 118.

Wellner, D., and Meister, A. (1960). *J. Biol. Chem.* **235**, 2013.

Wells, M. A. (1971). *Biochemistry* **10**, 4074.

Wells, M. A., and Hanahan, D. J. (1969). *Biochemistry* **8**, 414.

Williams, E. J., Sung, S. C., and Laskowski, M. (1960). *J. Biol. Chem.* **236**, 1130.

World Health Organization. (1971). *World Health Org., Tech. Rep. Ser.* **463**, Annex 1.

Yang, H. C. (1960). *Yamaguchi Igaku* **9**, 862.

Yang, C. C. (1965). *Toxicon* **3**, 19.

Yang, C. C., Chang, C. C., Hamaguchi, K., Ikeda, K., Hayashi, K., and Suzuki, T. (1967). *J. Biochem.* (*Tokyo*) **61**, 272.

Yang, C. C., Chang, C. C., Hayashi, K., Suzuki, T., Ikeda, K., and Hamaguchi, K. (1968). *Biochim. Biophys. Acta* **168**, 373.

Yang, C. C., Yang, H. J., and Huang, J. S. (1969). *Biochim. Biophys. Acta* **188**, 65.

Zeller, E. A. (1947). *Experientia* **3**, 375.

Zeller, E. A. (1948). *Advan. Enzymol.* **8**, 459–479.

Zeller, E. A. (1950). *Helv. Chim. Acta* **33**, 821.

Zeller, E. A. (1951). *In* "The Enzymes" (J. B. Sumner and K. Myrback, eds.), Vol. 1, Part 2, pp. 986–1013. Academic Press, New York.

Ziegler, F. D., Colon, L. V., Elliott, W. B., Gans, C., and Taub, A. (1967). *In* "Animal Toxins" (F. E. Russell and P. R. Saunders, eds.), pp. 236–243. Pergamon, Oxford.

Zwisler, O. (1965). *Hoppe-Seyler's Z. Physiol. Chem.* **343**, 178.

Author Index

Numbers in italics refer to the pages on which the complete references are listed.

A

Abbrecht, P. H., 42, *46*
Abdel-Messeih, G., 194, *215,* 365, *375*
Abe, H., 18, 19, *19*
Abel, J. J., 167, 172, *180*
Abrahamson, Y., 255, 257, 258, *270*
Abramoff, P., 204, *213*
Abramson, R. K., 82, *118*
Acher, R., 85, 103, 104, 105, 106, 108, *118, 119,* 141, *152,* 393, *396*
Adams, P. H., 30, 39, 40, *47*
Adolph, E. F., 51, 52, 54, *64,* 251, *270*
Adolph, P. E., 51, *64*
Afroz, H., 363, *374*
Agee, J. R., 82, 86, 96, 101, 103, *121,* 378, 393, *397*
Aggarwal, S. J., 86, 87, 88, 94, 96, 97, 100, 102, 114, *118*
Agid, R., 358, *374*
Albuquerque, E. X., 167, 176, *180*
Aleksiuk, M., 371, *374*
Algard, F. T., 115, *119*
Algauhari, A. E. I., 261, *270*
Ali, S. S., 363, *374*
Allen, E. R., 250, 252, *273*
Allen, J. E., 61, *65*
Allen, R., 173, *183*
Allen, R. F., 252, *270*
Allen, W. B., 250, 261, *270*
Allfrey, V. G., 14, *20*
Aloj, S., 439, *447*
Alonso, D., 38, 41, *46, 47*
Aloof-Hirsch, S., 467, *474*
Altamirano, M., 45, *46*
Altland, P. D., 80, *118,* 378, 394, *396*
Altman, P. L., 246, *246*
Alvarado, R. H., 54, *64,* 141, *152*
Amakasu, O., 167, *182*
Ambrosius, H., 201, 202, *213*
Amiconi, G., 89, 91, 99, 100, 104, 107, *118*
Amimoto, K., 71, 75, *76,* 338, 339, 347, *350*
Amin, A., 192, *215*
Amsterdam, A., 266, *270*
Anastasi, A., 174, 175, 178, 179, *180, 181*
Ancel, P., 7, *19*
Anderson, C. G., 457, 469, 470, *474*
Anderson, N. G., 253, 254, 261, 262, *270*
Andreoletti, G. E., 139, *158*
Angeletti, R. H., 467, *474*
Angelini, F., 417, *445*
Anthony, J., 251, *270*
Anthony, W. C., *182*
Antonini, E., 79, 89, 91, 92, 94, 95, 99, 100, 102, 103, 104, 106, 107, *118, 119,* 385, *396*
Anzalone, M. R., 129, *156*
Apell, G., 82, *120, 121*
Arai, H., 455, *478*
Arakawa, W., 84, 86, 88, 89, *120*
Araki, T., 95, 97, 107, *118*
Arata, H., 348, *351*
Aravindakshan, I., 467, 471, *474*
Armstrong, P., 6, *21*
Arnberg, H., 457, *475*
Arnone, A., 107, *118*
Ashley, H., 144, *152*
Atkinson, B. G., 86, 88, 115, 116, *120*
Attleberger, M. H., 200, *214*
Auerbach, R., 205, *216*
Auffenberg, W., 198, *213,* 250, *270*
Augustinsson, K., 211, *213*
Austin, S., 213, *216*
Avery, R. A., 363, *374*
Axelrod, J., 129, *159,* 167, *182*
Ayaki, Y., 343, *350*
Azari, P. R., 196, *213*

B

Bachmayer, H., 167, 179, *182*
Baglioni, C., 86, 87, 88, *118*
Bagnara, J. T., 129, 150, 151, *152*
Bailey, D., 379, 393, *397*
Baillien, M., 263, 268, 269, *270*
Bajandas, F. J., 36, 37, 42, *46*, *48*
Baker, C. M. A., 85, 101, *121*, 198, *215*
Baker, P. C., 150, *152*
Balagura, S., 300, *334*
Baldwin, E., 52, *64*
Baldwin, R., 58, *65*
Baldwin, T. O., 105, 106, 108, *118*
Balinsky, B. I., 6, 8, *19*
Balinsky, J. B., 52, *64*, 144, *152*
Ballester, A., 469, *477*
Ballmer, G. W., 252, *270*
Baltac, M., 406, *448*
Baltus, E., 4, 9, *19*
Bani, G., 148, *159*
Banks, B. E. C., 467, *474*
Bannister, W. H., 38, *46*
Banthorpe, D. V., 467, *474*
Barber, A. A., 190, 197, 198, 213, *213*, *216*
Baril, E. F., 189, 193, 194, *213*
Barker, S. B., 139, 143, *158*
Barnard, E. A., 256, 257, 258, 259, 260, *270*, *271*, *273*, *275*
Barne, M., 469, *474*
Barr, W. A., 136, 148, *152*
Bartel, A. H., 189, 192, 193, 194, 205, *213*, *215*, *216*
Bartels, H., 90, 91, 95, *119*
Barth, L. G., 5, 8, *19*
Barth, L. J., 5, 8, *19*
Bartholomew, G. A., 354, 370, *374*
Barton, A. J., 250, 261, *270*
Barwick, R. E., 218, 241, *246*
Bascom, W. D., 321, *336*
Bashirelahi, A., 30, 36, *48*
Battaglia, P. A., 99, *121*
Baumgardner, F. W., 93, *120*, 396, *397*
Bayne, C. G., 407, 414, 415, 424, 427, *444*
Baze, W. B., 300, 302, 303, 307, *334*
Beatty, B. R., 13, 14, *21*
Beaven, G. H., 81, *120*
Becker, R. O., 111, *120*
Beechwood, E. C., 304, *334*
Behleem, Z., 357, *376*
Behringer, H., 69, *76*, 172, *183*
Belkin, D. A., 261, *270*
Bellairs, A., 250, 261, *270*, 450, *474*
Bellamy, D., 140, *153*, 321, *336*, 429, *447*
Benbassat, J., 113, 115, 116, *118*
Benedict, F. G., 249, 261, *270*, 354, *374*
Benesch, R., 107, *118*
Benesch, R. E., 107, *118*
Benraad, T. J., 136, 137, *156*
Bentley, P. J., 51, 53, 54, 56, 57, *64*, 140, 141, 143, *152*, 251, 270, *270*, *274*, 278, 282, 288, 309, 310, 311, *334*, *336*, 354, 357, 365, 366, 368, *374*, *376*, 431, *444*
Bentzel, C. J., 45, *48*
Beraldo, W. T., 467, *477*
Berchtold, J.-P., 135, *152*
Berger, A., 467, *474*
Bergerhoff, K., 135, 138, *155*, *156*
Bergström, S., 341, *350*
Berkowitz, J. M., 45, *46*
Berman, R. H., 146, *152*
Bern, H. A., 123, 133, 146, 147, *152*, *157*, *159*, 252, 253, *272*, 403, 409, 410, 411, *444*, *447*
Berndt, W. O., 62, *64*, 304, *334*
Berridge, M. J., 27, *48*
Berry, A. R., 467, *474*
Berséus, O., 343, *350*
Bertaccini, C., 174, 179, *180*, *181*
Bertini, F., 83, 84, *118*
Bertrand, G., 170, *182*
Bessler, H., 467, *476*
Betke, K., 113, *118*, *120*
Betsuki, S., 69, 72, 75, *76*
Bewley, T. A., 403, *444*
Bharucha, M., 172, *180*
Bhide, N. K., 150, *155*
Bhown, A. S., 85, *118*
Bianchi, A., 269, *272*, *273*
Biber, T. U. L., 62, 63, *64*
Biemann, K., 175, *180*
Binyon, E. J., 357, 358, 361, 365, *374*
Birnstiel, M. L., 12, 13, 14, 18, *19*, *20*, *21*, *22*

Bishoff, G., 455, *476*
Bishop, W. R., 55, 57, *64, 65*
Bitensky, M. W., 150, *152*
Björk, W., 469, *474, 477*
Blackler, A. W., 11, 17, *19, 20*
Blackwell, R. Q., 108, *118*
Blain, A. W., 252, 261, *270*
Blaschko, H., 174, *180*
Blatt, W., 209, *215*
Bloch, E., 253, *274*
Blödorn, H. K., 164, *182*
Bloom, B., 267, *270*
Blumberg, B. S., 212, *214*
Bock, A. V., 388, *396*
Boell, E. J., 8, *19, 22*
Bogdanski, D. F., 174, *182*
Bogenschütz, H., 150, *152*
Bogert, C. H., 353, *374*
Bogert, C. M., 450, *474*
Bogert, C. W., 187, *213*
Bogoroch, R., 60, *64, 65*
Bohr, V. C., 467, *477*
Bommer, P., 175, *180*
Bonaventura, J., 107, 108, *118*
Bond, G. C., 201, 202, *213*
Bondi, K., 93, *120*, 396, *397*
Bonilla, C. A., 463, *474*
Bonting, S. L., 262, *271, 272*
Boos, J. O., 195, 198, 200, *214*, 383, *397*
Booth, B., 200, *214*
Boquet, P., 201, 205, *215*, 253, 255, *270*, 357, *375*, 461, 466, *474*
Borchert, P., 471, *477*
Borst, P., 10, *19*
Borut, A., 303, 313, 315, 330, *336*
Borysenko, M., 205, *213*
Boschwitz, D., 132, *152*
Bossert, W. H., 295, *334*
Botes, D. P., 455, 457, 460, 461, 462, *474, 477*
Botte, V., 136, *153*
Bouquegneaux-Tarte, C., 364, *375*
Bourne, G. H., 6, 7, *20*
Bouver, N. G., 82, *121*
Bovet, D., 470, *474*
Bovet-Nitti, F., 470, *474*
Bowman, H. G., 469, *474*
Bowman, R. I., 250, *274*
Bownds, D., 102, *121*
Boyd, E. M., 82, *120*
Brace, K. C., 80, *118*, 378, *396*
Brachet, J., 4, 8, 9, 10, 17, *19, 20*
Bradford, N. M., 27, *46*
Bradshaw, S. D., 212, *213*, 368, *374*, 402, 430, 432, 433, 434, 435, 436, 437, *444, 446*
Brady, K. L., 403, *445*
Brady, T. G., 38, *46*
Braganca, B. M., 466, 467, 471, *474*
Bragdon, D. E., 425, *444*, 469, *474*
Bragg, A. N., 7, *19*
Brambel, C. E., 210, *213*, 361, *374*
Brame, A. H., 90, *118*
Brandt, E. M., 111, 112, *119*
Bratcher, J. B., 423, *444*
Braun, W., 164, *182*
Braysher, M., 308, *334*
Brazaitis, P., 250, *270*
Brazda, F. G., 358, *374*
Brazil, V., 210, *213*
Breder, R. B., 250, *270*
Breuer, H., 136, *152*
Bricker, N. S., 307, *335*
Bridgwater, R. J., 343, *350*
Brimhall, R., 82, *118*
Brinkhous, K. M., 210, *216*
Brockman, H. L., 250, *273*
Brocq-Rousseu, D., 189, *213*
Brodsky, W. A., 307, *334, 335, 336*
Brooks, G. R., 83, *119*
Brooks, G. T., 83, 115, *120*
Brown, C. R. Jr., 213, *216*
Brown, D. D., 12, 13, 14, 15, 17, 18, 19, *19, 20, 21, 22*
Brown, E. R., 90, 91, *118*
Brown, G. S., 233, *246*
Brown, L. E., 84, *118*
Brown, P. S., 146, 147, *152, 153, 154*
Brown, R. D., 17, *19*
Brown, S. C., 147, *152*
Broyles, R. H., 115, *118*
Brunn, F., 56, *64*
Bruno, R., 139, *158*
Brunori, M., 79, 89, 91, 92, 94, 95, 99, 100, 104, 106, 107, *118*, 385, *396*
Bryan, J. A., 212, *215*, 255, 257, 258, *273*

Bryant, C., 218, 241, *246*
Bryant, R. E., 200, *214*
Buchwald, H. D., 167, *182*
Buckingham, M. B., 440, *444*
Budowski, E. I., 469, *476*
Büchmann, N. B., 127, *153*
Buess, F. W., 454, 467, 469, 470, 472, *477*
Bunn, H. F., 107, *118*
Bunt, A. H., 135, *153*
Burdick, C., 8, *21*
Burgers, A. C. J., 436, *444*
Burgess, W. W., 286, *334*
Burgos, M. H., 135, *157*
Burns, T. A., 378, 389, *397*
Burnstein, S. R., 150, *152*
Bustard, H. R., 261, *270*
Butcher, R. W., 150, *158*
Butler, C. F., 45, *48*
Butler, D. G., 431, 432, 433, 436, *444*

C

Caffin, J. P., 103, *118*
Cagle, F. R., 354, *374*
Caldwell, P. C., 39, *46*
Callan, H. G., 13, 14, 15, *19*
Callard, I. P., 136, *153*, 402, 407, 408, 414, 415, 424, 427, 432, 434, 436, *444*
Campbell, K. N., 252, 261, *270*
Campbell, M. L., 359, *374*
Camus, J., 266, *274*
Capraro, V., 269, *273*
Carballeira, A., 134, *156*
Cardeza, A. F., 407, *444*
Carmena, A. O., 80, 114, *118*, *119*
Carmichael, E. B., 358, 359, 360, *374*
Caro, L. G., 266, *271*
Carr, A., 250, *271*
Carta, S., 85, 94, 95, 100, 102, 103, 106, *118*, *122*
Carver, F. J., 115, *121*
Casper, J., 467, *475*, *477*
Cei, J. M., 174, 175, *180*, *181*
Cereijido, M., 63, *65*
Cha, W. K., 261, *273*
Chadwick, A., 436, *446*
Chaikoff, I. L., 267, *270*
Chalumeau-Le Foulgot, M. T., 136, *154*
Chan, D. K. O., 140, *153*, 408, 432, 434, 436, *444*
Chan, K. E., 467, *477*
Chan, S. K., 138, *153*
Chan, S. T. H., 127, 134, *153*
Chan, S. W. C., 134, *153*
Chandler, C., 30, 36, *48*
Chang, C. C., 255, *271*, 458, 459, 460, 461, 462, *474*, *476*, *477*, *478*
Channing, C. P., 423, *444*
Chase, J. W., 8, 10, 11, 12, 19, *20*
Chasis, H., 285, *336*
Chatterjee, A., 467, *475*
Chauvet, J. P., 85, 103, 104, 105, 106, 108, *118*, *119*, 141, *152*, 393, *396*
Chauvet, M. T., 141, *152*
Chen, H-C., *474*
Chen, L. C., 61, *65*
Chernoff, A., 379, 393, *397*
Chesley, L. C., 255, 256, 257, 258, *271*
Chester Jones, I., 140, 143, *153*, 312, 314, 318, 321, *335*, *336*, 402, 408, 409, 413, 430, 432, 434, 435, 436, *444*, *447*, *448*
Chiancone, B., 102, 103, 104, *119*
Chieffi, G., 130, 136, *153*, *157*, 424, *444*, *446*
Chiu, K. W., 406, 425, 426, *444*, *446*
Chiu, P. J. S., 460, *476*
Chiu, T. H., 460, *476*
Chopra, R. N., 469, 470, *475*
Choritz, E. L., 144, *152*
Chowdhurry, D. K., 255, *272*, 455, *475*
Christensen, C. U., 141, *153*
Christensen, P. A., 457, 467, 469, 470, *474*
Christomanos, A. A., 85, 102, 103, *119*, 394, *396*
Christophe, J., 266, *274*
Chua, A., 255, *274*, 469, *478*
Cieslak, E. S., 408, 413, *444*
Ciuchta, H. P., 467, *478*
Civan, M. M., 62, *64*
Clark, L. C., 24, *48*
Clark, N. B., 148, *153*, 205, 207, *213*, 298, 299, *334*
Clark, W. C., 195, *213*
Clarkson, T. W., 61, *65*

Clem, L. W., 201, 202, *216*
Clerici, P., 139, *158*
Cline, M. J., 115, *119*, 378, *396*
Coates, M. L., 98, 104, 108, 109, 110, *119*
Coe, C. G. L., 144, *152*
Cohen, E., 192, 194, *213*, *214*
Cohen, P. P., 138, *153*, 233, *246*
Cohen, S., 467, *474*, *476*
Colby, R. W., 213, *214*
Cole, C. J., 425, *444*
Cole, L. C., 354, *374*
Coleman, R., 132, 133, *153*
Collings, B. G., 339, *350*
Colon, L. V., *478*
Combescot, C., 413, 415, 425, *444*, *445*
Compher, M. K., 126, *153*
Condrea, E., 253, *271*, 466, 467, 469, 470, 471, *474*
Conte, F. P., 252, *272*
Conway, E. J., 38, *46*
Copeland, D. L., 137, *153*
Copp, D. H., 131, *153*
Cortelyou, J. R., 132, *153*, *156*, 299, *335*
Cosenza, L., 417, *445*
Cotlove, E., 29, *46*
Cott, H. B., 250, *271*
Couch, J. R., 213, *214*
Coulson, R. A., 188, 210, 211, *214*, 218, 219, 220, 221, 222, 223, 224, 225, 226, 227, 228, 229, 230, 231, 232, 233, 234, 235, 236, 237, 239, 240, 242, 243, *246*, *247*, 286, 298, 299, 302, 303, *334*, 356, 358, 367, 371, *374*, 406, 407, *445*
Covelli, I., 439, *447*
Cowles, R. B., 353, *374*
Cox, J. M., 105, 106, *121*
Crabbé, J., 60, 63, *64*, *65*, 142, *153*
Cragg, M., 52, *64*
Crane, E. E., 38, 43, *46*
Crane, R. K., 267, *271*
Crawford, E., 303, 313, 315, 330, *336*
Cremaschi, D., 269, *273*
Crenshaw, J. W., Jr., 85, *119*, 189, 192, 194, 195, *214*, *215*, 379, *397* *214*
Crippa, M., 15, 17, *19*, *21*
Crockett, A. T. K., 463, *475*
Crowther, R. A., 106, *121*
Csaky, T. Z., 45, *48*
Curran, P. F., 44, *46*, 63, *64*
Curti, B., 468, *474*
Cuthbert, A. W., 61, *64*
Czopek, J., 80, *122*
Czordás, A., 180, *180*

D

Daly, J. W., 167, 175, 176, 177, 178, *180*, *182*
Dance, N., 81, *120*
Dandrifosse, G., 256, 257, 258, 263, 266, 267, *271*
Danielsson, H., 71, 73, *76*, 341, 343, *350*
D'Anna, T., 8, *19*
Dantzler, W. H., 278, 279, 281, 283, 285, 286, 287, 288, 289, 290, 291, 292, 293, 296, 297, 298, 300, 301, 302, 303, 304, 305, 306, 307, 308, 310, *334*, 372, *374*
Darnell, J. E., 12, *21*
Darwin, C. R., 53, *64*
Davenport, H. W., 38, 42, *46*, 262, *271*
Davidson, E., 14, *20*
Davies, H. G., 79, *119*, *122*
Davies, R. E., 27, 29, 30, 38, 39, 40, 43, 44, *46*, *47*
Davis, H. S., 467, *474*
Davis, J. O., 143, *155*
Davis, L. E., 279, 281, 283, 284, 286, 290, 294, 295, *334*, *336*
Davis, T. L., 30, 36, 37, 42, *46*, *48*
Davis, W. J., 150, *157*
Dawbin, W. H., 300, *335*
Dawid, I. B., 8, 9, 10, 11, 12, 13, 14, 15, 17, *19*, *20*, *21*, *22*
Dawson, W. R., 284, 294, 297, 313, 315, 330, *335*, *336*, 368, *376*, 389, 396
Dayhoff, M. O., 79, 101, 104, *119*
De, S. S., 455, *475*
Deamer, D. W., 6, *21*
Dean, P. D. G., 342, 350, *351*
De Caro, G., 178, *180*, *181*
Deetjen, P., 285, *334*
de Knecht, A., 145, *155*
De Kok, A., 468, 469, *474*

Del Campo, R. M., 450, *474*
Del Conte, E., 434, *445*
Della Corte, F., 417, *445*
Delrio, G., 424, *446*
Denis, H., 5, 12, 15, 17, 19, *20, 21, 22*
Dennis, W. H., 41, *48*
Denson, K. W. E., 467, *474*
Dent, J. N., 125, 126, *153*
de Oliveira, A. V., 467, *478*
De Pont, J. J. H. H. M., 262, *271, 272*
Derby, A., 145, 146, *153, 154*
Derechin, M., 256, 260, *271*
deRoos, R., 135, 137, *153, 157*, 435, 436, *445*
DeSalle, R., 9, *21*
Dessauer, H. C., 83, 84, 86, 89, *119*, 187, 188, 189, 190, 191, 192, 193, 194, 195, 196, 197, 198, 199, 200, 201, 205, 206, 207, 208, 209, 210, 211, 212, 213, *214, 215*, 218, 234, *247*, 261, *271*, 300, 301, *334*, 355, 356, 358, 363, 370, *374*, 377, 378, 379, 381, 383, 384, 393, 394, *396, 397*
Detrait, J., 201, 205, *215*, 357, *375*, 469, *474*
Dettbarn, W. D., 470, *477*
Deuchar, E. M., 8, *20*
Deulofeu, V., 350, *351*
Deutsch, H. F., 192, *214*
Devi, A., 466, 467, *474, 477*
Devis, R. J., 6, 8, *19*
de Vries, A., 253, *271*, 462, 466, 467, 469, 470, 471, *474, 475, 476, 477*
de Vries, J., 145, *155*
DeWitt, W., 90, 91, 103, 106, 111, 112, 113, 115, 116, *118, 119, 120*
Dhyse, F. G., 321, *335*
Diamond, J. M., 44, *46*, 294, 295, *334*
Dicker, S. E., 142, *153*
Diener, E., 193, 201, 202, 203, 204, 205, 213, *215*
Dierickx, K., 125, 127, *153*
Dietrich, H., 455, *476*
Dill, D. B., 388, 391, *396*
DiMaggio, A., 404, *445*
Dimitriades, A., 102, 103, *119*
Diniz, C. R., 467, *474*
Dittmer, D. S., 188, *214*, 246, *246*
Dixon, G. H., 80, *119*
Djaldetti, M., 467, *476*
Dobson, C., 201, 202, *216*
Dodd, J. M., 133, *153*, 414, 415, 425, 427, 441, 442, *445*
Dodd, M. H. I., 441, 442, *445*
Doe, R. P., 212, *216*
Doerr-Schott, J., 124, 130, 131, *153, 154*
Doery, H. M., 469, 472, *475*
Donaldson, E. M., 419, 423, *444, 446*
Doonan, S., 467, *474*
Dornfeld, E. J., 252, *272*
Dozy, A. M., 82, *120, 121*, 378, 393, 394, *396*
Dubnoff, J. W., 454, 463, *475, 477*
Dubois, M. P., 130, 131, *153*
Dufaure, J.-P., 413, *445*
Duguet, M., 393, *396*
Duguy, R., 357, 358, 360, 361, 363, 365, *374*, 378, *396*
Dujarric de la Rivière, R., 201, *214*
Dull, D. L., 178, *181*
Dumont, J. N., 6, *22*
Dunlop, D., 146, *157*
Dunson, W. A., 313, 314, 315, 316, 317, 323, 325, 326, 327, 328, 329, 330, *335, 336*, 357, 367, *374*
Dupourque, D., 379, 393, *397*
Durbin, R. P., 24, 27, 29, 30, 32, 33, 34, 35, 38, 39, 40, 41, 45, *46, 47, 49*
Dutt, P. K., 255, *272*
Du Vair, G., 469, *477*

E

Eaker, D., 457, *475*
Eaker, D. L., 455, 457, 461, 462, *475*
Ealey, E. H. M., 193, 201, 202, 203, 204, 205, 213, *215*
Eaton, J. E., 144, *154*
Edelman, I. S., 30, 35, 37, *47*, 60, 62, 63, *64, 65*, 142, *154*
Edgren, R. A., 208, *214*
Edlund, Y., 266, *271*
Edwards, B. R., 127, 134, *153*
Edwards, E., 142, *156*
Edwards, H. T., 388, 391, *396*

Edwards, J. A., 91, *119*
Efrati, P., 394, *396*
Eggena, P., 64, *64*
Ehrensing, R. H., 211, *214*
Eik-Nes, K. B., 136, *154*
Eisentraut, M., 354, *374*
Ekholm, R., 266, *271*
Elias, V. E., 469, 472, *478*
Eliassen, E., *64*
Elizondo, R. S., 292, 293, 297, *335*, 430, 431, 432, *445*, *446*
Elli, R., 102, 103, 104, *119*
Elliott, A. B., 142, *153*
Elliott, W. B., 450, 467, *475*, *476*, *478*
Elsdale, T. R., 18, *20*
Elvehjem, C. A., 469, *478*
Elzinger, M., 86, 87, 88, 103, *119*
Emerson, C. P., 18, *20*
Emilio, M. G., 93, *119*
Endean, R., 178, *180*, *181*
Endo, Y., 457, 458, *475*
Enemar, A., 124, 146, *154*, 410, *445*
Engelbert, V. E., 394, *397*
Epp, O., 79, *120*
Eppig, J. J., 133, *155*
Epple, A., 133, *154*
Erdös, E. G., 210, *214*
Ernst, T., 189, *215*
Erspamer, V., 171, 174, 175, 178, 179, *180*, *181*
Esser, W., 167, *181*
Essner, E. S., 5, *20*
Essvik, B., 146, *154*
Etkin, W., 124, 127, 132, 144, 145, 146, *153*, *154*
Evans, D., 14, *20*
Evans, D. H., 52, *65*
Evans, E. A., Jr., 172, *181*
Evans, E. E., 200, 203, 204, *214*
Evennett, P. J., 130, 133, 149, *153*, *154*, *158*
Eventov, R., 454, 470, *477*
Eyeson, K. N., 402, 409, 411, 413, 414, 415, 418, 426, *445*
Eyquem, A., 201, *214*

F

Fabing, H. D., 171, *181*
Fänge, R., 312, 313, 314, 318, *336*
Faiman, L., 422, *445*
Falck, B., 124, *154*
Fantl, P., 208, 210, *214*
Farer, L. S., 212, *214*
Farina, R., 467, *478*
Faust, E. G., 164, *181*
Faust, E. S., 171, *181*
Feeney, R. E., 196, *213*
Fellows, R. E., 403, *445*
Ferguson, F. M., 441, 442, *445*
Ferguson, G. W., 418, *445*
Ferreira, S. H., 467, *475*
Ferreri, E., 143, *154*
Ferri, S., 251, *271*
Ficq, A., 10, 17, 20
Fiero, M. K., 463, *474*
Fimognari, G. M., 62, *64*
Findley, T., Jr., 288, *336*
Fine, J., 201, *214*
Finland, M., 198, *215*
Fiorindo, R. P., 402, *447*
Fischberg, M., 18, *20*
Fischer, G. A., 454, *475*
Fischer, H. G., 167, *181*, *182*
Fischer, R. C., 373, *376*
Fitch, H. S., 250, 252, *271*
Fitzgerald, J. B., 463, *475*
Fitzsimons, V. F. M., 250, *271*
Flickinger, R. A., 4, 5, 6, 8, *20*, *21*
Flores, G., 111, 112, *119*
Florey, H. W., 252, 253, 254, 257, 261, 262, 263, *275*
Florkin, M., 355, 372, *374*
Flury, F., 178, *181*
Follenius, E., 124, *153*, *154*
Follett, B. K., 4, 5, 6, 7, *20*, *21*, 136, 140, *153*, *154*, *157*, 400, 422, *445*, *447*
Fontaine, Y.-A., 442, *445*
Fontaine-Bertrand, E., 212, *213*, 430, *444*
Forbes, M. S., 441, *445*
Forbes, T. R., 426, *445*
Ford, P. J., 12, 15, 17, 19, *20*
Formanek, H., 79, *120*
Forster, R. P., 285, 289, *335*, *336*
Forte, G. M., *47*
Forte, J. G., 23, 24, 26, 28, 30, 35, 38, 39, 40, 43, *46*, *47*, *49*
Fowler, J. A., 250, *271*

Fox, A. M., 254, 261, 267, 268, *271*
Fox, K. R., 43, *48*
Fox, W., 83, 84, 86, 89, *119*, 190, 191, 192, 194, 195, 196, 197, 198, 200, 205, 206, 207, 208, 210, 212, *214*, 370, *374*, 378, 379, 381, 383, 384, *396*
Foxon, G. E. H., 93, *119*
Frair, W., 192, 194, 201, 202, 203, *214*
Frank, H., 45, *47*
Fränkel-Conrat, H., 455, 462, *477*
Fraser, R., 379, 393, *397*
Frazier, H. S., 307, *335*
Frieden, E., 82, 85, 87, 88, 96, 101, 102, 111, 112, 115, *118*, *119*, *120*, *122*, 144, *152*, *154*, *156*
Friedmann, G. B., 115, *119*
Frye, B. E., 146, *153*, *154*
Fuhrmann, F. A., 141, *154*, 167, 178, *181*, *182*
Fuhrmann, G. J., 167, 178, *181*, *182*

G

Gabe, M., 251, 252, *271*, 429, 434, *445*
Gahlenbeck, H., 90, 91, 95, *119*
Gall, J., 14, *20*
Gallien, L., 136, *154*
Gans, C., 194, *214*, 251, *272*, 450, 467, *475*, *476*, *478*
Gardiki, V., 102, 103, *119*
Gatten, R. H., Jr., 83, *119*
Gatzy, J. T., 61, 62, *64*, *65*
Gelb, A. M., 267, *272*
Genze, J. J., 133, *157*
George, D. W., 190, 196, 198, 200, *214*
Geschwind, I. I., 420, 436, *445*
Gessner, O., 164, 167, *181*
Gessner, P. K., 171, *181*
Geyer, G., 134, *154*
Ghiretti-Magaldi, A., 8, *21*
Ghosh, B. N., 255, *272*, 455, 467, *475*
Giebisch, G., 296, 297, *335*
Gilbert, L. I., 4, 5, *20*
Gilbert, N. L., 205, *214*
Gilles-Baillien, M., 268, 269, *272*, 355, 356, 357, 358, 359, 360, 361, 362, 364, 366, 367, 370, 371, *374*, *375*
Gillespie, J. H., 85, *119*
Giordana, B., 269, *272*, *273*
Gist, D. H., 408, 433, 435, 436, *445*
Gitter, S., 462, 467, 471, *474*, *475*
Giuliani, A., 102, 103, 104, *119*
Giunta, C., 139, *158*
Glaeser, A., 174, *180*, *181*
Glaser, R., 250, *272*
Glass, T. G., Jr., 463, *475*
Glenn, W. G., 463, *475*
Glynn, I. M., 39, *47*
Goaman, L. C. G., 105, 106, *121*
Godse, D. O., 171, *181*
Gohmann, E., Jr., 30, 36, *48*
Goldman, J. M., 150, *154*
Goldring, W., 285, *336*
Goldstein, D. A., 44, *48*
Gona, A. G., 145, 146, *154*
Gonçalves, J. M., 462, *475*
Gonzalez, C. F., 307, *335*
Gonzalez, J. D., 467, *477*
Good, R. A., 200, *216*
Goodman, D. B. P., 61, *65*
Goodnight, C. J., 357, 358, 359, 360, *375*
Goos, H. J. T., 124, 125, 127, 145, *155*, *158*, *159*
Goossens, N., 127, *153*
Gorbman, A., 123, 125, 132, 133, *155*, *157*, 252, 253, *272*, 437, 438, 439, 440, 441, 442, *445*, *446*, *447*
Gordan, A. S., 197, *215*, 394, *397*
Gordon, M. S., 52, 55, *65*
Gorman, G. C., 192, 195, 198, 200, 212, 213, *214*, 383, *397*
Gottfried, H., 424, 425, 429, *445*
Gould, H., 12, *20*
Graham, J. D. P., 150, *155*
Graham, W. J., 210, *214*
Graham-Smith, G. S., 192, *215*
Grant, P., 4, *20*
Grant, W. C., 147, *155*
Grassé, P. P., 253, *272*
Grasso, J. A., 116, *119*
Graul, C., 137, *155*
Green, B., 308, *334*
Green, H. H., 307, *335*
Green, K., 143, *155*, *156*
Green, N. D., 30, *47*
Greenfield, P., 8, *19*

Greenwald, L., 53, *65*, 141, *152*
Greenwald, O. E., 391, *397*
Greer, J., 106, *121*
Gregg, J. R., 9, *20*
Grégoire, C., 208, *215*
Gregosek, J. C., 439, *446*
Greslin, J. G., 422, *445*
Grey, H. M., 201, 204, 205, *215*
Grignon, G., 409, *445*
Gross, P. R., 4, 5, 12, *20*, *21*
Grossman, M. I., 23, *47*
Grundhauser, W., 359, 364, 369, *376*
Grunstein, M., 13, *19*
Guardabassi, A., 139, *158*
Guest, G. M., 387, 391, *397*
Guibé, J., 251, 261, *272*, 353, 363, 368, *375*
Guimond, R. W., 93, *119*
Gulland, J. M., 469, *475*
Gupta, B. J., 125, *153*
Gupta, I., 150, *155*
Gurdon, J. B., 17, 18, *19*, *20*, *21*
Guttman, S. I., 84, 85, 89, *118*, *119*, 381, 382, *397*

H

Haaf, G., 165, *181*
Habermann, E., 471, *475*
Habermehl, G., 164, 165, 167, 177, 180, *180*, *181*, *183*
Hackett, E., 208, 210, *215*
Hadley, M. E., 129, 150, 151, *152*, *154*, *155*
Hadley, N. F., 378, 389, *397*
Hagenbüchle, O., 144, *159*
Haggag, G., 300, 301, 302, 303, *335*, 357, 358, 359, 360, 363, *375*
Hahn, W. E., 205, 207, *215*, 427, *445*
Hajjar, J. J., 270, *272*
Hall, F. G., 93, *120*
Hallman, L. F., 321, *335*
Halmagyi, D. F. J., 467, *475*
Hamada, K., 86, 87, 88, 94, 97, 102, 103, 115, *120*
Hamaguchi, K., 462, *478*
Hampe, O. G., 467, *477*
Hanahan, D. J., 470, 471, *477*, *478*
Handler, J. S., 62, *65*
Handovsky, H., 170, *181*
Hanke, W., 127, 134, 135, 137, 138, 139, 143, 148, *155*, *156*, *157*, *158*
Hann, C. S., 208, 209, 210, *215*
Hanocq-Quertier, J., 4, *19*
Hansen, T. D., 251, 262, *271*, *272*
Hanson, D., 93, *120*, 246, *247*
Hansson, H. P. J., 40, *47*
Hardy, R., 368, 372, *376*
Hargens, A. R., 308, *336*
Harley-Mason, J., 170, *181*
Harrington, D. B., 111, *120*
Harris, J. A., 115, *120*
Harris, J. B., 30, 35, 37, 38, 41, *46*, *47*
Hartree, A. S., 418, 419, 441, 442, *446*
Hartwig, Q. L., 190, 196, 197, 198, 200, *214*
Harumiya, K., 135, *159*
Harvey, A. M., 286, *334*
Haselkorn, R., 9, *21*
Haslewood, G. A. D., 67, 70, 75, *76*, 253, *272*, 337, 339, 340, 341, 342, 347, 348, 349, 350, *350*, *351*
Hatta, K., 413, *447*
Hawkins, J. R., 171, *181*
Hayakawa, S., 75, *76*
Hayashi, K., 457, 462, *476*, *478*
Hayashida, T., 410, 412, 413, *445*, *447*
Hayes, M. B., 468, 469, *478*
Hearing, V. J., 133, *155*
Hecht, E., 81, *120*
Hedenius, A., 101, 104, *122*, 393, *398*
Heinz, E., 24, 29, 30, 32, 33, 34, *47*
Helander, H. J., 266, *272*
Heller, A., 141, *152*
Heller, H., 56, *65*, 141, *155*
Hellerström, C., 133, *155*
Hellman, B., 133, *155*
Hemme, L., 130, *155*, *157*
Henderson, I. W., 140, *153*
Henderson, R. W., 251, 261, *272*
Hendrickson, W. A., 79, *120*
Hennig, W., 13, *19*
Herbert, J. D., 231, 234, 235, 236, 240, *247*
Herlant, M., 409, *445*
Hermansen, B., 138, *155*
Hernandez, T., 188, 210, 211, *214*, 218, 219, 220, 221, 222, 223, 224, 225, 226,

227, 228, 229, 230, 231, 232, 233, 234, 235, 236, 237, 239, 240, *246*, *247*, 286, 298, 299, 302, 303, *334*, 356, 358, 367, 371, *374*, 406, 407, *445*
Herner, A. E., 88, *120*
Hertz, R., 321, *335*
Hesse, G., 171, *183*
Hiatt, H. H., 299, *335*
Hildemann, W. H., 200, *215*
Hill, L., 300, *335*
Hillyard, S. D., 378, 389, *398*
Hirofuji, S., 68, 75, *76*
Hirschfeld, W. J., 197, *215*, 394, *397*
Hirschowitz, B. I., 42, *48*, *49*
Hisaw, F. L., 425, *444*
Hodgkin, A. L., 39, *46*
Hoff-Jørgensen, E., 9, *20*
Hoffman, J. F., 44, *48*
Hoffman, R. E., 62, *64*
Hofmann, A., 171, *182*
Hogben, C. A. M., 24, 28, 29, 30, 37, 38, 40, 43, *46*, *47*, *48*
Holcomb, C. M., 212, *215*
Hollan, S., 82, *118*
Holler, B., 467, *477*
Hollyfield, J. G., 113, 114, *120*
Holmes, R. S., 212, *215*
Holmes, W. N., 312, 314, 318, 319, 320, 321, 322, 324, 329, *335*, 430, 431, *444*, *445*
Holtfreter, J., 5, *20*
Hong, B., 458, *478*
Honjin, R., 6, *20*
Hope, J., 6, 7, *20*
Hope, W. C., 256, 259, 260, *270*
Hopping, A., 358, 361, *375*
Horne, F. R., 300, 302, 303, 308, *334*
Horner, G. J., 467, *475*
Horning, E. C., 347, *351*
Horning, M. G., 347, *351*
Horton, B., 379, 393, *397*
Hoshino, T., 170, *181*
Hoshita, T., 68, 69, 75, *76*, 338, 339, 344, 347, *350*, *351*
Howards, S. S., 143, *155*
Howell, B. J., 93, *120*, 396, *397*
Hoyer, H. E., 405, 407, 426, 434, *446*
Huang, D. P., 424, 425, 429, 435, *445*
Huang, J. S., 457, 459, *478*
Huber, R., 79, *120*
Hudson, C. L., 288, *336*
Huehns, E. R., 81, *120*
Hüttel, R., 69, *76*
Huggins, S. E., 205, *215*
Huisman, T. H. J., 79, 82, *120*, *121*, 378, 393, 394, *396*
Hull, E., 111, *121*
Humphreys, T., 18, *20*
Humphries, A. A., 6, 7, *20*
Hunter, N. W., 138, *155*
Hunter, R. B., 441, 442, *445*
Hurley, T. W., 403, *445*
Hussein, M. F., 255, 257, *272*
Hutchinson, J. H., 142, *155*
Hutchinson, V. H., 93, *119*
Hutton, K. E., 357, 358, 359, 360, *375*
Hviid-Larsen, E., 63, *65*

I

Iizuka, K., 469, *478*
Ikeda, K., 462, *478*
Ikuma, S., 167, *182*
Imamura, H., 348, 349, *351*
Inai, Y., 350, *351*
Ingram, V. M., 79, 80, 81, 86, 87, 88, 103, 106, 111, 112, 113, 114, 115, 116, *119*, *120*, *121*
Irvine, F. R., 250, *272*
Isenberg, M. T., 285, *336*
Ishaq, M., 363, *374*
Ishii, S., 457, 458, *475*
Ishii, T., 212, *216*, 438, *447*
Ito, S., 24, 25, 26, *47*, 266, *272*
Iturriza, F. C., 124, 125, 131, *154*, *155*
Iwanaga, S., 457, 469, 470, *475*, *476*, *477*
Iwato, M., 348, *351*
Iyengar, N. K., 469, 470, *475*
Izard, Y., 201, 205, *215*, 357, *375*, 461, *474*

J

Jackson, A. H., 171, *181*
Jackson, C. G., Jr., 212, *215*

Jackson, M. M., 212, *215*
Jacobs, D., 426, *447*
Jacobson, A., 43, *47*, *48*
Jacques, F. A., 209, *215*, 362, 370, *375*
Jaeger, H., 172, *180*
James, G. P., 255, *274*
Jamieson, J. D., 266, *272*
Jandl, J. H., 198, *215*
Janowitz, H. D., 45, *46*
Janssens, P. A., 138, 139, *155*
Jaques, R., 469, *475*
Jard, S., 141, *157*
Jared, D. W., 6, 7, *22*, 140, *159*
Jeauniaux, C., 255, 256, 257, 258, 263, *271*, *272*
Jenkins, N. K., 205, 208, *215*
Jensen, H., 171, 172, 179, *181*
Jiménez-Porras, J. M., 460, 461, 462, 463, 465, 466, 469, 470, *475*
Johansen, K., 93, 98, *120*, 246, *247*, 386, *397*
Johnson, C. E., 138, *155*
Johnson, D. R., 250, *272*
Johnson, S. R., 141, *152*
Johnson, T. S., 252, *272*
Johnston, C. I., 143, *155*
Joly, J., 136, *155*
Jones, D. R., 93, *120*
Jones, K. W., 12, 13, *19*, *21*
Jones, M. E., 233, 234, *247*
Jones, R. E., 404, 405, 407, 408, 415, 418, 427, *445*, *446*
Jones, R. T., 82, 104, 108, 109, 110, *118*
Jordan, E., 13, *19*
Jordan, H. E., 114, *120*
Jørgensen, C. B., *64*, *65*, 123, 124, 125, 127, 133, 138, 141, 142, 148, *153*, *154*, *155*, *156*
Josephson, B., 253, *272*
Joshua, H., 467, *476*
Jouannet, M., 461, *474*
Junqueira, L. C. U., 270, *272*, 308, *335*
Juratsch, C. E., 450, *475*
Jurd, R. D., 83, 86, 90, 91, 111, 113, 115, 116, *120*
Just, J. J., 86, 88, 111, 115, 116, *119*, *120*
Justus, J. T., 91, *119*

K

Kabara, J. J., 454, *475*
Kachkarov, D. N., 368, *375*
Kaiser, E., 466, 467, *475*
Kajita, A., 95, 97, 107, *118*
Kalas, J. P., 210, *216*
Kaletta, U., 469, *474*
Kaltenbach, J. C., 148, *153*
Kaneko-Mohammed, S., 30, *47*
Kanemitu, T., 338, 347, *351*
Kaplan, H. M., 204, 209, *215*, 359, *375*, 394, *397*
Kara, J., 469, *475*
Karasaki, S., 5, 6, 8, *20*, *22*
Karle, I. L., 175, 177, *180*, *182*
Karle, J., 175, *182*
Karlsson, E., 457, 461, *475*
Kasbekar, D. K., 23, 26, 30, 35, 40, 41, 43, *47*
Kashiwa, H. K., 43, *48*
Kastin, A. J., 149, *156*
Katorski, B., 358, 361, *376*
Katsuno, A., 193, *215*
Katti, P., 144, *152*
Kawachi, S., 469, 470, *475*, *477*, *478*
Kawamura, M., 167, *182*
Kaye, M. A. G., 469, *475*
Kaye, N., 144, *156*
Kayser, C., 355, *375*
Kaziro, K., 86, 87, 94, 97, 102, 103, *120*
Kazuno, T., 68, 69, 73, 75, *76*, 338, 339, 341, 347, *350*
Kearney, E. B., 469, *477*
Keegan, H. L., 467, *475*
Keesee, D. C., 36, 37, 42, *46*, *48*
Kehl, R., 425, 427, *445*
Kelen, E. M., 467, *477*
Kelly, D. E., 129, *159*
Kelly, H. M., 52, *64*
Kemp, N. E., 8, *20*
Kendrew, J. C., 105, *121*
Kennedy, J. P., 250, *272*, *273*
Kent, G. C., 423, *444*
Kent, S. P., 200, *214*
Kenyon, W. A., 254, 255, 256, 257, *273*
Kern, H., 133, *156*

Kerr, T., 129, *156*
Kessler, J. I., 267, *272*
Keynes, R. D., 39, *46*
Khalil, F., 194, *215*, 300, 301, 303, *335*, 357, 358, 359, 360, 363, 365, *375*
Kidder, G. W., III, 47
Kikuyama, S., 145, *154*
Kilmartin, J. V., 106, *121*
Kim, C. H., 338, 347, *351*
Kim, D. W., 261, *273*
Kim, I. Y., 261, *273*
Kim, S., 468, *476*
Kim, Y. S., 146, *154*
King, H. W. S., 12, *20*
Kirchberger, M. A., 61, *65*
Kirschmann, C., 471, *474*
Kiss, G., 179, *181*
Kitahara, S., 30, 43, *47, 48*
Kitchen, H., 462, *476*
Klahr, S., 307, *335*
Klang, R., 146, *154*
Klauber, L. M., 250, 251, 252, 261, *273*, 450, *476*
Kleemann, C. R., 142, *156*
Kleihauer, E. F., 82, 113, *118, 120, 121*
Klein, D., 164, *182*
Klibansky, C., 467, *476*
Klicka, J., 425, *446*
Klitansky, C., 471, *474*
Klose, R. M., 296, *335*
Kmetová, S., 192, *215*
Knight, E., 12, *21*
Knobil, E., 427, 434, *446*
Knowlton, G. F., 250, *273*
Knox, W. H., 431, 432, 433, 436, *444*
Kobayashi, H., 437, 438, *446*
Kobayashi, S., 251, *273*
Koch, B., 429, 434, 435, *446*
Koch, K., 164, 165, *182*
Kochva, E., 450, *476*
Kochwa, S., 462, 467, *475*
Koerner, J. F., 469, *477*
Koler, R. D., 82, *118*
Komadina, S., 279, 281, 289, 290, 293, *335*
Komatsubara, T., 75, *76*
Konijn, T. M., 150, *158*
Konz, W., 170, 171, *183*
Kornalik, F., 467, *476*
Korovine, E. P., 368, *375*
Kosmos, A., 79, *122*
Kowell, A. P., 251, 261, *273*
Krakauer, T., 194, *214*
Kramer, M. F., 266, *273*
Krauss, J., 201, 202, *216*
Krogh, A., 54, *65*
Kroon, A. M., 10, *19*
Krull, A. H., 171, *181*
Kruse, H., 40, *49*
Kupchella, C. E., 362, 370, *375*
Kurata, Y., 84, 86, 88, 89, 102, *120*
Kurfees, J. F., 36, 42, *48*
Kuriki, Y., 9, *21*
Kuroda, M., 348, *351*

L

Lagios, M. D., 251, 252, 266, *273*
La Grange, R. G., 467, *476*
Lamont, D. M., 467, *474*
Landesman, R., 12, *21*
Lane, C. D., 17, *20, 21*
Lange, R., 132, *156*
Langley, J. N., 257, *273*
Langridge, W. H. R., 17, *22*
Lannom, J. L., Sr., 418, *447*
Lanzavecchia, G., 6, *21*
La Pointe, J. L., 427, *446*
Larsen, L. O., 124, 127, 129, 148, *155, 156*
Larson, M. W., 368, *375*
Larson, P. R., *476*
Laskowski, M., 205, *215*, 469, 472, *477, 478*
Latif, S. A., 268, *273*, 366, *375*
Latifi, M., 192, *215*
Lauritzen, L., 467, *477*
LaVia, M. F., 204, *213*
Lazo-Wasem, E. A., 425, *444*
Leaf, A., 59, 60, 61, 62, *65*, 142, *158*
Leathem, J. H., 136, *153*, 424, *444*
LeBrie, S. J., 285, 292, 293, 297, *335*, 430, 431, 432, *445, 446*
Lee, C.-Y., 255, *271*, 457, 458, 459, 460, 461, 467, *474, 476, 477*
Lee, P., 303, 313, 315, 330, *336*
Lee, S. Y., 460, *476*
Lefevre, M. E., 30, 31, 39, *48*

Leffkowitz, M., 462, *475*
Leggio, T., 93, 99, *120*, *121*
Leist, K. H., 135, 138, 139, 143, *155*, *156*
Leloup-Hatey, J., 436, *446*
Lenfant, C., 93, 98, *120*, 246, *247*, 396, *397*
Leonard, R., 6, *21*
Leone, C. A., 194, *215*
Lerch, E. G., 205, *215*
Levedahl, B. H., 439, 440, *446*
Levi, G., 467, *475*
Levi-Montalcini, R., 467, *474*, *476*
Levine, R., 406, *447*
Lewis, J. H., 192, 201, *215*
Lewy, J. E., 309, *335*
Lhotka, J. F., Jr., 252, *270*
Li, C. H., 403, *444*
Liang, C., 194, 210, *215*
Licht, P., 147, *159*, 195, 198, 200, *214*, 258, *273*, 284, 294, 297, *336*, 370, *375*, 383, *397*, 402, 404, 405, 406, 407, 408, 409, 410, 411, 412, 413, 414, 415, 417, 418, 419, 420, 421, 422, 423, 426, 427, 430, 433, 434, 435, 436, 437, 438, 439, 440, 441, 442, *444*, *446*, *447*
Limlomwongse, L., 26, 35, 43, *47*
Linde, H., *182*
Lindstedt, S., 343, *350*
Lipmann, F., 233, 234, *247*
Lippe, C., 269, *272*, *273*
Lipton, P., 63, *65*
Lisk, R. D., 402, *446*
Littna, E., 12, 14, 15, 17, 18, *19*
Liu, C. S., 108, *118*
Lodi, G., 128, 148, *156*, *157*, *159*
Loening, U. E., 12, 13, *19*, *21*
Loeschke, K., 45, *48*
Lofts, B., 130, 136, 137, 148, *156*, *159*, 355, *375*, 424, 425, 426, 429, *444*, *445*, *446*
Logier, E. B. S., 250, *273*
Longmuir, N. M., 38, 44, *46*
Love, W. E., 79, *120*, *121*
Løvtrup, S., 9, *20*
Low, B. S., 131, *153*
Lowry, P. J., 436, *446*
Luck, D. J. L., 9, *21*
Lueth, F. X., 261, *273*
Lupo di Prisco, C., 136, *153*, 424, *446*
Luppa, H., 252, *273*
Lustig, B., 189, *215*
Luzzio, A. J., 195, *215*
Lykakis, J. J., 192, 194, 201, 204, *215*, 379, *397*
Lynn, W. G., 437, 439, 440, 441, *446*
Lyons, W. J., 467, *476*

M

Mabuti, H., 75, *76*
McBean, R. L., 314, 319, 320, 321, 322, 324, 329, *335*, 431, *445*
McCarty, K. S., 12, *21*
McClanahan, L., 58, *65*, 310, *335*
McConnell, W. F., 407, 414, 415, 424, 427, *444*
McCormick, J. J., 439, *446*
McCurdy, H. M., 115, *119*
McCutcheon, F. H., 78, 94, *120*, 384, *397*
McGeachin, R. L., 212, *215*, 255, 257, 258, *273*
McGill, D., 437, 438, 439, 440, *447*
MacGregor, H. C., 14, *21*
Macht, D. J., 167, 172, *180*
McKennee, C. T., 402, *447*
MacKnight, A. D. C., 62, *65*
Mackrell, T. N., 29, *48*
Maclean, N., 83, 86, 90, 91, 111, 113, 115, 116, *120*
McMahon, E. M., Jr., 115, *120*
McManus, J. F. A., 469, *474*
McMenamy, R. H., 194, *215*
McMullan, J. M., 171, *181*
McNabb, F. M. A., 303, *335*
McNabb, R., 139, 140, *156*
McShan, W. H., 192, *214*
McWhinnie, D. J., 132, *153*, *156*, 299, *335*
Maden, B. E. H., 12, *21*
Maderson, P. F. A., 406, 426, *446*
Maeno, H., 469, 473, *476*
Maetz, J., 55, 57, *65*, 140, 141, 143, *156*, *158*
Märki, F., 167, 171, 175, 176, *182*
Mager, J., 467, *474*

Maggio, R., 8, *21*
Mahdavi, V., 17, *21*
Maher, M. J., 255, 257, 258, *270*, 439, 440, *446*
Mahmoud, I. Y., 425, *446*
Mahy, B. W. J., 369, *375, 376*
Mairy, M., 12, 15, *20, 21*
Makhlouf, G. M., 24, 34, 37, 44, 45, *48, 49*
Maldonado, A. A., 192, *215*
Mallette, W. G., 463, *475*
Malnic, G., 270, *272*, 296, 308, *335*
Maniatis, G. M., 86, 88, 112, 113, 114, 115, *120, 121*
Manthei, J. H., 467, *478*
Manwell, C. P., 85, 86, 87, 88, 94, 101, *121*, 198, *215*, 381, 384, 391, 394, *397*
Mao, S. H., 196, 198, 199, *215*
Marbaix, G., 17, *20, 21*
Marchalonis, J. J., 193, 201, 202, 203, 204, 205, 213, *215*
Marchlewska-Koj, A., 83, 84, 85, 89, 91, *121*
Maren, T. H., 262, *273*, 296, *335*
Markel, R. A., 52, *65*
Marques, M., 407, *446*
Marshall, E. K., 286, *334*
Martin, C. M., 198, *215*
Martinez, R., 139, *156*
Martinoya, C., 45, *46*
Martoja, M., 434, *445*
Marusic, E., 139, *156*
Masat, R. J., 189, 193, 194, 196, 201, *215*, 359, 369, *375*
Massey, V., 468, *474*
Massover, W. H., 6, *21*
Master, R. W. P., 473, *476*
Masters, C. J., 212, *215*
Masui, T., 75, *76*
Masui, Y., 149, *156*
Matson, J. R., 354, *375*
Matty, A. J., 123, 132, 139, 143, *155, 156, 158*
Maumus, L. T., 84, *119*
Maung, R. T., 192, 201, 203, 204, 205, 213, *215*
Maxwell, G. R., 467, *475*
Mayhew, W. W., 354, 355, 356, 369, 370, 371, *375*
Mazzarella, L., 106, *121*
Mazzi, V., 126, 128, 129, 139, 143, 145, 147, 148, *154, 156, 157, 158, 159*, 403, *446*
Meadows, P. E., *477*
Meaume, J., 461, *474*
Mebs, D., 457, *476*
Medda, A. K., 144, *156*
Medica, P. A., 418, *447*
Mehdi, A. Z., 134, *156*
Meier, A., 147, *156*, 405, *446*
Meints, R. H., 111, 115, *121*
Meischer, P. A., 200, *216*
Meister, A., 468, *478*
Meldrum, B. S., 467, 471, *476*
Mendelsohn, D., 342, 350, *351*
Mendelsohn, L., 342, 350, *351*
Mennega, A. M. W., 254, *273*
Menzies, R. A., 205, 206, 207, 213, *216*
Mersmann, H. J., 363, *376*
Mertens, R., 250, *273*
Metcalfe, J., 98, 110, *119*
Meyer, D. E., 250, *273*
Meyer, K., 172, 173, 174, *180, 182*
Miahle, C., 429, 434, 435, *446*
Micha, J. C., 257, 258, *273*
Micheel, F., 455, *476*
Michels, N. A., 361, *375*
Michl, H., 167, 179, 180, *180, 181, 182*, 466, 467, 469, *475, 476*
Middler, S. A., 142, *156*
Miller, J. E., 307, *336*
Miller, L., 13, *21*
Miller, M. R., 133, *157*, 251, 252, 266, *273*, 434, *447*
Miller, O. L., 13, 14, *21*
Miller, S. L., 308, *336*
Milstead, W. W., 250, *274*
Minnich, J. E., 300, 301, 302, 303, 309, 310, *335*
Minton, M. R., 450, *476*
Minton, S. A. Jr., 450, 454, 467, *476*
Mira-Moser, F., 130, *157*
Mirsky, A. F., 14, *20*
Mitidieri, E., 213, *216*
Mitropoulos, K. A., 343, *351*
Mitsuhashi, S., 469, 473, *476*
Mittasch, H., 170, 171, *183*
Miwa, I., 210, *214*

Moar, V. A., 17, *21*
Moberly, W. R., 356, *375*, 388, 389, *398*
Möckel, W., 256, 257, 259, 260, *271*, *273*
Möllenhoff, P., 167, *181*
Mohiuddin, A., 261, *274*
Molzer, H., 469, *476*
Monge, C., 270, *272*, 308, *335*
Monier, R., 17, 19, *22*
Moody, A., 54, *64*
Moody, F. G., 38, *48*
Morel, F., 141, *157*
Morgareidge, K. R., 187, *215*
Morimoto, K., 69, *76*
Moroz, C., 462, 467, *476*, *477*
Morpurgo, G., 93, 99, *120*, *121*
Morray, J., 17, *22*
Morrison, L. R., 467, *476*
Morrison, P. R., 209, *215*
Morrissey, S. M., 23, 24, *48*
Mosbach, E. H., 344, 347, *351*
Mosher, H. S., 167, 178, *181*, *182*
Moss, B., 86, 87, 88, 103, 113, 115, 116, *121*
Motulsky, A. G., 81, *120*
Moyle, V., 300, *335*
Müller, O. W., 164, *182*
Mugaas, J., 432, *447*
Muirhead, H., 105, 106, *121*
Mukerji, B., 469, 470, *475*
Mumbach, M. W., 55, 56, 57, 59, *64*
Munday, K. A., 369, *375*, *376*
Munjal, D., 467, *476*
Munsick, R. A., 141, *157*, 286, 288, *335*, *336*
Murakawa, S., 193, *215*
Murata, Y., 469, *478*
Murdaugh, H. V., Jr., 307, *336*
Murdock, D. S., 466, 469, *478*
Murphy, R. C., 210, *215*
Murray, A. W., 173, *182*
Murrish, D. E., 270, *273*, 308, *335*, 432, *447*
Musacchia, X. J., 189, *215*, 252, 254, 261, 267, *271*, *273*, *275*, 359, 362, 369, *375*, *376*, 378, *397*
Musacchia, X. L., 359, 364, 369, *376*
Myant, N. B., 343, *351*
Myer, J. S., 251, 261, *273*
Myers, C. W., 177, 178, *180*
Myers, R. M., 55, 57, *65*

N

Nace, G. W., 5, *20*
Nakagawa, T., 68, 75, *76*
Nakai, K., 457, *476*
Nakai, Y., 125, *157*
Nakamura, M., 134, *157*
Nakamura, S., 193, *215*
Nakamura, T., 6, *20*
Nakanishi, M., 192, 195, 213, *214*
Nandi, J., 123, 133, *152*, 429, *447*
Narita, K., 457, 460, *476*
Nass, S., 8, *21*
Nauss, A. H., 43, *47*
Nedegaard, S., 40, *49*
Needham, J., 9, *19*
Neelin, J. M., 463, *476*
Neill, W. T., 250, 252, *273*, 317, *335*
Neisser, C., 171, *182*
Nelson, B. L., 7, *22*
Neumann, U., 127, 137, 138, *155*, *157*
Nevo, E., 83, *119*
Newcomer, R. J., 89, 91, *121*, 192, *215*, 379, *397*
Nicholls, T. J., 136, 140, *154*, *157*
Nichols, C. W., 146, *157*, 409, 410, 411, *447*
Nickerson, M. A., 90, 100, 104, *122*
Nickol, J. M., 7, *22*
Nicoll, C. S., 146, *152*, *157*, 402, 403, 409, 410, 411, 412, 413, *444*, *446*, *447*
Nicolson, G. L., 62, *65*
Nigon, K., 41, *47*
Nikander of Kolophon, 164, *182*
Nikol'skaya, I. I., 469, *476*
Nir, E., 394, *396*
Nitoma, C., 174, *182*
Norris, K. W., 313, 315, 330, *335*
Notenboom, C. D., 142, *157*
Novales, B. J., 149, 150, 151, *157*
Novales, R. R., 149, 150, 151, *157*
Noyes, D. H., 33, 34, *48*
Nussbaum, N., 440, *447*
Nute, P. E., 95, *121*

O

Ochsman, J. L., 27, *48*
O'Dor, R. K., 131, *153*
Oestlund, E., 167, *182*
Ogston, A. G., 38, *46*
Ohad, I., 266, *270*
Okada, S., 88, 89, 102, *120*, 350, *351*
Okasaki, R., 9, *21*
Okasaki, Y., 68, 75, *76*
Okonogi, T., 469, *476*
Okunda, K., 75, *76*, 347, *351*
Olcese, O., 213, *214*
Oliver, J. A., 250, *273*
Olivero, M., 139, *158*
Orlando, M., 90, 91, 92, 104, *121*
Orloff, J., 62, *65*
Ortiz, E., 192, *215*
Oshima, K., 125, *157*
Ostrom, J. H., 250, *273*
Ouyang, C., *474*
Overton, E., 54, *65*
Ozawa, H., 394, *397*
Ozcan, E., 467, *476*
Ozon, R., 136, *152*, *157*

P

Pacifico, A. D., 29, 44, *48*
Padlan, E. A., 79, *121*
Paganelli, C. V., 194, *214*
Paik, W. K., 468, *476*
Painter, E., 61, *64*
Palade, G. E., 266, *271*, *272*, *273*, *274*
Palmer, J. L., 189, 193, 194, *213*
Pandha, S. K., 413, *447*
Panigel, M., 415, 427, *447*
Panijel, J., 4, 6, *21*
Papkoff, H., 412, 413, 415, 417, 418, 419, 420, 421, 422, 423, 441, 442, *444*, *446*, *447*
Parkes, C. O., 131, *153*
Parks, H. F., 266, *273*
Parnas, I., 467, *476*
Parsons, J. A., 402, *447*
Parsons, T. S., 313, *336*
Pasini, C., 174, *181*
Passey, R. B., 466, *478*
Pasteels, J., 7, *21*
Pataski, S., 172, *182*
Patlak, C. S., 44, *48*
Paulov, S., 192, *215*
Pavlopulu, C., 394, *396*
Peaker, M., 250, *274*
Pearson, A. K., 402, 413, 414, 415, 417, 418, 419, 426, *446*
Pearson, D. D., 192, *215*, *216*
Pearson, J. E., 469, 472, *475*
Pearson, O. P., 245, *246*
Pehlemann, F. W., 125, 130, 135. 152, *155*, *156*, *157*
Pell, S. M., 250, *274*
Pène, J. J., 12, *21*
Peng, M. T., 459, *476*
Pengally, E. T., 373, *376*
Penhos, J. C., 406, *447*
Perkowska, E., 14, *21*
Perschmann, C., 300, 301, *336*
Perutz, M. F., 105, 106, *121*
Petcher, P. W., 358, 359, 360, *374*
Petersen, J. A., 386, *397*
Petrucci, D., 8, *19*, *21*
Peute, J., 124, 125, *157*, *159*
Peyer, J., 171, *182*
Peyrot, A., 126, 128, 129, 145, *156*, *157*, *159*
Pfleiderer, G., 469, *476*
Phillips, J. E., 312, 314, 318, *335*
Phillips, J. G., 134, 140, *153*, 321, *336*, 406, 424, 425, 426, 429, 430, 435, *444*, *445*, *446*, *447*
Philpot, V. B., 194, *216*
Phisalix, C., 170, *182*
Piezzi, R. S., 124, 135, *157*, *158*
Piper, G. D., 135, 138, *157*
Pitts, R. F., 232, *247*, 300, *334*, *336*
Platner, W. S., 369, *376*
Plenck, H., 249, 251, 252, *274*
Plinius, C., 164, *182*
Poort, C., 133, *157*, 266, *273*
Pope, C. H., 250, 255, 261, *274*
Porath, J., 457, 461, 462, *475*
Porter, G. A., 60, 62, 63, *64*, *65*, 142, *155*, *157*
Porter, K. R., 82, 107, *121*, 377, *397*
Porter, P., 302, 303, *336*
Posner, H. S., 174, *182*

Potts, W. T. W., 52, *65*
Pough, F. H., 196, 198, 200, 212, *214*, 361, *376*, 384, 385, 389, 391, *397*
Poulson, T. L., 303, *335*, 389, *396*
Prado, J. L., 213, *216*, 358, *376*, 385, *397*
Prange, H. D., 278, 282, 311, *336*
Prasad, M. R. N., 413, 415, 418, 425, 426, 427, *447*
Prescott, J. M., 469, *478*
Preston, A. S., 62, *65*
Preusser, H. J., 180, *181*
Price, R. P., 111, 112, *119*
Privitera, C. A., 363, *376*
Prosser, R. L., III, 207, *216*
Puffer, H. W., 467, *477*
Purdom, I. F., 13, *19*
Putnam, F. W., 187, 193, 197, *216*

Q

Quadri, M., 366, *376*
Quastel, J. H., 471, *474*
Quay, W. B., 129, 150, *157*
Quintana, G., 54, *65*

R

Rabinowitz, J., 342, 350, *351*
Rabinowitz, M., 9, *21*
Raheem, K. A., 357, 358, 359, 360, 363, *375*
Rahman, M. A., 357, *376*
Rahn, H., 93, *120*, 396, *397*
Rall, J. E., 212, *214*
Ramaswami, L. S., 426, 434, *447*
Ramirez, J. R., 83, 86, 89, *119*, 378, 379, 381, 383, 384, 393, *396*, *397*
Ramsey, H. J., 384, *397*
Ramsey, H. W., 462, *476*
Randall, E., 432, *447*
Ranges, H. A., 285, *336*
Rao, C. A. P., 191, 204, *216*
Rao, S. S., 473, *476*
Rapatz, G. L., 369, *376*
Rapola, J., 135, *157*
Rapoport, S., 387, 391, *397*
Rasmussen, H., 61, *65*
Rastogi, R. K., 130, *157*
Rathe, G., 83, 84, *118*
Rathe, J., 266, *274*
Raudonat, H. W., 467, *477*
Rawitch, A. B., 468, 469, *474*
Rebbert, M., 11, *20*
Rechnic, J., 467, *475*, *477*
Reddy, P. R. K., 413, 415, 418, 426, *447*
Redshaw, M. R., 4, 5, 6, 7, *20*, *21*, 136, 140, *154*, *157*
Reeder, R. H., 13, 17, *20*
Reese, A. M., 252, *274*
Reeves, R. B., 93, *121*
Rehm, W. S., 27, 29, 30, 31, 32, 33, 34, 35, 36, 37, 38, 39, 40, 41, 42, 43, 45, *46*, *47*, *48*, *49*
Reich, E., 9, *21*
Reichert, E., 261, *274*
Reichert, L. E., Jr., 422, *445*
Reichstein, T., 172, *180*, *182*
Reid, H. A., 467, *477*
Reinhardt, W. O., 267, *270*
Reitman, M., 406, *447*
Reynolds, C. A., 378, 393, 394, *396*
Ribeiro, L. P., 213, *216*
Richards, A. N., 288, *336*
Richards, G. M., 469, *477*
Rick, C. M., 250, *274*
Rider, J., 192, 193, *216*
Ridley, A. R., 52, 57, *65*
Ridley, H. A., 23, *47*
Riggs, A., 82, 86, 87, 88, 94, 96, 97, 100, 101, 102, 103, 105, 106, 108, 114, *118*, *121*, *122*, 378, 379, 380, 385, 387, 393, 394, *397*, *398*
Rinaudo, M. T., 139, *158*
Ringle, D. A., 4, 5, *21*
Robberson, B., 82, *121*
Robbins, J., 212, *214*
Roberts, J. S., 281, 283, 284, 290, 293, 294, *336*
Roberts, R. C., 189, 193, *216*
Robertson, A. V., 171, *182*
Robertson, D. R., 131, 132, *158*
Robertson, R. N., 38, *48*
Robin, E. D., 307, *336*
Robison, G. A., 150, *158*

Rocha e Silva, M., 467, *477*
Roche, J., 439, *447*
Rodriguez, E. M., 124, *158*
Rogers, D. C., 132, *158*
Rogers, L. J., 301, 302, *336*, 372, *376*
Romer, A. S., 93, *121*, 188, *216*
Roodyn, D. B., 9, 10, *21*
Root, H. D., 261, *274*
Rose, F. L., 91, *122*
Rose, W., 250, *274*
Roseghini, M., 174, 175, 178, *180, 181*
Rosenberg, L. L., 402, 414, 415, 437, 438, 439, 440, *446*
Rosenberg, M., 113, *121*
Rosenberg, P., 467, 470, *477*
Rosenbluth, J., 145, *154*
Rosenfeld, G., 321, *336*, 467, *477*
Rosenkilde, P., 126, 133, 142, 148, *154, 156, 158*
Rosenquist, J. W., 205, 207, 208, 209, 213, *216*
Rosenthal, H. L., 213, *216*
Rosse, W. F., 111, *121*
Roth, S. C., 467, *477*
Rothschild, A. M., 467, *477*
Rothman, S. S., 266, *274*
Rotta, G. P., 139, *158*
Rotunna, C. A., 63, *65*
Rounds, D. E., 4, *21*
Rousseau, G., 60, *65*
Roussel, G., 189, *213*
Rucknagel, D. L., 82, *118*
Rudack, D., 6, *21*
Rueff, W., 204, *215*, 359, *375*, 394, *397*
Ruibal, R., 54, 56, *65*
Russell, F. E., 253, *274*, 450, 451, 453, 454, 463, 466, 467, 468, 469, 470, 471, 472, *475, 476, 477*
Rutledge, J. R., 36, 37, 42, *46, 48*
Ryan, R. J., 422, *445*
Ryffel, G., 144, *159*

S

Saave, J. J., 82, *118*
Sacerdote, M., 147, 148, *159*
Sachs, C., 37, *48*
Sachs, G., 24, 34, 42, 44, *48, 49*
Sage, M., 409, 411, *447*
Saint-Girons, H., 251, 252, *271*, 355, 358, 365, *374, 376*, 400, 401, 402, 414, 434, *445, 447*
Saint Girons, M.-C., 378, *397*
Saito, K., 470, *477*
Sakai, K., 167, *182*
Sakai, Y., 86, 87, 94, 97, 115, *120*
Saland, L. C., 125, *158*
Salimaki, K., 132, 144, *158*
Salmon, F., 250, 261, *274*
Saluk, P. H., 201, 202, *216*
Salvati, A. M., 85, 102, *122*
Salvatore, G., 439, *447*
Salvatore, M., 439, *447*
Samejima, Y., 457, *476*
Sampietro, P., 139, *158*
Sandberg, A. A., 212, *216*
Sanders, A. S., 252, 253, 254, 257, 261, 262, 263, *275*
Sanders, S. S., 29, 33, 36, 39, 42, 45, *48*
Sandler, R., 427, 434, *446*
Sandor, T., 134, *158*, 429, 430, *447*
Sanyal, M. K., 425, 427, *447*
Sarkar, N. K., 466, *477*
Sasaki, T., 69, *76*, 455, 457, 459, 461, *476, 477*
Satake, K., 394, 397
Sato, S., 455, 457, 458, *475, 477, 478*
Sawai, Y., 469, *476*
Sawyer, M. K., 286, 288, *336*
Sawyer, W. H., 285, 286, 288, *336*
Saxén, E., 132, 144, *158*
Saxén, L., 132, 144, *158*
Scalenghe, F., 139, 143, *154, 158*
Schaeffer, W. H., 434, *447*
Schally, A. V., 149, *156*
Scharffenberg, R. S., 253, *274*, 468, 470, *477*
Schatzlein, F. C., 437, 438, 439, 440, *447*
Scheer, B. T., 52, 55, 57, *64*, 65, 134, 140, *158*
Schilb, T. P., 307, *334, 336*
Schindler, O., 172, *180*
Schjeide, A. O., 205, 207, *216*, 427, *448*
Schlesinger, C. V., 381, 394, *397*

Schlesinger, H. S., 41, *48*
Schmid, W. D., 141, *158*
Schmidt, E., 242, 243, *247*
Schmidt, F. W., 242, 243, *247*
Schmidt, K. P., 250, *274*
Schmidt-Nielsen, B., 278, 279, 281, 283, 284, 285, 286, 289, 290, 291, 292, 293, 294, 295, 296, 298, 300, 301, 302, 303, 306, 307, 308, 309, 310, 311, *334*, *336*, 372, *374*
Schmidt-Nielsen, K., 52, *65*, 251, *270*, *273*, *274*, 278, 282, 303, 308, 309, 310, 311, 312, 313, 314, 315, 318, 330, *334*, *335*, *336*, 354, 365, 368, *374*, *376*, 386, *397*
Schöpf, C., 164, 165, *182*, *183*
Schoffeniels, E., 52, *65*, 256, 257, 258, 263, 266, 268, 269, *270*, *271*, *272*, *274*, 355, 357, 359, 365, 372, *374*, *375*, *376*
Scholander, P. F., 308, *336*
Schonberger, C. F., 250, *274*
Schramm, M., 266, *270*
Schricker, J. A., 321, *335*
Schroeder, W. A., 79, 82, *120*, *121*
Schröter, H., 172, *182*
Schuetz, A. W., 149, *158*
Schultheiss, H., 143, *158*
Schwantes, A. R., 383, *397*
Schwartz, M., 29, 43, *47*, *48*
Scott, W. N., 307, *336*
Scudder, C., 209, *215*
Seal, U. S., 189, 193, 205, 207, 210, 212, *216*
Sedar, A. W., 24, 26, *49*
Seegers, W. H., 210, *215*
Sehra, K. B., 469, 470, *475*
Seitz, G., 164, *182*
Sela, M., 462, *476*
Selander, R. K., 383, *398*
Sells, B. H., 13, *19*
Selman, G. G., 9, *21*
Sembrat, K., 439, *447*
Semple, R. E., 356, 357, 360, 361, *376*, 378, *397*
Sena, L., 439, *447*
Seniów, A., 192, *216*
Shaffer, B. M., 146, *158*
Shah, P. P., 342, 350, *351*
Shamloo, K. D., 192, *215*
Shamoo, Y. E., 307, *335*, *336*
Sharp, G. W. G., 60, 61, 62, *65*, 142, *158*
Shaw, T. I., 39, *46*
Sheeler, P., 190, 197, 198, 213, *213*, *216*
Shefer, S., 344, 347, *351*
Shellabarger, C. J., 437, 438, 439, 440, *447*
Shelton, G., 93, *119*
Shelton, J. B., 82, *120*, *121*
Shelton, J. R., 82, *120*, *121*
Sherwood, N. P., 202, *213*
Shimasaki, S., 6, *20*
Shimizu, T., 253, *274*
Shimodaira, K., 170, *181*
Shindelman, J., 178, *182*
Shipolini, R., 467, *474*
Shmerling, D. H., 257, *275*
Shoemaker, R. L., 24, 34, 35, 37, 42, 43, *48*, *49*
Shoemaker, V. H., 142, *158*, 284, 294, 297, *336*, 368, *374*
Shontz, N. N., 90, *121*
Shreffler, D. C., 82, *118*
Shukuya, R., 86, 87, 88, 94, 95, 97, 102, 103, 107, 115, *118*, *120*
Sidky, Y. A., 205, *216*
Sieberg, C., 200, *214*
Siekevitz, P., 266, *274*
Sievers, M. L., 369, *376*, 378, *397*
Sigsworth, D., 356, 357, 360, 361, *376*, 378, *397*
Simkiss, K., 205, 207, 208, *215*, *216*
Sinclair, J., 9, *21*
Singer, S. J., 62, *65*
Singer, T. P., 469, *477*
Sinsheimer, R. L., 469, *477*
Siperstein, M. D., 173, *182*
Sjövall, J., 350, *351*
Skadhauge, E., 281, 284, 285, 290, 294, 298, 301, 309, 311, *336*, 400, *447*
Skoczylas, R., 252, 254, 261, *274*
Slaunwhite, W. R., Jr., 212, *216*
Slegers, J. F. G., 262, *272*
Slotta, C. H., 171, *182*

Slotta, K., 455, 462, 467, 469, 471, *477*
Smith, D. D., 250, *274*
Smith, H. P., 210, *216*
Smith, H. W., 285, *336*
Smith, M., 325, *336*
Smith, R. G., 194, *216*
Smith, R. T., 200, *216*
Smith, S., 18, *20*
Smoller, C. G., 125, *158*
Snart, R. S., 63, *65*, 140, *153*
Snyder, B. W., 146, *154*
Snyder, G. K., 462, *476*
Sobotka, H., 253, *274*
Socino, M., 143, *154*
Sodero, E., 406, *447*
Sokol, O. M., 250, *274*
Solie, T. N., 458, *478*
Solinger, R. E., 307, *335*
Solomon, A. K., 45, *47*
Solomon, S., 279, 281, 289, 290, 293, *335*
Sonnenberg, H., 285, *334*
Sorcini, M., 85, 89, 90, 91, 92, 99, 100, 102, 104, *118*, *121*, *122*
Sormova, Z., 469, *475*
Southern, E. M., 17, 19, *20*
Spangler, S. G., 29, 30, 34, 35, 45, *48*, *49*
Sparks, C. E., 86, 87, 88, *118*
Speeter, M. E., 171, *182*
Speirs, J., 13, *19*
Spiedel, C. C., 114, *120*
Spiegelman, S., 8, *21*
Spies, I., 127, *153*
Sprent, J. F. A., 201, *216*
Stahlmann, K., 30, *47*
Staley, H. F., 252, 253, *274*
Stanić, M., 467, *477*
Stanley, J. G., 432, *447*
Staple, E., 342, 350, *351*
Starzecki, B., 467, *475*
Steen, W. B., 56, *65*
Steigemann, W., 79, *120*
Steinbach, H. C., 8, *21*
Steiner, L. A., 88, 115, *121*
Steinmetz, P. R., 307, *335*, *336*
Stenger, C., 413, *448*
Stevenson, O. R., 219, 220, 221, *247*
Stickler, G. B., 194, *214*
Still, J. M., 378, 393, 394, *396*
Stitt, J. T., 356, 357, 360, 361, *376*, 378, *397*
Stoll, A., 171, *182*
Stoner, J. A., 151, *157*
Strassberg, J., 467, *477*
Stratton, L. P., 86, 87, 88, *121*
Streb, M., 132, *158*
Strittmatter, C. F., 8, *21*
Strittmatter, P., 8, *21*
Strohman, R. C., 146, *152*, *157*
Strydom, A. J. C., 457, 461, *477*
Strydom, D. J., 457, 460, 462, *474*
Su, C., 458, 459, *477*
Sulkowski, E., 469, *477*
Sullivan, B., 78, 82, 86, 95, 96, 100, 101, 103, 107, 108, *121*, *122*, 378, 379, 380, 381, 383, 384, 385, 387, 388, 390, 391, 392, 393, 394, *397*, *398*
Sumyk, G., 469, *476*
Sung, S. C., 472, *478*
Sutherland, E. W., 150, *158*
Sutherland, I. D. W., 285, *335*
Sutton, D. E., 394, *396*
Suzuki, H. K., 207, *216*
Suzuki, T., 462, 469, *477*, *478*
Svedberg, T., 101, 104, *122*, 393, *398*
Swanson, R. F., 11, *20*, *21*
Swift, H. H., 9, *21*
Swigart, R. H., 27, *49*
Sy, J., 12, *21*
Szarski, H., 79, 80, *122*, 250, *274*
Sze, L. C., 9, *21*
Szelenyi, J. G., 82, *118*

T

Taborda, A. R., 469, *478*
Taborda, L. C., 469, *478*
Tachikawa, R., 167, *182*
Tait, N. N., 202, 203, 204, 213, *216*
Taketa, F., 90, 100, 104, *122*
Takewaki, K., 413, *447*
Talbott, J. H., 388, *396*
Tam, W. H., 424, 425, 429, *445*, *447*

Tamaki, Y., 212, *216*
Tamiya, N., 455, 457, 458, *475, 477, 478*
Tamm, C., 172, *182*
Tammar, A. R., 347, *351*
Tamura, C., 167, *182*
Tanabe, Y., 212, *216*, 438, *447*
Tanaka, S., 69, *76*
Tanaka, Y., 75, *76*
Tang, T. E., 113, *120*
Tanner, G. A., 285, *336*
Tata, J. R., 144, *158*
Taub, A. M., 313, 315, 323, 325, 329, *335, 336, 478*
Tayeau, F., 253, *274*
Taylor, J. D., 151, *152, 158*
Taylor, R. E., 139, 143, *158*
Taylor, W. J., 462, *476*
Taynon, H. J., 208, *215*
Tchen, T. T., 71, *76*, 343, *350*
Templeton, J. R., 284, 313, 315, 329, 330, 332, *336*, 353, 369, *376*, 432, *447*
Tentori, L., 85, 89, 90, 91, 92, 94, 95, 99, 100, 102, 103, 104, 106, *118, 119, 121, 122*
Teorell, T., 27, 29, 45, *49*
Tercafs, R. R., 52, *65*, 365, *376*
Terlou, M., 124, *158, 159*
Terner, C., 40, 45, *46, 49*
Thapliyal, J. P., 413, *447*
Thean, P. C., 467, *477*
Theil, B. C., 115, 116, *122*
Thein, M., 213, *216*
Thesleff, S., 310, *336*
Thinès, G., 371, *376*
Thiruvathukal, K. V., 252, *274*
Thomas, C., 14, *21*
Thomas, W. L., 250, *273*
Thompson, D. C., 299, *335*
Thompson, E. C., 394, *396*
Thornburn, C. C., 139, *158*
Thornton, V. F., 149, *154, 158*
Thrasher, J. D., 418, *447*
Thurmond, W., 127, *158*
Ti Ho, 7, *22*
Timmerman, W. F., *477*
Timourian, H., 201, 202, *216*
Tipton, S. R., 439, *447*
Titius, E., 173, *182*
Tocchini-Valentini, G. P., 17, *19*
Toh, C. C., 261, *274*
Toivonen, S., 132, 144, *158*
Tokuyama, T., 175, 177, *180, 182*
Tominaga, S., 193, *215*
Toom, P. M., 466, 469, *478*
Tooze, J., 79, *122*
Torretti, J., 139, *156*
Toscano, C., 129, *156*
Trachtenberg, P., 467, *477*
Trader, C. D., 82, 85, 87, 96, 101, 102, *122*
Trandaburu, T., 406, *448*
Trethewie, E. R., 261, *275*
Trobec, T. N., 432, *447*
Troxler, F., 171, *182*
Tsai, F. F., 469, *478*
Tseng, T. C., 460, *476*
Tsuda, K., 167, *182*
Tsui, H. W., 425, 426, *444*
Tsushima, K., 115, *120*
Tu, A. T., 255, *274*, 458, 466, 469, *478*
Tucker, V. A., 354, *374*
Tullner, W. W., 321, *335*
Turbeck, B. O., 40, *49*
Turner, A. H., 359, *374*
Turner, F. B., 418, *447*
Turner, J. E., 439, *447*
Twigg, G. I., 357, 358, 361, 365, *374*
Twining, H., 250, 252, *271*
Tyler, A., 201, *216*

U

Udenfriend, S., 174, *182*
Uranga, J., 54, *65*
Urban, G., 167, *181*
Urist, M. R., 205, 207, *216*, 427, *448*
Ussing, H. H., 141, *154*

V

Vallee, B. L., 260, *274*
Van Brunt, J., III, 205, 206, 207, 213, *216*

Vandermeers, A., 266, *274*
Vandermeers-Piret, M. C., 266, *274*
van de Veerdonk, F. C. G., 150, *158*
van der Schans, G. S., 144, *152*
van Dongen, W. J., 129, 130, 134, *156, 159*
van Dyke, H. B., 286, 288, *336*
Van Handel, E., 212, *216*
van Kemenade, J. A. M., 127, 129, 134, 135, *156, 158, 159*
Vankin, G. L., 111, 112, *119*
Van Oordt, P. G. W. J., 124, 125, 129, 130, 134, 137, 145, *155, 157, 158, 159,* 426, *446*
van Tienhoven, A., 413, 427, *448*
Varma, M. M., 135, *159*
Vasilenko, S. K., 469, *478*
Velani, S., 85, 102, 103, *122*
Vellano, C., 126, 128, 145, 147, 148, *156, 157, 159*
Vellard, J., 210, *213*
Vernon, C. A., 467, *474*
Vester, J. W., 307, *336*
Vialli, M., 171, 179, *181*
Vick, J. A., 467, *477, 478*
Vidal Breard, J. J., 469, 472, *478*
Vijayakumar, S., 127, *153*
Villalonga, F. A., 63, *65*
Villegas, L., 34, 35, 36, 38, 43, 45, *49*
Villela, G. G., 213, *216*
Villers, A., 250, *274*
Villers Pienaar, U., 360, *376*
Vinegar, A., 378, 389, *398*
Vinson, G. P., 134, *153*, 429, 435, *445*
Vintemberger, P., 7, *19*
Viscidi, E., 439, *447*
Vital Brazil, O., 467, *478*
Vitali, R., 174, 175, *181*
Vitali, T., 174, *181*
Vitanen, P., 355, *376*
Vivaldi, G., 85, 94, 95, 100, 102, 103, 106, *118, 122*
Vivien, J., 413, *448*
Vladescu, C., 406, 433, *448*
Vocke, F., 171, *183*
Vogel, G., 165, *181*
Volk, T. L., 135, *159*
Volsoe, H., 355, 363, *376*
von Brehm, H., 132, *156*
Vonck, H. J., 249, 251, 255, 257, 260, *274*
von Euler, U. S., 170, *182*
von Mecklenberg, C., 410, *445*
Voris, H. K., 192, 195, *213, 216*
Vullings, H. G. B., 128, *159*

W

Waddington, C. H., 7, *21*
Wade, M., 91, *122*
Wagner, E. M., 409, 413, *448*
Wagner, F. W., 469, *478*
Wakely, J. F., 167, *182*
Wakui, K., 469, *478*
Waldmann, T. A., 111, 115, *119, 121,* 378, *396*
Walker, A. M., 288, *336*
Walker, R. F., 439, *448*
Wallace, H., 13, 17, 18, 21, 22
Wallace, R. A., 4, 5, 6, 7, *21, 22,* 140, *159*
Walsh, E. O., 469, *475*
Wan, B., 23, 24, *48*
Wang, C. L., 108, *118*
Ward, R. T., 6, *22*
Waring, H., 142, *158*
Warner, E. D., 210, *216*
Wartenberg, H., 6, 8, *22*
Watanabe, K., 348, *351*
Waterman, A. J., 440, *448*
Waters, J. A., 177, *180*
Watson, F., 194, *215*
Watson, H. C., 105, *121*
Watts, D. C., 81, *122*
Watts, R. L., 80, *122*
Way, L. W., 30, *47*
Webb, E. C., 212, *215*
Weber, C. S., 13, 14, 19, *19*
Weber, K. M., 127, 134, *155*
Weber, R., 8, *19, 22,* 144, 146, *159*
Webster, T. P., 383, *398*
Wegnez, M., 15, 17, 19, *20, 22*
Weil, F. J., 171, 173, *183*
Weitz, G., 164, *183*
Wellen, J. J., 136, 137, *156*
Wellner, D., 468, 469, *478*

Wells, M. A., 471, *478*
Welsh, J. H., 179, *182*
Wensink, P. C., 13, 19, *22*
Wersall, R., 27, *49*
Weymouth, R. D., 357, 367, *374*
White, F. H., 201, *216*
White, F. N., 187, *215*, 396, *398*
White, R., 406, *447*
White, T. D., 27, *49*
Whitehouse, M. W., 342, 350, *351*
Whittemore, F. W., 467, *475*
Wieland, H., 170, 171, 173, *183*
Wilbur, K. M., 253, 254, 261, 262, *270*
Wilhoft, D. C., 439, *448*
Wilkie, D., 9, 10, *21*
Wilkinson, R. F., Jr., 427, *448*
Willard, E., 402, *444*
Willem, R., 17, *20*
Williams, E. J., 472, *478*
Williams, J., 8, *22*
Williams, J. N., Jr., 469, *478*
Wilson, A. C., 192, 195, 213, *214*
Wilson, F. E., 194, *215*
Wilson, J. W., 385, 391, 392, *398*
Wilson, T. H., 267, *271*
Wilt, F. H., 114, *122*
Winchester, R. J., 266, *272*
Windhager, E. E., 309, *335*
Wingstrand, K. G., 124, 142, *156, 159*, 400, *448*
Wintrobe, M. M., 80, *122*
Wirz, H., 288, 289, *336*
Wischnitzer, S., 6, 8, *22*
Wise, P. T., 134, *158*
Wise, R. W., 86, 87, 88, *121, 122*
Witkop, B., 167, 171, 175, 176, 177, *180, 182*
Witschi, E., 423, *448*
Witten, P. W., 213, *214*
Wölfel, E., 164, *183*
Wolff, J., *476*
Wolstenholme, D. R., 9, *22*
Wolvekamp, H. P., 255, 256, 257, *274*
Woo, M. L., 454, 470, *477*
Woo, M. Y., 469, 472, *477*
Wood, S. C., 80, 89, 91, 92, 98, 104, 111, 113, *122*, 388, 389, *398*
Woodard, J. W., 116, *119*
Woodbury, A. M., 368, 372, *376*
Woodland, H. R., 17, *20*
Woolley, P., 288, *336*
Wootton, V., 339, 340, 347, 348, 349, 350, *351*
Wortham, J. S., 101, *122*
Wright, A., 409, 413, 432, 434, *448*
Wright, A. A., 250, *274*
Wright, A. H., 250, *274*
Wright, E. M., 268, *275*
Wright, F. S., 143, *155*
Wright, R. D., 252, 253, 254, 257, 261, 262, 263, *275*
Wu, C. H., 406, *447*
Wurth, M. A., 252, 273, *275*
Wurtman, R. J., 129, *159*
Wyman, J., 94, 95, 100, 106, *118*
Wyssbrod, H. R., 307, *335*

Y

Yaari, A., 394, *396*
Yamana, K., 18, 19, *19*
Yamasaki, K., 337, 343, 347, *350, 351*
Yang, C. C., 457, 459, 462, *478*
Yang, H. C., 469, *478*
Yang, H. J., 457, 459, *478*
Yang, S. Y., 383, *398*
Yanni, M., 358, *375*
Yaron, Z., 427, *448*
Yau, W. M., 45, *48*
Yoshida, L., 467, *478*
Yoshimura, F., 135, *159*
Yoshinori, T., 438, *447*
Young, A. D., 394, *397*
Yu, C. I., 107, *118*
Yuasa, C., 88, *120*
Yuuki, M., 337, 347, *351*

Z

Zain, B. K., 268, *273*, 366, *375, 376*
Zain-Ul-Abedin, M., 268, *273*, 357, 358, 361, 366, *375, 376*
Zalesky, S., 164, *183*
Zamecnik, P. C., 467, *476*
Zarafonetis, C. J. D., 210, *216*
Zarrow, M. X., 425, *444*

Zelander, T., 266, *271*
Zeller, E. A., 255, *275*, 466, 468, 469, 471, *478*
Zendzian, E. N., 256, 257, 258, 259, 260, *275*
Zeuthen, E., 9, *20*
Ziegler, F. D., *478*
Ziegler, H., 407, 414, 415, 427, *444*
Zinner, S. H., 151, *157*
Zipf, J. B., 179, *182*
Zipser, R. D., 147, *159*
Zito, R., 103, *122*
Zmudka, M., 468, *474*
Zoppi, G., 257, *275*
Zwanenbeek, H. C. M., 145, *155*
Zwisler, O., 469, *478*

Subject Index

A

Acetazoleamide,
  bicarbonate excretion and, 233
  urate excretion and, 303
  urine pH and, 296, 298
Acetoacetate, aldosterone effect and, 62
Acetylcholine,
  chitinase secretion and, 264
  snake venoms and, 469, 470
6-*O*-Acetylglucose, metabolism of, 223
N-Acetylhistamine, occurrence of, 174
O-Acetylsamandarine, chemical properties, 165
Acid phosphatase, snake venom, 469
Acris crepitans, hemoglobins of, 83–84
Acris gryllus, hemoglobins of, 83
Acrochordus javanicus, bile salt of, 348
Actinomycin D, aldosterone effects and, 60, 142
Adenohypophysis,
  sodium transport and, 57
  structure, 129
Adenosine, envenomation and, 451
Adenosine 3′,5′-monophosphate,
  batrachotoxin and, 176
  gastric secretion and, 41–42
  melanocyte-stimulating hormone and, 150
Adenosine polyphosphates, hibernation and, 363
Adenosine triphosphatase,
  gastric, secretion and, 40–41
  snake venom, 469, 472
  sodium-potassium-activated, hydrochloric acid secretion and, 262
Adenosine triphosphate,
  erythrocyte content, 93, 95
  gastric acid secretion and, 38–40
  hemoglobin and, 97, 98, 111, 387, 388, 391, 392
Adenylcyclase, chromatophores and, 150
Adrenalectomy,
  potassium excretion and, 297
  sodium transport and, 292
Adrenaline,
  bufonid toxins and, 167, 170
  chitinase secretion and, 264
Adrenocorticotropic hormone, *see* Corticotropin
Agama, prolactin of, 411
Agama agama,
  gonadotropin effects in, 418
  hypophysectomized, 409, 432
  testosterone and, 426
Agama stellio, hemoglobin synthesis in, 394
Agamodon anguliceps, bile salts of, 347
Agkistrodon,
  haptoglobin in, 210
  hemoglobin of, 211
Agkistrodon acutus, transferrin of, 196
Agkistrodon contortrix, venom, potency, 452
Agkistrodon piscivorus,
  bile salt of, 350
  digestive enzymes, 255, 257, 258
  erythrocytes, phosphates in, 387
  hemoglobin, function, 390
  venom, potency, 452
Agkistrodon rhodostoma, venom, potency, 453
Alanine,
  ammonia excretion and, 232
  bicarbonate synthesis and, 232
  intestinal transport, 366
  metabolism in reptiles, 237
  reptile tissues and, 224, 225, 227
  synthesis of,
    alligator, 229
    chameleon, 236
  transamination and, 229
  uptake of, 268–269
Albumin,
  reptilian plasma, 190, 191
    anion binding, 194–195
    evolutionary aspects, 195–196
    isolation and identification, 193
    volume expander function, 193–194

Aldosterone,
action, cellular basis, 59–64
biosynthesis of, 429–430
blood glucose and, 138
corticotropin and, 135
interrenal and, 133
molting and, 148
osmoregulation and, 431, 432
sodium transport and, 57, 58, 59, 142–143, 292–293
Alkaline phosphatase, snake venom, 469, 472
Alligator,
amino acid deamination in, 244
blood chloride, 188
glycogenolysis in, 218–219
nitrogen transport in, 227
pituitary, size of, 400
tissues, glycogen content, 218
transcortin of, 212
Alligator,
Hageman factor in, 210
plasma proteins, comparative aspects, 192
Alligator mississippiensis,
albumin,
anion binding, 194
evolutionary aspects, 195
bile salts of, 342, 350
blood seasonal variations,
glucose, 358
oxygen capacity, 361
calcium and phosphate excretion in, 299
erythrocytes, phosphates in, 387
esterases of, 211
glucose utilization, season and, 367
hemoglobin, 384
function of, 392
photoperiod and, 371
plasma vitellin of, 206
transferrin of, 196
urine, carbonic anhydrase and, 298
vitamin B_{12}-binding protein of, 213
Allochenodeoxycholic acid, occurrence of, 68, 75
Allocholic acid,
biosynthesis of, 71–72
occurrence of, 68, 75, 347, 349, 350
Amanita muscaria, toxin, activity, 163
Amberlite IRC-50, gonadotropins and, 421
Amblyrhynchos cristatus, nasal gland secretion, 313, 314, 327–330
Amblystoma mexicanum,
eggs, size of, 3
mitochondrial deoxyribonucleic acid, 9
nucleolar organizer in, 12–13
oocytes, nucleoli in, 14
Ambystoma, pancreatic islets of, 133
Ambystoma annulatum,
hemoglobin of, 89
polymorphism, 91, 92
Ambystoma maculatum,
erythrocytes of, 80
larval hemoglobin, 91
hemoglobin of, 89–90
functional properties, 95, 99–100
polymerization, 103–104
metamorphosis and, 116
Ambystoma opacus, hemoglobin, polymorphism, 91
Ambystoma tigrinum, larval hemoglobin, 91
Ambystoma tigrinum tigrinum,
hemoglobin of, 89–90
functional properties, 99–100
Ameiva ameiva, bile salts of, 347
Amino acid(s),
blood seasonal variations, 359
deamination, species and, 244
hemoglobin composition, 88, 102, 104
intestinal absorption of, 268–269
luteinizing hormone composition, 421
metabolism,
alligator, 224–234
ammonium bicarbonate synthesis, 231–234
chameleon, 234–237
deamination rate, 225–226
insulin and, 226–227
interconversions, 228–231, 236–237
α-keto acid metabolism, 228
nitrogen transport, 227–228
protein digestion and fate of products, 224–225
summary, 237–241
neurotoxin composition, 454

sequences, snake venom polypeptides, 456–457
viper hemoglobin composition, 394, 395
yolk proteins and, 4, 5
L-Amino acid oxidase,
snake plasma and, 213
venom and, 255, 468, 469, 470, 473
α-Aminobutyrate, methionine and, 239
p-Aminohippurate, transport, glomerular filtration rate and, 285–286
Ammonia,
blood seasonal variations, 359
chitinase secretion and, 265, 266
excretion, 239, 298, 312
regulation of, 299–301
synthesis, amino acids and, 232
Ammonium bicarbonate,
synthesis in crocodilia, 231–232
ammonia synthesis, 232
bicarbonate synthesis, 232
carbamyl phosphate synthesis, 233–234
cabonic anhydrase and, 233
Amphenone, salt gland function and, 320, 321, 322, 431
Amphibians,
ionic fluxes in, 52–56
osmotic ranges and tolerance, 51–52
ribosomes, structure and synthesis, 12–13
Amphibolurus barbatus, bile salts of, 347
Amphibolurus ornatus, sodium retention by, 368
Aphiuma means, erythrocytes, size, 79, 80
Amphiuma tridactylum, hemoglobin, functional properties, 98
Amphotericin B, intestinal permeability and, 269
Amyda japonica, bile salts of, 338, 347
Amylase,
reptilian digestive juice, 255–259
snake serum, 212
Amytal, potassium transport and, 44
Anaerobiosis, chitinase secretion and, 265
Anaetulla nasuta, bile salt of, 348
Anaspida, bile salts of, 337–339, 347
Androgens, physiological actions, 425–427
Androstenedione, biosynthesis of, 424
Anguis fragilis, digestive enzymes, 257
Anions, binding, albumin and, 194–195
Anolis,
esterases of, 212
gonadotropin, neuraminidase and, 420
hemoglobin of, 383
plasma albumins, 195
transferrins of, 200
Anolis aeneus, hybrids, transferrin of, 198
Anolis carolinensis, see also Chameleon
blood seasonal variations, glucose, 358
corticosterone in, 430, 435
digestive enzymes, 257
food intake, 261
gonadotropins and, 426
growth hormone and, 404–405
hypophysectomized,
growth of, 408
hepatic glycogen in, 433
testicular regression in, 414, 415
thyroidal activity and, 440
iodine metabolism in, 437–439
liver,
glycogen content, 218
seasonal variations in, 363
males, yolk synthesis in, 207, 208, 209
melanocyte stimulating hormone of, 436–437
metabolic rate, season and, 356
ovary, gonadotropins and, 418, 419
photoperiod and, 370
pituitary, gonadotropins in, 423
plasma vitellin of, 205
progesterone and, 427
prolactin and, 404–405, 407–408
spermatogenesis, temperature and, 417
tail regeneration, prolactin and, 406
testis, gonadotropins and, 419
testosterone and, 426
urea excretion by, 301
weight, seasonal variation, 356
Anolis garmani, bile salts of, 347
Anolis grahami, bile salts of, 347
Anolis lineatopus, bile salts of, 347
Anolis vichardi, bile salts of, 347
Anolis trinitatus, hybrids, transferrin of, 198
Anoxia, gastric secretion and, 39–40
Antibody,
growth hormones and, 410

luteinizing hormone and, 421
Antigens, reptilian immunoglobulins and, 202–205
Anura,
bile salts of,
bufomidae, 69–70
other anurans, 70
ranidae, 70
erythrocytes, size of, 79, 80
hemoglobin,
components, 82–89
function, 93–98
sequence studies, 104–108
structure, 101–103
neurohypophysis, 124
pancreatic islets of, 133
testis, steroid-producing cells, 136
toxins,
atelopotidae, 178
bufonidae, 167–174
dendrobatidae, 175–178
discoglossidae, 179
hylidae, 178–179
leptodactylinae, 174–175
pipidae, 179
ranidae, 179
Arabinose,
metabolism of, 223
uptake of, 268
Arginine,
glutamine synthesis and, 235
metabolism in reptiles, 239
L-Arginine-ester hydrolases, snake venom, 469
Arginine vasotocin, *see also* Vasotocin
cloaca and, 308
glomerular filtration rate and, 286–288
intestinal permeability and, 270
potassium excretion and, 297
renal tubular function and, 290–291
sodium transport and, 292
water balance and, 56
Arizona, hemoglobin of, 211
Arsenate, chitinase secretion and, 265
Asparagine, transamination and, 229
Aspartate,
metabolism in reptiles, 238
deamination of, 226
synthesis of, 236
transamination and, 229
Aspartate residues, amphibian hemoglobins and, 106
Atelopotidae, toxins of, 178
Atelopus ambulatorius, toxin of, 178
Atelopus cruciger, toxin of, 178
Atelopus planispina, toxin of, 178
Atelopus varius, toxin of, 178
Atelopus zeteki, toxin, activity, 163, 178
Atheris squamiger, bile salt of, 349
Autopharmacology, snake venoms and, 467
Azide, glucose uptake and, 267

B

Barium ions, chloride transport and, 29
Batrachoseps attenuatus, erythrocytes of, 79
Batrachotoxin,
occurrence and activity, 163, 175–176
structure, 175
Batrachotoxinin A, occurrence of, 175
Bee, stings, fatal, 451
Behenate, uptake of, 267
N-Bonzoyl-L-arginine ethyl ester, reptilian enzymes and, 259
Bicarbonate,
chitinase secretion and, 264, 266
chloride transport and, 29
excretion of, 232, 298, 312
gastric adenosine triphosphatase and, 40, 41
gastric serosa and, 35
intestinal secretion, 270
reptilian blood, 188
transport, bladder and, 307
Bile salts,
anaspid, 337–339, 347
anuran, 69–70, 75
caudata, 68, 75
crocodilian, 342–343
distribution,
amphibia, 71–74
reptiles, 343–346
nature of, 67
saurian, 339–340
snake, 340–342
Biogenic amines, bufonid toxins and, 167, 170

Biticholic acid, occurrence of, 341–342, 349, 350
Bitis, bile salts of, 341–342
Bitis gabonica, bile salt of, 349
Bitis lachesis,
bile salt of, 349
venom, potency, 453
Bitis nasicornis, bile salt of, 349
Bitis worthingtoni, bile salt of, 349
Bladder,
body water supply and, 56, 58
inorganic ions, season and, 364
interrelations with other excretory functions, 309–312
short circuit current, aldosterone and, 60, 62
sodium flux and, 53–54
sodium space and, 54, 55
uric acid in, 303, 310
water and ion excretion and, 306–308
Blastoporal lip, yolk and, 7
Blood,
glucose, growth hormone and, 406–407
seasonal variations in,
cells, 359–362
glycemia, 358
inorganic ions, 357–358
nitrogenous compounds, 358–359
Blood cells, seasonal variations in, 359–362
Blood flow, metabolic rate and, 244–246
Boa constrictor amarali, hemoglobin of, 384
Boa constrictor constrictor, bile salt of, 348
Boa constrictor imperator, bile salt of, 348
Boa constrictor occidentalis, bile salts of, 348
Bohr effect, amphibian hemoglobins and, 92–100, 106, 111
Boiga blandingi, bile salt of, 348
Boiga dendrophila, bile salt of, 348
Bombina bombina,
hemoglobins of, 83
toxin of, 179
Bombina variegata,
hemoglobins of, 83
toxin of, 179
Bone marrow, erythropoiesis in, 114
Bothrops alternatus,
bile salt of, 350
venom, phospholipase of, 472
Bothrops atrox, venom, potency, 452
Bothrops jararaca, glycemia in, 358
Bothrops phylodrias, glycemia in, 358
Botulinus toxin, occurrence and activity, 163
Bradykinin,
envenomation and, 451
structure and occurrence of, 179
Branchiogenic glands, structure and function, 131–133
Bromide, acid secretion and, 30
Bufo,
ovarian hormones, 136
toxins, activity, 163
Bufo alvarius, toxin of, 171
Bufo americanus,
erythrocytes of, 80
hemoglobins of, 84
Bufo arenarum,
hemoglobin of, 84
interrenals of, 135
toxin of, 171
Bufo bocourti, hemoglobin of, 84
Bufo boreas, hormone effects, food consumption and, 147
Bufo bufo,
adenohypophysis, cell types, 130
bile salts of, 69, 75
blood glucose levels, 138, 139
corticotropin secretion in, 127
gonads, hypothalamus and, 126
hemoglobin of, 84
end groups, 103
functional properties, 93
molting, endocrine effects, 147–148
neurohypophysectomy in, 142
oocytes, meiosis, 149
pancreatic islets of, 133
thyroid gland regulation in, 126
vasotocin in, 141
Bufo calamites, hemoglobin of, 84
Bufo canaliferus, hemoglobin of, 85
Bufo coccifer, hemoglobin of, 84
Bufo cognatus, hemoglobin of, 85
Bufo coniferus, hemoglobin of, 85
Bufo formosus, toxin of, 171
Bufo fowleri, hemoglobin of, 84

Bufo garmani, hemoglobin of, 84, 85
Bufogenins, isolation and structure, 171–172
Bufo granulosus fermandezae, hemoglobin of, 84
Bufo granulosus major, hemoglobin of, 84
Bufo hemiophrys, hemoglobin of, 84
Bufo houstonensis, hemoglobins of, 84
Bufo kellogi, hemoglobin of, 85
5α-Bufol,
occurrence of, 68, 69, 75
structure, 68
5α(or β)-Bufol, formation of, 71, 73
5-β-Bufol, occurrence of, 69, 75
Bufo leutkeni, hemoglobin, 84
Bufo marinus,
bile salts of, 69, 75
erythrocytes, life span, 80
hemoglobin of, 84
functional properties, 93
hormone effects, food consumption and, 147
neurohypophysectomy in, 142
nucleolar organizer in, 13
toxin of, 171
Bufo marmoreus, hemoglobin of, 85
Bufo microscaphus, hemoglobins of, 84
Bufonidae,
bile salts of, 69–70, 75
hemoglobins of, 84–85
toxins,
biogenic amines, 167, 170
bufagenins, 171–172
bufotoxins, 173–174
indolalkylamines, 170–171
Bufo paracnemis, hemoglobin of, 84
Bufo perplexus, hemoglobin of, 84, 85
Bufo punctatus, hemoglobin of, 85
Bufo rangeri, hemoglobin of, 84
Bufo regularis,
bile salts of, 69, 75
hemoglobin of, 84
water metabolism, 141
Bufo speciosa, hemoglobin of, 84, 85
Bufo spinosus, hemoglobins of, 84
Bufo spinulosus, hemoglobin of, 84, 85
Bufotaline,
isolation and structure, 171–172
Bufotenidine,
occurrence of, 174, 179
structure, 171
Bufotenine,
occurrence of, 174, 178
structure and synthesis, 170–171
Bufo terrestris, hemoglobins of, 84
Bufothionine, structure, 171
Bufotoxin,
occurrence and activity, 163, 174
structure, 173
Bufo valliceps,
hemoglobin of, 84, 85
polymerization of, 101
Bufoviridine, structure of, 171
Bufo viridis,
hemoglobin of, 84
polymerization, 101
plasma, urea in, 52
Bufo viridis viridis, toxin of, 171
Bufo vulgaris,
hemoglobin of, 84
tadpoles, hemoglobins of, 86, 88
toxin, activity, 163
Bufo vulgaris japonicus,
bile salts of, 69, 75
embryos, hemoglobins of, 89
Bufo woodhousei, hemoglobins of, 84
Bungarus caeruleus, venom, potency, 453
Bungarus fasciatus,
bile salt of, 349
digestive enzymes, 257
Bungarus multicinctus,
bile salt of, 349
venom polypeptides,
amino acid sequence, 456–457
biological activity, 461

C

Caerulein, occurrence of, 178
Caffeine, phosphorylase and, 221
Caiman, adrenocorticotropic hormone in, 435
Caiman crocodilus,
bile salts of, 342, 350
digestive enzymes of, 257
Caiman latirostris,
bile salts of, 350
hemoglobin of, 384

metabolic rate, season and, 356
Caiman sclerops,
bile salts of, 350
cloaca, function, 309
extraurinary water loss, 282
Calcitonin, ultimobranchial glands and, 131
Calcium,
acid secretion and, 42–43
blood clotting and, 208–209
blood, seasonal variations, 358
melanophores and, 150
plasma vitellin and, 205, 206, 208
reptilian blood, 188
storage, reproduction and, 208
transport, regulation of, 298–299
ultimobranchial glands and, 131, 132
Calliphora erythrocephata, salivary gland, secretory cells, 27
Calyptocephalella gayi, hemoglobin of, 83
Campesterol, occurrence of, 69
Candicin, 174
occurrence and activity, 163
Carbachol, digestive juice pH and, 254, 261
Carbamylcholine, chitinase secretion and, 264, 266
Carbamyl phosphate,
conversion to ammonium bicarbonate, 233–234
renal synthesis, 233
Carbohydrate(s),
metabolism,
glycogenolysis and glycolysis, 218–221
hyperglycemic agents, 221
regulation of, 137–140, 433
sugars and insulin effects, 222–224
reptilian plasma proteins and, 189
Carbonic anhydrase,
ammonia excretion and, 301
bicarbonate synthesis and, 233, 234
bladder and, 307
gastric secretion and, 24, 40, 262
hydrogen ion excretion and, 296–297, 298
Carbon monoxide, chitinase secretion and, 265
Carboxypeptidase,
amphibian hemoglobins and, 102–103, 106
reptilian digestive juice and, 256, 257, 260
Cardiovascular system, snake venoms and, 467
Caretta caretta,
salt gland, 318
ion concentration in, 314
Carnivores,
reptilian, 250
digestive enzymes, 255
Carotenoids,
proteins binding, 212–213
xanthophores and, 151
yolk and, 4
Casein, hydrolysis of, 256–257, 260
Catecholamine(s),
hypothalamus and, 124
melanophores and, 150
Cathepsin, yolk platelets and, 8
Cellobiose, chitinase secretion and, 264
Central nervous system,
batrachotoxin and, 176
salamandra alkaloids and, 166–167
Cerastes cerastes, bile salt of, 349
Ceruloplasmin, reptilian plasma and, 210
Cesium, chitinase secretion and, 265, 266
Chameleo muelleri, bile salts of, 347
Chameleon,
amino acid deamination in, 244
nitrogen transport in, 227
Chameleo vulgaris, digestive enzymes, 257
Chelodina,
Hageman factor in, 210
plasma albumin, 194
Chelodina longicollis,
dehydration effects, 372
urea excretion by, 301–302
Chelone midas, bile salts of, 347
Chelonia,
hemoglobins, 378–381
function, 385–388
immunoglobulins of, 201
interrelations among various excretory functions, 309–311
plasma proteins and, 190

Chelonia, gonadotropin, neuraminidase and, 421
Chelonia mydas,
glomerular filtration rate, 279, 280
salt gland, 318
corticoids and, 431
ion concentration in, 314, 319–323, 329
sodium chloride excretion, 279, 286
Chelydra,
fibrinogen of, 209
gonadotropin, neuraminidase and, 420
Hageman factor in, 210
progesterone synthesis in, 425
Chelydra serpentina,
blood seasonal variations,
inorganic ions, 357
blood cells, 360, 361
plasma volume, 361
coagulation, 361
digestive juices,
enzymes, 256, 258, 259–260
pH, 254
erythrocytes, phosphates in, 387
follicle stimulating hormone, biological tests, 423
growth hormone of, 412–413
naphthylamidase in, 211
pituitary, size of, 400
weight, seasonal variation, 356
Chelys fimbriata, hemoglobin, function, 386
Chersydrus granulatus, bile salt of, 348
Chironomus, hemoglobin of, 79
Chitinase, reptilian digestive juice and, 255, 256, 257, 258, 263–267
Chitobiase, reptilian digestive juice and, 257
Chitobiose, chitinase secretion and, 264, 266
Chlamydomonous kingii, esterases of, 211–212
Chlorate, thyroid and, 133
Chloride,
amino acid uptake and, 269
carriers of, 28–30
chitinase secretion and, 264, 266
seasonal variations,
blood, 357
tissues, 364
urine, 362
Chloride transport,
components of, 28
coupling to hydrogen, 31–34
historical background, 27–28
intestinal, 269, 366
Chloromercuribenzoate, chitinase secretion and, 265, 266
Chloromethyl ketones, chymotrypsins and, 259–260
Chlorophis, see *Philothamnus*
Chlorosoma, see *Philodryas*
5α-Cholestane-3α,7α,12α,26,27-pentol,
formation of, 72
occurrence of, 68
5α-Cholestane-3α,7α,12α,26-tetrol,
allocholate synthesis and, 72
occurrence of, 69
5β-Cholestane-3α,7α,12α,26-tetrol, occurrence of, 69
5α-Cholestane-3α,7α,12α-triol, cholesterol modifications and, 71, 72
5α(or β)-Cholestane-3α,7α,12α-triol, hydroxylation of, 71, 73
5β-Cholestane-3α,7α,12α-triol,
cholesterol modifications and, 71, 72
transformations of, 72
5α-Cholestan-26-oic acid, occurrence of, 68
Cholesterol,
bile salts and, 67, 69, 71, 72, 337, 343–346
bufotoxins and, 173
interrenal steroids and, 429
reptilian plasma, 212
salamandra alkaloids and, 165–166
seasonal variations, 363
testis and, 137
Cholic acid,
biosynthesis, 71, 72, 73, 343–346
occurrence of, 68, 75, 347–350
Choline chloride, acid secretion and, 42
Cholinesterase, venoms and, 255, 466, 469, 470, 473
Choline sulfate, gastric potential difference and, 37
Chondodendron tomentosum, toxin, activity, 163
Chrysemys, prothrombin of, 210

Chrysemys belli,
digestive juices,
enzymes, 255, 256
pH, 254
Chrysemys cinera,
digestive juices,
enzymes, 255
pH, 254
Chrysemys d'Orbignyi, hypophysectomized, 409
Chrysemys elegans, blood magnesium content, 369
Chrysemys picta,
blood seasonal variations,
blood cells, 359
mucopolysaccharides, 362, 370
nitrogenous compounds, 359
digestive juices,
enzymes, 256, 258
pH, 254
hemoglobin, 381
structure, 393
immunoglobulins of, 204
intestine
absorption in, 267–268
mucosa, 252
mitochondria, seasonal variation in, 363
parathyroid hormone effects, 299
renal function, corticosteroids and, 430, 431
tissue water content, season and, 364, 369
transferrin of, 198
Chymotrypsin, reptilian digestive juice, 256, 257, 259–260
Citrulline, metabolism of, 236–237, 239
Clemmys caspica, hemoglobin of, 394
Clemmys caspica, rivulata, digestive enzymes, 256
Clemmys guttata, hemoglobin, function, 385
Clemmys insculpta, bile salts of, 347
Clemmys leprosa,
digestive enzymes of, 256, 263
leukocytes, seasonal variation, 361
Cloaca,
interrelations with other excretory functions, 309–312
uric acid in, 303
water and ion excretion and, 308–309
Clostridium botulinum, toxin, activity, 163
Clostridium tetani, toxin, activity, 163
Clotting components, reptilian plasma, 208–210
Cnemidophorus lemniscatus, pituitary-interrenal axis in, 434
Cnemidophorus tigris,
esterases of, 212
transferrins of, 200
Coagulation defects, snake venoms and, 467, 468
Cobra toxin, occurrence and activity, 163
Color change, endocrine regulation, 149–152
Coluber,
hemoglobin of, 211
transferrin of, 196
Coluber constrictor,
hemoglobin of, 393
transferrins of, 200
Coluber constrictor mormon, bile salt of, 348
Coluber ravergieri, bile salt of, 348
Coluber viridiflavus, bile salt of, 348
Complement, reptilian, 202
Conolophus subcristatus,
salt gland, 330
ion concentration in, 314, 329
Corallus canina, bile salts of, 348
Corallus enhydris, bile salt of, 348
Cordylus giganteus, erythroblasts, seasonal variation, 360
Cordylus vittifer, erythroblasts, seasonal variation, 360
Corticosteroids, carbohydrate metabolism and, 138
Corticosterone,
biosynthesis of, 429–430
blood glucose and, 138
corticotropin and, 135, 435–436
interrenal and, 133
molting and, 148
nitrogen metabolism and, 139
osmoregulation and, 431, 432
protein binding, 212
salt gland and, 320, 321
sodium transport and, 142
water metabolism and, 143

Corticotropin,
adenohypophysis and, 129–130
biochemistry of, 436–437
blood glucose and, 138
corticosterone metabolism and, 435–436
definition of, 428–429
interrenal organs and, 134–135
lipids and, 139
molting and, 148
pars distalis and, 127–128
regulation of, 434–435
secretion, hypothalamus and, 127
water metabolism and, 143
Cortisol,
liver glycogen and, 138
nitrogen metabolism and, 139
occurrence of, 430
sodium retention and, 143
water metabolism and, 143
Cortisone,
antibodies and, 204
growth hormone and, 407
Corynebacterium diphteriae, toxin, activity, 163
Creatine phosphate, hibernation and, 363
Crocodilia,
albumins of, 193
amino acid interconversions in, 228
glutamine and glutamate from non-nitrogenous precursors, 229–231
glycine, alanine, glutamate and glutamine, 229
transamination, 229
bile salts of, 342–343
glucose and glycogen levels in, 218
hemoglobins, 384
functions of, 391–392
immunoglobulins of, 201
plasma proteins of, 190
urinary function in, 311–312
Crocodylus, plasma proteins, comparative aspects, 192
Crocodylus acutus,
bile salts of, 342, 350
cloaca, function, 309
extrarenal excretion in, 326–327
glomerular filtration rate, 281, 284–285
hemoglobin of, 384
potassium excretion in, 297–298
salt gland, ion concentration in, 316, 317
sodium transport in, 294
urinary function in, 311–312
urine, osmolality, 290
Crocodylus johnsonii,
bile salts of, 350
glomerular filtration rate, 281, 284, 286
Crocodylus niloticus, bile salts of, 342, 350
Crocodylus porosus,
glomerular filtration rate, 281, 284, 286
salt gland, ion concentration in, 316, 317
Crocodylus siamensis, habitats, 317
Crotalus,
cloaca, function, 308
gonadotropin, neuraminidase and, 420
water reabsorption in, 270
Crotalus adamanteus,
bile salt of, 350
venom,
albumin and, 195
electrophoresis of, 463
phospholipases of, 470–471
potency, 452
Crotalus atrox,
amylase of, 255
bile salt of, 350
digestive enzymes, 255, 257
neurohypophysis, arginine vasopressin in, 286
venom,
electrophoresis of, 463
phospholipases of, 471
potency, 163, 452
Crotalus cerastes, venom, potency, 452
Crotalus confluentes, bile salt of, 350
Crotalus durissus terrificus,
hemoglobin of, 384
venom,
fractionation of, 462–463
potency, 452
Crotalus horridus,
bile salt of, 350
blood seasonal variation,
blood cells, 360
inorganic ions, 358
nitrogenous compounds, 359
hemoglobin, function, 390

Crotalus horridus atricaudatus, venom, fractionation of, 463
Crotalus horridus horridus,
venom,
fractionation of, 463
potency, 452
Crotalus oregonus, bile salt of, 350
Crotalus ruber ruber, venom, potency, 452
Crotalus scutulatus, venom, potency, 452
Crotalus terrificus, bile salt of, 350
Crotalus viridis helleri,
venom,
amino acid composition, 454
fractionation of, 463
pharmacological activities, 464
potency, 452
Crotalus viridis lutosus, venom, potency, 452
Crotalus viridis viridis,
venom,
fractionation of, 463
potency, 452
Crotaphopeltis natamboeia, bile salt of, 348
Crotaphytus, progesterone and, 427
Cryptobranchus,
hemoglobin, 104
functional properties, 100
neurohypophysis, 124
Cryptobranchus alleganiensis,
erythrocytes of, 80
hemoglobin of, 90
Cryptodira, plasma proteins, comparative aspects, 192
Ctenosaura pectinata, nasal gland secretion, 313, 315, 329, 330, 331
Curare, occurrence and activity, 163
Cyanide,
chitinase secretion and, 265, 266
glucose uptake and, 267
toxicity of, 163
Cycloheximide,
aldosterone effects and, 142
amino acid interconversion and, 229
Cycloneosamandaridine, chemical properties, 165
Cycloneosamandione, chemical properties, 165
Cyclura carinata, bile salts of, 347
Cylindrophus rufus, bile salt of, 348
5α-Cyprinol, occurrence and structure, 68, 75
5β-Cyprinol, occurrence of, 75
Cysteine, chitinase secretion and, 266
Cysteine residues, amphibian hemoglobin and, 102, 108, 114
Cystine, metabolism in reptiles, 239
Cystine residues, venom polypeptides and, 461
Cytochrome(s), egg mitochondria and, 8
Cytochrome oxidase, embryonic development and, 8, 9

D

Dasia smaragdinum, transferrins of, 198
Dehydration,
glomerular filtration rate and, 279–281
urea synthesis and, 302
Dehydrobufotenine,
occurrence and activity, 163
structure, 171
Dendroapsis angusticeps, bile salt of, 349
Dendroaspis angusticeps, venom, potency, 453
Dendroapsis jamesoni kaimosae, bile salt of, 349
Dendroapsis viridis, bile salt of, 349
Dendrobates auratus,
toxin, 177
activity, 163
Dendrobates histrionicus, toxin of, 177
Dendrobates pumilio, toxin, activity, 163, 177
Dendrobatidae, toxins of, 175–178
Deoxycholic acid, occurrence of, 68, 350
27-Deoxy-5α(or β)-cyprinol,
formation of, 71, 73
oxidation of, 71, 73
26-Deoxy-5β-ranol, formation of, 72, 73
Deoxyribonuclease, snake venom, 469
Deoxyribonucleic acid,
mitochondrial, oocytes and embryos and, 9–12
synthesis, hemoglobin and, 116
Dermatemys, plasma albumin of, 194

Dermatemys mawii, hemoglobin, function, 385–386, 387, 392
Dermochelys coriacea, salt gland, 318
Desmognathus fuscus, hemoglobin of, 90
Desmognathus monticola, hemoglobin of, 90
Desmognathus ochrophaeus, hemoglobin of, 90
Desmognathus quadramaculatus,
  erythrocytes of, 80
  hemoglobin of, 90
Dexamethasone, corticosterone and, 435
Diamox, amphibian gastric secretion and, 24
Dibenamine, melanophores and, 150
Dicamptodon ensatus,
  erythrocytes of, 80
  hemoglobin of, 89 ,104
    functional properties, 98
    larval, 91
    polymorphism, 91–92
Dicoumarol, chitinase secretion and, 265
Diemyctylus pyrrhogaster, bile salts of, 68, 75
Diemyctylus viridescens, molting, hormones and, 148
Diethylaminoethyl cellulose, gonadotropins and, 421–422
Digestion,
  anatomical considerations,
    gross, 251
    microscopic, 251–252
  mechanism, 260–261
    gastric secretion, 261–267
    intestinal absorption, 267–270
25,26-Dihomo-5α-cholane, 3α,7α,12α,24ξ, 26-pentol, occurrence of, 68
Dihydroisohistrionicotoxin, structure, 177
3α,7α-Dihydroxy-5β-cholestan-26-oic acid, occurrence of, 350
7α,12α-Dihydroxy-5α-cholestan-3-one, formation of, 71, 72
7α,12α-Dihydroxy-5β-cholestan-3-one, formation of, 71, 72
7α,12α-Dihydroxy-4-cholesten-3-one, transformation of, 71, 72
3α,12α-Dihydroxy-7-oxo-5-β-cholan-24-oic acid, occurrence of, 348
Diisopropyl fluorophosphate, proteases and, 260
N,N-Dimethylhistamine, occurrence of, 174
2,4-Dinitrophenol,
  chitinase secretion and, 265
  glucose uptake and, 267
  potassium transport and, 44
  sodium transport and, 39
2,3-Diphosphoglycerate,
  erythrocyte content, 93, 95
  hemoglobin and, 96, 97, 107
Diphtheria toxin, occurrence and activity, 163
Dipsas variegata trinitas, bile salt of, 348
Dipsosaurus, hemoglobin of, 384
Dipsosaurus dorsalis,
  amylase of, 255, 257
  antibody synthesis by, 203, 204
  bile salt of, 347
  blood oxygen capacity, 361
  cloaca, function of, 308–309
  estrogen effects, 427
  hemoglobin, function, 388, 389
  metabolic rate, seasonal variation, 356
  osmoregulation, prolactin and, 408
  ovary, prolactin or growth hormone and, 407
  salt gland, 330
    corticoids and, 432
    ion concentration, 315, 331, 332
  urate excretion by, 302
  water reabsorption in, 270
Discoglossidae,
  hemoglobins of, 83
  toxins of, 179
Discoglossus pictus, bile salts of, 70, 74, 75
Dispholidus typus, bile salt of, 348
Dog,
  amino acid deamination in, 244
  nitrogen transport in, 227
Dopamine,
  bufonid toxins and, 167
  hypothalamus and, 124, 125
Drought, seasonal variation and, 371–372
Drymarchon corais couperi, bile salt of, 348
Dryophis, see *Anaetulla*
Duberria lutrix, bile salt of, 348

E

Echis carinatus,
bile salt of, 349
venom, potency, 453
Egernia, complement of, 202
Egernia cunninghami,
amino acid metabolism, 237–241
immunoglobulins of, 202, 204
liver, glycogen content, 218
Elaphe moellendorfi, bile salt of, 348
Elaphe quadrivirgata, bile salt of, 348
Elaphe quatorlineata, bile salt of, 348
Elaphe situla, bile salt of, 348
Elaphe taeniurus, bile salt of, 348
Elaphe varinata, bile salt of, 348
Electrophoresis,
albumins, 195
hemoglobin components and, 81–82
pituitary extracts and, 410
plasma proteins and, 191–192
Eleutherodactylus discoidalis, hemoglobins of, 83
Embryo(s), mitochondrial nucleic acids, 9–12
Embryonic development, ribosome synthesis and accumulation, 17–19
Emys europaea, digestive enzymes, 256
Emys orbicularis,
adrenocorticotropic hormone effects, 433
bile salts of, 338, 347
digestive enzymes of, 256, 263
immunoglobulins of, 204
Endocrine system,
annual cycle and, 365
color change regulation,
cellular action, 150–152
hormone release, 149–150
gonadal function and reproduction and, 148–149
molting induction and, 147–148
regulation of metabolic processes,
energy metabolism, 137–140
growth, 146–147
osmomineral regulation, 140–144
regulation of metamorphosis, 144–146
structure and function,
branchiogenic glands, 131–133
hypothalamohypophysial system, 123–131
mesodermal glands, 133–137
pancreatic islets, 133
Endoplasmic reticulum, oxyntic cells and, 25–27
Energy,
anaerobic production of, 220
gastric secretion and, 37–38
Energy metabolism, regulation of, 137–140
Enhydrina schistosa,
bile salt of, 349
venom, potency, 453
Enhydris enhydris, bile salt of, 349
Enhydris pakistanica, bile salt of, 349
Enhydris plumbea, bile salt of, 349
Environment,
arid terrestrial, salt glands and, 330–334
kidney function and, 278, 280–282
marine, salt glands and, 318–330
Enzymes, venoms and, 465–474
Epicrates cenchria, bile salt of, 348
Epinephrine, blood lactate and, 219–220
Epinine, bufonid toxins and, 167, 170
Eretmochlys imbricata, salt gland, 318
Ergosterol, bile salts and, 70
Eristocophis macmahoni, bile salts of, 350
Erucate, uptake of, 267
Eryx conicus, bile salt of, 348
Eryx jaculus, bile salt of, 348
Eryx johnii, bile salt of, 348
Esophagus, glands in, 252
Esterases, reptilian plasma, 211–212
Estradiol-17β,
biosynthesis of, 424
lipid and, 140
yolk synthesis and, 7, 207–208
Estrogens,
physiological action, 427
plasma proteins and, 191, 205, 207
Estrone, biosynthesis of, 424
Ethylenediaminetetraacetate, carboxypeptidase and, 260
Ethyleneimine, reptile hemoglobin and, 387
Eumeces obsoletus,
amylase of, 255, 257

hemoglobin, function, 389
Eunectes murinus, bile salts of, 348
Eurycea bislineata, erythrocytes of, 80
Evolution,
reptilian albumins and, 195–196
snake venoms and, 449–450
transferrins and, 198–200
Excretion,
extrarenal, 312–318
arid terrestrial environment, 330–334
marine environment, 318–330
Exopeptidases, snake venom, 469
Eye(s), melatonin and, 150

F

Fasting, digestive juice pH and, 254
Fat body,
seasonal variations, 363
yolk synthesis and, 207
Fatty acids,
absorption of, 267
seasonal variations, 363
Feeding, digestive juice pH and, 254
Fibrin, formation, reptilian plasma and, 209
Fluoride, chitinase secretion and, 265
Follicle stimulating hormone,
action of, 417–419
subunits and, 419
lipid and, 140
localization of, 414
Foods, electrolyte content, 324
Formiminoglutamate, formation of, 240
Fowlea, see *Xenochropis*
Fructose,
intestinal transport, 366
metabolism of, 223
uptake of, 268

G

D-Galactosamine, metabolism of, 223
D-Galactose,
metabolism of, 223
uptake of, 267–268
Garter snake,
digestive juices, pH, 254
transcortin of, 212
Gastric mucosa,
osmotic properties, 45–46
structure and function, 24–27
Gastric potential difference, 34
ionic gradients and,
mucosal permeability, 36–37
serosal permeability, 35–36
Gastric secretion,
cellular origin, 27
electrical aspects,
ionic gradients and gastric potential difference, 35–37
site of gastric potential difference, 34–35
energetic aspects,
oxygen consumption, 38
phosphate metabolism, 38–42
requirements, 37–38
hydrochloric acid production, 261–262
in vivo and *in vitro*, 24
ionic requirements,
calcium, 42–43
potassium, 42
sodium, 42
pH of, 254
Gastrin, amphibian response to, 23
Gavialis gangeticus, bile salts of, 350
Genes,
human hemoglobins and, 80–81
ribosomal, amplification of, 14
Geochelone carbonaria, hemoglobin of, 379
Geochelone denticularia, hemoglobin of, 379
Geophagy, function of, 250
Gerrhonotus multicarinatus,
hemoglobin, 381
function of, 389
Gillichthyes, prolactin assay and, 409, 411
Glomerular filtration, salt acclimation and, 53–54
Glomerular filtration rate, regulation of water and ion excretion and, 278–288
Glomerulus, interrelations with other excretory functions, 309–312
Glucagon,
blood glucose levels and, 137–138
effects in reptile, 220–221
Gluconate, hydrogen secretion and, 30
3-*O*-Glucopyranose, uptake of, 268

D-Glucosamine, metabolism of, 223
Glucose,
  aldosterone effect and, 62, 63
  blood,
    growth hormone and, 406–407
    seasonal variations, 358, 369
  chitinase secretion and, 264
  ion secretion and, 44
  metabolism of, 223
  uptake of, 267, 268, 366
Glucose-6-phosphatase, thyroxine and, 139
Glucuronate, hydrogen secretion and, 30
Glutamate,
  deamination of, 226, 229
  glutamine synthesis and, 231
  metabolism in reptiles, 238
  proline and, 236
  synthesis, chameleon and, 235
  uptake of, 268
Glutamate-oxaloacetate transaminase, concentrations, 242, 243
Glutamine,
  ammonia excretion and, 232
  bicarbonate synthesis and, 232
  reptile tissues and, 224, 225, 227
  salamandra alkaloids and, 165
  synthesis of,
    alligator, 229, 231
    chameleon, 235–236
  transamination and, 229
Glycera, hemoglobin of, 79
Glycine,
  ammonia excretion and, 232
  bicarbonate synthesis and, 232
  bile salts and, 67
  deamination *in vivo*, 243–244
  metabolism in reptiles, 237
  reptile tissues and, 224, 225, 227
  synthesis of,
    alligator, 229
    chameleon, 236
  uptake of, 268
Glycogen,
  embryonic development and, 9
  hepatic,
    corticoids and, 433
    levels, 138, 139, 218
  seasonal variation, 363
Glycosidases, reptilian plasma, 212
Gonad(s),
  endocrine secretions, 423–428
  hypothalamus and, 125–126
Gonadotropin(s),
  chemical fractionation, 421–422
  estrogen synthesis and, 136
  growth hormone and, 407
  nonreptilian, actions of, 415–419
  prolactin and, 407–408
  reptilian,
    biological characterization, 422–423
    chemical studies, 420–422
    endogenous, 413–415
  species differences, 419, 420, 443
  testis and, 137
Gonglyophis, see *Eryx*
Gopherus, estivation in, 368
Gopherus agassizii,
  ammonia, urea and urate excretion by, 300, 301, 302
  bladder, function, 306
  dehydration in, 372
  extraurinary water loss, 282
  potassium excretion by, 296
  renal function, 278, 310
    glomerular filtration rate, 279, 280
  sodium transport in, 292
  urine, osmolality, 289
Gopherus berlandieri, urea excretion by, 302
Gopherus polyphemus, hemoglobin, function, 385
Graptemys geographica,
  blood seasonal variations,
    blood cells, 360, 361
    inorganic ions, 357
    plasma volume, 361
  weight, seasonal variation, 356
Grass snake, digestive juices, pH, 254
Growth, hormone, *see* Somatotropin
Gymnophiona, neurohypophysis, 124

H

Haemachatus haemachatus,
  bile salt of, 349
  venom polypeptides,
    amino acid sequences, 456–457, 460–461
    purification of, 460–461

Hageman factor, reptilian plasma and, 210
Haptoglobin, reptilian plasma and, 210–211
Harderian glands, *see* Salt gland
Heart,
  bufotoxins and, 174
  ranid toxins and, 179
Helicops angulatus, bile salt of, 349
Heloderma, estivation in, 368
Heloderma horridum, bile salt of, 339, 347
Heloderma suspectum,
  copper-containing protein of, 210
  erythrocytes, phosphates in, 387
  venom, autoantibodies and, 201
Hematin, binding, albumin and, 194
Hemidactylus,
  glomerular filtration rate, 281, 284
  sodium transport in, 293–294
  urine, osmolality, 290
Hemidactylus flaviviridis,
  gonadotropin effects, 418
  testosterone and, 426
Hemoglobin(s),
  adult,
    amphibian, 82–86, 89–91
    reptilian, 378–384
  amino acid composition, 102
  association, dissociation and polymerization, 101–102
  comparative studies, value of, 77–79
  components, 79–82
    anura, 82–89
    apoda, 92
    urodela, 89–92
  embryonic, 88–89, 384
    components, 81
  end group studies, 102–103, 104
  function, 92–93
    anura, 93–98
    chelonia, 385–388
    crocodilia, 391–392
    squamata, 388–391
    urodela, 98–100
  haptoglobins and, 210–211
  larval, 86–88, 91
    functional properties, 94
  sequence studies, structure-function correlations, 104–111
  structure,
    anura, 101–103
    reptilia, 393–394
    sequence studies, 104–111
    urodela, 103–104
  synthesis, 393–394
    larval to adult transition, 115–116
    seasonal variation and phenylhydrazine treatment, 111–113
    sites of, 113–115
  transferrin and, 197
Hemolysis, venom phospholipases and, 471
Hemorrhage, snake venoms and, 467–468
Heparin, blood clotting and, 209
Herbivores,
  colon length, 251
  reptilian, 250
    digestive enzymes, 255
Heterocholic acid, occurrence of, 347
Heterodon, haptoglobin in, 210
Hexose monophosphate shunt, aldosterone and, 61
Hibernation, reptiles and, 354–355
Hippuryl-L-phenylalanine, carboxypeptidase and, 260
Histamine,
  digestive juice and, 254, 261, 262, 263
  envenomation and, 451
  gastric osmotic permeability and, 45
  hydrogen secretion and, 33
  occurrence of, 174, 178
  phospholipase and, 471
Histidine,
  metabolism of, 237, 240
  uptake of, 268
Histidine residues, amphibian hemoglobins and, 106, 107, 110
Histrionicotoxin, structure, 177
Homalopsis buccata, bile salt of, 349
Homobatrachotoxinin, structure, 175
Hormones, water balance and, 56–59
Human, hemoglobin, components of, 80–82
Hyaluronidase, venom and, 255, 466, 469, 472, 473
Hybrids, bufonids, hemoglobins of, 84–85
Hydrochloric acid, secretion of, 253, 261–262

Hydrocortisone,
growth hormone and, 407
hyperglycemia and, 221
Hydrogen ions,
gastric serosa and, 35
transport,
coupling to chloride, 31–34
regulation, 295–298
separate mechanism, 30–31
Hydrophis cyanocinctus, bile salt of, 349
Hydroxide ions, gastric serosa and, 35
β-Hydroxybutyrate, aldosterone effect and, 62
18-Hydroxycorticosterone,
biosynthesis of, 429, 430
interrenal and, 133
Hydroxyproline, metabolism of, 240
5-Hydroxytryptamine, derivatives, toxins and, 174
5-Hydroxytryptophan, formation of, 240
Hyla arborea,
hemoglobins of, 83
toxin of, 178
Hyla arenicolor, xanthophores, regulation of, 151
Hyla caerulea, toxin of, 178
Hyla cinerea, hemoglobins of, 83
Hyla crucifer, erythrocytes, size, 79, 80
Hyla pearsoniana, toxin of, 178
Hyla peroni, toxin of, 178
Hyla trachytorax, hemoglobins of, 83
Hylidae,
hemoglobins of, 83–84
toxins of, 178
Hynobius tsuensis, hemoglobin of, 90
Hyperglycemic agents, carbohydrate metabolism and, 221
Hyperphagia, prolactin and, 405
Hypophysectomy,
growth and, 408–409
interrenals and, 434
iodine uptake and, 440
molting and, 147–148
Hypothalamohypophysial system,
structure and function, 123
adenohypophysis, 129–131
neural portion, 124–129
Hypothalamus,
function of, 124–127
metamorphosis and, 144–145
pituitary secretions and, 402
Hypsirhina, see *Enhydris*

I

Iguana,
bile acid of, 344
marine, temperature fluctuations, 187
Iguana, fibrinogen of, 209
Iguana iguana,
albumin, evolutionary aspects, 195
bile salt of, 347
carotenoid-binding protein of, 213
digestive enzymes, 257
hemoglobin, function, 389
plasma osmotic pressure, 194
salt gland, ion concentration in, 315
transferrin of, 196
Immunoglobulins,
reptilian,
induction, 202–205
isolation and identification, 200–202
Immunology,
reptilian plasma proteins and, 192
snake venoms and, 467
Indolalkylamines, bufonid toxins and, 170–171
Indole propionate, binding, albumin and, 194
Inorganic ions, blood, seasonal variations, 357–358
Inositol hexaphosphate,
erythrocyte content, 95
hemoglobin and, 97, 387–388
Insulin,
amino acid deamination and, 226–227
ammonia excretion and, 232
growth hormone and, 406–407
sodium uptake and, 63
sugar metabolism and, 137, 223–224
Interrenal corticoids,
physiological action,
carbohydrate regulation, 433
water and electrolytes and, 430–433
Interrenal glands,
definition of adrenocorticotropic hormone, 428–429
products of, 133–134
steroidogenesis, chemistry of, 429–430
structure and function, 133–135
Intestinal juice, pH of, 254

Intestine,
absorption in,
amino acids, 268–269
fatty acids, 267
other substances, 269–270
sugars, 267–268
inorganic ions, season and, 364
permeability, season and, 366–367
sodium elimination and, 54–55
urate in, 303
Invertase, reptilian digestive juice, 256, 257
Iodide,
acid secretion and, 30
chitinase secretion and, 264, 265, 266
Iodine, metabolism of, 437–439
Iodoacetamide, hemoglobins and, 96, 390–391
Iodoacetate,
chitinase secretion and, 265
glucose uptake and, 267
reptile hemoglobin and, 387
Iodosobenzoate, chitinase secretion and, 265
Iridophores, hormonal regulation, 151
Iron, transport, transferrins and, 196–198
Isethionate,
hydrogen transport and, 30
potential differences and, 36
Isocitrate dehydrogenase, concentrations, 242
Isoleucine, metabolism in reptiles, 238
Isomaltase, reptilian digestive juice and, 257

K

α-Keto acids, metabolism of, 228
α-Ketoglutarate,
metabolism of, 228
transamination and, 229
uptake of, 235
Kidney, erythropoiesis in, 114, 115
Kinosternon, plasma albumin of, 194
Kinosternum, gonadotropin, neuraminidase and, 420

L

Lacerta, estrogen synthesis in, 424
Lacerta agilis, estivation in, 368
Lacerta sicula,
growth hormone and prolactin effects, 405
pituitary-interrenal axis in, 434
reproduction, prolactin and, 407
testosterone and, 426
Lacerta viridis,
digestive enzymes of, 257, 258, 263–267
estivation in, 368
hydrochloric acid secretion by, 262
urea excretion by, 301
Lacerta vivipara,
fat bodies, seasonal variations, 363
hemoglobin of, 393
Lachesis muta, bile salt of, 350
Lachesis mutus, venom, potency, 452
Lactate,
intestinal permeability to, 269
reptilian blood, 188, 219
Lactate dehydrogenase, concentrations, 242, 243
Lactose, chitinase secretion and, 264, 266
Lamprey, hemoglobin of, 79
Lampropeltis, haptoglobin in, 210
Lampropeltis getulus,
bile salt of, 348
erythrocytes, phosphates in, 387
Lampropeltis getulus holbrooki, haptoglobin of, 211
Laticauda colubrina,
glomerular filtration rate, 280, 283, 286
sodium chloride excretion, 279
Laticauda laticauda, venom, toxin of, 455
Laticauda semifasciata,
extrarenal excretion in, 313, 315, 329
venom, toxins of, 455–458
Lepidochelys olivacea
salt gland, 318
ion concentration in, 314, 323, 325
Lepidosauria, bile salts of, 339–342, 347–350
Leptodactylidae, hemoglobins of, 83
Leptodactylin,
biosynthesis, 174
occurrence and activity, 163, 174
Leptodactylinae, toxins of, 174–175
Leptodactylus bufonius, hemoglobin of, 83

Leptodactylus chaquensis, hemoglobins of, 83
Leptodactylus fallax, hemoglobins of, 83
Leptodactylus luticeps, hemoglobins of, 83
Leptodactylus ocellatus,
hemoglobins of, 83
toxins of, 174
Leptodactylus pentadactylus,
hemoglobin of, 101
toxins, activity, 163, 174
Leptodactylus pentadactylus labyrinthicus, skin gland, imidazolylalkylamines in, 174
Leptodactylus podicipinus, hemoglobins of, 83
Leptospira, reptilian immunoglobulins and, 201
Leptotyphlops, hemoglobin of, 211
Leucine, metabolism in reptiles, 238
Leukocytes, seasonal variations in, 361
Lialis burtonis, bile salt of, 347
Lichnura, hemoglobin of, 211
Linoleate, uptake of, 267
Lipid,
estradiol and, 140
lipovitellin and, 4
metabolism, aldosterone and, 61
plasma vitellin and, 206
prolactin and, 139, 405
seasonal variations in, 363, 369
water loss and, 365
yolk protein and, 207
Lipovitellin, yolk and, 4–6
Lithophagy, function of, 250
Liver,
bile salts and, 67
erythrocyte maturation and, 112, 114, 115
glycogen content, 218
seasonal variations in, 363, 369
yolk proteins and, 6
Lizards,
body temperature, 187
herbivorousness of, 250
venomous, 449
Luteinizing hormone,
action of, 418–419
amino acid composition, 421
localization of, 414
oocyte meiosis and, 149
reptilian, tests for, 423
Lysine, metabolism in reptiles, 240
Lysine residues, frog hemoglobin, 105, 107
Lysolecithins,
binding, albumin and, 195
hemolysis and, 471
Lysosomes, enzymes, metamorphosis and, 146
Lytic factors, snake venoms and, 467

M

Macrochelys temmincki, bile salts of, 347
Magnesium, blood, seasonal variations, 358, 369
Malaclemys centrata,
blood seasonal variations,
blood cells, 360–361
inorganic ions, 357–358, 367
nitrogenous compounds, 359
weight, seasonal variation, 356
Malaclemys centrata centrata, digestive enzymes, 256, 258, 259, 263
Malaclemys geographica,
blood seasonal variation,
nitrogenous compounds, 359
Malaclemys terrapin, salt gland, ion concentration in, 314, 329
Malaclemmys terrapin centrata,
hemoglobin of, 381, 384
oxygen affinity, 388
Malate dehydrogenase, concentrations, 243
Malayemys subtrijuga, hemoglobin, function, 385
Malonate, glucose uptake and, 267
Malpolon monspesselana, bile salt of, 348
Maltase,
reptilian digestive juice, 256, 257
turtle serum, 212
Maltose, chitinase secretion and, 264
Mannitol, acid secretion and, 42
D-Mannoheptulose, metabolism of, 223
D-Mannose, metabolism of, 223
Marine environment, salt glands and, 318–330
Marinobufagin, structure, 172
Mecholyl, potential differences and, 35

Megalobatrachus japonicus, bile salts of, 68, 71, 75
Melanocyte-stimulating hormone,
  adenosine 3′,5′-monophosphate and, 150
  adrenocorticotropic hormone and, 436–437
  cells producing, 131
  inhibition of, 124, 125
  release, stimulus, 149
Melanophores, division rate, 151–152
Melatonin, effects, 149–150
Membranes, seasonal variations and, 365–367
Mercaptoethanol,
  hemoglobin and, 82, 83, 101, 387, 393
  immunoglobulins and, 201
Mesodermal glands,
  structure and function,
    gonads, 136–137
    interrenal organs, 133–135
Mesotocin, water balance and, 56
Metabolic rate,
  body size and, 246
  body temperature and, 241
  enzyme concentration and,
    in vitro, 242–243
    in vivo, 243–244
  substrate concentration and, 242
  substrate delivery as a function of blood flow, 244–246
Metamorphosis,
  hemoglobin synthesis and, 115–116
  regulation, endocrine system and, 144–146
Metapyrone, corticosteroids and, 434
β-Methasone, hepatic glycogen and, 433
Methemoglobin,
  oxygen affinity studies and, 97, 99, 101
  reptile blood and, 385, 389
Methionine,
  deamination of, 225, 227
  interconversion of, 237, 239
O-Methylbufotenine, occurrence and activity, 163, 171
3-*O*-Methylglucose,
  metabolism of, 223
  uptake of, 267–268
6-*O*-Methylglucose, metabolism of, 223
α-Methylglucoside, uptake of, 268
N-Methylhistamine, occurrence of, 174
N-Methyl-serotonin, occurrence of, 174
Methylxanthines, gastric secretion and, 42
Metopirone, corticotropin and, 130, 135
Microorganisms, bile salts and, 67, 337
Micrurus fulvius,
  venom,
    fractionation of, 462
    potency, 452
Micrurus nigrocinctus, venom, 470
Mitochondria,
  deoxyribo- and ribonucleic acids in oocytes and embryos, 9–12
  interrenal tissue, 135
  renal tubules and, 293–294
  structure and activity in oocytes and embryos, 8–9
  yolk platelets and, 6
Molting,
  induction, endocrine system and, 147–148
  prolactin and, 406
Morelia spilotes, immunoglobulins of, 201
Mucopolysaccharides, blood, seasonal variations, 362, 370
Mucous cells, gastric mucosa, 24
Muscarine, occurrence and activity, 163
Muscle,
  glycogen, 218
    conversion to lactate, 219
Myoglobins, hemoglobins and, 79
Myosin-adenosine triphosphatase, lizard, 258

N

Naja goldi, bile salt of, 349
Naja haje,
  bile salt of, 349
  venom polypeptide, amino acid sequence, 456–457
Naja haje haje, venom polypeptide, purification of, 460
Naja melanoleuca, bile salt of, 349
Naja naja,
  bile salt of, 349
  venom polypeptides,
    amino acid sequences, 456–457

antigenicity, 461
potency, 163, 453
Naja naja atra,
venom polypeptides,
amino acid sequences, 456–457
pharmacological effects, 459, 460
purification of, 458–459
Naja naja naja,
neurotoxin,
amino acid sequence, 456–457
composition, 455
Naja naja siamensis,
neurotoxin,
amino acid sequence, 456–457
composition, 455
Naja nigricollis, 470, 472
bile salt of, 349
venom polypeptide,
amino acid sequence, 456–457
purification of, 461
Naja nivea,
bile salt of, 349
venom polypeptides, amino acid sequences, 456–457
Naphthylamidases, reptilian plasma, 211
Nasal glands, *see* Salt glands
Natrix,
excretion of,
ammonia, 301
calcium and phosphate, 298–299
potassium, 296, 297
urate, 303–304, 305
fibrinogen of, 209
progesterone in, 425
transferrin of, 196, 199, 200
Natrix cyclopion, renal function, corticosteroids and, 430–431
Natrix cyclopion cyclopion, sodium transport in, 292
Natrix cyclopion floridana, sodium transport in, 292
Natrix natrix,
bile salt of, 348
blood seasonal variations,
blood cells, 361
glucose, 358
inorganic ions, 357
corticosterone in, 430
digestive juices,
enzymes, 257
pH, 254
hypophysectomized, growth of, 409
thyroid gland, annual cycle, 365
Natrix rhombifera, naphthylamidase of, 211
Natrix sipedon,
glomerular filtration rate, 279, 280, 286–287
renal tubular function, 290–291
urinary function in, 311
urine, osmolality, 289
Natrix taxispilota,
erythrocytes, phosphates in, 387
extraurinary water loss, 282
hemoglobin,
function, 390–391, 392
structure, 393
Necturus,
gastric mucosa, potential difference, 34
hemoglobin, 104
functional properties, 100
neurohypophysis, 124
Necturus maculosus,
aldosterone and, 143
erythrocytes of, 80
hemoglobin of, 90
functional properties, 98
insulin effects, 137
mitochondrial deoxyribonucleic acid, 9
testosterone in, 136
Neostigmine, cobra neurotoxin and, 459
Nerve growth factors, snake venom, 467
Nervous system, snake venoms and, 467
Neuraminidase, follicle stimulating hormone and, 420–421
Neurohypophysis, water balance and, 56, 140–141
Neurotoxins, properties of, 454
Newt, prolactin assay and, 409
Nitrate, acid secretion and, 30
Nitrogen,
regulation of excretion,
ammonia, 299–301
urate, 302–306
urea, 301–302
transport of, 227–228
Nitrogenous compounds, blood seasonal variations, 358–359

Nonprotein nitrogen, blood seasonal variation, 358
Noradrenaline,
  melanocyte-stimulating hormone and, 125
  occurrence and activity as toxin, 163, 167, 170
Notechis, prothrombin of, 210
Notechis scutatus,
  bile salt of, 349
  venom, potency, 453
Notophthalmus viridescens, growth, hormonal effects, 147
Nuclei, amphibian erythrocytes and, 79–80
Nucleolus, ribosomal nucleic acids and, 12–13
5′-Nucleotidase, snake venom, 469, 472–473
Nucleotide pyrophosphatase, snake venom, 469
Nux vomica, toxin, activity, 163

O

Oleate, uptake of, 267
Omnivores, reptilian, 250
Oocytes,
  meiosis, hormones and, 148–149
  mitochondria, 8
    nucleic acids, 9–12
Oogenesis,
  ribosome synthesis and accumulation, 14–17
  yolk accumulation and, 6–7
Ophidia, interrelation among various excretory functions, 311
Ornithine,
  glutamine synthesis and, 235
  metabolism in reptiles, 239
Osmoregulation,
  corticosteroids and, 430–433
  hormones and, 140–144
  prolactin and, 408
Osmotic pressure,
  body fluids and, 242
  reptilian plasma, 193–194
Osteolaemus tetraspis, habitat, 317
Ouabain,
  chitinase secretion and, 265
  glucose uptake and, 267
  hydrochloric acid secretion and, 262
  sodium transport and, 43, 295
Ovary,
  endocrine function, 136
  gonadotropins and, 418
  hypophysectomy and, 414
  steroid biosynthesis in, 424–425
Oviduct, estrogen and, 427
Oxaloacetate,
  alanine synthesis and, 236
  aldosterone effect and, 62
  glutamine synthesis and, 235
  transamination and, 229
11-Oxycorticosterone, interrenal and, 430
Oxygen,
  consumption,
    acid secretion and, 38
    aldosterone and, 62
    amino acid deamination and, 244
    embryos, 8
Oxyntic cells,
  gastric mucosa and, 24, 27
  potential difference and, 34
Oxytocin,
  chitinase secretion and, 264, 265
  glomerular filtration rate and, 288
  hydrolysis of, 211

P

Pancreas, growth hormone and, 407
Pancreatic islets, structure and function, 133
Parathyroidectomy, calcium and phosphate excretion and, 298–299
Parathyroid gland, structure and function, 132
Pelamis platurus, extrarenal excretion in, 313, 323, 325, 326, 329
Pelobates fuscus, hemoglobins of, 83
Pelusios sinuatus, bile salts of, 347
Pepsin, reptilian digestive juice, 256, 257, 262–263
Peptidase, venom and, 255
Peptide(s),
  hemolytic, 178
  hypotensive, 174–175
  toxic, 179

pH,
gastric potential difference and, 37
plasma vitellin and, 206
reptilian blood, 187–188
reptilian digestive juice, 253, 254
transferrins and, 198
urate excretion and, 302
Pharmacology, snake venoms and, 467
Phenylacetate, intestinal permeability to, 269
Phenylalanine,
bufonid toxins and, 167
interconversion of, 237, 241
p-Phenylenediamine oxidase, reptilian plasma and, 210
Phenylhydrazine, amphibian hemoglobin and, 111–113
Philodryas olfersi, bile salt of, 348
Philothamnus hoplogaster, bile salt of, 348–349
Phlorizin, glucose uptake and, 267
Phosphate, transport, regulation of, 298–299
Phosphate metabolism,
acid secretion and,
adenosine triphosphate, 38–40
adenosine triphosphate and thiocyanate, 40
cyclic adenosine monphosphate, 42
gastric adenosine triphosphatase, 40–41
Phosphodiesterase, venom and, 255, 469, 472
Phospholipase,
cobra toxin and, 461
venoms and, 462, 469, 470–472, 473
Phospholipid,
membrane permeability and, 61, 62
yolk protein and, 207
Phosphomonoesterase, venom and, 255, 469, 472
Phosphoprotein, plasma vitellin and, 206
Phosphoprotein phosphatase, yolk and, 8
Phosphorus,
blood seasonal variation, 358
phosvitin and, 4, 7
Phosphorylase, methylxanthines and, 221
Phosvitin,
plasma vitellin and, 206
yolk and, 4–6
Photoperiod, hibernation and, 370–371
Phrynosoma cornutum,
glomerular filtration rate, 281, 283
sodium transport in, 293
urine, osmolality, 290
Phrynosoma m'calli,
metabolic rate, seasonal variation, 356, 369
photoperiod and, 370
Phyllobates aurotaenia,
toxin,
activity of, 163
structure, 175
Phyllobates lugubris, toxin of, 176
Phyllobates vittatus, toxin of, 176
Phyllokinin, structure of, 179
Phyllomedusa, toxin of, 175
Phyllomedusa rhodei, toxin of, 179
Phyllomedusa sauvagei, hemoglobins of, 83
Phylodria,
cloaca, function, 308
water reabsorption in, 270
Physalaemus bresslaui, toxin of, 175
Physalaemus centralis, toxin of, 175
Physalaemus cuvierii, toxin of, 175
Physalaemus fuscumaculatus, toxin of, 175
Physignathus lesueri, bile salts of, 347
Pigeon, prolactin assay and, 409, 411
Pilocarpine, digestive juice pH and, 254, 261
Pineal gland,
gonads and, 129
light reception and, 125
melatonin and, 149, 50
Pipidae,
hemoglobins of, 83
toxins of, 179
Pituitary gland, *see also* Hypophysis, *etc.*
extracts, hyperglycemia and, 221
Pituitary-gonadal axis,
evidence for,
endogenous gonadotropic factors, 413–415
nonreptilian gonadotropins and, 415–419
Pituitary-interrenal axis,
evidence for,

in vitro, 435–436
in vivo, 433–435
Pituitary system,
reptilian,
biochemistry of hormones, 402–403
components of adenohypophysial system, 401–402
gross morphology and cytology, 400
Pituitary-thyroidal axes, reptilian, 440–441
Pituitrin, intestinal permeability and, 270
Pituophis, transferrin of, 196
Pituophis catenifer, hemoglobin, function of, 391
Pituophis catenifer affinis, extraurinary water loss, 282
Pituophis melanoleucus,
glomerular filtration rate, 279, 280
urinary function in, 311
urine, osmolality, 289–290
Pituophis sayi,
bile salt of, 349
digestive juices,
enzymes, 255, 257
pH, 254
Plasma, amino acid concentration, 224–225
Plasma protein(s),
poorly characterized,
ceruloplasmin and haptoglobin, 210–211
clotting components, 208–210
hydrolytic enzymes, 211–212
transport proteins, 212–213
reptilian,
comparative aspects, 192
composition and fractionation, 189–192
Plasma vitellin,
reptilian,
identification and properties, 205–206
vitellinogenesis and, 207–208
Plethodon cinereus, erythrocytes of, 80
Pleurodema bibroni, hemoglobin of, 83
Pleurodema bufonina, hemoglobins of, 83
Pleurodema nebulosa, hemoglobin of, 83
Pleurodema tucumana, hemoglobins of, 83
Pleurodira,
gonadotropins, neuraminidase and, 420–421
plasma proteins, comparative aspects, 192
Podocnemis unifillis, digestive enzymes, 256, 258
Polychrus marmoratus, bile salts of, 347
Polymorphism,
hemoglobins, 84
bufonid, 84
urodele, 91–92, 103–104
Pores, gastric mucosa, 45
Potassium,
acid secretion and, 42
amino acid uptake and, 269
chitinase secretion and, 265, 266
corticosteroids and, 431, 432
gastric mucosa and, 36–37
gastric serosa and, 35–36
membrane ionic conductance and, 30
salt glands and, 312, 313, 317–333
seasonal variations,
blood, 357–358
urine, 362
tissues, 364
transport, 43–44
intestinal, 269–270, 366
regulation, 295–298
urate excretion and, 302, 304–305
Potassium chloride, secretion of, 27
Pregnancy,
corpus luteum and, 428
progesterone synthesis and, 424–425
Pregnenolone,
corticosteroids and, 134
estrogen synthesis and, 424
interrenal steroids and, 429
steroid hormones and, 136
Proaccelerin, reptilian plasma and, 210
Progesterone,
androgen synthesis and, 424
biosynthesis of, 424
interrenal steroids and, 134, 430
oocyte meiotic division and, 148–149
ovary and, 136
physiological action, 427
Prolactin,
assay and biochemical characterization, 409–411, 413

definition, 403–404
glycogen and, 139
hypothalamus and, 128
metamorphosis and, 144–145, 146
physiological action,
body growth, 146–147, 404–406
osmoregulation, 408
reproduction, 407–408
skin shedding and, 148
water metabolism and, 143
Proline, interconversion of, 236, 240
Propionate, esters, hydrolysis of, 211
Propranolol, melanophores and, 150
Propylthiouracil, metamorphosis and, 145
Protease,
reptilian digestive juices, 256, 257, 260
venom and, 255, 466–468, 469, 473
Protein(s),
amphibian ribosomes, 12, 15
blood seasonal variations, 359, 369
catabolism of, 139
digestion and fate of products, 224–225
liver, season and, 363
mitochondrial deoxyribonucleic acid and, 11
synthesis, metamorphosis and, 146
Prothrombin, reptilian plasma and, 210
Psammophis condanarus, bile salt of, 349
Psammophis sibilans, bile salt of, 349
Pseudoechis colletti, venom, 470
Pseudechis porphyriacus, bile salt of, 349
Pseudemys,
fibrinogen of, 209
gonadotropin, neuraminidase and, 420
Hageman factor in, 210
Pseudemys concinna, blood, coagulation, 361
Pseudemys elegans,
blood, coagulation, 361
digestive enzymes, 256–257, 259–260
Pseudemys floridana, iodine metabolism in, 437
Pseudemys ornata, bile salts of, 347
Pseudemys scripta,
ammonia excretion by, 299–300
arginine metabolism in, 239
bladder, function of, 306–307
blood seasonal variations,
blood cells, 359–360
glucose, 358
inorganic ions, 357, 358
mucopolysaccharides, 362, 370
nitrogenous compounds, 359
digestive juices, pH, 254
extraurinary water loss, 282
gonadotropin,
biological test, 423
effects, 418
fractionation, 421
hemoglobins, 378, 381
fingerprints, 394
hydrogen ion excretion by, 296
hypophysectomized, 409
plasma vitellin of, 205
renal function, 278, 309–310
glomerular filtration rate, 279, 280, 286
sodium transport in, 292
transferrin of, 197, 198
urea excretion by, 301
urine, osmolality, 289
vitamin B_{12}-binding protein of, 213
Pseudemys scripta, troosti, hydrochloric acid secretion, 261, 262
Pseudobatrachotoxin, breakdown of, 175
Pseudoboa cloelia, bile salt of, 349
Pseudoboa nuwiedi, bile salt of, 349
Pseudoboa pelota, bile salt of, 349
Pteridines, pituitary and, 151
Ptyas, gonadotropin, neuraminidase and, 420
Ptyas mucosus, bile salt of, 349
Pumiliotoxin(s), occurrence and activity, 163
Pumiliotoxin C, structure, 177
Puromycin, aldosterone effects and, 60
Pyruvate,
alanine synthesis and, 236
aldosterone effects and, 62
blood levels, 219
glutamine synthesis and, 231, 235, 237
transamination and, 229
Pyruvate kinase, concentrations, 243
Pythocholic acid,
biosynthesis, 346
occurrence of, 340–341, 348
Python molurus, bile salt of, 348
Python reticulatus, bile salts of, 348
Python sebae, bile salt of, 348

R

Rana,
adenohypophysis, cell types, 129
pars intermedia, innervation of, 125
Rana cancrivora,
habitat, 140
osmotic tolerance, 51
plasma, urea in, 52
skin, water permeability, 141
vasotocin in, 142
Rana catesbeiana,
bile salts, 70, 75
biosynthesis of, 72–73
carbohydrate metabolism, 138
drinking by, 54
erythrocytes of, 80
hemoglobins of, 85–86
end groups, 103
functional properties, 94, 95, 96
larval and adult, 78, 113, 114, 115, 116
phenylhydrazine and, 111
polymerization of, 101, 108
sequence studies, 105, 106–107
tadpole, 87–88, 94, 97–98, 101–102
hypophysectomy in, 57
pancreatic islets of, 133
pituitary-interrenal axis in, 135
urine, tonicity, 52
water metabolism, 141
Rana clamitans,
erythrocytes, Bohr effect and, 94
tadpoles, hemoglobin of, 97
Rana esculenta,
adenohypophysectomy in, 57
bile salts of, 75
hemoglobin of, 85, 102
end groups, 103
functional properties, 94–95
sequence studies, 104–106, 108
insulin effects, 137
interrenals and, 133
neurohypophysis, structure, 124
osmotic tolerance, 51–52
pancreatic islets of, 133
reproductive cycle, adenohypophysis and, 130
sodium efflux by, 55
toxins of, 179
urine, tonicity of, 52
Rana grylio,
habitat, 140
hemoglobins of, 85, 114
polymerization, 101
tadpole, 87–88, 97
Rana montezumae, hemoglobin, synthesis of, 114
Rana nigromaculata,
bile salts of, 70, 75
toxin of, 179
Rana pipiens,
aldosterone levels, environment and, 59
bile salts of, 75
drinking by, 54
gastric secretion in, 24
growth, prolactin and, 146–147
hemoglobins of, 85, 101
metamorphosis and, 113, 114, 115
phenylhydrazine and, 111
tadpole, 86–87
iridophores, regulation of, 151
melanophores, cyclic adenosine monophosphate and, 150
melatonin effects, 149
metamorphosis, prolactin and, 145
mitochondrial deoxyribonucleic acid, 9
oocytes, miosis, 149
osmotic tolerance, 51–52
parathyroid glands of, 132
preoptic nucleus, salt and, 57
skin, sodium transport and, 55–56
sodium efflux in, 53
thyroxine effects, 139
ultimobranchial glands, 131, 132
urea excretion, 140
urine, tonicity, 52
yolk proteins of, 4, 5
Rana ridibunda,
bile salts of, 74, 75
hemoglobins of, 85
end groups, 103
Rana temporaria,
adenohypophysis,
annual cycle and, 130
function, 129–130
adrenal cortex, hypophysis and, 127
bile salts of, 75
carbohydrate metabolism in, 137–138

eggs, size of, 3
gonads, hypothalamus and, 125
hemoglobin of, 85, 101
interrenals of, 134–135
molting in, 148
photoperiod, hypothalamus and, 128
testes, lipoidal material in, 137
thyrotropin-producing cells of, 131
toxin of, 179
Rana terrestris, hemoglobin of, 85
Rana tigrina,
hemoglobins of, 85
melanophores, catecholamines and, 150
Ranidae,
bile salts of, 70, 75
hemoglobins of, 85–86
toxins of, 179
5α-Ranol,
occurrence of, 68, 75
structure, 69
5-β-Ranol,
cholic acid and, 73
formation of, 72, 73
occurrence of, 75
Rat,
amino acid deamination in, 244
growth hormone assay and, 410
nitrogen transport in, 227
reptilian follicle stimulating hormone and, 422
Regeneration, prolactin and, 406
Regina, transferrin of, 199
Renal tubular function,
regulation,
calcium and phosphate transport, 298–299
coupling of sodium and water transport, 294–295
nitrogen excretion, 299–306
potassium and hydrogen ion transport, 295–298
sodium transport, 291–294
tubular permeability to water, 288–291
Renal tubules, interrelations with other excretory functions, 309–312
Reproduction, growth hormone and prolactin and, 407–408
Reptiles,
digestive juices, 252–253
bile, 253
enzymes, 253–260
pH, 253
evolutionary lines, divergence of, 188
feeding habits, 249–251
hemoglobins,
adult, 378–384
fetal, 384
numbers and distribution, 377
Reserpine, catecholamines and, 124, 125
Reticulocytes, transferrin and, 197, 198
Rhacophorus sehlogelis var. *arborea*, embryos, hemoglobin of, 89
L-Rhamnose, metabolism of, 223
Rhamphiophis rostratus, bile salt of, 349
Rhynchocephalia, albumin of, 193
Riboflavin, snake plasma and, 213
Ribonuclease,
reptilian digestive juice, 256, 258–259
snake venom and, 469
Ribonucleic acid,
mitochondrial, oocytes and embryos and, 11–12
ribosomal, accumulation in oocyte, 14–15
synthesis, aldosterone and, 60
yolk and, 4
D-Ribose, metabolism of, 223
Ribosome(s),
mitochondrial, 11
structure and synthesis in amphibia, 12–13
synthesis and accumulation,
embryonic development and, 17–19
oogenesis and, 14–17
Rivanol,
immunoglobulins and, 201
transferrin and, 196
Rubidium ions, acid secretion and, 42

S

Salamanders, numbers and distribution, 90
Salamandra,
ovarian hormones, 136
pancreatic islets of, 133
Salamandra maculosa,
hemoglobin, polymerization, 101, 104

toxin,
activity, 163, 166–167
biosynthesis, 165–166
historical background, 164
isolation and structure, 164–165
Salamandra salamandra,
bile salts of, 68, 73, 75
hemoglobins of, 89
interrenals of, 135
Salamandridae,
hemoglobins of, 89
venoms of, 164–167
Salivary glands, feeding habits and, 251
Salt glands,
corticoids and, 431–432
excretion and, 312–313, 317
marine environment and, 318–330
Samandaridine, chemical properties, 165
Samandarine,
chemical properties, 165
occurrence and activity, 163
Samandarone, chemical properties, 165
Samandenone, chemical properties, 165
Samandinine, chemical properties, 165
Samanine,
chemical properties, 165
synthesis of, 168–169
Sauria,
bile salts of, 339–340
plasma proteins, 190
comparative aspects, 192
Sauromalus obesus,
antibody synthesis by, 203, 204
arterial pressure, temperature and, 284
extraurinary water loss, 282
nasal gland secretion, 313, 315, 329, 330
Scaphiopus couchi,
melanophores, catecholamines and, 150
moisture and, 51
Sceloporus cyanogenys,
ovary, prolactin and, 407
progesterone and, 427
salt gland, 330
ion ratio, 315, 331, 333
urine, osmolality, 290, 291
Sceloporus floridanus, testosterone and, 426
Sceloporus magister, urate excretion in, 304, 305
Sceloporus occidentalis,
critical thermal maximum, 368
hemoglobin, function, 388–389
intestinal mucosa, 252
iodine metabolism in, 437, 439
Sceloporus undulatus, hemoglobin of, 381
Seasonal variations,
cold-climate reptiles, 356–357
blood, 357–362
membranes, 365–367
tissues, 362–365
urine, 362
origin,
drought, 371–372
photoperiod, 370–371
summary, 372–373
temperature, 368–370
reptiles from dry-summer regions, 367–368
Secretin, reptile intestine and, 253
Sephadex G-100, gonadotropins and, 421
Serine,
deamination of, 225
metabolism in reptiles, 237–238
synthesis of, 236
Serine residues, phosvitin and, 4, 5, 206
Serpentes,
bile salts of, 340–342
plasma proteins, 190
comparative aspects, 192
Shrew, blood flow in, 245
Sialic acid,
gonadotropins and, 420–421
reptilian plasma proteins and, 189
Siren intermedia nettingi, hemoglobins of, 89
Siren intermedia texana, hemoglobins of, 89
Siren lacertina, hemoglobins of, 89
Serotonin, occurrence and activity as toxin, 163, 178, 179
β-Sitosterol, occurrence of, 69
Skin,
aldosterone effects, 63–64
sodium transport and, 55–56, 59
water influx and, 54, 56
Snakes, venomous, families represented, 449
Sodium,
acid secretion and, 42

batrachotoxin and, 176
blood, prolactin and, 408
chitinase secretion and, 265
corticosteroids and, 431, 432
efflux, urine and, 53–54
flux, aldosterone and, 60
gastric mucosa and, 36
gastric serosa and, 35
melanin dispersion and, 151
reptilian blood, 188
salt glands and, 312, 313, 317–333
seasonal variations,
blood, 357, 368
tissues, 364
urine, 362
transport, 43
cloaca and, 308–309
corticosteroids and, 142
coupling to water transport, 294–295
intestinal, 269, 270, 366, 367
neurohypophysial hormones and, 57, 141
regulation, 291–294
urate excretion and, 302
Sodium chloride,
anuran plasma and, 52
loading, dehydration and, 278–279
Somatotropin,
assay and biochemical characterization, 410–413
definition, 403–404
glycogen and, 139
physiological action,
blood glucose, 221, 406–407
body growth, 146–147, 404–406
osmoregulation, 408
reproduction, 407–408
L-Sorbose, metabolism of, 223
Sphenodon punctatus,
ammonia, urea and urate excretion by, 300, 301
Sphoeroides rubripes, toxin of, 167
Spinaceamine, occurrence of, 174
Spleen, erythropoiesis in, 114
Squamata,
albumins of, 193
hemoglobins, 381–384
function of, 388–391
plasma proteins, 190
Staurotypus, plasma albumin of, 194
Stearate, uptake of, 267
Stenotherus, plasma albumin, 194
Stenotherus odoratus, reproduction, calcium and, 208
Steroids,
gonadal, 423–424
biosynthesis, 424–425
interrenal gland, 128
physiological action, 425–428
Steroid dehydrogenase,
hypothalamus and, 128
interrenals and, 134, 135
ovary and, 136
Steroidogenesis, temperature and, 417
Sterols, yolk protein and, 207
Stilling cells, interrenals and, 134
Stomach, glands of, 252
Storeria, transferrin of, 199
Strychnine, occurrence and activity, 163
Suberylarginine, bufotoxins and, 173
Subunits, amphibian ribosomes, 12
Sucrase, *see* Invertase
Sucrose,
chitinase secretion and, 264, 266
metabolism of, 223
Sugars,
intestinal absorption of, 267–268
metabolism of, 222–223
Sulfanilamide, hydrochloric acid secretion and, 262
Sulfate,
bile alcohols and, 67, 68, 74
chitinase secretion and, 264, 265
hydrogen transport and, 30, 32, 40–41
potential differences and, 36, 37

T

Tachydosaurus, fibrinogen of, 209–210
Tachysaurus rugosus,
digestive juices,
enzymes, 257, 262–263
hydrochloric acid secretion, 261
pH, 254
Tadpole,
hemoglobins of, 86–88
metamorphosis and, 115–116
oxygen binding properties, 78
stomach, potential difference and, 35
D-Talose, metabolism of, 223

Tarbophis, see *Telescopus*
Taricha granulosa,
erythrocytes of, 80
hemoglobin,
amino acids and, 104, 109
functional properties, 98
sequence studies, 108–111
toxin of, 167
Taricha rivularis,
hemoglobin, functional properties, 98
toxin of, 167
Taricha torosa,
interrenals of, 135
toxin, activity, 163, 167
Taurine,
bile salts and, 67, 74, 337
hibernation and, 364
Tauroallocholate, occurrence of, 339–340, 342
Taurocholate,
biosynthesis of, 71
occurrence of, 70, 339–340, 342
Taurovaranate, occurrence of, 339
Telescopus fallax, bile salt of, 349
Telmatobius hauthali, hemoglobin of, 83
Telocinobufagin, structure, 172
Temperature,
digestion and, 261
origin of seasonal variation and, 368–370
regulatory mechanisms, 353–354
sodium transport and, 294
thyroxine synthesis and, 439
Terrepene carolina,
amylase of, 255, 257
blood seasonal variations,
blood cells, 360
glucose, 358
inorganic ions, 357, 358
nitrogenous compounds, 359
erythrocytes, life span, 377–378
extraurinary water loss, 282
hydrochloric acid secretion, 261, 262
transferrin of, 196
Testis,
endocrine function, 136–137
gonadotropins and, 418
hypophysectomy and, 414
steroid biosynthesis in, 424
Testosterone,
biosynthesis, 424
occurrence of, 136
physiological action, 425–427
prolactin secretion and, 128
skin sloughing and, 425–426
Testudo graeca,
bile salts of, 347
digestive juices,
enzymes, 257, 262–263
hydrochloric acid production, 261
pH, 254
esterase of, 211
intestinal absorption in, 268
thyroglobulin of, 439
Testudo hermanni,
blood seasonal variations,
inorganic ions, 357, 364
nitrogenous compounds, 359
digestive enzymes, 257
immunoglobulins of, 202
intestine, permeability, 269
photoperiod and, 370–371
thyroglobulin of, 439
torpor in, 370
urine, seasonal variation in, 362, 365–366
weight, seasonal variation, 356
Testudo hermanni hermanni, amino acid uptake in, 268–269
Testudo ibera,
antibody synthesis by, 203–204
temperature effects, 369
Testudo leithii, urea excretion by, 301–302
Testudo sulcata, urea excretion by, 301–302
Testudo tabulata, hemoglobin, function, 386
Testudo vicina, bile salts of, 347
Tetanus toxin, occurrence and activity, 163
$3\alpha,7\alpha,12\alpha,24$-Tetrahydroxy-$5\alpha$(or β)-cholestan-26-oic acid, metabolism of, 71, 73
$3\alpha,7\alpha,12\alpha,22\xi$-Tetrahydroxy-$5\beta$-cholestan-26-oic acid, occurrence of, 347, 349
$3\alpha,7\alpha,12\alpha,24$-Tetrahydroxy-$5\beta$-cholestan-26-oic acid, decarboxylation of, 72

Tetrahydroxyisosterocholanic acid, occurrence of, 347
Tetrahydroxysterocholanic acid, oxidation of, 338
Tetraiodothyronine, pituitary thyrotropes and, 441
Tetrodotoxin,
 batrachotoxin effects and, 176
 occurrence and activity, 163, 167
Thalassochelis caretta, thyroglobulin of, 439
Thamnophis,
 haptoglobin in, 210
 hemoglobin of, 211
 transferrin of, 196, 199
 urate excretion by, 304, 305, 306
Thamnophis sauritus,
 gonadotropin effects in, 418
 plasma vitellin of, 205–206
 vitellinogenesis in, 207, 208
Thamnophis sirtalis,
 gonadotropin effects in, 418
 hemoglobin of, 384
 hypophysectomized, growth of, 408–409
Thamnophis sirtalis sirtalis, hemoglobin, function of, 391
Thelotornis kirtlandi, bile salt of, 349
Theophylline, phosphorylase and, 221
Thiocyanate,
 adenosine triphosphate and, 40, 41
 hydrogen transport and, 29, 30, 31, 32, 33
 potassium transport and, 44
 thyroid and, 133
Thiouracil, thyrotropin-producing cells and, 131
Thiourea,
 intestinal permeability to, 269
 thyroid and, 133
Threonine,
 metabolism in reptiles, 238
 serine synthesis and, 236
Thrombin, reptilian plasma and, 209
Thrombocytopenia, snake venoms and, 467
Thyroid gland,
 annual cycle and, 365
 hypothalamus and, 126–127
 structure, 132–133
Thyrotropin,
 adenohypophysis and, 130
 biochemistry of, 441–442
 thyroxine biosynthesis and, 437–440
Thyrotropin releasing factor, metamorphosis and, 144, 145
Thyroxine,
 amino acid metabolism and, 139–140
 biosynthesis of, 437–440
 blood glucose and, 139
 molting and, 148
 osmomineral regulation and, 143–144
 proteins binding, reptilian plasma, 212
Tiliqua, Hageman factor in, 210
Tiliqua nigro-lutea,
 digestive juices,
  enzymes, 257
  hydrochloric acid secretion, 261
  pH, 254
Tiliqua rugosa,
 potassium transport in, 297
 sodium transport in, 294
Tiliqua scincoides,
 glomerular filtration rate, 280, 283–284, 286, 288
 urine, osmolality, 290
Trachysaurus rugosa, see also Tiliqua
p-Toluene sulfonyl-L-arginine methyl ester, reptilian trypsin and, 259
Trachysaurus rugosus,
 blood seasonal variations, inorganic ions, 357
 dry summers and, 368
Transaminase, snake venom, 469
Transamination, amino acids involved, 229
Transcortin, reptilian plasma, 212
Transferrin,
 reptilian plasma and, 190, 191
  evolutionary aspects, 198–200
  iron transport, 196–198
  isolation and identification, 196
Transport proteins, reptilian plasma, 212–213
Trehalase,
 reptilian digestive juice and, 257
 turtle serum, 12
Trehalose, chitinase secretion and, 264
3α,7α,12α-Trihydroxy-5α-cholestan-26-oic acid,

formation of, 72
occurrence of, 68, 75
3α,7α,12α-Trihydroxy-5α(or β)-cholestan 26-oic acid, formation of, 71, 73
3α,7α,12α-Trihydroxy-5β-cholestan-26-oic acid,
formation of, 72
occurrence of, 70, 350
3α,7α,12α-Trihydroxy-5β-cholest-22-ene-24-carboxylic acid, occurrence of, 69, 70, 75
3α,7α,12α-Trihydroxy-5β-cholest-23-en-26-oic acid, occurrence of, 69, 75
Trihydroxyisosterocholanic lactone, occurrence of, 338
Trihydroxysterocholanic lactone, occurrence of, 338
Triiodothyronine,
hemoglobin synthesis and, 116
metamorphosis and, 144–146
phenylhydrazine treatment and, 112
Trimeresurus flavoviridis, bile salts of, 350
Trimeresurus rhombeatus, bile salts of, 350
Trimeresurus wagleri, bile salts of, 350
Trionyx, gonadotropin, neuraminidase and, 421
Trionyx ferox, hemoglobin, function, 385
Trionyx phayrei, bile salts of, 347
Trionyx sinensis, see *Amyda japonica*
Trionyx spinifer, blood seasonal variations, inorganic ions, 357, 367
Trionyx triunguis, bile salts of, 347
Triturus,
hemoglobins of, 89
neurohypophysis, 124
pancreatic islets of, 133
Triturus alpestris,
sodium transport in, 141
toxin of, 167
vasotocin in, 141
Triturus cristatus,
fat body weight, 139
hemoglobin of, 90, 104
functional properties, 99
polymorphism, 91–92
interrenals of, 135
thyroid gland, regulation of, 126
toxin, activity, 163, 167
Triturus cristatus carnifex,
growth, hormonal effects, 147
prolactin,
effects, 145
release, 128
Triturus marmoratus, toxin, activity, 163, 167
Triturus viridescens,
hemoglobin of, 90
larval, 91
metamorphosis and, 116
Triturus vulgaris, toxin, activity, 163, 167
Tropiduras,
glomerular filtration rate, 283
sodium transport in, 293
urine, osmolality, 290
Trypsin,
chymotrypsinogens and, 259, 260
erabutoxins and, 458
reptilian digestive juice, 256, 257, 259
Tryptophan,
binding, albumin and, 194
metabolism in reptiles, 240
Tuatara,
body temperature, 187
immunoglobulins of, 202, 203, 204
d-Tubocurarine, cobra neurotoxin and, 459
Turanose, chitinase secretion and, 264, 266
Turtles,
albumins of, 193–194
amino acid metabolism, 237–241, 244
nitrogen transport in, 227
Typhlops, hemoglobin of, 211
Typhlops jamaiciensis, bile salt of, 348
Tyrosine,
formation of, 241
leptodactylin and, 174

U

Uma notata inornata, hemoglobin, function, 389
Uncoupling, snake venoms and, 467
Urate,
blood seasonal variations, 359
dehydration and, 372
excretion,
regulation of, 302–306, 310

reptiles and, 239
urea excretion and, 301–302
Urea,
anuran plasma and, 52
corticosteroids and, 143
excretion,
regulation of, 301–302
reptiles and, 239
thyroxine and, 140
formation from arginine, 239
hemoglobins and, 88
immunoglobulins and, 201
intestinal permeability to, 269
seasonal variations, 359, 364, 372
urinary, hibernation and, 362
Uric acid, salts, solubility of, 302
Uridine, uptake, aldosterone and, 60
Urine,
pH, ammonia excretion and, 300
seasonal variations in, 362
sodium efflux in, 52–53
tonicity, plasma and, 52
Urinary excretion,
water and ions, 277–278
bladder and cloacal function, 306–309
glomerular filtration rate, 278–288
interrelations among various functions, 309–312
renal tubular functions, 288–306
Urodeles,
erythrocytes, size of, 79, 80
hemoglobin,
components, 89–92
function, 98–100
sequence studies, 108–111
structure, 103–104
neurohypophysis, 124
pancreatic islets of, 133
testis, steroid-producing cells, 136
venoms of, 164–167
Uromastix,
progesterone and, 427
testosterone and, 426
Uromastix acanthinurus, digestive enzymes, 257
Uromastyx aegyptia,
amylase of, 255, 257
blood seasonal variations, glucose, 358
salt gland, 330
ion concentration in, 315
Uromastyx hardwickii,
amino acid transport, season and, 366
blood seasonal variations,
coagulation, 361
inorganic ions, 357
lipids, seasonal variation, 363
sugar uptake by, 268
Uromastix thomasi, bile salts of, 347

V

Vagus, stimulation, digestive juice and, 254, 261, 262
Valine, metabolism in reptiles, 238
Varanic acid,
biosynthesis, 346
occurrence of, 347
Varanus, testosterone and, 426
Varanus gouldi,
bile salts of, 347
cloaca, function, 308
Varanus griseus,
bile salts of, 347
blood seasonal variations, 357
blood cells, 360
glucose, 358
inorganic ions, 358
nitrogenous compounds, 358
liver, seasonal variations in, 363
muscle, glycogen in, 363
Varanus niloticus, bile salts of, 339, 347
Varanus salvator, bile salts of, 347
Varanus varius, bile salts of, 347
Vasotocin, *see also* Arginine vasotocin
antidiuretic action, 57–58, 59
hibernation and, 365–366
skin water permeability and, 141–142
Venom(s),
amphibian, pharmacological activities, 163
commercial, reliability of, 451, 454
enzymes in, 255, 465–474
functions of, 161, 180
general characteristics of, 450–454
low molecular weight components, 455–465
physiopharmacological activities, 465
plasma albumin and, 194–195

Vipera ammodytes, bile salts of, 350
Vipera aspis,
blood composition, season and, 357, 358, 360
fat bodies, seasons and, 363
hemoglobin, structure, 394, 395
hypophysis, season and, 365
venom, autoantibodies and, 201
Vipera berus,
annual metabolic cycle, 355
bile salts of, 350
venom, potency, 452
Vipera palaestinae,
bile salts of, 350
venom, fractionation of, 462
Vipera russelli,
bile salts of, 350
venom, potency, 453
Vitamin B_{12}, reptilian plasma and, 213
Vitamin K, prothrombin synthesis and, 210
Vitellin, plasma, estrogen and, 191
Vitellogenesis,
estrogen and, 427
reptiles and, 207–208
Vitellogenin, composition of, 6

W

Water,
balance, endocrine control, 56–59
elimination, urinary tract and, 57–58
flux, season and, 365–366
ingestion by reptiles, 250–251
metabolism, neurohypophysis and, 140–141, 143
permeability regulation, renal tubular function and, 288–291
reabsorption of, 53, 270
respiratory and cutaneous losses, 282
tissue content, season and, 364
transport,
bladder and, 307
cloaca and, 308, 309
coupling to sodium transport, 294–295
mucosal osmotic properties, 45–46
skin and, 56
water of secretion and, 44–45
yolk content, 5

X

Xanthophores, hormonal regulation, 151
Xantusia vigilis,
gonadotropin effects, 418
progesterone and, 427
spermatogenesis, temperature and, 417
Xenochrophis piscator, bile salt of, 349
Xenodon,
cloaca, function, 308
water reabsorption in, 270
Xenopeltis
haptoglobin in, 210
hemoglobin of, 211
Xenopus,
adenohypophysis,
corticotropin and, 130
structure, 129
corticosteroid effects, 139
insulin and, 137
interrenal tissue of, 135
liver glycogen, 138
oocytes, meiotic division, 148–149
ovarian hormones, 136
parathyroid glands, 132
thyroid gland, inhibition of, 133
thyrotropin-producing cells of, 130–131
thyrotropin-releasing factor in, 145
vasotocin in, 142
Xenopus laevis,
anucleolate, development of, 18
corticosteroid production in, 127
eggs,
mitochondrial nucleic acids, 9, 11
size, 3
embryo, ribosomal ribonucleic acids, 18–19
habitat, 140
hemoglobins of, 83
metamorphosis and, 113, 116
phenylhydrazine treatment and, 112–113
polymerization, 101
tadpole, 86
hypothalamic function in, 124, 125
melanophores,
catecholamines and, 150
division rate, 151–152
melatonin in, 150
oocyte,

ribosomes of, 14
ribosomal ribonucleic acid, 17
yolk deposition in, 6
pancreatic islets of, 133
ribosomal genes, 13
toxin of, 179
water metabolism, 141, 143
yolk proteins of, 4, 5
Xylose, uptake of, 268
D-Xylose, metabolism of, 223
L-Xylose, metabolism of, 223

Y

Yolk,
accumulation during oogenesis, 6–7
breakdown during embryonic development, 7–8
structure of, 4–6

Z

Zootoca, progesterone and, 427
Zymogen granules, gastric mucosa, 24

A 4
B 5
C 6
D 7
E 8
F 9
G 0
H 1
I 2
J 3